W9-BRU-910

**Library Patrons,
You may initial this sheet
to indicate you have read
this book.**

			PRINTED IN U.S.A.

AMERICAN
CARNAGE

Ambition must be made to counteract ambition. The interest of the man must be connected with the constitutional rights of the place. It may be a reflection on human nature, that such devices should be necessary to control the abuses of government. But what is government itself, but the greatest of all reflections on human nature? If men were angels, no government would be necessary. If angels were to govern men, neither external nor internal controls on government would be necessary. In framing a government which is to be administered by men over men, the great difficulty lies in this: you must first enable the government to control the governed; and in the next place oblige it to control itself.

—JAMES MADISON, *FEDERALIST*, NO. 51

AMERICAN CARNAGE

ON THE FRONT LINES OF THE
REPUBLICAN CIVIL WAR
AND THE RISE OF PRESIDENT
TRUMP

Tim Alberta

HARPER

An Imprint of HarperCollinsPublishers

To Abraham and Lewis and Brooks:

Reject false choices. Think for yourselves.

HarperCollins books may be purchased for educational, business, or sales promotional use. For information, please email the Special Markets Department at SPsales@harper collins.com.

FIRST EDITION

Library of Congress Cataloging-in-Publication Data has been applied for.

ISBN 978-0-06-289644-5

19 20 21 22 23   LSC   10 9 8 7 6 5 4 3

CONTENTS

AUTHOR'S NOTE

HOW DID DONALD TRUMP BECOME PRESIDENT OF THE UNITED STATES? Since the early morning hours of November 9, 2016, game attempts to solve this riddle have been subject to the same disorienting forces that came to define his political ascent: ideological bias and tribal loyalty, social alienation and demographic transition, institutional breakdown and political polarization.

There is a temptation not only to associate these things with Trump, but to blame him for the havoc they hath wrought on America and the world. Throughout his campaign for the presidency and in his first two years in office, Trump defied every law of gravity while shattering societal conventions that will prove difficult to repair. In so doing, he nurtured narratives of his own centrality to a bruising reconfiguration of modern American life.

Trailblazing as he might be, Trump is not the creator of this era of national disruption. Rather, he is its most manifest consequence.

IN PRESIDENTIAL POLITICS, THE CANDIDATE MUST MEET THE MOMENT. Barack Obama could not have won the White House with his dovish foreign policy platform in 2004, an election decided on the question of whom Americans wanted as their wartime president. It was not until 2008, with the country weary from intervention, and his heavily favored Democratic rival tainted with a vote for the Iraq invasion, that the electorate was primed for his candidacy.

Similarly, Trump's appeals to America's darker impulses would have fallen flat in 2000. The nation was too peaceful, too prosperous, too cohesive. All throughout history the world over, efforts to exploit anxiety have succeeded when there is anxiety to be found: times of war, financial despair, national disunity. The United States circa 2000 did not fit the bill. Even after winning the most disputed race in presidential history, with a grueling thirty-six-day Florida recount splintering the nation before the Supreme Court intervened on his behalf, George W. Bush entered office with a 57 percent approval rating—higher than the incoming marks of both his father, George H. W. Bush, and the previous Republican president, Ronald Reagan.

Eight years later, the scenery had changed. The country was trapped in two deeply unpopular military conflicts. It was shedding jobs at an alarming rate, particularly in the manufacturing hubs of middle America. And its electorate was increasingly bifurcated, with partisans estranged from one another not just ideologically but geographically and culturally as well.

Into this breach stepped Obama. He promised to heal these wounds, to reject the labels of "red states" and "blue states" and unify the country. It would prove an impossible task.

To understand Trumpism is to understand the dramatic national makeover of the previous ten years. When Obama was sworn in as president in January 2009, he opposed same-sex marriage. Facebook lagged behind MySpace in monthly Web traffic. Brad Pitt and Angelina Jolie were the most famous couple on earth. Airbnb had just launched its website. Tiger Woods was the world's most dominant athlete. Uber did not exist. Nobody had heard of the word *gluten*. The Twitter account @realDonaldTrump had not yet been created.

The ensuing decade saw a convergence of phenomena—economic displacement, technological innovation, civic upheaval—that fundamentally reshaped our politics. And nowhere was that change felt more acutely than inside the Republican Party.

Having moved to Washington in the twilight of Bush's presidency, I found myself less fascinated by Obama and his incoming Democratic government than by the power vacuum in the GOP. Bush was leaving office with record-low approval and Republicans were headed into the wilderness. There was no vision, no new generation of leaders, no energy in the conservative base. One of America's two major parties had gone politically bankrupt.

In the ten years since, covering Congress and campaigns, I have interviewed more than a dozen Republican presidential candidates, scores of GOP congressmen and senators and governors, hundreds of party activists and strategists, and countless voters all across the country.

I watched as the Republican Congress went to war with Obama; as the Tea Party rebranded the American right; as platoons of insurgent lawmakers came to Washington intent on its demolition; as the government was shuttered and pushed to the brink of default; as a group of renegades established themselves as an effective veto threat over their own party; and as the mutiny overthrew the most powerful man in the GOP, having already swallowed up his heir apparent.

A wave of revolution was cresting in the Republican Party. The only question was who would ride it.

THE NARRATIVE I AM SETTING FORTH IN THESE PAGES IS NOT MEANT TO be comprehensive. What I have attempted to construct is an account guided by my own coverage of what, until the spring of 2016, was commonly known as the Republican civil war, as told through the eyes of the combatants on the front lines.

Although this project was conceived in the middle of 2016, it was not until early 2018 that I began researching and reporting the new material that would form this book's foundation. In the year since, I have interviewed more than three hundred people specifically for the purposes of this recounting. Those interviews, joined with my work over the previous decade, comprise the spine of the story you are about to read.

Wherever I have drawn from other sources of information, proper attribution and/or citation is provided.

Everything else is rooted in my own reporting. Most of the interviews were conducted on the record; any quote that appears in the present tense ("says") is drawn from original reporting, as are the past-tense quotes ("said") that do not include outside attribution.

This book is fashioned not as a traditional work of journalism, but as a storytelling narrative. Portions of some interviews were conducted on deep background, giving me license to use the material but not attribute the source. In those instances, the only reporting I present in these pages is taken from sources with firsthand knowledge of the events and conversations in question.

At a time when truth is under attack, I have labored to verify every quote, story, and circumstance down to the smallest detail. In addition to hundreds of hours of taped interviews, I have also drawn from emails, audio files, contemporaneous memos, and other private records provided by sources whose identities I agreed to safeguard.

It is my hope that by reading this account of the political and cultural turbulence that rattled the nation during the first two decades of the twenty-first century, you will gain a more textured understanding of Donald Trump's rise to the presidency—and of its implications for America.

TIM ALBERTA
January 2019

PROLOGUE

THE RESOLUTE DESK IS CLEAN AND UNCLUTTERED, ITS WALNUT VEneer glimmering as rays of late afternoon sun dart between the regal golden curtains and irradiate the Oval Office. There are two black telephones situated at the president's left elbow, one line for everyday communications within the White House and the other for secured calls with the likes of military leaders, intelligence officials, and fellow heads of state. There is a small, rectangular wooden box adorned with a presidential seal and a red button so minuscule that it escapes my eye until he pushes it, prompting only momentary distress over the impending possibility of nuclear apocalypse before a tidily dressed butler appears moments later and places a tall, hissing glass of Diet Coke in front of the commander in chief. Finally, there is a single sheet of paper, white dotted with red and blue and yellow ink, that the president continues to nudge in my direction. It proclaims the results of a CBS poll taken two nights earlier, immediately following his 2019 State of the Union address: 97 percent of Republican viewers approved of the speech.

"Look," Donald Trump says, shoving the document fully in front of me, reciting the 97 percent statistic several times, "It's pretty wild."

It certainly is.

Pretty wild that Trump, a thrice-married philanderer who paraded mistresses through the tabloids and paid hush money to pinups and porn stars, could conquer a party that once prided itself on moral fiber and family values.

Pretty wild that Trump, a real estate mogul and reality TV star with no prior experience in either the government or the military, could win the presidency in his first bid for elected office.

And yes, pretty wild that Trump, who spent his first two years as president conducting himself in a manner so self-evidently unbecoming of the office—trafficking in schoolyard taunts, peddling brazen untruths, cozying up to murderous tyrants, tearing down our national institutions, weaponizing the gears of government for the purpose of self-preservation, preying on racial division and cultural resentment—finds himself in a stronger standing with the Republican base heading into the 2020 election than he did upon winning the presidency in 2016.

To be clear, the 97 percent statistic is meaningless: Presidents have long utilized the State of the Union tradition as a pep rally, vision-casting in ways designed to draw standing ovations and project the awesome power of the bully pulpit, leaving little for the party faithful not to love.

But Trump is not wrong when he boasts of the unwavering support he enjoys within the GOP. In fact, according to Gallup, he is more popular with his party two years into office than any president in the last half century save for George W. Bush in the aftermath of 9/11. Unlike with Bush, there is no connective national crisis to which Trump can attribute the allegiance of the Republican rank and file; to the contrary, Trump celebrated the two-year anniversary of his inauguration by presiding over the longest government shutdown in U.S. history, a climax of the absolutism that came to infect the body politic around the turn of the century and now appears less curable with each passing day. Even in this valley, with his overall approval rating sliding to an anemic 37 percent, Gallup showed Trump maintaining the support of 88 percent of Republicans. Sitting in his office in early February, less than two weeks since the reopening of the federal government, his overall approval rating has shot back up to 44 percent. Among Republicans, it's steady at 89 percent.

That he has executed a hostile takeover of one of America's two major parties—the one he never formally belonged to before seeking its presidential nomination—will astonish generations of intellectuals, think-tankers, and political professionals. But it does not surprise Trump.

A decade earlier, as a transition of political power coincided with a jarring suspension of America's economic stability, Trump saw what many Republicans refused to acknowledge: Their party was "weak." Not just weak in the sense of campaign infrastructure or policy positions, but weak in spirit, weak in manner, weak in appearance. The country was hurting. People were scared. What they wanted, Trump realized, was someone to channel their indignation, to hear their grievances, to fight for their way of life. What they got instead was George W. Bush bailing out banks, John McCain vouching for Barack Obama's character, and Mitt Romney teaching graduate seminars on macroeconomics.

"Nobody gave them hope," Trump says of these anxious Americans, enumerating the deficiencies of those three Republican standard-bearers who came before him. "I gave them hope."

TO CONSIDER TRUMP'S RISE IS TO RECOGNIZE THE ESSENTIAL INGREDI-ents of American life: instinctual outrage and involuntary contempt, geographic clustering and clannish identification, moral relativism and self-victimization.

These conditions began to shape the modern political era dating back to the mid-1990s. There was the zero-sum warfare practiced by then-Speaker Newt Gingrich, who encouraged Republicans to use words such as "traitors" and "radicals" to describe their political opponents. There was the impeachment of then-President Bill Clinton, whose deception in the face of extramarital scandal might have angered more Democrats had Republicans not brimmed with hypocritical and opportunistic indignation. There was, after the brief interlude of 9/11, the Bush administration's disastrous handling of the Iraq War and Hurricane Katrina. And there was, of course, the election of Obama, whose unifying rhetoric met with the cold realities of governing in a bitterly divided nation, stoking the flames of polarization that had come to inform seemingly every issue, even those where once considerable agreement existed.

Trump understands—intuitively, if not academically—how this atmosphere invited his emergence.

In an interview inside the Oval Office, he misses no opportunity to highlight the failures of the governing class, alternatively delighted and vexed that no politician proved capable of identifying or exploiting the opportunity he did.

The president seems particularly gleeful in swiping at his most immediate predecessors, one of each party, a merry violation of etiquette inside the world's most exclusive fraternity. Bush, he says, "caused tremendous division . . . tremendous death, and tremendous monetary loss" by focusing on nation-building abroad instead of fortifying a wobbly domestic economy. Obama, the president argues, was even more clueless when it came to the economy, standing idly by as the country hemorrhaged blue-collar jobs, more concerned about preserving political norms than protecting American workers.

Whether fair, or accurate, or nuanced, these sentiments were undeniably shared by a sizable chunk of the American electorate in the waning years of Obama's presidency. By stepping into the arena and presenting himself as a brawler—someone unbeholden to any special interest, someone unencumbered by the conventions of Washington, someone willing to burn down the government on behalf of the governed—Trump returned the Republican Party to power.

He also understands the relationship between his victory and the GOP's previous two losses.

The 2008 campaign was "a very rough time for the Republican Party," Trump says. He recalls how McCain embraced Bush's floundering war in Iraq; how he resorted to gimmicks as the global banking system teetered on the edge of the abyss; how he repeatedly told laid-off midwestern voters that some of their jobs wouldn't be coming back. "I gave him money—believe it or not, because I wasn't a huge fan, then or now, but I raised money for him," Trump says of McCain. "And then he just gave up on an entire section of the country."

The 2012 defeat was worse, because to Trump, it was avoidable. Whereas McCain had been hamstrung four years earlier by Bush's deep unpopularity as well as a dramatic financial crisis, Romney faced a vulnerable incumbent presiding over a sputtering economy. Yet, instead of appealing to the primal instincts of the right, running a bloody campaign against a president whose team was showing no mercy to Romney, the Republican nominee played by the rules. "Romney's problem was he had too much respect for Obama," Trump says. "And he shouldn't have, because Obama didn't deserve it."

In both cases, Trump says, these Republicans lost because they acted like, well, *Republicans*. McCain and Romney defended the merits of free trade, promoted the exporting of American military force, and advocated the importing of cheap labor—all while adhering to a code of conduct that pacified the graybeards of the GOP establishment.

What Trump does *not* understand is that his populist, inward-facing "Make America Great Again" mantra is less a revelation than a resurrection. For generations, his ideological forebears—from Ohio senator Robert A. Taft, to the leaders of the John Birch Society, to "Pitchfork" Pat Buchanan, who challenged George H. W. Bush in the 1992 GOP primary—have peddled a version of conservatism, known commonly as *paleoconservatism*, anchored by an intense skepticism of international commerce, military adventurism, and foreign immigration.

As a political philosophy, this brand of right-wing nationalism was crushed under the heel of William F. Buckley's *National Review*, the Ronald Reagan revolution, and the Bush dynasty. In pursuing a vision of expansionist, growth-oriented *neoconservatism*, these forces transformed the post–World War II Republican Party into a champion of global markets, global policing, and global citizenry.

But it was unsustainable. Between the lives and treasure lost fighting a pair of Middle East quagmires, the devastation to America's manufacturing sector wreaked by automation and outsourcing, the hoarding of corporate profits and the widening chasm of income equality, the declining faith in institutions ranging from the media to organized religion, the recklessness of the financial class, and the nation's rapidly darkening ethnic complexion, voters had cause to question whether the *Wall Street Journal*'s editorial board truly had their best interests at heart.

The revolt was near. Not everyone could see it—and not all those who did took it seriously. Trump saw it. He took it seriously. And he became its voice, as the unlikeliest of insurgents, the commercial tycoon who cheated the little guy, who employed illegal workers, who made his products overseas, and who enhanced his inherited fortune through scams and fixers and lawsuits, railing against a shredded social contract from the gilded penthouse of his Manhattan skyscraper.

With the people taking to the streets, hoisting flags and chanting against the government—ostensibly in protest of spending, taxation, infringement of liberty—Trump stepped to the front lines of the cultural conflict. By securing a regular place on the airwaves of Fox News, by propagating lies about the president's birthplace, by attacking the left in ways no leader of the Republican Party would dare consider, he adopted a movement, and a movement adopted him.

"The Tea Party was a very important event in the history of our country. And those people are still there. They haven't changed their views," Trump says. "The Tea Party still exists—except now it's called Make America Great Again."

His conquest of the right was sufficient to win the presidency. Yet the circumstances of his victory were freakish if not fluky. And the truth is, presidents of the United States come and go. Most are transitional, as historians have long observed. Only a select few are transformational. To be transformational, to durably alter the American identity, Trump would have to do more than bend conservatism to his will. He would have to redefine the Republican Party.

THE PRESIDENT IS RACING TOWARD A MINEFIELD.

His longtime lawyer Michael Cohen is preparing to testify in front of the House Oversight Committee, having already implicated Trump in several criminal conspiracies.

A special counsel's dual-track investigation—of possible collusion between Trump's campaign and the Russian government in 2016, and of whether the president obstructed that inquiry—is drawing to a close, with potentially ruinous consequences for himself, his administration, and his political and personal associates.

Finally, having reopened the government on a short-term basis, allowing congressional negotiators to go through the motions of pursuing a nonexistent compromise on border funding, he faces the specter of another shutdown in one week's time. Sensing that a deal is improbable, Trump is preparing to take the unprecedented step of declaring a national emergency, redirecting other pools of money toward the construction of his promised wall along the U.S.-Mexico border. But the plan is fraught with peril; myriad legal challenges are certain, and congressional opposition could thwart the president's end run around the legislative branch.

Whispers of impeachment gust through every corridor of the Capitol. Having retaken control of the House of Representatives in November, Democrats confront daily pressure, from the party's base and from within their own ranks, to vote on Trump's eviction from office.

It is the Republicans, however, who control the Senate. And only with a two-thirds vote of the upper chamber can Trump be removed. The president's fate is in the hands of his own tribe.

"The Republican Party was in big trouble," Trump says. "I brought the party back. The Republican Party is strong. The Republican Party is strong."

He takes a long pause.

"They've got to remain faithful. And loyal."

CHAPTER ONE

FEBRUARY 2008

"These isms are gonna eat us alive."

SHE WAS SHAKEN WITH DISBELIEF, THEN ABLAZE WITH DEFIANCE, the churn of emotion ultimately yielding a pure, righteous fury. It was mere minutes until the pinnacle of her professional life: introducing her favored presidential candidate, in the heat of a contested Republican primary, at the Conservative Political Action Conference. This felt like an inflection point for the party, and Laura Ingraham, the acid-tongued radio host, had prepared accordingly. Worried that Republicans would succumb to nominating John McCain, the ideologically autonomous senator who had betrayed the right on everything from tax policy to campaign finance reform, Ingraham readied a blistering attack on the Grand Old Party's front-runner and a final plea for voters to rally behind the true conservative in the race: Mitt Romney.

It mattered not that Romney was a Mormon, nor that he'd piloted a controversial program in Massachusetts requiring individuals to buy health insurance, nor that he was an opportunistic centrist with shape-shifting views on abortion and gay rights. The conservative movement, and the loudest voices on talk radio—Rush Limbaugh, Sean Hannity, Mark Levin, and Ingraham herself—had fallen for the former governor.

Everything about Romney screamed presidential: his elegant wife, his five strapping sons, his vintage jawline, his flawless coif of black with flecks of silver distinguishing his temples. He was a wealthy business guru who had created tens of thousands of jobs. He was a technocrat and a turnaround specialist, having rescued the 2002 Salt Lake City Olympics from financial ruin. He was a fiscal messiah who preached the gospel of free markets and low taxes and deregulation.

And for whatever his faults, Romney was strong when it came to Mc-
Cain's greatest weakness: immigration.

McCain had long been viewed warily by the right wing of the Republican Party. While celebrated for his Vietnam heroism—he spent five and
a half years as a prisoner of war in the Hanoi Hilton, refusing the early
release offered due to his father's rank as a four-star admiral[1]—the Arizona senator reveled in deviating from party orthodoxy. He opposed his
onetime rival George W. Bush's tax cuts after losing to him in the GOP
primary of 2000. He decried the administration's use of torture and advocated for closing the detention facility at Guantanamo Bay. He teamed
with Wisconsin Democrat Russ Feingold to rewrite the nation's campaign finance laws, undermining the GOP's structural cash advantages.
On a personal level, McCain could be gruff and churlish, prone to angry
outbursts that left colleagues questioning his steadiness. "The thought
of his being president sends a cold chill down my spine," Senator Thad
Cochran of Mississippi, a Republican, told the *Boston Globe* in 2008. "He
is erratic. He is hotheaded. He loses his temper, and he worries me."[2]

But McCain's unforgivable sin came in 2007. Along with the "Liberal
Lion," Ted Kennedy, he led the charge in Congress to pass Bush's comprehensive immigration reform plan—including a path to citizenship for
millions of illegal residents. The fallout was devastating. McCain took a
beating from the right, which, combined with the early mismanagement
of his 2008 campaign, nearly ended his second bid for the White House
before it began.

Even as his pirate ship of a campaign steadied, and McCain climbed
back into contention, his vulnerability was all the more exposed. Republican candidates had expected the 2008 primary fight to revolve around
two issues: the domestic economy and the wars in Iraq and Afghanistan.
There was no shortage of discussion and debate on these topics. Yet
more visceral for GOP voters, especially those in the lily-white early-
nominating states of Iowa and New Hampshire, was immigration. The
fear was that by offering "amnesty" to millions of foreign-born intrud-
ers, Republicans threatened to destabilize the economy while staining
the American social fabric.

During a town hall meeting in New Hampshire in the late summer of 2007, McCain grew exasperated upon hearing yet another voter raise concerns about Mexican immigrants endangering her community. "Ma'am, you live in New Hampshire. We're two thousand miles from the southern border," the senator said. "What are you worried about, a bunch of angry French Canadians?"

McCain's traveling staff, which due to financial troubles had been pared back to campaign manager Rick Davis and a few local organizers, howled at the remark. But as they spoke afterward, McCain warned Davis that the issue could derail his candidacy. "If we're going to get this in New Hampshire," McCain said, "we're going to get this everywhere."

The rival campaigns reached a similar conclusion and began to recalibrate, seeking tougher tones to channel the ire of their electorate. Fortunately for McCain, many of his opponents were ill equipped to attack him on the issue: Rudy Giuliani, the former New York City mayor, was a longtime friend and an immigration dove himself; former Arkansas governor and onetime Baptist minister Mike Huckabee had a similarly soft record; Texas congressman Ron Paul's libertarian worldview called for a free flow of goods and people; and former senator Fred Thompson, like the rest of the field, lacked the viability to inflict damage on McCain.

The exception was Romney. He methodically chiseled away at McCain's immigration record, painting him as a career politician oblivious to the plight of working Americans. The irony, in retrospect, is that Romney now realizes that the churning resentment among voters had far less to do with people coming in than it did with jobs going out—something he and other Republicans spent little time discussing, having accepted as canon the political infallibility of free markets.

"It was evident that certain industries would be substantially affected and harmed on a disproportionate basis—the auto industry, mining, metals—[and] the argument was, well, in the long run this is all good for the country, and as a nation we'll do better," Romney says. "I think the evidence is that as a nation we *did* do better. But if you're working in an auto factory in Detroit you're not doing better, and if you're working in Ohio or Indiana for a car factory or a steel factory, you're not doing better. Your life has been devastated. You had a home, a community. Suddenly the community becomes almost a ghost town. You can't sell your home because who wants to buy a home in Lordstown, Ohio, when GM pulls out and there's just no one else moving in? These people are very angry, and the elites, Republicans and Democrats in power, didn't do anything about it, and didn't really think about what the implications would be for those disproportionately affected. So, people were very angry—and continue to be angry."

Convinced that immigration was the galvanic issue of the race, Romney didn't limit his attacks to McCain. He savaged Huckabee as well, hoping to undermine the preacher's down-home populism. The former Arkansas governor had fought for illegal minors in the state to qualify for college scholarships, inspired by the story of a local high school valedictorian who was brought to the United States when he was four years old. "If a cop pulls over a car for speeding, he gives the ticket to the dad, not to the kid in the backseat," Huckabee says. "I wanted this kid to go to college and become a doctor and pay taxes, rather than just have him pick tomatoes while the government subsidizes him for the rest of his life."

Things got especially testy in Iowa. With McCain skipping the evangelical-laden caucuses to train his efforts on New Hampshire, where he had legendarily revived his 2000 primary bid, Romney and Huckabee escalated their attacks on one another down the home stretch, each man sensing that a victory in the Hawkeye State was their only springboard to capture the party's nomination. Romney's operation was cutthroat: Several former staffers recall handing out flyers in Iowa with a picture of the Mexican consulate in Little Rock, Arkansas, asking why Governor Huckabee had permitted so many Mexicans to work illegally in poultry factories.

"There were a lot of things Romney's people should have apologized for. We were constantly putting out fires that his people were starting," Huckabee says. "I had never seen a more disingenuous campaign in trying to portray somebody who was anything but conservative as more conservative than everyone else on the stage. It was truly laughable."

While losing ground thanks to Romney's sustained assault, Huckabee reached into his own bag of tricks. Speaking with a reporter from the *New York Times Magazine* in December, Huckabee asked, "Don't Mormons believe that Jesus and the devil are brothers?"[3]

This was the last election in which social media would not play a dominant role—and yet even still, Huckabee's quote went viral, dumping kerosene on the fire already burning in Iowa. He quickly apologized to Romney; ten years later, Huckabee still insists the comment was born out of ignorance rather than animus. Either way, it played into a whisper campaign that sought to cast his rival's Mormon faith in a suspicious light. And in the closing days of the Iowa race, Huckabee ran a now-famous television ad in which he spoke directly to the camera in front of a Christmas tree, framed by the corner of a white bookshelf that gave the unambiguous appearance of a cross.[4] This, paired with his remark to the *Times*, struck no one as coincidental, given the outsize influence Christian voters held over the outcome of the Iowa caucuses.

Ultimately, the sequence played out perfectly for McCain: Huckabee won Iowa, weakening Romney, whose loss of momentum allowed McCain to win New Hampshire and South Carolina. With Romney starved of an early-state victory, Huckabee strapped for cash, Giuliani exiting after a poor performance in Florida, and the other fringe candidates a nonfactor, it was McCain's nomination to lose.

But Romney would not quit. Pouring millions of his own fortune into the campaign, he hung around, amassing enough delegates to remain mathematically alive and stirring an eleventh-hour optimism on the right that McCain could be defeated. Romney's speech to CPAC on February 7, then, was meant to commence a last stand, rallying the conservative troops against the reviled front-runner.

Instead, with the endgame increasingly apparent, and his political future to consider, Romney sat down on the eve of the event and crafted a withdrawal speech.

Nobody told Ingraham. When Romney's director of conservative outreach, Gary Marx, picked her up for the event, the radio host still believed she was calling for the storming of McCain. Informed that Romney was quitting, Ingraham looked puzzled at first, the news not fully registering. She then became agitated, promising Marx and other

Romney staffers that she would change his mind. "If none of you can convince him, I will," she huffed. Confronting Romney backstage at the event, Ingraham pleaded with him to reconsider. To no avail: The candidate said his mind was made up. Ingraham looked volcanic. Romney's aides wondered whether it was a good idea to send her out onstage.

Minutes later, Ingraham strode to the lectern inside the Omni Shoreham Hotel in Washington. She appeared visibly torn between delivering the scorched-earth speech she had prepared and giving the decaffeinated version the party now needed. "I don't think it's enough to say that you were a foot soldier in the Reagan revolution," she said, mocking McCain's claim to the Gipper mantle. "I think the question is, what have you been doing for conservatism lately?"

The crowd roared lustily. With both McCain and Huckabee set to speak after Romney at CPAC, and attendees still unaware that their champion was about to quit, Ingraham, dressed for a funeral in her black jacket, dark blouse, and cross necklace, announced, "He is a national security conservative. He is a proud social conservative. And he is a fiscal conservative. In other words, Mitt Romney is the *conservative's* conservative."

The boy band–shrieking welcome of Romney onto the stage quickly dissolved into a bad breakup melancholy at the realization of the news he'd come to share. "If I fight on, in my campaign, all the way to the convention"—he was drowned out by cheers. Romney looked as though he were putting the family dog to sleep. It was a pained expression, his lips pursed, his eyes betraying the fact that he was sincerely concerned about the state of the country and had a whole lot more to say about it.

"I want you to know, I have given this a lot of thought," Romney said. "I would forestall the launch of a national campaign, and, frankly, I would make it easier for Senator Clinton or Obama to win." Cries of protest now filled the ballroom. "Frankly, in this time of war, I simply cannot let my campaign be a part of aiding a surrender to terror."

When McCain arrived at the event two hours later, his coronation as the now-presumptive Republican nominee was interrupted by boos and jeers. "Many of you have disagreed strongly with some positions I have taken in recent years. I understand that," McCain said. "I might not agree with it, but I respect it for the principled position it is. And it is my sincere hope that even if you believe I have occasionally erred in my reasoning as a fellow conservative, you will still allow that I have, in many ways important to all of us, maintained the record of a conservative."

The next morning, the two biggest news stories in America were McCain clinching the GOP nomination and pop star Britney Spears leaving a mental hospital. Both were subjects of popular psychoanalysis; in the case of McCain, the question was whether he could possibly placate conservatives. It wouldn't be easy. As right-wing author and provocateur Ann Coulter had told Fox News host Sean Hannity days earlier, "If he's our candidate, then Hillary is going to be our girl, Sean, because she's more conservative than he is."[5]

When Bill O'Reilly hosted Ingraham on his Fox News show, hours after her performance at CPAC, he asked whether the hard-liners on talk radio would now cease their bludgeoning of McCain. Ingraham protested, saying their critiques had been substantive and demanding evidence to the contrary.

"They call him 'Juan,'" O'Reilly replied.

IF THE RIGHT'S LOVE AFFAIR WITH ROMNEY MIGHT HAVE BEEN FOR lack of a better option, then its loathing of McCain owed in part to frustrations with the man they called "Dubya."

Increasingly, Bush's legacy was feeling treasonous to small-government Republicans: prodigious spending, endless war overseas, rising debt and deficits, a massive federal intrusion into K–12 education, the biggest entitlement expansion since the Great Society. "Who in the end prepared the ground for the McCain ascendency? Not Feingold. Not Kennedy. Not even Giuliani. It was George W. Bush," conservative pundit Charles Krauthammer wrote in the *Washington Post* after Romney quit. "Bush muddied the ideological waters of conservatism."[6]

Perhaps most irritating to the base, Bush had proclaimed upon winning reelection that he had "political capital" and would spend it on two things he barely mentioned during the 2004 campaign: privatizing Social Security and reforming the country's immigration system.

Paul Ryan, a young congressman from Wisconsin, had distinguished himself as the most outspoken advocate in the House of Representatives for making changes to Social Security. He found himself at once excited and perplexed when Bush suddenly pledged to tackle the issue. "He didn't really talk about Social Security at all," Ryan says. "The campaign was more about security—you know, 9/11, three Purple Hearts, the swift boat stuff. So, I remember him declaring what he wanted to do, which I thought was spectacular. But I also thought, gosh, he didn't really till the field for this."

Soon, Ryan was aboard Air Force One with Bush, flying to Wisconsin for an event aimed at selling the Social Security overhaul. Of all the Republicans in the state's congressional delegation, he was the only one to accompany the president. "They all thought it was too risky," Ryan recalls.

His colleagues were smart to sense trouble: Bush's plan to create personalized accounts, while fashionable among the conservative intellectual class, was a nonstarter with much of the party's base, particularly blue-collar workers, middle-class earners, and the elderly. The backlash was so harsh, in so many congressional districts, that Speaker Denny Hastert and his GOP leadership team refused to give the president's bill a committee hearing.

The failed Social Security push was crucial—not just for what it foretold about the party's fraught relationship with entitlement programs, but for its dooming of immigration reform as well. White House officials had vigorously debated which initiative to lead the second term with; Social Security reform, the heavier lift, ultimately won out. By late May, when House GOP leadership apprised members of the summer schedule, Bush's proposal was dead.

The time and momentum lost proved critical: Democrats won back majorities in both chambers of Congress in 2006, leaving Bush weakened to sell an immigration deal to Republicans and the Democrats emboldened to hold out for something better. "The sequencing was off," Karl Rove, Bush's chief strategist, admits. "If we had done immigration first, it would have passed."

The effort came tantalizingly close nonetheless: For much of 2007, Kennedy and McCain strong-armed their Senate colleagues while Democrat Luis Gutierrez and Republican Jeff Flake whipped support for a companion bill in the House.[7] When it failed, some Republicans could argue honestly that they had done their part. Democrats were in control of Congress, after all, and given the rivalries within that party's coalition (Blue Dog labor versus Green-minded progressives, Charlie Rangel's Congressional Black Caucus versus Gutierrez's Congressional Hispanic Caucus), the collapse of immigration reform looked to be a bipartisan feat.

But the GOP's struggle with immigration was, and would remain, something more existential than any one legislative outcome might suggest.

The 2000 campaign was meant to signal a new era of Republican politics: warm, aspirational, inclusive. Bush spoke Spanish at events and promised to champion diversity as a core American quality. He fumed when his body man, a Texan of Hispanic descent, was pulled over in Iowa and hassled about his citizenship. Bush's deployment of the term "compassionate conservatism" wasn't merely an electoral ploy. The Texas governor, known to converse at length with hotel maids and interrupt dinner parties with Luke 12:48—"From everyone who has been given much, much will be wanted"—envisioned a presidency founded on principles of equality and charity. Tending to war-torn nations with medical assistance and refugee resettlement would be a priority, as would tending to war-torn American communities with education reforms and prisoner reentry programs.

What Bush couldn't foresee was September 11, 2001, and how the attacks would not only alter his own priorities but also contribute to a changing psyche on the American right.

"The term 'compassionate conservatism' ticked off conservatives," recalls Jim Towey, the director of faith-based initiatives in Bush's White House. "I'd go into meetings on Capitol Hill with members who didn't have any African Americans in their districts and talk about prisoner reentry or drug treatment with them. They could not have cared less. Their eyes would glaze over, like, 'Why are you talking to me about this?'" When Bush would engage those lawmakers on the issues himself, Towey recalls, "The president would be told, 'You sound like a

Democrat. All this stuff about poor people, immigrants, refugees—this is how Democrats talk.'"

Much of this dissension within the Republican ranks boiled below surface level. Although racism was alive and well, and fears of ethnic targeting spiked after 9/11 (prompting Bush to visit a mosque and declare Islam a peaceful religion), xenophobia did not dominate the national political environment. That began to change after 2004. Bush believed his reelection was a mandate to do big things, including immigration reform. But his base saw it very differently. "The 2004 campaign was about who's keeping us strongest and who's keeping us safe, not who's making for a more just society," Rove says.

Pete Wehner, who led Bush's Office of Strategic Initiatives, says "there was a shift" after the president announced his intention to change U.S. immigration policy. "If you read the books by Rush and Hannity prior to 2005, you don't find anything on immigration. There was nothing on talk radio," Wehner says. "But our base was changing. Some of it was tied to 9/11. Some of it was tied to economic insecurity. Some of it was tied to a sense of lost culture for a lot of people on the right. Those attitudes on immigration were proxies for a lot of other things."

Bush saw it sooner than most. Early in his second term, while meeting with a few trusted advisers, the president confided that he was worried about some interconnected trends taking root in the country—and most acutely within the Republican Party. There was protectionism, a belief that global commerce and international trade deals wounded the domestic workforce. There was isolationism, a reluctance to exert American influence and strength abroad. And there was nativism, a prejudice against all things foreign: traditions, cultures, people.

"These isms," Bush told his team, "are gonna eat us alive."

THE WHEELS WERE COMING OFF BUSH'S WHITE HOUSE BY THE TIME the 2008 primary got under way.

Any two-term president's party craves fresh perspective after eight years, but the administration's snowballing ineptitude—the half-baked nomination of Harriet Miers to the Supreme Court, the abysmal handling of Hurricane Katrina, the gross negligence at the Walter Reed Army

Medical Center, the vice president sending a buckshot into his hunting buddy's face, among other misadventures—imposed a unique urgency on Republican candidates seeking distance from the incumbent.

For Ron Paul, the iconoclastic libertarian from Texas, this meant denouncing foreign intervention and relinquishing the role of global policeman. For Huckabee, the key was merging folksy Christian appeal with folksier populism, arguing that changes to Social Security or Medicare would unfairly punish the people whose hard work funded the programs in the first place. For Romney, the darling of the right, it was all about fiscal responsibility: balancing budgets, reducing spending, streamlining government agencies.

Then there was McCain. His candidacy was built less on policy specifics than on biography, implicit to which was a message of preparedness. McCain was hardly a Bush ally; the two battled ferociously for the Republican nomination in 2000, a race that climaxed with a smear campaign in South Carolina alleging that McCain had fathered a black child out of wedlock.[8] Yet the senator had tied his presidential hopes to the troop surge in Iraq. And despite its decided unpopularity when it was announced in January 2007, the surge worked, stabilizing a country that had been ravaged by sectarian violence since the U.S. invasion in 2003.

As it became clear that the party would nominate someone who, on two defining issues (immigration and Iraq), was in lockstep with a deeply disliked president, Republicans couldn't help but take out their frustration on McCain.

Not that he didn't have it coming. Years of bucking the party line and antagonizing the right had finally caught up with McCain by the time he became the party's presumptive nominee. That spring, when the Arizona senator dispatched his longtime colleague and respected fiscal hawk Phil Gramm to meet with House conservatives on his behalf, the former Texas senator (and would-be treasury secretary in a McCain administration) got an earful.

"Tell me something," Patrick McHenry, a young Republican from North Carolina said to Gramm. "Why should I not be physically ill at the prospect of a President John McCain?"

Taken aback, Gramm gave the only reply that passed the smell test: "Conservative judges."

The party's new standard-bearer wanted to show a united Republican front as he pivoted to the general election. But it was proving elusive. When McCain met with conservative movement leaders throughout the summer, seeking to soothe their concerns and coalesce their support, he was hammered with a constant refrain: He needed a conservative, pro-life running mate to offset the right's reservations about him.

The problem was, none of the conventional "short list" choices was appealing to McCain. He didn't care for Romney. He couldn't relate to Huckabee. He thought Minnesota governor Tim Pawlenty and Florida governor Charlie Crist were boring. In a contest against Barack Obama, the dynamic young Illinois senator who had outlasted Hillary Clinton and was marching toward history as the nation's first black president, McCain would not win by being conventional.

What McCain *wanted* to do—what he told senior aides he *would* do— was pick his close friend Joe Lieberman.

A former Democrat turned independent who shared McCain's hawkish foreign policy views, Lieberman would accentuate the Republican nominee's strengths: independent-minded, muscular on national security matters, experienced, and ready to govern on day one.

It wouldn't be that simple, however. Lieberman was pro-choice, an apostasy that McCain's advisers warned could shatter his uneasy alliance with the GOP base. McCain learned this lesson for himself in mid-August after suggesting to the *Weekly Standard* that former Pennsylvania governor Tom Ridge was under consideration, noting that Ridge's pro-choice position wasn't a deal breaker.[9] If this were a trial balloon, its pop registered on the Richter scale. The blowback—delegates threatening to defeat his choice at the convention, activist leaders saying their constituencies might sit out the election entirely—convinced McCain to heed the advice of his brain trust. Lieberman could not join the ticket.

It was time to panic. With the convention weeks away, the presumptive Republican nominee had no clue whom to choose as his running mate. Obama was consistently leading in the polls, and all the fundamentals—

unpopular president, dragging economy, change election—were in his favor. The circumstances called for a bold stroke, a daring maneuver from the maverick. But the base would revolt over Lieberman. McCain felt helpless, irritated by the quandary and underwhelmed by his options. Who could possibly shake up the race *and* excite conservatives?

CONGRESSIONAL REPUBLICANS SPENT THE SUMMER ATTEMPTING TO shift the national argument from two wars and a looming recession to friendlier terrain: energy. With gas prices soaring and Democrats opposing new means of exploration, the GOP tried to frame an exaggerated contrast between the parties: one looking out for the far-left environmental lobby, one looking out for working families struggling to fill their tanks.

As part of the initiative, Ohio congressman John Boehner, who after the 2006 election had succeeded Hastert as the top House Republican, led a delegation of GOP legislators to tour energy sites in Colorado and Alaska. When the group arrived on the remote North Slope, they were met by a forty-four-year-old first-term governor most had never heard of. By the time they left, the men in the group were remarking to one another: Sarah Palin had a *great* physique.

Nobody returned to Washington raving about Palin as a prodigy. To the extent that she was recalled by the congressmen to their friends and staff members, it was *that governor of Alaska is a knockout*. The thought of Palin as national timber simply did not compute. She had spunk and self-confidence by the barrelful, but was far removed, physically and intellectually, from the debates driving Washington policymaking. The only prominent Republican hawking Palin for VP was *Weekly Standard* editor Bill Kristol, who had met her on his own trip to Alaska. When Kristol called Palin his vice-presidential "heartthrob" on *Fox News Sunday*, just days after the delegation trip concluded, the congressmen sneered at the prospect of a pinup following in the footsteps of Dick Cheney, Nelson Rockefeller, and Hubert Humphrey.

What they didn't know was that McCain's campaign had been in contact with Palin.

It seemed unserious at first; McCain had met Governor Palin just once, and briefly, at a meeting of the National Governors Association. But with time running out, and the door suddenly slammed on the prospect of either Lieberman or Ridge, McCain's team flagged Palin as a last-second Hail Mary.

Intrigued, McCain spoke with Palin by phone on August 24—eight days before the GOP convention was set to convene in the Twin Cities of Minnesota.[10] The call was ostensibly to quiz the Alaska governor on her answers to McCain's vetting questionnaire. But for Steve Schmidt, the campaign's pugilistic senior strategist and its staunchest pro-Palin advocate, it was meant to get McCain comfortable with the only checks-all-the-boxes option available. Four days later, after a last-ditch plea from McCain to his staff to let him pick Lieberman, Palin arrived at the senator's Sedona ranch for a face-to-face meeting. The next afternoon, inside a rowdy arena in Dayton, Ohio, John McCain introduced Sarah Palin, a woman with whom he'd spent a sum total of a few hours, as his choice for the vice presidency of the United States.

"WHO?" OBAMA ASKED, FLYING TO CHICAGO FROM DENVER FOLLOWING his own party's convention.

David Axelrod, the Democratic nominee's chief strategist, was at a loss. His team had performed extensive opposition research on eight potential VP selections, and Palin wasn't one of them. He told Obama, sheepishly, that he didn't know much about the Alaska governor.

"Well, why did he pick her?" Obama asked.

"Because she's a woman," Axelrod replied, "and because he wants his campaign to be about change, too."

Obama thought for a moment. "Yeah, but this whole running for national office thing is really hard," he said. "She may be the greatest politician since Ronald Reagan. Maybe she comes out of Alaska and handles this whole shitshow. But let's give her a month and then we'll see."

Back in the nation's capital, Bush had a similar reaction. "So, what do you make of the news?" the president asked Dana Perino, his press secretary, as they watched Palin's introduction on television in the Oval Office.

"For a Republican woman, it's really exciting," Perino said, beaming.

"It is exciting," Bush agreed. He took a lingering pause. "But she's got no idea what's about to hit her."

Nor did the Republican Party itself. Practically overnight, Sarah Palin came to embody the most disruptive "ism" of them all, one that would reshape the GOP for a decade to come: populism.

THE ANNOUNCEMENT OF PALIN WAS SO RUSHED THAT PARTY OFFICIALS could not offer basic biographical talking points to inquiring reporters. Several of McCain's own staffers mispronounced the Alaska governor's name on conference calls the day of her rollout.[11] The famously disorganized campaign was already ill-prepared for the convention, and now his staff was tasked with selling the American public on someone they themselves knew nothing about.

Amid this inferno, the coolest performer was Palin herself. If she was daunted by the expeditious leap from Alaska obscurity to international celebrity, it didn't show. The governor was a born performer: warm, funny, charismatic, and devastatingly common. A union member's wife and a mother of five, including an infant son with Down syndrome, Palin oozed relatability to middle America. She was not merely a breakout star; she was a political phenomenon.

Both in her Dayton introduction speech and in her address to the convention a few days later, Palin sent tremors through the Republican Party. Her combination of homespun magnetism and theatrical fearlessness was breathtaking. Inside the suite belonging to Boehner, who was serving as the convention's chairman, party heavyweights whispered that she was Reagan reincarnate. Boehner was taking painkillers to manage sciatica and thought he might be hallucinating; this could not be the same small-state governor he and his colleagues had met just a few months earlier.

It was her, all right, and Palin was proving even more telegenic than any of those House Republicans could have imagined. After watching her onstage in Dayton, Roger Ailes, the chairman and CEO of Fox News, phoned one of McCain's senior advisers. Ailes explained a process he used for scouting talent at other networks: He would flip through news channels and stop when a female anchor held his attention. "That's what she just did for the Republicans," Ailes said.

All eyes were on Palin. With the GOP's new sensation dominating the headlines and turbo-charging the convention audience, it was easy to overlook who wasn't in Minnesota: Bush.

McCain had kept a strategic distance from the White House since winning the primary, not wanting to aid Obama's argument that he represented a third term for the current president. Bush was not offended, having embraced a gallows humor regarding his own unpopularity. But the McCain team's decision to keep him away from the convention, with Hurricane Gustav approaching the Gulf and the wounds of Katrina still fresh, enraged some Republicans. "It was disgraceful. He's the sitting president, the leader of the Republican Party for eight years, and he doesn't get to speak?" says Sara Fagen, the White House political director under Bush. "It was a disgraceful moment for John McCain and for the Republican Party."

Bush would later dismiss the snub, telling friends he wasn't bothered by it. But watching the convention on television one evening, he turned to Perino. "Do you think they know they're insulting me?" he asked.

Hurt feelings were an afterthought: For the first time all year, as they departed the Twin Cities, Republicans believed they could hold the White House. The polling suggested as much: After trailing Obama in almost every major national survey since May, McCain was either tied or ahead in eleven of the fourteen polls conducted the week after the convention. The impact on both campaigns, tactically and psychologically, was immediate.

The McCain team had gotten into Obama's head with its TV ad, "Celebrity," comparing the Democratic nominee to hotel heiress Paris Hilton, someone famous for being famous, and Spears, the troubled singer. Sensitive to the perception of hero worship, Obama's campaign scaled back, swapping rock star rallies for intimate town halls. Now, suddenly, it was McCain and Palin drawing the enormous crowds. Axelrod and Obama agreed it was time they revert to the bigger events, concerned that enthusiasm—and momentum—had switched teams.

The Palin bump proved to be a sugar high—stimulating but unsustainable, loaded with flavor but lacking in substance. Her unmasking was brutal: the yawning void of knowledge, the allergy to nuance, the lack of discipline and restraint. These deficiencies burst into view right around the one-month anniversary of her selection, just as Obama guessed, during a now-infamous interview with CBS's Katie Couric.

Among the other lowlights of the CBS segment, Palin could not name a single newspaper she read. Facebook and Twitter were in their infancies, but video of the interview rocketed around the internet thanks to YouTube, the blogosphere, and Tina Fey's devastating *Saturday Night Live* impersonation. It was the beginning of the end for Palin. Her polished facade was dissolving to reveal someone plainly unprepared to be a heartbeat away from the presidency.

"Everyone began to realize this woman is vacuous. She's got fifteen minutes of pop phrases and slogans, but she's not a deep thinker," Rove recalls. "We went from wanting people who were experienced and qualified to wanting people who would throw bombs and blow things up. The ultimate expression of that was Donald Trump, but Sarah Palin was the early warning bell."

Palin's ascendance, and the warring intratribal assessments of her, foreshadowed the party's struggle in foundational ways. To some on the right, her anti-elite, anti-intellectual shtick was hostile to that which makes government effective. "The Palin candidacy is a symptom and expression of a new vulgarization in American politics. It's no good, not for conservatism and not for the country," Peggy Noonan wrote in the *Wall Street Journal* in October.[12]

To others, however, Palin's swaggering one-woman insurgency represented the first blow struck against a corrupt league of career politicians who were hopelessly disconnected with America. "We have a polarization in this country not just of conservative-liberal, Democrat-Republican. We also have a polarization in this country of elites versus average people," Rush Limbaugh said on Fox News that same month. "The elites think they're the smart people, and they think Sarah Palin's a hayseed hick. . . . It's an amazing divide to watch the condescension, the arrogant condescension, for average, ordinary people, the people who make this country work."

The GOP's time in power was drawing to a close. But its cultural schism was only just emerging.

THEIR EXPRESSIONS WERE GRAVE, THESE GUARDIANS OF THE FINAN-cial universe, as they settled around the rectangular conference table in the Roosevelt Room. It was Tuesday the sixteenth of September. McCain was still riding a postconvention bounce, and Palin had not yet marginalized herself with self-inflicted wounds. But politics was the last thing on President Bush's mind.

Evidence had mounted over the past year suggesting an economic powder keg: wildly fluctuating bond markets, panic over bad mortgages, sluggish economic growth, risky bets by big banks. On "Black Monday," September 15, it exploded: Lehman Brothers and Merrill Lynch, two pillars of American finance, collapsed. Meeting with the president a day later, Treasury Secretary Henry Paulson and Federal Reserve chairman Ben Bernanke explained to Bush and his senior staff what had just transpired—and urged immediate federal intervention. They said he could not afford to wait: AIG, the world's largest insurance company, was also about to go under, threatening a run on the banks and a global economic meltdown.[13] Savings accounts would be wiped out; deflationary spirals would crush small businesses; banks would have no credit to extend; home mortgages and car loans would default by the millions.

Bush was sickened if not shocked. Paulson had kept in regular contact since the summertime failure of mortgage subsidizers Fannie Mae and Freddie Mac, and after their federal takeover in early September, Paulson had warned the president that the worst might be yet to come. For his blue-blooded heritage, Bush had never trusted the financial establishment and was known to sometimes question aloud its contribution to an equitable society. But this was not the time for such recriminations. "I don't give a shit about Wall Street right now," he told aides during the crisis. "This is about moms and dads getting money from ATMs."

The bankruptcy of AIG would be calamitous, Paulson and Bernanke told him. A federal rescue of the insurance giant was their only recourse. "You have my blessing," Bush said.

But it wasn't enough. Wall Street continued to crumble over the ensuing forty-eight hours, leading the same cast to reassemble in the Roosevelt Room on Thursday the eighteenth of September. This time, Paulson and Bernanke had a steeper ask. They believed a historic infusion of capital was needed to stabilize the financial system, and that it should be appropriated by Congress rather than authorized by government agencies.

This was heresy: Two weeks earlier, the GOP's convention platform had stated unequivocally, "We do not support government bailouts of private institutions." But Bush, in the twilight of his presidency, had long since spurned the electoral ramifications of his policymaking. When Senator Mitch McConnell had suggested to him two years earlier that he draw down troops in Iraq as a Democratic wave built in 2006, Bush told him, politely, to take a hike.

The president studied the two men. He knew Paulson's perspective better than Bernanke's. "What do you think will happen," Bush asked the Fed chairman, "if we don't ask for the money?"

Bernanke did not hesitate. "Mr. President, before I came into government, I was an economic historian. I studied the Great Depression. If we don't act, we could be facing another Great Depression—this one worldwide."

Bush grimaced. He told Paulson and Bernanke that he needed an hour. Back in the Oval Office, he called in his senior staff. "I want you to tell me exactly what you just heard him say," the president said. One by one, they repeated Bernanke: "Another Great Depression."

Bush nodded. "I'm sure as hell not gonna be that guy."

That evening, Paulson and Bernanke led a delegation to Capitol Hill to meet with the bicameral leadership of both parties. Once gathered in Speaker Nancy Pelosi's suite, Paulson wasted no time: The situation was dire and getting worse by the minute, he declared. They had a plan, but for it to succeed they needed prompt bipartisan cooperation. If anyone was tempted to call the treasury secretary's bluff, all they had to do was steal a glance at the Federal Reserve chairman. "I'll never forget it as long as I live," Boehner, the House GOP leader, recalls. "As Paulson was talking, I looked over at Bernanke and his lips were quivering. That's when I knew we were in big, big trouble."

Concerns were aired by leaders in both parties. Democrats wanted the root causes of the crisis addressed, Republicans wanted as light a government footprint as possible, and everyone pressed for a specific price tag. "Several hundred billion," Paulson said, eliciting groans and anxious looks. Still, the meeting was an overall success: Everyone left agreeing that timely action was necessary.

The next morning, Paulson announced the bipartisan consensus in favor of taking a "comprehensive approach," and Bush delivered remarks from the Rose Garden.

"Our system of free enterprise rests on the conviction that the federal government should interfere in the marketplace only when necessary," he said. "Given the precarious state of today's financial markets, and their vital importance to the daily lives of the American people, government intervention is not only warranted; it is essential."

The stock market rallied. Washington breathed a sigh of relief.

And then things went south. Paulson's subsequent unveiling of a $700 billion Troubled Asset Relief Program (TARP, for short) couldn't have gone over much worse inside the Capitol. The three-page proposal was intentionally scant on details, which the treasury secretary hoped would facilitate speedy passage. Instead, it had the opposite effect: With the narrative fast taking hold of Wall Street getting saved and Main Street getting screwed, constituents flooded their representatives' phone lines. House members revolted; every congressional office suddenly had prescriptions to fill the void of legislative specifics. The TARP proposal was snowed under. After two days of dead-end negotiations in Congress, Bush decided to use his bully pulpit.

In the middle of the afternoon on Wednesday the twenty-fourth of September, Bush stood in the White House theater practicing a primetime address he would deliver that evening urging Congress to approve TARP. Josh Bolten, the president's chief of staff, kept looking down at his phone. McCain's campaign manager, Rick Davis, was calling repeatedly.

When the rehearsal ended, an aide burst into the theater and announced that Senator McCain was trying to reach the president. Bolten and Bush exchanged glances and retreated to their respective quarters to return the calls, reconvening in the Oval Office a short while later. They had both been told the same thing: McCain was suspending his campaign and returning to Washington to focus on the financial crisis. He wanted a summit with Bush, Obama, and the congressional leadership at the White House.

The president frowned. "We don't *have* to say yes to this," Bush asked Bolten. "Do we?"

THE NEXT AFTERNOON, BOEHNER AND THE HOUSE GOP LEADERSHIP huddled in his suite on the second floor of the Capitol. It had been a miserable seventy-two hours. Uncertainty gripped all of Washington, but nobody felt more heat than Boehner. While the rescue package had critics in both parties and both chambers, the emerging silhouette of a deal was expected to ultimately have sufficient support among Senate Democrats, Senate Republicans, and House Democrats. All bets were off, however, when it came to House Republicans.

Many of them flatly rejected the concept of a bailout, arguing that free markets must be allowed to fail and self-correct. Even those who were open to Paulson's plan disliked what few specifics they had been given.

The wild card was the party's presidential nominee. Those Republicans opposed to the bailout hoped that McCain was returning to DC to smash it with a populist hammer, while those Republicans inclined to support it hoped McCain was coming to give them cover.

Suddenly, there he was, the maverick himself, parachuting into Boehner's suite unannounced and flopping down on the couch. The Republicans around the room exhaled in unison. The big White House summit was hours away, and nobody knew what McCain had up his sleeve.

"So, Boehner," the Republican nominee said, leaning forward. "What's the plan?"

Boehner's face turned a special shade of crimson. "What the hell do you mean, what's the plan?" he said, coughing cigarette smoke. "You're running for president. You suspended your campaign. You called the meeting. You tell us!"

McCain's return to Washington was a politically tactical move, not a legislatively strategic one. He had no plan. Republicans in the room knew it, and soon so would everyone else.

The economy had never been McCain's area of expertise, nor had he ever mastered the art of projecting empathy to the masses. As the son of a renowned Navy admiral and the husband of a multimillionaire beer heiress, the senator was unfamiliar with financial discomfort. He teased Huckabee during the primary for sounding off on corporate CEOs and global trade deals. He told voters in the Rust Belt, on more than one occasion, that some of their lost jobs were gone for good. He joked that his knowledge of economics boiled down to reading Alan Greenspan's book. He became flummoxed during an interview when asked how many houses he and his wife owned.[14] And just weeks earlier, on the day Lehman Brothers and Merrill Lynch went down, he had responded, "The fundamentals of our economy are strong."[15]

This was not what voters wanted to hear, particularly those whose livelihoods depended on an industrial vibrancy that was diminishing before their eyes. According to the Bureau of Labor Statistics, more than two million manufacturing jobs were eliminated between December 2007 and June 2009—15 percent of the entire manufacturing workforce, vanished, in just eighteen months.[16]

McCain's staff was unfamiliar with the bailout negotiations, so Boehner tasked one of his lieutenants, Mike Sommers, with accompanying the Republican nominee to the White House meeting. Sommers was supposed to brief McCain en route, but the candidate spent most of the car ride talking by phone with his wife, Cindy, about the first general election debate, scheduled for the following day. When they arrived at 1600 Pennsylvania Avenue, McCain turned to Sommers as they climbed out of their black SUV: "Okay, what do I need to know?"

Things weren't going much better in the Oval Office. Bush had arranged for some of the top Republicans—Boehner, McConnell, and Vice President Dick Cheney—to game-plan with Paulson prior to the larger gathering. But the powwow blew up when Bush, a close friend and golfing buddy of Boehner's, kept needling the House GOP leader about his inability to corral Republican votes. "That son of a bitch pissed me off," Boehner recalls. "I said, 'Well, if your treasury secretary had any fucking ears, we wouldn't be in this position!'" Paulson leapt from his chair to protest, and Bush ended the meeting before further damage could be done.

Things were more disciplined on the Democratic side. As the procession moved from Bush's office to the Cabinet Room, the Republicans noticed that Obama was huddled with Speaker Nancy Pelosi and Senate Democratic leader Harry Reid in a hallway, the three of them plotting in hushed tones. Boehner walked into the room and found his seat next to McCain. "Look," he whispered, "I think the less you say, the better."

As everyone found their chairs, Bolten pulled Bush aside. "Mr. President, I never sent you into a meeting of any consequence where I could not tell you what I expected or hoped to have happen," the chief of staff said. "I just want to apologize in advance."

By protocol, after his introductory remarks and a quick summary from Paulson, the president turned to his right and recognized Pelosi, the Speaker of the House, to lead off the discussion. "Senator Obama will be speaking for the Democrats," she replied.

The Republicans gulped hard. Their counterparts had coordinated. Bush turned to Obama, three seats to his left, and gave him the floor. What followed, according to six Republicans in the room, was a flawless diagnosis of the moment—the policy failures responsible, the solutions at hand, the political complexities of passing TARP through a discordant Congress.

Before he finished, Obama made sure to emphasize that a bipartisan agreement had been within reach *before* McCain called this meeting. That wasn't entirely true, but it was brilliant gamesmanship: If McCain hoped to be credited with saving the day by putting politics aside, Obama's counternarrative was that McCain's actions had actually *jeopardized* the negotiations at a moment of maximum delicacy.

"Well," Bush said, turning to McCain, "it's only fair that I call on you next."

McCain shook his head. "I'll wait my turn."

Republicans in the room were mortified. Boehner, now regretting having told McCain to say as little as possible, jumped in. He explained his members' qualms and suggested that tweaks would be necessary to deliver a respectable chunk of them. The room began to buzz with side conversations, talk of whipping votes and calling hearings and changing legislative text. Amid the chaotic cross talk, Obama finally raised his voice. "I'd like to hear what Senator McCain has to say."

The table fell silent. Everyone turned toward the GOP nominee. Clearing his throat and sizing up a single index card, McCain delivered a few boilerplate talking points about Republicans' reasonable concerns with the plan and his hope that a bipartisan consensus would emerge. Reid rolled his eyes, while Obama let escape an audible half laugh, half sigh.

"It was totally embarrassing," Boehner recalls of McCain's commentary. "He was unprepared. He had no message. He knew, at that point, he was going to lose."

As if things couldn't get uglier, Spencer Bachus, the ranking member on the House Financial Services Committee, seized the moment to boast that Republicans had incorporated strong taxpayer protections into Paulson's proposal. This infuriated Pelosi, who snapped back that Democrats had insisted on such provisions. Chaos engulfed the room once more, with its screaming matches heard down the hallway.

"Well," Bush said, pushing his chair away from the table, "I've clearly lost control of this meeting."

THINGS WORSENED IN A HURRY. DESPITE MCCAIN'S TEPID ENDORSE-ment of TARP during the next day's presidential debate, Republicans led the charge in routing the bill when it came to the House floor the following week. The total was 205 yeas and 228 nays; among Republicans, it was 65 in favor and 133 opposed.

As the clerk tallied the votes, GOP congressman Fred Upton of Michigan stood in the back of the chamber, phone to his ear, announcing to colleagues that the Dow Jones Industrial Average was plummeting. "Two hundred points . . . three hundred points . . . five hundred points . . . seven hundred points."

Boehner, watching the markets on a television in the cloakroom, said a prayer. McHenry, the feisty young conservative who'd voted against the bill, approached Upton and felt a wave of nausea. "I was a hard-core no," he says, "but as I listened to Fred, this sinking feeling came over me."

Back at the White House, Bush looked for the positives. "Hopefully," he told a pair of staffers, "this will scare some people straight."

It did and it didn't. A number of on-the-fence Republicans were sufficiently spooked to back the next version of the bill, no matter what it included. But for many on the right, the vote was cathartic. After failing for eight years to break the GOP's addiction to big government, conservatives had defied their party's president and its congressional leadership on the most urgent legislation in modern American history.

It had been a long time coming. Jim DeMint, the conservative South Carolina senator, had pledged to be a "team player" when he jumped from the House to the Senate, working on behalf of the party's senatorial committee to raise money for upcoming elections. But by the summer of 2008, his feuds with the establishment having escalated, DeMint broke away and started his own group, the Senate Conservatives Fund, meant to recruit and support conservatives who could restore the party's credibility. "The basic Republican platform of limited government was not evident in any of the things we were doing," DeMint recalls. "It just felt like Republicans had nothing to run on anymore."

A parallel sense of revolution was percolating in the lower chamber. As the White House scrambled to make recommended changes to the package, with Bolten taking over for Paulson as the point man on passing TARP, House conservatives called an emergency meeting in the Budget Committee's hearing room. On one side, Wisconsin congressman Paul Ryan and another conservative favorite, Kevin Brady of Texas, implored their colleagues to reconsider. They argued that $700 billion was nothing compared to the cleanup that would be necessary if hundreds of thousands of people lost their jobs overnight.

But Ryan and Brady were outnumbered. A murderer's row of House conservatives—Mike Pence of Indiana, Jeff Flake of Arizona, Jim Jordan of Ohio, and Jeb Hensarling of Texas—argued that free-market principles meant nothing if they could be jettisoned at the first sign of crisis. "The question is, are we Republicans or are we conservatives?" Hensarling asked the group.

"We're Americans," Ryan replied angrily, "And if we don't do some-thing, this economy is going to crash." In truth, Ryan feared not just the crash itself. If Democrats wiped out Republicans that November, with the economy in ruins, he warned his comrades, "this will be FDR on steroids. It'll be another New Deal, run through Alinsky in Chicago," he said, referring to the legendary community organizer and left-wing boogeyman Saul Alinsky.

Ultimately, a revised bill passed both chambers a few days later and was signed into law. The bleeding stopped. The financial sector stead-ied. The program, by any objective metric, worked. But the political repercussions suggested otherwise. Institutional mistrust and class divisions were exploding in real time; Republicans who backed TARP would be punished, while those who rejected it benefited.

"Not enough of our members bought into the gravity of the situation—and they were rewarded for not buying into it," says Eric Cantor, the House GOP's chief deputy whip, who struggled to secure the votes needed to pass TARP. "That mentality would come back to haunt us."

A decade later, Flake, who touted his TARP opposition while jump-ing from the House to the Senate in 2012, is the only such Republican to express regret for how he voted in September 2008. "In the House, it's much easier to vote no and hope yes," he explains. "When you're one of 435, it's easier to cast an ideological vote and force someone else to carry your water. But it's irresponsible. When I got to the Senate, I decided I couldn't do that anymore."

TARP is quite possibly the most successful government program of its generation. All the money was paid back, with interest, and experts believe that the intervention almost certainly staved off a Depression-like catastrophe.[17] But the entire episode was scarring for millions of Americans who became convinced that Washington and Wall Street were playing by a different set of rules; that the economy was rigged against them; that professional politicians had sold them out.

"McCain came back to bail out the banks. He had a chance. I was hoping he wouldn't vote for it," says Jordan. "That was when the populist sentiment started to take root around the country. I think that was probably laying the groundwork for what happened in 2016."

THE FINAL MONTH OF THE CAMPAIGN WAS ANTICLIMACTIC. THOUGH the financial rescue package had finally passed, the Republican Party's management of the affair had hardly inspired confidence. After a second term plagued by volatility, it was yet another crisis on the GOP's watch. This, combined with his own economic amateurism and his running mate's slow-motion implosion, was too much for McCain to overcome.

The candidate came to peace with this. But many Republicans could not.

Palin thrashed wildly in the campaign's final weeks. She alleged that Obama was "palling around with terrorists." She also invoked his former pastor, Rev. Jeremiah Wright, whose controversial sermons at a black church McCain had declared out-of-bounds for criticism, determined that, win or lose, he would not be remembered for injecting race into the contest.

Several of McCain's top aides spent the home stretch bad-mouthing Palin to the press, attempting to pin the imminent defeat on her. Steve Schmidt, the senior strategist who had insisted on picking Palin (and who, ten years later, would announce his departure from the Republican Party due to Trump's takeover) suffered what friends described as a nervous breakdown and left the campaign for three weeks in October, returning just before the election to begin shoveling blame onto Palin.

On Hannity's Fox News show, a guest described the nature of Obama's community organizing work in Chicago as "training for a radical overthrow of the government." In battleground Pennsylvania, Bill Platt, chairman of the Lehigh County GOP, warmed up a McCain rally by mentioning how Obama didn't wear an American flag lapel pin. "Think about how you'll feel on November 5 if you wake up in the morning and see the news that Barack Obama—Barack *Hussein* Obama—is the president-elect of the United States of America," Platt warned.[18]

McCain's closing argument—"Who is the real Barack Obama?"— aimed to contrast the Democratic senator's ultraliberal voting record with his centrist rhetoric. Obama promised to deliver comprehensive immigration reform, for instance, but the Illinois senator had helped torpedo the McCain/Kennedy effort by supporting a "poison pill" labor amendment. Obama railed against money in politics, but he became the first presidential nominee ever to reject public financing for his campaign, reversing an earlier pledge and triggering an avalanche of outside spending.[19] Even Obama's opposition to same-sex marriage, Republicans felt, was insincere, aimed at mollifying white moderates and black churchgoers.

Yet more than drawing attention to these issues, McCain's approach was unwittingly successful in eliciting ugly responses from the right. Shouts of "terrorist!" echoed at Republican events nationwide. Conservative websites exploded with last-minute allegations that Obama had been born overseas; that he was a Muslim; that he was a Manchurian candidate. Rock bottom was reached at an October 10 rally in Minnesota, where McCain was repeatedly booed for telling his town hall audience that they should not be scared of Obama. At one point, a woman named Gayle Quinnell stood to speak. "I can't trust Obama," she told McCain. "I have read about him and he's not . . . he's not . . . he's an Arab."

McCain shook his head and took the microphone from her hands. "No, ma'am," he replied. "He's a decent family man, citizen, that I just happen to have disagreements with on fundamental issues. And that's what this campaign is all about."

Watching the news coverage with Obama and their team at Chicago headquarters, Axelrod says he was stunned—not at McCain's honorable defense of his opponent, but at the reactions to it. "I remember when John McCain was the 'it' guy among the young Reaganite class in Congress," Axelrod says. "To see him shouted down at his own rally for showing a modicum of civility, I just said out loud, 'My God, I've never seen anything like this.'"

Obama brought out the worst in the Republican base. The seeds of anger and resentment, of nativism and victimization, were sown by forces outside his control long before his ascent. But he harvested them in a way no other Democrat could. The succession of liberal policies; the ostensive shaming of patriotism ("I believe in American exceptionalism, just as I suspect that the Brits believe in British exceptionalism and the Greeks believe in Greek exceptionalism"); the imperiously lecturing tone; the hints of class condescension ("They cling to guns or religion"); perhaps most critically, the dark skin and the African roots and the exotic name—any of these elements, on their own, might not have been so provocative. But in this era of convulsion and cultural dislocation, Obama was a perfect villain for the forgotten masses of flyover country.

DAYS BEFORE THE ELECTION, RICH BEESON, THE REPUBLICAN NAtional Committee's political director, told a pair of junior staffers that they were about to witness history. The youngsters perked up, having heard nothing but doom and gloom for the past month with regard to their party's prospects. "The way we lose this election," Beeson told them, "is going to be historic."

Barack Obama won the presidency in a landslide, carrying the Electoral College by a margin of 365 to 173 and winning the popular vote by nearly ten million—the biggest spread since Ronald Reagan's forty-nine state reelection romp in 1984. There was no silver lining for the GOP: Democrats expanded their majorities in both houses of Congress, giving the incoming president and his party unified control of the government and a mandate to make wholesale changes to Washington and the rest of the nation.

More concerning for Republicans than the scope of Obama's victory were the fundamentals behind it: The Democratic nominee had turned out huge numbers of minorities, young voters, and women with college degrees. This "coalition of the ascendant," as journalist Ron Brownstein described it, represented the fastest-growing segments of an electorate undergoing a rapid, far-reaching makeover. While McCain captured 57 percent of white men and 55 percent of whites overall, he won just 43 percent of women, 31 percent of Hispanics, and 32 percent of voters under age thirty.

The implications were chilling. Republicans weren't just heading into political hibernation; they were at risk of entering a demographic death spiral. "Things looked pretty bleak," Boehner recalls. "You've got this young, dynamic African American who rebuilt the Democratic

Party in one fell swoop. There was no way out. We were going to be in the minority for one hell of a long time."

The morning after Obama's victory, a senior RNC official handed down orders to his communications staff. They were to plant a story in the media about grassroots support for a new party chairman, a black Republican by the name of Michael Steele.

AS THE NEW PRESIDENT PREPARED TO TAKE OFFICE, REPUBLICANS faced a moment of reckoning. For the past generation, the party had promoted a set of principles colloquially described as a "three-legged stool": fiscal responsibility, social conservatism, and strong national defense. Sifting through the wreckage of 2008, they found that the stool had collapsed. Republicans had spent recklessly while exposing their military's limitations after fighting a two-front war for the better part of a decade.

Only the social conservatism had been strictly adhered to, and even within that foundational conception, cracks were showing. The Bush administration's effort to pass a constitutional amendment banning same-sex marriage, for instance, had rankled many in the Republican professional class. This foretold of the growing disconnect between the party's elite and its base on many other issues that transcended the divide between secularists and religious voters. Whereas the questions of immigration, trade, and entitlement spending were understood by upscale, white-collar Republican moderates through a prism of macroeconomics, they were processed by the party's working-class conservatives through a filter of societal insecurity.

The GOP had once been a country club party, drawing its life force from the discipleship of affluent suburbanites. But that was changing. As America's wealthier, better-educated voters grew more progressive in their social views, they had begun drifting into the Democrats' column. At the same time, the Democratic Party's rejection of Bill Clinton's centrism—and its abandonment of big labor's focus on protecting American jobs—was beginning to push its blue-collar, less-educated voters rightward into the Republican camp.

Into this moment of realignment had stepped Obama, the urbane citizen of the world, and Palin, the tough-talking hockey mom whose husband worked oil rigs and raced snow machines.

Even as Democrats ran away with the election of 2008, Palin's appeal was a revelation. She was connecting with portions of the electorate in ways that nobody had since Reagan. But unlike the Gipper, she was not channeling their hopes and ambitions and highest aspirations. Instead, she was provoking their fears, fanning their anxieties, inciting their animosities. And it worked.

This, more than any botched interview or off-the-cuff comment, fueled the rift over Palin within the GOP: She was doing that which horrified the party's establishment, and doing it well.

"She was the early embodiment of some of the problems that would plague the party: mediocrity, anger, resentment, populism, proudly anti-intellectual, and increasingly bitter. And she was a *rock star* for it," says Wehner, the Bush White House official. "That was a sign that something was going on in the Republican base. We went from glorifying excellence and achievement to embracing this anger and grievance and contempt."

It was a long time coming. Palin's resonance with Republican voters was, above all, an indictment of the party's tone-deaf arrogance. Having catered to the aristocrat caste atop the GOP for decades, winning far more elections than they lost along the way, Republicans were blissfully ignorant of the discontent simmering below the surface. When it boiled over, the defensiveness of the elites—reproaching Palin, for example— only made things worse.

"I really think what created the problem we have today in the party was the donor class and the intellectual class blaming that loss on Sarah Palin," says David McIntosh, a former Indiana congressman and longtime leader in the conservative movement. "We felt the establishment guys blew it—they were the ones in charge under Bush. They lost; they were out of power. So, the effort to scapegoat Palin fell on deaf ears."

In the aftermath of the 2008 election, McIntosh and several other heavyweights on the right launched a group called the Conservative Action Project.[20] Its mission was to bring together under one roof the leaders of prominent activist groups, hoping to pool their ideas and leverage their numbers to rebuild a Republican Party that sounded more like Palin than Bush. They began holding weekly meetings in early January of 2009, their first one in a conference room at the Family Research Council, plotting the ways in which they could steer the new GOP farther to the right.

The energy these conservatives saw and tapped into might have been lost on the Republican establishment, but it did not escape the Democrats. Indeed, while many GOP leaders worried about a permanent tilt in the country's political axis, Obama knew, by virtue of the huge expectations for him, that backlash was inevitable. The only question was its size and strength.

Six weeks after the election, the incoming president and his advisers met with Bush's team for an official transition briefing on the economy. The updates were brutal: While the bank bailouts had prevented a systemic collapse, families and communities were being pounded. Thousands of jobs were being shed, waves of homes were going into foreclosure, and all indicators pointed to things getting much worse before they got better.

When the meeting ended, David Axelrod walked out of the room and looked at his boss. "We're going to get our asses kicked in the midterms," he told Obama.

WHATEVER WAS TO HAPPEN BETWEEN THE TWO PARTIES, OR WITHIN them, was no longer Bush's concern. He was taking a vow of political silence, he told friends, eager to extract himself from the public glare after eight enormously trying years.

He would not spend his life as a private citizen consumed by partisan wins and losses. He would be rooting for the country; he would be rooting for his successor. If there was one thing Bush worried about as his tenure closed, it was the "isms" he saw infecting America's mind-set—and how they might animate the GOP's opposition to Obama.

On Wednesday, January 14, six days before he left office, the president convened a group of conservative talk radio hosts in the Oval Office. The firebreathers, such as Rush Limbaugh and Sean Hannity, were not invited; they were hopeless cases. Bush wanted to speak to the "reasonable right-wingers," including Dennis Prager, Hugh Hewitt, and former secretary of education Bill Bennett.

"Look, I asked you here for one reason," Bush told the group in a solemn tone. "I want you to go easy on the new guy."

CHAPTER TWO

JANUARY 2009

"He could have annihilated us for a generation."

THE CANNONS THUNDERED AT FIVE MINUTES PAST NOON. THE FORTY-fourth president of the United States had just taken his oath of office, and the faux artillery fire merged with the roar of some two million people[1] on the National Mall to create a spectacle befitting the momentous occasion. Stepping to the podium, surveying the record-setting crowd braving a subfreezing chill to witness the inauguration of America's first black president, Barack Obama itemized our national crises: protracted wars abroad, economic hardships at home, rising health care costs, failing schools, flawed energy policies, and a reluctance to recognize the changes inherent to a new century.

"Today I say to you that the challenges we face are real. They are serious, and they are many. They will not be met easily or in a short span of time," Obama said. "But know this, America: They will be met."

He continued: "On this day, we gather because we have chosen hope over fear, unity of purpose over conflict and discord. On this day, we come to proclaim an end to the petty grievances and false promises, the recriminations and worn-out dogmas that for far too long have strangled our politics. We remain a young nation. But in the words of Scripture, the time has come to set aside childish things. The time has come to reaffirm our enduring spirit; to choose our better history; to carry forward that precious gift, that noble idea passed on from generation to generation, the God-given promise that all are equal, all are free, and all deserve a chance to pursue their full measure of happiness."

Seated on the stage risers behind him, Eric Cantor, the second-ranking House Republican, felt a twinge of panic. This wasn't just a quadrennial shift of power in Washington; it might prove a tectonic disturbance in the trajectory of the country. Obama and the Democrats, it seemed, could rule for as far as the eye could see.

"I had the best seat in the house at that inauguration. I was sitting against the rail and looking out across that sea of people, all the way to the monument, and it was just staggering," Cantor says. "We had elected a black president, and here he was, talking about changing America and certainly acting as if he wants to incorporate us into the end product. He had a seventy-some-percent approval rating[2] and these Democratic supermajorities. We were up against it."

Patrick McHenry, the North Carolina congressman, who'd never seen such a crowd in his life, had a more visceral reaction. "I thought we were completely, permanently screwed."

He wasn't the only one. Two weeks earlier, the president-elect had returned to Washington for the first time since Election Day and called a meeting with congressional leaders of both parties. As they gathered in the Lyndon Baines Johnson Room, just off the Senate floor, Obama made brief comments to a gaggle of reporters. "We are in one of those periods in American history where we don't have Republican or Democratic problems, we've got American problems," he said. "My commitment as the incoming president is going to be to reach out across the aisle, to both chambers, to listen and not just talk, to not just try to dictate but to try to create a genuine partnership, so that we are actually doing the people's business at this time of extraordinary difficulty."

The Republicans lawmakers in the room exchanged smirks—putting on a show for the press, they figured. But once the journalists were shooed away, and the players got to work discussing the framework of an economic stimulus package, the incoming president's tenor remained the same: earnest, approachable, even humble. This was not the Obama they'd expected. He listened intently to Republicans' ideas. He acknowledged their concerns. He hinted that he was receptive to their biggest priority, making tax relief central to the stimulus.

By the time the meeting adjourned, those Republicans present were somewhere between delirious and devastated. They had never bought Obama's campaign rhetoric, his promises to transcend partisanship and heal a fractured body politic. They believed him to be a hardened progressive with velvety eloquence, and they were counting on the emergence of his true colors for their survival. They hoped to use his sky-high potential against him: Once voters realized that Obama wasn't a great compromiser, his astronomical numbers would fall back to earth and Republicans would begin their journey out of the wilderness. Masterminding this theory was Mitch McConnell, the GOP Senate leader, who told allies after the election that the key to regaining power would be shattering the mystique of Obama's post-partisan image.

The January 5 meeting was hardly the start McConnell or his colleagues had envisioned. There was nary a negative word to say about Obama as Republicans confronted a waiting horde of media outside the LBJ Room. "I think this bill is going to start out, and hopefully end, as an example of very significant bipartisan cooperation," McConnell said.

Ducking the cameras, Cantor hustled across the Capitol complex. He glanced at his chief of staff, Steve Stombres. "What did you think?" Stombres shook his head. "I was inspired," he replied. Brad Dayspring, Cantor's communications director, was somewhat less diplomatic. "If he governs like that," Dayspring told his boss, "we are all fucked."

Cantor knew as much. So did Boehner. But they were in no position to sabotage the incoming president. There would be plenty of opportunities to draw lines in the sand; for the time being, with the economy on life support and Washington under tremendous pressure to produce, they would take Obama's promise of cooperation at face value. Back on the House side of the building, Boehner popped into Cantor's office with a request: Put together an outline of some core Republican suggestions for the stimulus bill. "And none of the right-wing stuff," he added. "We want broad support."

Sitting on the inauguration stage two weeks later, Cantor fretted. He had done what Boehner asked, drafting a list of five items to share with Obama when they reconvened after his swearing-in. But now he wondered what good it would do. Staring out at the National Mall, Cantor would recall to friends, he wondered: Was the GOP going extinct?

THAT NIGHT, MORE THAN A DOZEN CONSERVATIVE LAWMAKERS, INcluding Cantor, Paul Ryan, and Kevin McCarthy from the House and

Jim DeMint, Bob Corker, and Jon Kyl from the Senate, gathered inside the Caucus Room, a stylish Washington steakhouse. Parties were jumping across the capital city in honor of the new president, but Republicans weren't in any mood to celebrate. Organized by focus group guru Frank Luntz, and featuring a guest appearance by former Speaker Newt Gingrich, the dinner would gain infamy as the inception of the GOP's coordinated resistance to Obama.

And yet, opposition was not the only item on the menu. The dinner also was an exercise in wound-licking and soul-cleansing for members of a Republican Party that had strayed far from its principles, as the journalist Robert Draper, who first reported on the dinner, wrote in his book *Do Not Ask What Good We Do: Inside the U.S. House of Representatives.*[3] History will remember the GOP's obstructionism as its organizing principle during the Obama years, and appropriately so. But the backdrop of Bush's presidency and the pall it cast over conservatism often goes ignored in understanding the mentality of Republicans circa 2009. Cooperating with the new president was dangerous not just because it handed him a victory, but because it fed a perception on the right that there was no longer any meaningful distinction between the two parties.

The dilemma within the GOP, of course, was that many rank-and-file Republicans *were* moderates. They *didn't* diverge sharply from the Democrats. They *hadn't* objected to the big-government policies of the Bush administration. And they *weren't* keen to tangle with a dauntingly popular new president. This was especially true for the thirty-seven House Republicans whose districts Obama had just won.[4]

Cantor understood this better than most. As minority whip, the House GOP's designated vote-counter, he was tasked with knowing his members the way a husband knows his wife: likes and dislikes, goals and motivations, verbal tics and personality quirks. Raised in Southern Virginia as an observant Jew, Cantor had long since learned to straddle disparate worlds. (There weren't many Jewish lawmakers to be found in the GOP, though Cantor always got a kick out of hearing Mike Pence, his evangelical colleague from Indiana, refer to Jews as "our people.")

Despite being more conservative than most in his party, the forty-five-year-old Cantor had skillfully worked his way into its leadership, intuitively harmonizing the dueling instincts of pragmatism and purity. Now he was confronting intraparty dynamics that were seemingly impossible to balance: Conservatives had incentive to fight Obama, while moderates had reason to work with him.

The stimulus offered a fascinating first case study.

Describing the economy as a "very sick" patient who needed to be stabilized, Obama set a deadline of February 16, Presidents' Day, for the stimulus package to arrive on his desk. Democrats weren't waiting around. By the time Obama met with congressional leaders at the White House three days after his inauguration, Pelosi and her colleagues had already drafted legislation. This irritated House Republicans, whose ideas Obama had promised to consider. Pelosi scoffed at those concerns. "Yes, we wrote the bill. Yes, we won the election," she told reporters at the Capitol.

The next morning, when Obama hosted a meeting with lawmakers inside the Roosevelt Room, Cantor promptly handed out copies of the five-point priority list he had crafted. It was more than a tad presumptuous, and laid the foundation for Obama's dislike of Cantor, but the president played it cool. "Nothing on here looks outlandish or crazy to me," he remarked.

The list was heavy on tax relief: for families, small businesses, home buyers, the unemployed. This surprised no one. (*Republicans and Tax Cuts: A Love Story.*) More telling was what it omitted: infrastructure.

Boehner and Cantor knew the one thing that could buy off their members was big spending on roads and bridges; Republican voters, whether in busy commuter suburbs or neglected rural communities, love few things more. What they didn't know was whether *Obama* knew this. By not proposing a massive investment in infrastructure, GOP leadership was both testing the new president and carving out a potential escape hatch if the negotiations went south.

This bit of chicanery crystallized the GOP's quandary in dealing with the unified Democratic government. Obama had the votes to pass laws with or without them; the trick for Republican leaders was influencing legislation in a way that made it appealing to conservatives, not just moderates, so they wouldn't be accused of selling out their right flank. A more experienced Democratic president might have recognized this and reacted accordingly. But having served less than one full term on Capitol Hill before winning the White House, Obama seemed not to fully grasp the ideological fault lines within the congressional GOP—or how to exploit them.

Their own suggestions aside, Republicans were puzzled by many of the Democrats' priorities. Obama had conceptualized the legislation as a shot in the arm to the flatlining economy—instant help in the form of shovel-ready jobs, tax cuts, and funding for state governments. Yet the emerging bill looked more like a liberal grab bag of programs years in the wishing: increased Pell Grants, expanded broadband internet, investments in green energy companies. These proposals had merit, certainly, but Republicans were justified in questioning why billions of dollars should be spent on projects that paid no immediate dividend when *urgency* was the buzzword inside the Beltway.

As this debate intensified in the January 23 meeting, with Cantor and Kyl, the Senate minority whip, pressing Obama on why the White House was favoring certain programs, the president lost his sense of humor. "Elections have consequences," he told Republicans around the table. "And I won."[5]

It was an unforced error by Obama—and an immeasurable gift to the GOP.

Boehner, Cantor, and McConnell had already seized on Pelosi's quote to impress upon their colleagues that the House Speaker, a San Francisco progressive, was pushing Obama leftward, persuading him not to waste time playing footsie with Republicans. Not everyone believed that. But now Obama had echoed Pelosi's sentiment in a way that seemed dismissive at best and hostile at worst.

Back on Capitol Hill, Boehner and Cantor convened their members inside a sprawling conference room in the House basement. They relayed the details of the meeting, including Obama's quote and the disagreements over spending in the package. In that moment, the process surrounding the stimulus package changed in fundamental ways. Not only were members bothered by Obama's remark, but they were dismayed at the relative pittance being allocated to infrastructure.

Cantor, before that morning's White House visit, had counted at least thirty Republicans whom he expected Democrats to pick off, especially if the final product featured significant spending on shovel-ready projects that would be visible in their states and districts. But the Democrats, to his disbelief, weren't prioritizing transportation infrastructure. They weren't doing anything to court his most easily converted members. Soon, Cantor's number of susceptible Republicans was cut in half. Within a few days, it had dwindled into the single digits, and then even lower.

Boehner was stunned. He knew there would be a renewed emphasis on fiscal restraint among conservatives hoping to turn the page on Bush's legacy. And he suspected that, sooner or later, if the economy didn't show signs of life, moderate Republicans would feel emboldened to do battle with Obama as well. But the stimulus was an unlikely showdown. The minority leader had felt certain that at least a few dozen of his members, particularly those in Obama districts, would support the president's first initiative.

On January 28 the American Recovery and Reinvestment Act passed the House—and remarkably, not a single Republican voted for it.

Because of Cantor's well-publicized confrontation with Obama, and the fact that one-third of the bill comprised tax breaks,[6] it was natural to blame the minority whip (or credit him) with imposing total discipline on his ranks. But this missed the bigger picture. By allowing the stimulus to become larded with pet projects, by not pressing for massive infrastructure investments, and by saying, "I won," however benign the intent, Obama had given Boehner and Cantor just what they needed to lock up a House Republican Conference that was primed for a jailbreak. It also played right into McConnell's master plan of puncturing the president's bipartisan aura.

"We came back in here at the beginning of 2009, we were on the way down to forty, which is the irrelevant number in the Senate. And the question was, is there a way back?" McConnell recalls. "My view was we needed to test whether the American people were simply frustrated in 2008 by the war, and the financial meltdown, or whether they really wanted to go hard left. . . . We had to draw a bright line of distinction between us and what the Democrats in full control of [government] were trying to achieve. And that meant keeping our fingerprints off things."

The Democrats walked right into this trap. They had been out of power for eight years—twelve, really, when considering the turbulent second term of Bill Clinton—and Republicans had run roughshod over them during that period, doing "nothing to encourage bipartisanship," as Boehner admitted. Now, imbued with absolute authority over Washington and feeling compelled to act quickly, congressional Democrats had convinced the new president that he didn't owe Republicans anything. The result was a flawed bill hustled onto the House floor just eight days after the inauguration and approved by the Senate less than two weeks later.

"If it's passed with 63 votes or 73 votes, history won't remember it," Dick Durbin, the Illinois senator and a mentor to Obama, told the *Washington Post* the day of the contentious White House meeting.

In fact, the vote totals are about the only thing history remembers about the stimulus.

Three Republicans broke with McConnell when the bill passed the Senate, giving Obama's legislation the faintest whiff of bipartisanship. But it was the House GOP's blanket opposition that stole the headlines and set the tone for eight years of escalating polarization.

"That was the beginning of the end for Obama," Boehner says of the stimulus fight. "If he had reached across the aisle in a meaningful way, he would have found a lot of Republicans ready to work with him—whether Eric and I liked it or not. He could have annihilated us for a generation."

TWO DAYS AFTER HOUSE REPUBLICANS UNIFORMLY REJECTED THE stimulus package, another tense vote was under way—this one inside a Washington hotel ballroom. It was time for Republicans to choose a party chairman. Mike Duncan was Bush's handpicked choice to lead the RNC during the final two years of his presidency, but the winds of change were gusting after the party's drubbing in November, and Duncan dropped out after the third round of balloting. That left four remaining candidates. The insiders were Katon Dawson and Saul Anuzis, chairmen of the state parties in South Carolina and Michigan, respectively. The outsiders were Ken Blackwell and Michael Steele. The RNC is composed of three members—a chairperson, a committeeman, and a committeewoman—from each of the fifty states and five territories, plus Washington, DC. Blackwell and Steele were not part of "the 168," as the RNC's membership was known. But something more conspicuous set them apart: Both candidates were black.

Diversity had never been a strength for the Republican Party, yet the homogeneity of its national leadership was especially striking: Only three of the 168 were black. With the post-Bush GOP suddenly leaderless, and Obama dramatizing the racial chasm between the two parties, there was a groundswell among the party elite to choose a nonwhite chairman. And despite being a non-RNC member, Steele was an obvious fit. Not only did the former lieutenant governor of Maryland have establishment cred (Johns Hopkins undergrad, Georgetown Law, longtime insider, and a onetime state party chairman), but he was also a regular on the cable news circuit, exuding a charismatic media savvy rarely associated with Republican politics.

Not everyone was sold. The GOP's most glaring vulnerability was organization; whatever remained of the Rove/Bush machine had been wiped out by Obama's historic grassroots army. Steele was selling himself as an optical counterbalance to the new president, someone who could lead the GOP's messaging and public relations operation. A party chairman's work, however, is done primarily behind the scenes, raising money and strengthening state affiliates. For this task, Dawson, the South Carolina chief, was better equipped. But as the contest intensified, so, too, did the whispers about Dawson's membership in an all-white country club. It was enough to tip the scales: On the sixth ballot, Steele prevailed over Dawson to become RNC chairman.

Immediately following his victory, Steele was whisked two hundred miles southwest to Hot Springs, Virginia, where House Republicans were holding their annual retreat. The atmosphere there was ebullient. Fresh off their defiance of Obama, and for the first time in months, the GOP lawmakers felt a sense of optimism. Pence, the newly elected number three House Republican, planned a weekend-long pep rally. Boehner told his troops that the stimulus vote would be remembered as the party's return to fiscal responsibility. Cantor asked his lieutenants to autograph a bottle of wine that they would uncork after winning the majority in 2010. And Pence played a clip from *Patton* in which George C. Scott, portraying the famed World War II general, says, "We are advancing constantly, and we're not interested in holding onto anything—except the enemy. We're going to hold onto him by the nose, and we're gonna kick him in the ass. We're gonna kick the hell out of him all the time, and we're gonna go through him like crap through a goose!"

None of this seemed terribly realistic. Their rejection of the stimulus did not change the fact that House Republicans, by and large, were big spenders. Or that House Republicans, deep in the minority, stood little chance of retaking the majority in two years. Or that House Republicans, intimidated by Obama's approval rating, still believed they would be obligated to cooperate with him.

In fact, exhilarating as their stand against the stimulus was, many of Boehner's members had hoped to vote for it. Those representing districts in the industrial heartland were especially desperate to help their constituents. Paul Ryan, whose hometown of Janesville, Wisconsin, was being devastated by the closing of a General Motors plant, had expected the stimulus package to offer something closer to a fifty-fifty split between tax cuts and infrastructure spending, and was slack-jawed by the Democrats' final product.

Ryan still believed, however, that there was plenty of work to do with the new president. Having crafted a controversial bill in 2008 calling for a restructuring of entitlement programs—a proposal that the House GOP's campaign arm urged its incumbents to disown—Ryan was pleasantly surprised when Obama, after taking office, reached out privately to offer positive feedback. There seemed to be real potential for an alliance: A year later, in early 2010, Obama would praise Ryan as a "sincere guy" with a "serious proposal" for cutting the deficit.[7]

Ryan, an annoyingly earnest midwesterner who came to Washington straight out of college—he worked as a waiter, a think tank researcher, and a Capitol Hill staffer before winning his congressional seat at age twenty-eight—was under no illusions about Obama's liberalism. But unlike most of his comrades, Ryan believed the president was uniquely suited to pursue major bipartisan reforms to government in a way Bush had never been. While other Republicans spent the retreat talking tough about Obama, the congressman from Wisconsin's First District was more circumspect.

"The president did a good job laying the groundwork for the future," Ryan told *Politico*, referencing their recent conversations. "Speaking for myself, I think the president is showing us that he wants to collaborate."[8]

Steele, meanwhile, received a standing ovation when he stood to address the gathering. "My mom was a sharecropper's daughter with a fifth-grade education," the new party chairman said. "If my mother knew how to balance the budget without taking money out of my pocket, I'm sure that the rest of the folks out here on the other side should know how to do that as well."

Republicans were ecstatic at the new voice, and the new look, of their party. It wouldn't last.

Over the next few months, Steele put his foot in his mouth so many times it warranted a surgical relocation. He promised to bring conservatism to "hip-hop settings."[9] He said Democrats voting for the stimulus were trying to "get a little bling, bling."[10] He jokingly linked Bobby Jindal, the Indian American governor of Louisiana, to the film *Slumdog Millionaire*, a film set in the ghettos of Mumbai.[11] He called abortion "an individual choice" that should be left to the states.[12]

None of this should have been surprising. Steele was a known quantity in the party, an opinionated rabble-rouser with a penchant for provocation. Interestingly, while the RNC tried to tame its leader in some respects—members suggested, for instance, that his colorful suits and ties were a bit much—they were eager to use his *uniqueness* in other ways. When Steele went to give a major speech early in his tenure, he was handed prepared remarks that included several jokes about Obama's birthplace. He refused to read them.

"I said, 'Do you understand what it's like for a black man to stand up and say another black man is not born here?'" Steele recalls. "People in the party wanted me to go out there and start hitting Obama on his birth certificate to score points with the base and get them all fired up. I'd rather get people fired up about being right about policy and challenging the status quo that way, as opposed to playing the race card against the president."

Others weren't so reluctant. As Rush Limbaugh led conservative talk radio down a dim path of racial dog whistling, declaring himself in the summer of 2009 to be a full-fledged "birther"—one who doubted that Obama was born in the United States—Steele felt compelled to push back. Most memorably, he went on CNN with comedian D. L. Hughley and denounced Limbaugh when Hughley suggested the radio jock was the leader of the Republican Party. "Rush Limbaugh, his whole thing is entertainment," Steele said. "Yes, it's incendiary. Yes, it's ugly."

Predictably, El Rushbo devoted much of his next show to lampooning Steele.[13] The coverage of the incident, and the fury in the base, indicating mass allegiance to the talk radio host instead of the party chairman, suggested that the tail was wagging the dog.

Meanwhile, Steele was trying to back-channel with the White House to set up a meeting with the president. He wanted to get acquainted; to let Obama know that he would be civil and attempt to keep the party's nativist voices at bay. But the president never responded.

"Despite the public image of Barack Obama, he's very, *very* partisan. He does not like Republicans. He didn't like any of us," Steele says. "I don't think he really appreciated the roles that we were both in, at the same time, as black men. And in my estimation, there should have been some space for the two of us to get in a room together, just so we could say, 'Hey, can you believe these white people?'"

As Steele grappled with concerns over identity and image, the GOP's fund-raising sputtered and its state-by-state organization fossilized. The buyer's remorse was sudden. Steele had barely moved into his new office when the mutiny began. In early March, *The Hill* reported that North Carolina committeewoman Ada Fisher, one of the RNC's three black members, had emailed her colleagues calling for Steele's resignation.[14]

"I don't want to hear anymore language trying to be cool about the bling in the stimulus package or appealing to D. L. Hughley and blacks in a way that isn't going to win us any votes and makes us frankly appear to many blacks as quite foolish," Fisher wrote.

Steele survived the uprising, thanks in part to the rigorous defense of a baby-faced Wisconsin lawyer. He was the RNC's general counsel and the chairman's right-hand man, Reinhold Richard Priebus—"Reince" for short.

FROM THE FLOOR OF THE CHICAGO MERCANTILE EXCHANGE, A MAN flapped his arms and bellowed into the camera. It was February 19 and CNBC reporter Rick Santelli was irate—not just about the recently enacted stimulus, but about the Obama administration's plans to rescue homeowners from bad mortgages.

"How about this president and new administration," Santelli shouted. "Why don't you put up a website to have people vote on the internet as a referendum to see if we really want to subsidize the losers' mortgages? Or would we like to at least buy cars and buy houses in foreclosure, and give them to people that might have a chance to actually prosper down the road, and reward people that could carry the water instead of drink the water?"

Back in the studio, the hosts of *Squawk Box* chuckled at their correspondent's indignation. But Santelli grew only more animated.

"This is America!" he shouted, motioning to the traders surrounding him. "How many of you people want to pay for your neighbors' mortgage that has an extra bathroom and can't pay their bills? Raise your hand!" The traders cheered in solidarity. Santelli turned back to the camera. "President Obama," he cried, "are you listening?"

Santelli's rant blew up overnight. He had bottled the anger fermenting over Bush's bailout and Obama's stimulus, not to mention the inevitability of further government intervention into the automotive companies, the energy sector, and the health care industry. Worried that Democrats were exploiting the economic upheaval to make wholesale changes to the country—a suspicion fed by White House chief of staff Rahm Emanuel's comment "You never let a serious crisis go to waste"[15]—Republicans were already on the edge. By warning that he might organize a "Chicago Tea Party" to protest, Santelli provided the push.

Soon, they were everywhere: people in the streets, some wearing revolutionary-era costumes and toting Gadsden "Don't Tread on Me" flags, demonstrating against the president and his party and the skulking odor of socialism. These weren't the first protests since Obama was elected, and Santelli wasn't the first person to use the phrase "Tea Party" in opposition to Obama's policies, but the CNBC host had galvanized a movement.

The day after Santelli's outburst, roughly two dozen conservative activists joined a conference call to harness the sudden vigor in the grass roots. Many of them had never met in person; they were associated through the internet, particularly social media, having huddled under the hashtag #tcot (Top Conservatives on Twitter). The organizer of this online community, a business consultant named Michael Patrick Leahy, put together the conference call. "We need to strike while the iron's hot, while people are talking about Santelli," Leahy announced.

One of the participants on the call was Jenny Beth Martin, an Atlanta-based computer programmer who had become active in conservative forums online. Having lost her home to foreclosure and moved into a downsized rental just two weeks earlier, Martin and her husband were doing odd jobs to make ends meet. She says that they didn't need the government's help and were livid at its intervention on behalf of the powerful.

"People sensed that Washington was rigging the system against the average American and expecting the average American to pay for that rigged system," she says. "Big business and big labor and big technology teaming up with big government against the average American."

The voices on the call agreed to coordinate their first events one week later, on Friday, the twenty-seventh of February. They hoped to hold ten events and draw a few thousand demonstrators; instead, nearly fifty rallies popped up across the country, and attendance was five times what anyone had anticipated.

This had been achieved on short notice and with almost no money behind it. Watching with interest, the right's wealthiest benefactors saw an investment too good to pass up. Hoping to capitalize on this burst of momentum, some of the biggest donors in conservative politics, including the libertarian-minded industrialists Charles and David Koch, jumped into the action, spending heavily to build out expansive lists of activists and volunteers. Rare was a movement that came about so quickly; even rarer was an opportunity to shape it. By pumping untold millions of dollars into a network of right-wing organizations, deep-pocketed ideologues such as the Kochs aimed to build a machine capable of displacing a hollowed-out Republican Party.

Meanwhile, smaller, more organic groups were springing up across the country under the Tea Party banner, planning meetings and coordinating with sister outfits. One of them, cofounded by Martin, was the Tea Party Patriots, which in a matter of weeks had amassed thousands of members in its Atlanta chapter alone.

The potency of this combination, AstroTurf money and grassroots mobilizing, was realized on April 15, 2009. It was Tax Day, and the bill for eight years of big-government policies had finally come due. As if the bank bailouts, stimulus vote, and mortgage rescues weren't unpopular enough, Obama was now throwing some $55 billion at the Detroit automakers (on top of the $25 billion in TARP funding Bush had provided) in exchange for government-mandated restructurings inside General Motors and Chrysler. Hundreds of Tea Party rallies were held nationwide in protest of Washington's fiscal recklessness and its intrusion into the private marketplace.

It wasn't just rage against Uncle Sam animating the masses. The news in March 2009 of New York financier Bernie Madoff having defrauded thousands of people out of tens of billions of dollars stoked the same feelings of exploitation and unfairness that were increasingly being directed at the government over its policies on spending, immigration, and trade. Decades of a widening chasm in incomes, a diminishment of factory work, a shredded national identity, a dissipating sense of societal cohesiveness, a vanished sense of postwar unity—it was all blurring together in an abstract expression of outrage.

America was in open revolt, and Obama's honeymoon period was suddenly over.

The cover of *Time* magazine for May 18, 2009, showed a Republican elephant logo with the headline "Endangered Species." It didn't come across as hyperbole—not when party honchos had spent the past six months thinking the same thing, and certainly not when Arlen Specter, the senator from Pennsylvania, felt compelled to defect to the Democratic Party that spring, giving Obama a 60th vote in the Senate once Al Franken was seated after a protracted legal fight in Minnesota.

But all the while, something was happening—an authentic rebellion the likes of which the right hadn't seen since the days of the John Birch Society. The irony was subtle but significant: Republicans had failed for the past four years to mobilize their base, yet Obama had done for the party what it could not do for itself.

GOP LEADERS WERE KEEN TO CAPITALIZE. BOEHNER WAS SCHEDULED to be in Bakersfield, California, on April 15, headlining a fund-raiser in the district of Cantor's chief deputy whip, Kevin McCarthy. Boehner and McCarthy agreed to attend the Bakersfield Tea Party event on the condition that they not give any remarks. Boehner suspected with some justification that these crowds would be just as hostile to Republican politicians, especially leadership officials, as they were to Democrats.

"We're at this event, and there's some people who are really happy that we showed up," Boehner recalls. "But there were others that just looked at us with more disdain than you could ever imagine. They thought we were the enemy."

If anything, strange as it would seem given the events of the next several years, Boehner embodied the Tea Party before it existed. A self-made businessman who worked as a janitor to put himself through school, Boehner had earned a small fortune selling corrugated boxes and injection-molded plastics before turning his attention to politics. After winning a state legislative race, he defied the Ohio GOP establishment to win a congressional seat in 1990, overcoming a last name that was mispronounced by everyone, from talk radio hosts to a young volunteer for his campaign, a local college Republican named Paul Ryan. "I didn't know him," Ryan laughs. "I thought his name was *Boner*."

Boehner ("BAY-ner") quickly cut a reputation as a crusader against waste and corruption. Leading a group of young lawmakers known as the "Gang of Seven," he gained fame for investigations into the House Bank and the House Post Office that rattled Congress to its core. These triumphs earned him a spot at Gingrich's leadership table following the GOP Revolution of 1994, but he was later exiled to the rank and file. Clawing his way back to congressional relevance over the ensuing decade, Boehner sharpened his legislative skills, won a committee chairmanship, authored the No Child Left Behind Act, and forged alliances with members across the aisle—and across his party's ideological divide. By the time Denny Hastert had vacated his spot atop the House GOP, Boehner was positioned as the undisputed heir apparent.

He was a breathing paradox: the creature of K Street who rented his Capitol Hill apartment from a lobbyist, but who never requested an earmark in his career; the chain-smoking bullshitter who wept at the mere mention of schoolchildren; the midwestern Everyman who never left home without a clean shave and an ironed shirt; the bartender's son who grew up in a two-room ranch with his parents and eleven siblings only to become the Speaker of the House.

If Cantor served as the House GOP's head—calculating the angles, crunching the votes—Boehner was its heart. Having evolved from insurgent to institutionalist, he specialized in reading people, in building relationships, in giving tough love and getting respect in return. He trafficked in candor and humor, often at the same time, never hesitant to tell someone what they didn't want to hear or dispense pearls of wisecracking wisdom. Once, when Patrick McHenry was brand new to Congress, Boehner spotted him eating an ice-cream sandwich in the Republican cloakroom. "Don't do that," Boehner told the freshman, pointing to the frozen snack.

"Why?" McHenry asked.

"You're gonna be a fat-ass," Boehner replied. (Sure enough, McHenry says, his weight ballooned during his first term in Washington.)

Another time, after he had finished railing against Alaska-based earmarks on the floor, Boehner was confronted by the state's congressman, Don Young, who after a verbal skirmish shoved Boehner against a wall and held a ten-inch blade to his throat. "Fuck you," Boehner said, staring the Alaskan in the eye. Young would later ask Boehner to serve as best man at his wedding.

This political sixth sense, however, did nothing for Boehner when it came to the Tea Party. He understood the recoil against a growing federal government, but he wasn't sure what to make of grown men wearing tricornered hats. He agreed that Washington spent taxpayer money carelessly, but he wasn't convinced this movement was really about fiscal responsibility.

Boehner wasn't alone in this regard. The *tea* in Tea Party was an acronym for "Taxed Enough Already," yet the more time Republicans spent observing the nascent movement, the less certain they felt about its organizing philosophy.

"In theory, it was all about spending," Cantor says. "*In theory*. But I began to question that."

"The Kochs didn't like the social issues, so they tried to make it a small-government thing and put 'values' on the back burner," says Tony Perkins, president of the Family Research Council. "But if you actually looked at the survey data, the Tea Party were our people—and the cultural issues were the top priority for them. Moms raising their kids aren't thinking about tax rates; they're thinking about what kind of culture their kids are growing up in."

"It was a populist movement, rooted in conservative limited-government principles," recalls Jim Jordan, the Ohio congressman who would become Boehner's archnemesis. "But part of it was a reaction to what many Americans viewed as Obama apologizing for America.... It was about more than spending, as evidenced by what happened with President Trump."

"It *was* fiscal," says Jim DeMint, the South Carolina senator. "But it was also about lots of intrusion into our lives, control of your health care, and redefining marriage." (Iowa on April 3 became the third state to legalize same-sex marriage, after Massachusetts and Connecticut, causing disorientation for those who had once taken solace in believing that such developments would be limited to coastal blue states.)

More than anyone in Congress, DeMint heralded the Tea Party's arrival as the GOP's salvation. He had, after all, launched the Senate

Conservatives Fund out of a belief that the right would soon rise up against the Republican Party. Now it was actually happening. The only problem: The establishment was resisting it.

This became clear to DeMint in September, when tens of thousands of Tea Party protesters (perhaps more, depending on warring crowd-size estimates) marched in Washington for the "9/12" rally.[16] It was a powerful rebuke to Washington and its two tribes; only a handful of Republicans, including DeMint and Pence, were invited to address the event. But when the Senate GOP gathered for its weekly lunch a few days later, there was zero discussion of the march. It was as though nothing had happened.

DeMint was beside himself. "This is what we've been waiting for!" he told his colleagues. Years later, reflecting on that meeting, he is still upset. "They looked down on those people," DeMint says of his Senate colleagues at the time. "The Republican Party stiff-armed the Tea Party."

Actually, many Republicans, even the crusty establishment types, embraced the Tea Party in public even as they harbored reservations in private. There was reason for them to keep their distance. For one thing, much of the grassroots enmity was directed toward career politicians who had cheated conservative principles, campaigning in one way and then governing in another; in that regard, many GOP lawmakers on Capitol Hill would be found guilty as charged. The Bush years, stained with Medicare Part D, the unfunded overseas adventurism, and a host of other apostasies, had cemented that. Even onetime revolutionaries such as Boehner had fallen in line.

Beyond their own professional self-preservation, Republicans could be excused for viewing the Tea Party with a certain bewilderment. Conservatism in the tradition of Edmund Burke had been temperamental more than ideological, emphasizing prudence and deliberation. Plenty of individuals who identified with the Tea Party possessed those qualities, no doubt. Yet the collective offered little trace of either. This, combined with a not-infrequent renewal of the racially tinged atmospheres witnessed during the final month of the McCain campaign, gave Republicans justifiable pause about the Tea Party and whether to associate themselves with it.

It wasn't the last time an explosive movement would divide the GOP and vex its establishment.

Through the kaleidoscope of history, the Tea Party can be viewed most honestly as an early indication of the disquiet felt by many Americans regarding the changes sweeping the country—demographically, culturally, politically, and otherwise. This societal restlessness would manifest itself in many different ways, over many different issues, in the years that followed. But in 2009, its energy was drawn from one primary source: Obamacare.

THE PRESIDENT'S DECISION TO SEEK AN OVERHAUL OF THE AMERICAN health care system sparked a visceral resistance on the right that made opposition to the stimulus look like child's play.

Although the fight over Hillary Clinton's reform proposal was fifteen years old, combatants on both sides still nursed grisly wounds. Knowing this, and learning from his missteps with the stimulus, Obama moved methodically. He took a long view, allowing for months of committee hearings and casting a wide conceptual net that brought Republicans to the table during the drafting process. Much to the chagrin of progressives, Obama himself staked out centrist positions, refusing to give advocates of a single-payer system (later dubbed "Medicare for All") a seat at the negotiating table.

It made no difference. The stimulus showdown and the Tea Party's eruption, not to mention the fiercely partisan House vote to pass cap-and-trade legislation in June 2009, had crushed the president's approval rating in Republican congressional districts. If Obama's mistake in January with the stimulus was moving too quickly, too self-assuredly when Republican moderates were ripe for the picking, his mistake with health care for the remainder of 2009 was moving too slowly after those same moderates had already reached the conclusion that they could not afford to do business with him.

It wasn't just the anti-Obama intensity that drove GOP opposition to the health care overhaul. On the statistical whole, Democrats were aiming to protect certain populations (poor people, the unemployed, minorities) that did not heavily reside in Republican congressional districts. Years later, America would witness a surge of support in conservative parts of the country for Democrat-sponsored ideas; in 2018, voters in three red states, Idaho, Utah, and Nebraska, passed ballot referenda approving Medicaid expansion. But in 2009, no obvious political incentive existed for GOP lawmakers to mess with a system that worked reasonably well for many of their constituents.

Few issues in America can be demagogued as effectively as health care. Even though the "individual mandate" was championed by the right's favorite think tank, the Heritage Foundation, in the late 1980s (and enacted by the right's favorite presidential candidate, Romney, when he was governor), it suddenly became the paver of a pathway to socialism. The hysteria didn't stop there. Unfounded talk of the government subsidizing coverage for illegal immigrants was rampant by the summer of 2009, as was the wildly irresponsible speculation, spearheaded by Sarah Palin and her cronies in conservative talk radio, that the Democratic legislation would set up "death panels" to decide which patients would be deserving of lifesaving treatment.[17]

The right's frenzy climaxed over the August recess, with raucous town hall meetings from coast to coast scaring sense into any Republicans who might have considered working with Obama. Lawmakers who had never drawn a hometown crowd of more than one hundred people found themselves facing angry audiences of a thousand or more. The atmosphere was not just angry but fearful: Three and a half million jobs were lost in the last six months of Bush's presidency, and another 3.5 million would disappear[18] in the first six months of the Obama presidency. Unemployment, which registered at 7.6 percent the month Obama took office,[19] had spiked to 9.7 percent by August[20] and would clear 10 percent by year's end.[21] Meanwhile, the stock market—for those fortunate enough to still have stocks—was barely inching upward despite the taxpayer aid given to Wall Street.

The rapidly deteriorating economy, plus the party-line stimulus vote, followed by the rise of the Tea Party and the national panic over a government takeover of health care, had rendered Obama toxic to half the country less than a year after his taking office.

The president's only recourse for passing health care legislation was to execute another brutal, party-line vote with his Democratic super-majorities. But that option disappeared on August 25 when Ted Kennedy lost his battle with brain cancer. Kennedy's death, commemorated as the end of a dynasty that had entranced the country since the middle of the last century, cost Obama a pivotal vote and cast the fate of his health care bill into serious doubt. Whatever mandate the president had ridden into office seven months earlier was now departed; his approval rating had dropped from 68 percent in late January[22] to 50 percent in late August.[23]

With his health care plan languishing, the president convened a joint session of Congress on September 9 to reset the national conversation and dispel some of the more sinister myths about his proposal. It was an attempt to bring down the temperature. Instead, the fever spiked. When Obama reiterated that his bill did not provide coverage to illegal immigrants, Joe Wilson, a South Carolina congressman seated near the front of the House chamber, hollered, "You lie!" It was an atrocious breach of decorum. It was also erroneous: Obama was right on these facts, as health care experts and fact-checkers certified, and Wilson was wrong.

Not that it mattered. Wilson's online fund-raising exploded the next day. Talk radio hailed him as a hero. Conservative movement groups made him an honored guest at upcoming banquets. He was reamed out by Boehner behind closed doors and forced to apologize, but the lesson of the incident was clear. By disrespecting the president of the United States with a blatant, provable falsehood, Wilson had become right-wing royalty.

It was a promising blueprint.

CHAPTER THREE

APRIL 2010

"We've come to take our government back!"

THE NEWS WAS SHOCKING IF UNSURPRISING: CHARLIE CRIST HAD left the Republican Party.

Once upon a time, the Florida governor had seemed invincible. He was charming and handsome, a media darling with pragmatic instincts and a nose for mass appeal: When Obama visited Florida to promote the stimulus, Crist embraced him onstage. He could do no wrong, and with an approval rating hovering near 70 percent,[1] his future was brighter than a Palm Beach sunrise. After cruising to reelection in 2010, Crist would be ideally positioned, as the esteemed governor of America's biggest swing state, to seek the GOP presidential nomination in 2012.

All this changed when Republican senator Mel Martinez announced his retirement in 2009. Jeb Bush, the revered former governor who had built the Florida GOP into a powerhouse, was the odds-on favorite to replace him. If Bush wanted the seat, nobody in the party would stand in the way—certainly not Marco Rubio, a dynamic young Cuban American who had risen to become Speaker of the state House of Representatives thanks in part to Bush's mentorship.

But when Bush declined to run, and encouraged his pupil to enter the race, Rubio pounced. The thirty-seven-year-old was barely known beyond Miami; he had few statewide connections and even less money. But what he did have, as the son of an immigrant hotel bartender and a maid, was a rousing biography and the oratory to sell it. Launching his campaign with irrational optimism, he toured the state and tied his candidacy to the American dream: how he was living it, how government threatened it, how conservatives could preserve it.

Everything was on schedule until Crist jumped into the race in May 2009. It was like David versus Goliath—if Goliath had air support. The

National Republican Senatorial Committee, which serves as the Senate GOP's campaign machine in Washington, had traditionally observed a policy of not meddling in primaries where no incumbent was involved. Yet the NRSC under chairman John Cornyn, the Texas senator who took charge after the 2008 cycle, was far more interventionist. Within hours of launching his campaign, Crist won the NRSC's support. Making this especially excruciating for Rubio was that he was in Washington that very day. He had come to meet with Cornyn and other top party officials, hoping to find some support inside the Beltway. Instead, Rubio found Cornyn urging him to quit the campaign.

It was a no-brainer from the national party's perspective: Six incumbent Republican senators had announced their retirements early in 2009, and Crist, a well-liked governor and big-time fund-raiser, would deprive Democrats of a pickup. But Cornyn's heavy-handed approach infuriated the right. "The NRSC is actively trying to undermine the conservative Republican base with milquetoast establishment Republicans the nation rejected in 2006 and 2008," the influential activist Erick Erickson wrote on his blog, *RedState*. "We must not let them win."[2]

Sitting in his rental car after the Cornyn meeting, Rubio saw the emails and text messages arriving in rapid-fire fashion: Dozens and dozens of prominent Republicans, both in Florida and in Washington, were endorsing Crist. It was a shock-and-awe tactic meant to scare him from the race, and it very nearly worked.

"I thought for sure that was the end, because everything I knew about traditional politics had told me it was the end," Rubio recalls. "As it turned out, it was the greatest thing that could have happened. There was a possibility that had they not played it that way, I may have never been able to take off and nationalize my race and raise enough money to be competitive. But embedded in that was this resentment at being told from the top down, 'This is who you're going to elect.' And I think that wound up extending all the way into 2016."

The most essential ingredient to a political victory is timing. The story of Barack Obama's presidential conquest, and of Donald Trump's eight years later, cannot be told without the context of the mood and the moment they were uniquely suited to meet. Crist would have crushed

Rubio in any primary prior to 2010. Yet the conditions of that year on the right—militant opposition to Obama, lingering disillusionment with George W. Bush, antipathy toward Washington and its sovereign class— inverted the playing field. Rubio turned Crist's strengths into weaknesses: his centrist skill set, his big-donor connections, his support from the party establishment, his aura of inevitability. Despite carrying some baggage from his days in the statehouse and owning a less than ideologically pristine legislative record, Rubio crafted a contrast that proved irresistible, that of a conservative outsider versus a moderate insider.

He won the support of Jim DeMint as well as other Tea Party leaders and their grassroots groups. He utilized the internet and social media to raise millions in small-dollar donations from every state in the union. He earned glowing coverage as the star of a nascent movement, including a cover of *National Review* that played on Obama's 2009 slogan: "Yes *He* Can." Republican power brokers in Washington were impressed— and a little bit terrified. Rubio wasn't just beating their anointed candidate; he was threatening to expose them as obsolete.

"When I worked my first political campaign, for Bob Dole in '96, you needed some formal apparatus to conduct politics," Rubio says. "You don't need anything anymore. With social media and the internet, I can reach millions of people instantly without paying virtually anything. People started realizing that in 2010. You didn't *need* the party."

(It was, with the addition of social media, a next-generation realization of the "party in a laptop" strategy that Howard Dean had used in 2004 to test the Democratic establishment with few of the infrastructural assets brought to the race by a John Kerry or a Richard Gephardt.)

Once trailing Crist in the polls by 30 points, by April of 2010, Rubio was *ahead* by 30 points.[3] Crist's decision to abandon the GOP and run as an independent did nothing to alter the outcome; Rubio won the Senate seat easily. But the symbolism was striking. Crist had been one of the most popular Republicans in America, and Rubio, by tapping into the fury of the right (and by branding Crist as an Obama-hugging squish), had driven him from the party altogether.

Rubio's conquest of Crist was but the most visible ripple in a wave that grew throughout 2009 and 2010. No Republican was safe: All across the country, card-carrying members of the establishment were thrown overboard, victims of the party's complacency and of the base's exasperation with politics as usual.

Interestingly, the biggest contributor to this renaissance inside the GOP was the signature achievement of the Democratic president.

LOOKING BACK, THE ODDS OF A HEALTH CARE OVERHAUL WERE NEVER fully in Obama's favor. His administration's efforts to woo Republicans, particularly in the Senate, began wilting in the heat of conservative opposition in the summer of 2009. Meanwhile, the institutional troubles on the Democratic side—a Minnesota recount preventing Senator Al Franken from being seated until July, Ted Kennedy's death from cancer a month later—rendered their vaunted "filibuster-proof majority" somewhat meaningless. It wasn't until after the appointment of Kennedy's temporary replacement, Paul Kirk, that Democrats could pass a health care bill through the Senate. It happened on Christmas Eve 2009, in a party-line vote of 60 to 39, with one GOP senator absent.

But the Democrats' newfound momentum was crushed less than a month later.

The January special election to fill Kennedy's seat was expected to be a snoozer: Dark blue Massachusetts hadn't elected a Republican to the Senate since the Jimmy Carter administration. With health care reform hanging in the balance, and the liberal lion's dying wish awaiting fulfillment, Democrats couldn't possibly lose. This assumption was profoundly arrogant, given that Republicans had several months earlier won the governors' elections in Virginia and New Jersey—a purple and blue state, respectively, both of which Obama had carried comfortably.

The certainty Democrats felt about holding Kennedy's seat also ignored the one intangible that can transform any race: candidate quality. Scott Brown, the truck-driving, barn jacket–wearing, up-by-the-bootstraps Republican, was a superb candidate; Martha Coakley, the Democrat who mocked the idea of shaking voters' hands outside Fenway Park, was not. Brown won the race by 5 points, eliminating the Democrats' 60-vote majority and, it seemed very possibly, the prospects for enacting health care reform.

The problem for Obama was that the House and Senate had passed two different bills, and there was no longer a sufficient number of Democratic votes in the Senate to pass a merged version. Some top party officials, including White House chief of staff Rahm Emanuel, argued for a scaled-back product. Others, most notably Pelosi, rejected this approach. ("Kiddie care," she scoffed.) Harry Reid pressed for the House to pass his chamber's version, but Pelosi's members were cold to the idea: Moderate Democrats objected to the scope of abortion coverage, while liberals protested the lack of a government-sponsored insurance entity.

Still, there was no other way—and Obama knew it. As the president launched a public relations campaign, touring the country and citing news of insurance rate hikes to argue for the bill as a cost-containment measure, Reid and Pelosi agreed on a two-step legislative strategy. The House would pass the Senate bill; then the Senate, using a parliamentary tactic known as reconciliation, would pass a package of changes demanded by House Democrats in exchange for their votes.

As it became clear that their plan would work, it became equally clear that Pelosi was asking some of her centrist members to walk the plank. Dozens of House Democrats were already facing stiff reelection fights; voting for the president's polarizing bill was akin to nailing shut their own coffins. "She is a strong Speaker—there isn't any question about that," Boehner told reporters at the time. "So, you pass a very unpopular bill. You shove it down the throats of the American people, and you lose your majority. How good is that? How smart is that?"

Pelosi did nothing to tamp down Republican criticisms of the product, and the process, when she remarked in early March, "We have to pass the bill so that you can find out what is in it." The Speaker knew what was in the Affordable Care Act; it had been debated for many months, and her comment, in full context, was clearly meant to assure the public that they would like the bill once its benefits were realized. But Pelosi's verbal blunder was a political gift that would keep on giving in the years ahead.

"Look at how this bill was written," Boehner barked from the House floor. It was March 21, minutes before the House would pass the Senate bill in a landmark victory for the Democratic Party. "Can you say it was done openly?" Lawmakers in the chamber shouted in response. "With transparency and accountability?" The shouts grew louder. "Without backroom deals struck behind closed doors, hidden from the people?" Boehner mustered every ounce of righteous indignation. "Hell no, you can't!"

When Obama signed the Affordable Care Act into law a few days later, it was, in the immortal words whispered into his ear by Vice President Joe Biden, "a big fucking deal." Democrats had succeeded, where generations of their forebears had failed, in approving a sweeping reorganization of the American health care industry. That meant, among other things, the implementation of a requirement to own insurance; expanded Medicaid coverage for low-income individuals; the opening of a federal insurance marketplace with government subsidies for those who qualified; and revamped regulations governing how insurance companies provided or denied coverage, such as to people with preexisting medical conditions.

It was also a big deal for the GOP. Not a single Republican in either chamber had voted for the bill dubbed "Obamacare," even though a number of Senate Republicans had spent months negotiating the details and said privately that they found the legislation to be fair-minded. "Mitch did everything he could to keep a Republican from crossing over. We had meetings every Wednesday just to keep discipline," recalls Tennessee senator Bob Corker. "Mitch is really good at loosening lug nuts, and over time the wheels just fall off. That's what he did with Obamacare."

"It was the unifying effort for us, the definitive effort going into the 2010 elections," McConnell says. "It gave us a chance at a comeback. It set up a referendum in the country on whether or not [voters] were suffering from some buyer's remorse over the decision two years earlier. I think that would have been less likely had we signed on and a bunch of people had gone over to the other side. They would have claimed it was bipartisan."

Boehner was correct in saying the law did not have majority support. One day before the president signed it, CNN released a poll showing 59 percent of Americans opposed and just 39 percent in favor.[4] This was one of many surveys to suggest that the partisan exercise had aggravated the middle of the electorate. The Republican base was already on fire in opposition to Obama, and now there were signs of an independent exodus away from the Democratic Party as well.

Perhaps most significant, in the political short term, was the Democrats' decision to cut $500 billion from Medicare to help pay for the legislation. The cuts were not immediate; in fact, they were designed to slow the growth of Medicare over the next decade, a longtime goal of reform-minded Republicans. This introduced no small amount of irony to the 2010 campaign season. Two years earlier, Paul Ryan had been ostracized within the GOP for his entitlement-tweaking "Roadmap"; his cosponsors of the legislation numbered fewer than ten. Now, with the Tea Party rewriting the rules of the GOP, dozens of candidates nationwide ran for Congress endorsing Ryan's proposal.

It suggested a potential inflection point: Perhaps Republicans would become the party of hard truths and tough choices, after all.

Instead, party strategists, eyeing big gains among elderly voters and middle-income whites nearing retirement, made "Obama's Medicare cuts" the go-to attack of 2010. The maneuver reflected equal parts tactical brilliance and intellectual nihilism.

ONE DAY AFTER OBAMA'S SIGNING CEREMONY, THE INK STILL WET ON Democrats' historic legislation, Republicans filed bills in both chambers of Congress to repeal the Affordable Care Act.

This was a promise made by effectively every single Republican running for Congress in 2010: They would repeal Obamacare. Boehner and the House GOP leadership stated as much in their "Pledge to America," a document that itemized the reforms Republicans would make once back in power. Among the party's other vows: to slash $100 billion from the budget in year one; to reduce overall government spending to pre-2008 levels; to prohibit federal funding for abortions; and to post all bills online for three days before a vote could occur.

If repealing Obamacare was one pillar of the GOP's midterm platform, the other was arresting the president's fiscal profligacy. Obama could be excused for shaking his head in disbelief. For the previous eight years, Republicans had spent like teenagers with their first credit card, blowing a hole in the deficit and incurring unprecedented amounts of debt. Now their "Pledge" promised to get America's fiscal house back in order—with fuzzy math and unspecified budget cuts.

Seeking to claim the high ground and avoid being typecast as a big-spending progressive, Obama had in February 2010 created a National Commission on Fiscal Responsibility and Reform. Co-chaired by Republican Alan Simpson and Democrat Erskine Bowles, the eighteen-member group was tasked with producing a comprehensive plan for debt and deficit reduction.

The final passage of Obamacare a month later pushed Simpson-Bowles to the national back burner and ended whatever fleeting moment of ideological cease-fire it had created. With the midterm elections fast approaching and each party now bunkered down in their positions, any notion of transcending partisanship faded away. The Republicans who had bet against Obama's messianic promises were proved right: He *couldn't* break the impasse in Washington.

What they couldn't anticipate, however, was the deepening schism within their own tribe.

There was mounting buzz of a mutiny inside the Republican National Committee, with members complaining of Steele's odd management style and lavish spending habits. The juiciest rumor, percolating in the spring of 2010, was that Reince Priebus, the RNC's general counsel and Steele's consigliere, was plotting a coup. Priebus laughed it off when the chairman confronted him, claiming the rumor was totally fabricated. Steele believed him. He couldn't trust many people inside the RNC anymore, but Priebus had been in his corner through thick and thin.

Around that same time, in Arizona, a firestorm erupted when conservatives in the statehouse muscled through a bill, SB 1070, that required all residents to carry immigration paperwork on their person and required law enforcement to check the immigration status of anyone they suspected might be illegal, even if they were not stopped or detained for committing a crime. Republicans in the state overwhelmingly supported the effort, including John McCain, who faced a primary challenge that year from an immigration hard-liner. But the bill's sanctioning of racial profiling made national Republicans nervous, especially given the increasingly hostile tone struck on the immigration issue ever since the collapse of Bush's reform effort in 2007.

Meanwhile, in April, the Heritage Foundation, the GOP's academic bedrock for a generation, announced the creation of a spinoff lobbying organization. It would be called Heritage Action. Unlike its scholarly cousin, this new group favored baseball bats over bow ties. Republican politicians had been making big promises since Obama took office, Heritage officials reasoned. It was time to hold them accountable.

LESS THAN TWO YEARS REMOVED FROM BUSH LEAVING THE WHITE House with record-low approval ratings and Obama taking office with Washington at his feet, Republicans were beginning to entertain a scenario that had once seemed unfathomable: They could regain control of Congress in 2010.

It was a heavy lift. The Senate appeared out of reach; a net gain of 10 seats was needed to flip the chamber. And even the more realistic target, the House of Representatives, required a net gain of 39 seats for Republicans to retake the majority. Still, there was cause for bullishness. A first-term president's party traditionally takes a thumping in the midterms, and Obama, wounded by the stimulus and health care fights, suddenly seemed mortal. With the president's job approval sagging, the economy barely yawning to life, and the energy of the electorate squarely behind them, Republicans schemed to put Democrats on the defensive. They would expand the political map, targeting not just the Blue Dog moderates in coin-flip districts, but also the progressives who rarely faced general election challenges, forcing the Democrats to spread their resources thin.

But first, the GOP had to exorcise its own demons.

Despite the long-brewing disillusionment of the conservative base, there had never been a true intraparty bloodletting in 2008. Bush was still in office, McCain had been mostly within striking distance of Obama, and the Republicans who voted for TARP did so *after* the conclusion of primary season. Now, with a new day dawning for the GOP, it was imperative for many first-time congressional hopefuls to run not just against Obama and the excesses of the Democratic Party, but against Bush and the excesses of the Republican Party.

In Michigan's Third District, where the GOP incumbent was retiring, a thirty-year-old freshman state lawmaker named Justin Amash separated himself from a crowded primary field by blasting the policies of his own party. The approach was not without risk: Prominent Republicans in the state put a target on Amash's back, and some refused to support him even after he'd advanced to the general election. But Amash, a Ron Paul acolyte, won the West Michigan race anyway, thanks to support from Tea Party groups and a surge of participation from low-propensity voters attracted to his libertarian message. "I got active in politics in part because of what George W. Bush was doing," Amash recalls. "The Obama backlash of course started around the time of the Tea Party, but a lot of us blamed George W. Bush for Obama in the first place."

In South Carolina's Fourth District, a local prosecutor named Trey Gowdy shredded the Republican incumbent, Bob Inglis, for supporting Bush's bailout of Wall Street. It wasn't the congressman's only vulnerability: Inglis had denounced his South Carolina colleague Joe Wilson for shouting "You lie!" at Obama, and had further infuriated voters by asking them, at a town hall meeting, why they were so afraid of the president. His greater offense was criticizing the Fox News host Glenn Beck for "trading on fear" and urging his constituents to stop watching the program. Gowdy capitalized on all this, identifying Inglis as a part of the old Republican guard that had sullied the party's reputation. Propelled by considerable Tea Party support in one of America's most conservative districts, Gowdy trounced the GOP incumbent.

On the opposite end of the state, down in the Lowcountry of South Carolina's First District, the anti-establishment revolt proved metaphorically rich. In a sprawling primary contest, two dynastic giants towered above the field: Carroll Campbell III, son and namesake of the state's iconic ex-governor, and Paul Thurmond, son of Strom Thurmond, the famous senator and segregationist Dixiecrat. Lesser known was Tim Scott, an African American state lawmaker and a self-made insurance salesman from the hard neighborhoods of North Charleston.

Boosted by national Tea Party groups (and also by Eric Cantor, who was desperate to diversify the House Republican ranks), Scott built his candidacy around the populist crusades of eliminating earmarks and introducing term limits. After defeating Thurmond in the runoff, a contest thick with racial and historical subplots, Scott punched his ticket to Congress in November.

And in Idaho's First District, incumbent congressman Walt Minnick, a Blue Dog and a fiscal hawk, became the country's only Democrat to receive a national Tea Party endorsement. The only problem: Both Republicans running against him *also* had Tea Party backing. Raúl Labrador, a state lawmaker and former immigration attorney, had the support of numerous local activist groups, while Vaughn Ward, a Marine combat veteran, was endorsed by Sarah Palin. The weakness of Ward: He was also a prized recruit of the national party leadership, a fact that his GOP rival wielded mercilessly against him. "The Republican Party brought us the Obama administration because they couldn't get their act together in Washington," Labrador told the *Idaho Statesman*. He also vented frustration at Palin's attempt to establish herself as the Tea Party's figurehead. "This is a movement," Labrador said at the time, "not a party." He scored an upset over Ward and then defeated Minnick in the general election.

The pattern was inescapable: In congressional districts from sea to shining sea, self-styled insurgents found success by racing to the right and distancing themselves from the GOP's ruling class. These candidates were running as a new breed; they would legislate as conservatives first and Republicans second, prizing ideological purity over partisan achievement, coming to Washington not to climb the institutional ladder but to dismantle it rung by rung.

It wasn't limited to federal races. In Texas, popular U.S. senator Kay Bailey Hutchison launched a primary challenge to Governor Rick Perry on the grounds that he was too much of an absolutist. Campaigning as a pragmatic centrist, Hutchison was supported by a small army of political heavyweights, including former president George H. W. Bush, former secretary of state James A. Baker III, and former vice president Dick

Cheney. It didn't have the intended effect: Perry, running against Washington and the party's graybeards, crushed Hutchison by 21 points.

Steele, who visited more than one hundred cities that fall on the RNC's "Fire Pelosi" bus tour, saw the anti-establishment fanaticism everywhere he traveled. He recalls wondering how, if Republicans took back the House, Boehner would handle a mob of rookie revolutionaries. When they met in Washington, shortly before Election Day, Boehner's answer was simple: They would fall in line. Freshmen *always* fall in line.

But the party chairman was not convinced. "These guys are out there blowing up Republicans as much as they're blowing up Democrats," Steele told Boehner. "You mean to tell me you can't see that?"

Boehner could see it, all right. But after two years of being steamrolled by Pelosi, all he cared about was flipping seats and reclaiming the House majority. And so far, despite the emergence of some Republicans he would come to describe as "assholes" and "legislative terrorists," none of them had proved so crazy as to cost the GOP a winnable race.

McConnell was not so fortunate.

WHAT MADE THE TEA PARTY VIBRANT IS ALSO WHAT MADE IT UNSUS-tainable: a lack of organization. With thousands of groups springing up overnight—national, state, county—cohesion was a pipe dream. There could be no designated platform, no organizing doctrine, no shared sense of vision for what specifically they hoped to accomplish. In a sense, this seemed appropriate. Conservatism is by definition distrustful of top-down, one-size-fits-all thinking. But the movement's administrative void created a Wild West ecosystem in which supremacy belonged to whatever organization, or candidate, could push hardest and farthest to the right.

Palin was the de facto figurehead, hence her keynote address to the inaugural Tea Party Nation conference in February 2010. (Her six-figure speaking fee, and the exorbitant ticket prices,[5] invited a lasting skepticism of the "grassroots" leaders and their commercial incentives.) It hardly mattered that she had abruptly resigned as Alaska's governor, a move that validated the perceptions of her volatility. More than any elected official alive, Palin possessed a God-given capacity for channeling the forces of panic and populist grievance swirling throughout much of America—especially its older, whiter parcels. And now that she wasn't running Alaska, she was free to lead a much larger constituency.

At the same time, a chorus of conservative groups—the Club for Growth, FreedomWorks, and the Tea Party Express, among many others—fought for organizational dominance atop the movement. They jockeyed over donors, events, endorsements, and troops to populate their armies (and their lucrative email lists). It was an exciting time to be a conservative activist: After decades of being dictated to by party overlords, the commoners had wrested away their power.

Even so, the fundamental problem of parameters, or a lack thereof, remained. Having little regard for the practical considerations of winning a general election, activists and their allied groups often defaulted to backing the candidate farthest from the mainstream, empowered by DeMint's infamous observation that he would rather have thirty conservative Republicans in the Senate than sixty moderate Republicans. (To be clear, thirty senators, even if all reborn as Barry Goldwater, lack the capacity for passing legislation or confirming judicial nominees.)

The House map was too expansive for this pursuit of ideological purity to have a studied, concentrated effect. But a small batch of Senate races drew disproportionate amounts of money, energy, and attention from the nascent professional right—with decidedly mixed results.

First blood was drawn in Utah: Senator Bob Bennett was ousted at the state's May 8 GOP convention, finishing in third place behind two conservative challengers and thus failing to qualify for the runoff. Having voted for TARP and signaled his support for immigration reform, Bennett knew the activist-dominated convention could be treacherous. An endorsement and stay-of-execution plea from Mitt Romney failed to persuade the party faithful; a heartier ovation was reserved for De-Mint, who appeared via video to announce his endorsement of a young attorney named Mike Lee. Though he finished second in the convention voting, Lee went on to win the primary and the general election, later establishing himself as one of the more serious conservative voices in Congress. But the unceremonious exiling of Bennett was deeply unsettling to the GOP's ruling class and a harbinger of the disruption to come. "The political atmosphere, obviously, has been toxic," a weepy Bennett told the *Salt Lake Tribune*, "and it's very clear some of the votes that I have cast have added to the toxic environment."[6]

Less than two weeks later, in Kentucky, a libertarian Republican named Rand Paul, an ophthalmologist and the son of Ron Paul, crushed the national party's handpicked recruit, winning the Senate primary by 23 points. It was another blow to Cornyn and the NRSC, but it was especially humiliating for McConnell, the state's senior senator, who had been working against Paul behind the scenes. Knowing this, Paul felt a special satisfaction campaigning against the bailout vote McConnell engineered and calling for the overthrow of an establishment McConnell embodied. The result was another win for DeMint and the conservative groups that had pooled their resources behind Paul knowing full well the significance of beating McConnell in his own backyard. "I have a message from the Tea Party—a message that is loud and clear and does not mince words," Paul declared at his victory rally. "We've come to take our government back!"

In reality, it barely mattered whom Republicans nominated in Utah and Kentucky. No Democrat was going to carry either of those ruby red states circa 2010. Thus, an even bigger win for the Tea Party came in Wisconsin, where liberal icon Russ Feingold was expected to cruise to a fourth term. Priebus had other ideas. In addition to serving as the RNC's general counsel, Priebus was the Wisconsin GOP's chairman. Skilled at uniting the intraparty factions that warred in other states, he had set out looking for someone who could excite both the Tea Party and the establishment. What he found: Ron Johnson, a self-made manufacturing baron who was gaining renown among the state's grassroots for his rants against the advance of big government. Recruited into the race by Priebus, Johnson checked every box: He was an angry, business-minded outsider with deep pockets to fund a competitive campaign. The race turned into the biggest surprise of the election cycle: By the time the RNC bus tour pulled into Wisconsin in mid-October, Johnson was trouncing Feingold in the polls.

The mood was so jubilant that Priebus, who owned a gray suit that

Steele admired, invited his local tailor onto the bus and had him measure the chairman for an identical match. After the election was won, everyone joked, Steele and Priebus would wear them on the same day and
pose as twins—the towering black man and the diminutive Greek guy.

That Johnson poured $8 million of his own fortune into the Wiscon-
sin race, and was boosted by millions more in outside money, dripped
with irony. Early in 2010, the Supreme Court's ruling in *Citizens United
vs. FEC* had established that corporate political donations qualified as
protected speech, inviting an unprecedented deluge of "dark money"
into the midterm cycle. (Anonymous donations far predated *Citizens
United*; in fact, the justices ruled that lawmakers have the power to reg-
ulate campaign finance disclosures, something Congress has not done.)

Republicans could ask for nothing more than the eradication of the
McCain-Feingold law *and* its Democratic coauthor in one fell swoop.
Johnson beat Feingold by 5 points.

This was the reward of the Tea Party: uncorking an energy that had
simmered for decades, yielding fresh candidates who captured the
mood of the electorate.

It was also the risk.

REPUBLICANS HAD HARRY REID ON THE ROPES. THE SENATE MAJORITY
leader, part of the Democratic triad in Washington, was badly underwater in Nevada. Polling consistently showed a majority of voters disapproving of his performance, owing partially to tepid support from
his own base: A *DailyKos* survey in late 2009 reported that just 58 percent of Nevada Democrats viewed him favorably.[7] This, in concert with
booming enthusiasm on the right, should have spelled the end for Reid,
potentially altering the course of Obama's tenure by removing the man
who wielded the Senate to safeguard the president's legacy.

Instead, Republicans nominated Sharron Angle.

A former state assemblywoman, Angle operated out of her living
room with just two paid staffers, one of whom, her campaign manager,
was prone to going AWOL for weeks at a time. Angle's former statehouse
colleagues whispered that she was fit for a straitjacket; that she wanted

to outlaw alcohol, that she had strange associations with the Church of Scientology, that she once protested black football uniforms because they insinuated a satanic influence. None of this prevented the Tea Party Express from endorsing Angle—and then pumping a half million dollars into the primary. The Tea Party Express endorsement sparked a cascade of outside conservative support from the likes of Palin, radio host Mark Levin, prominent activist Phyllis Schlafly, and gospel singer Pat Boone. In its endorsement, weeks before the primary, the Club for Growth called Angle "Harry Reid's worst nightmare."[8]

In fact, she was Reid's dream come true. Having previously flirted with fringe positions such as abolishing Social Security and eliminating the Department of Education, Angle went completely off the reservation after winning the Republican primary. She said that Islam's Sharia law was being imposed on cities in Michigan and Texas. She suggested that the 9/11 hijackers came across the "porous" Canadian border. She spoke of using "Second Amendment remedies" to clean up Congress if elections failed to do the trick. When a group of Hispanic high school students questioned the tone of her immigration-themed attacks on Reid, she said to them, "I don't know that all of you are Latino. Some of you look a little more Asian to me."[9]

A race that should have been a referendum on Reid instead became a choice between the unpopular incumbent and his unhinged opponent. Having trailed the GOP establishment's preferred candidate, former state party chairman Sue Lowden, by double digits earlier in the year, Reid wound up beating Angle by 6 points in November.

It wasn't the only time conservatives would snatch defeat from the jaws of victory.

In Colorado's Senate race, the former lieutenant governor, Jane Norton, began the GOP primary as the prohibitive favorite until the local activist base exploded in opposition. Their vessel became Ken Buck, a local district attorney with a penchant for controversy. DeMint swooped into the race in support of Buck, funneling more money to him than any other candidate in 2010. Local right-wing groups rallied as well, viewing Buck's candidacy as a metaphorical middle finger to Washington.

The outside financial and organizational support proved critical: Buck scored a 4-point upset over Norton.

But his structural handicaps persisted:. The Democratic nominee, Michael Bennet, out-raised Buck by a three-to-one ratio. This triggered a massive influx of outside spending on Buck's behalf, from conservative groups who adored him and also from reluctant Republican donors who found him clownish but couldn't stomach the possibility of losing such a winnable race. Having waded clumsily into gender politics by mocking Norton's "high heels" during their primary duel, Buck was carpet-bombed with Democratic attacks focused on his weaknesses with women, specifically his failure to prosecute a rape case as district attorney[10] and his opposition to abortion in cases of rape or incest.

Bennet won the general election by fewer than 30,000 votes, and exit polls showed that Buck had lost women by 17 points.[11] In contrast, Republicans carried female voters by 1 point nationwide that November. Norton, in other words, would have coasted to victory as the GOP nominee.

The grand finale, and the weirdest Tea Party implosion of them all, came in the First State.

Delaware's race for U.S. Senate never really registered on the national radar. Mike Castle, a longtime moderate congressman, was the clear favorite both to capture the GOP nomination and to win the general election. He was known as a subduing voice within the party; the summer prior, Castle was booed and shouted down in a town hall meeting for claiming that Obama was an American citizen. "I want to know, why are you people ignoring his birth certificate?" a woman shouted at the congressman to raucous applause in a video that gained widespread attention. "He is not an American citizen. He is a citizen of Kenya. I am American. My father fought in World War II with the Greatest Generation in the Pacific theater for this country, and I don't want this flag to change. I want my country back!"

Castle's challenger, Christine O'Donnell, was a known gadfly with a checkered past. Even as she collected endorsements from conservative groups toward summer's end (the Tea Party Express, the Susan B. Anthony List, the Family Research Council), the campaign never felt competitive. Numerous polls showed Castle leading not only O'Donnell, but also Chris Coons, the Democratic nominee, by double digits. Despite mounting attacks on his voting record from the right, Castle's centrist brand seemed to suit Delaware just fine.

Everything changed on August 24. More than four thousand miles away, a Tea Party favorite named Joe Miller shook the political world by defeating Senator Lisa Murkowski in Alaska's GOP primary. It was thoroughly unexpected. Murkowski, the former governor's daughter, was seen as political royalty and immune to a primary challenge. That perception didn't scare off Miller's supporters: the Tea Party Express, the Club for Growth, FreedomWorks, Mark Levin, Laura Ingraham, and a number of antiabortion groups backed Murkowski's rival. Miller's eventual loss in the general election, to a write-in campaign staged by Murkowski, would diminish the significance of his primary win. Yet, in the moment, as they celebrated their triumph over the establishment, Tea Partiers turned to the final primary on the 2010 calendar: Delaware.

With the national spotlight blazing down on the state, both sides of the GOP civil war readied for a defining battle. O'Donnell savaged Castle as the most liberal Republican in Washington and mocked his lack of masculinity; her allies, in uncoincidental harmonization, spread rumors about the congressman's sexuality.[12] Castle, a country club gentleman, told allies he didn't want to run a negative campaign—so the party did the dirty work for him. Both the Delaware GOP and the NRSC went Dumpster-diving on O'Donnell, unearthing lethal opposition research on everything from her sloppy personal finances to potential illegalities in her use of campaign funds.

"She's not a viable candidate for any office in the state of Delaware," Tom Ross, the state's Republican chairman, told reporters. "She could not be elected dog catcher."[13]

The establishment attacks boomeranged. Days before the primary, Palin, a professional martyr and veteran victim of party pile-ons, swooped into the race on O'Donnell's behalf, completing the Tea Party's takeover of Delaware. Castle was helpless. Congressional primaries typically draw a fraction of the eligible voter pool—between 10 and 20 percent—and are therefore dominated by the most passionate, ideologi-cally motivated constituents. There was no disputing which side had the passion in this contest. Once the overwhelming favorite to be Delaware's next senator, Castle lost by 6 points to a primary opponent nobody in the state had ever taken seriously. The reaction inside the Republican Party was disgust: A top NRSC official told the *Wall Street Journal* min-utes after the race was called that the party committee wouldn't be sup-porting O'Donnell financially in the general election.[14] This, along with Karl Rove's instant analysis that O'Donnell was a lost cause, caused an uproar on talk radio. Eager to avoid additional bloody intraparty war-fare, Cornyn condemned his staffer's unauthorized comment and cut a $42,000 check to O'Donnell's campaign.

But that wasn't enough. Several weeks later, without warning, O'Donnell flew to Washington and confronted Cornyn at the NRSC's headquarters. She knew the $42,000 check had been an investment in crisis management; she knew the party wasn't planning to give her an-other dime. Now, sitting across from the committee chairman, O'Don-nell demanded that the national party spend several million dollars in Delaware or else she would go to Rush Limbaugh and Sean Hannity to tell of the GOP's internal sabotage.

"I don't respond well to threats," Cornyn replied, standing up from his chair. "This meeting is over."

The encounter confirmed what party officials had long suspected: O'Donnell was unreliable at best and unstable at worst. Two days before her DC visit, she had released a television ad in which, against a cloudy backdrop and wearing a dark jacket, she looked into the camera and de-clared, "I'm not a witch. I'm nothing you've heard. I'm you." The spot was meant to dismiss news coverage of an old television clip in which O'Donnell spoke of dabbling in witchcraft; instead, it became the punch line that would forever crystallize the GOP's Tea Party problem.

O'Donnell lost the general election to Chris Coons by 17 points.

THERE WAS NO SPINNING THE RESULTS OF ELECTION DAY 2010. IT WAS, as Obama told reporters the next afternoon, "a shellacking." The outcome served as a reminder of the economic unease still gripping much of America, and doubled as a swift rebuke of the Democrats' one-party rule in passing the stimulus, the cap-and-trade bill (in the House), the Affordable Care Act, and the Dodd-Frank financial regulatory law.

Thanks to mass mobilization on the right and a decided swing of independents away from the president's party, Republicans flipped an astonishing 63 Democratic seats in the House of Representatives, retaking the majority and positioning Boehner as the next Speaker. He would be dealing with the largest freshman class in modern congressional history: 87 House Republicans in total.

Republicans also gained 6 Senate seats, including the one formerly held by Obama in Illinois (which the state's governor, Rod Blagojevich, had attempted to sell to the highest bidder, later earning himself a lengthy prison sentence.) This increased the GOP's number in the upper chamber to 47—which should have been 50 if not for giveaways in Nevada, Colorado, and Delaware.

The scope of the Democratic wipeout extended far beyond DC. Prior to 2010, Republicans had unified control of the government, both legislative chambers plus the governorship, in nine states. That number more than doubled on Election Day. In total, Republicans picked up more than 700 state legislative seats from coast to coast. They regained the majority in twenty individual chambers. They also won back six governor's mansions, including a clean sweep of the Rust Belt (Pennsylvania, Ohio, Michigan, and Wisconsin) and the election of a rising star named Nikki Haley, who overcame a whisper campaign aimed at her Sikh background and Indian heritage to become South Carolina's first female governor.

Meanwhile, in addition to the hemorrhaging of legislative seats and governorships, Democrats lost dozens of offices (secretary of state, auditor, attorney general) that play various roles in regulating state elections. The implications were enormous. Republicans could now introduce tighter voting laws, which they said were necessary to combat fraud (and that critics alleged were aimed at suppressing poor and minority votes). Even more consequentially, with the decennial Census wrapping up, and states preparing to redraw their congressional lines based on the population shifts, the GOP could consolidate its majorities with gerrymandered district maps.

For senior Republicans, however, the thrill of victory carried unexpected consequences. Shortly after Election Day, a Minnesota congresswoman named Michele Bachmann walked into Boehner's office for a private meeting. Bachmann had built a sizable following as a right-wing instigator, accusing Obama in 2008 of harboring "anti-American views" and saying a year later that she found it "interesting" how swine flu only seemed to break out under Democratic presidents.[15] Bachmann told Boehner that she needed something from him: to be seated on the powerful House Ways and Means Committee.

"That's not going to happen," Boehner said.

"Oh, yes, it is," she replied. "Or else I'm going to go to Rush, and Hannity, and Mark Levin, and Fox News, and I'm going to tell them that John Boehner is suppressing the Tea Partiers who helped Republicans take back the House."

Unlike the similar threat made by O'Donnell to Cornyn during campaign season, this one carried real weight: Bachmann was a sitting member of Congress and a leading figure in the Tea Party movement, someone who could complicate Boehner's ascension to the speakership.

"I had never been put in a position like that before," he remembers. "She had me by the balls. She had all the leverage in the world, and she knew it."

Boehner scrambled for a quick solution. Ways and Means was going to be tedious in the next Congress, he told her. What about joining the Intelligence Committee?

Bachmann liked the idea, and Boehner paid a visit to Mike Rogers, the Michigan congressman who chaired the Intel panel. "No, no, no," Rogers said when Boehner broke the news. "You can't do this to me."

"Listen," Boehner told him. "The Tea Party is going to raise me to the top of the flagpole naked if that woman doesn't get what she wants."

Rogers acquiesced; Bachmann joined the committee. She soon became regarded as a diligent, hardworking member of the panel; Boehner's handpicked new appointee, a leadership ally from California named Devin Nunes, was widely viewed as a policy deadbeat.

BEYOND THE INTRATRIBAL WARFARE THAT HAD DOMINATED THE CAMpaign cycle, and that would soon spawn rival factions in Congress, something else felt out of place inside the party. Republican officials worried that the energy of 2010 had masked fundamental deficiencies that Obama would exploit in 2012. Democrats lost because the president's organization was garaged, party leaders whispered anxiously, and because the president's voters had stayed home for the midterms. If Republicans were going to take down Obama, they needed to build a machine capable of competing with his. That would require money, technology, discipline—and leadership.

Priebus had reached those same conclusions. Although he denied to Steele on multiple occasions that he was interested in running for his position, Priebus had spent the summer warning the chairman against seeking reelection. The membership was deeply unhappy, he explained, and the committee was engulfed by controversy. It would benefit everyone if Steele graciously stepped aside after his two-year term.

Steele refused to listen. "If they want me gone, they're gonna have to throw me out," he said.

Priebus finally decided that he had no choice. Just before the midterms, he knocked on Steele's door at the RNC. "I think the party needs to go in another direction," he told the chairman.

"You know what, Reince?" Steele responded. "Keep the fucking suit."

CHAPTER FOUR

JANUARY 2011

"Shakespeare has got nothing on this shit."

H E STOOD BEFORE THEM IMPERIALLY DRESSED, SKIN FRESHLY KISSED a bright citrus by the Florida sun, his two-pack-a-day baritone rumbling with alternating notes of caution and aggression. It was the first week of Congress, and John Boehner was the Speaker of the House. Two decades spent collecting favors, dialing donors, and hustling votes had finally paid off. Now, as he addressed the enormous new class of House Republicans—eighty-seven of them—Boehner needed to set a few things straight. He was delighted that reinforcements had arrived. And he was looking forward to fighting alongside them in presenting a muscular check on the Obama administration. But he was going to need their cooperation. And their trust.

Boehner had heard the rhetoric from Republican candidates in 2010, and he wanted to make one thing clear. "Campaigning," he told the freshmen, "is different than governing." Republicans now controlled the House, but Democrats were still running the Senate and the White House. If the House GOP was to accomplish anything—if they were to make gains for conservatism in a divided government—compromise and incrementalism would be necessary.

Many of the new members nodded their heads. Although they had all hugged the Tea Party, plenty of the incoming lawmakers were factory produced: noncontroversial, corporate-friendly, mechanical-mannered Republicans whose twin objectives were scoring victories for the team and winning reelection for themselves (not in that order). They listened to Boehner and heard a leader worth following; he was savvy, experienced, disciplined.

The true believers had a different reaction. If half the freshmen were standard fare Republicans who expected symmetrical partisan combat with the Democrats, the other half were Tea Party guerrillas. These law-makers felt, with some justification, that they had been sent to Washington not to trade chess moves with Obama, but to flip over the board and send the pieces scattering. They looked at Boehner, the back-slapping, cabernet-swirling, country club connoisseur, and saw that which they had come to destroy. As these rookies stewed, watching their fellow classmates pledge allegiance to Boehner, the earliest seeds of discord were sown within the House majority.

"I thought it was a revolution. I thought we were going to completely change the way that Washington worked," Raúl Labrador, the new Idaho congressman, recalls. "Within one week—I'm not exaggerating—I saw a large majority of my class saying, essentially, 'Whatever you need us to do, we will do.' And I was sick inside."

THIS INTERNAL TENSION WENT UNDETECTED AT FIRST. AFTER ALL, THE closing months of 2010 had set the stage for Armageddon between the parties. The week before Election Day, Boehner had told Sean Hannity that it was "not a time for compromise," promising specifically that Republicans would do everything possible to keep the Obamacare law from being implemented. "We're going to do everything—and I mean everything—we can do to kill it, stop it, slow it down," he said.

Perhaps eager to one up his counterpart, Mitch McConnell told *National Journal* that same week, "The single most important thing we want to achieve is for President Obama to be a one-term president."[1] This quote would become legend, neatly encapsulating the GOP's obstructionist outlook, even as it was blown slightly out of proportion. (Minority party leaders have always schemed to prevent the other party's president from winning a second term; furthermore, McConnell had said in the next breath that Republicans would cooperate with Obama "if he's willing to meet us halfway on some of the biggest issues.")

More consequential than any verbal jousting had been the collapse of the Bowles-Simpson commission. After nearly eight months of deliberation, the commission released its final report on December 1, 2010. It prescribed a combination of spending cuts and revenue increases that would slash the deficit by nearly $4 trillion by 2020. This would be achieved by slaughtering sacred cows left and right: raising the Social Security age, reducing spending on Medicare and military operations, eliminating a trillion dollars in tax loopholes, and capping government spending at 21 percent of GDP.

Congress would vote on the plan if a supermajority of the commission, fourteen of its eighteen members, approved it. But only eleven did. And while there was opposition from liberal members, the more conspicuous resistance came from the three House Republicans appointees who voted as a bloc: Paul Ryan, Jeb Hensarling, and Dave Camp.

Ryan's vote came across as particularly cynical. The Budget Committee's senior Republican had spent years warning of America's unsustainable fiscal course, yet when handed a concrete, bipartisan proposal to help correct it, Ryan balked, arguing that the plan would drive Obamacare's price tag even higher while hindering economic growth with considerable new tax hikes.

This sparked a feud between Ryan and Obama that dashed whatever prospects once existed for a partnership. Months after the Bowles-Simpson vote, when Ryan had taken over as Budget Committee chairman and released a new version of his entitlement-cutting "Roadmap," the White House invited him to a speech Obama was giving on fiscal solubility. With Ryan sitting in the front row at George Washington University and "shaking his head in disgust," as the *Wall Street Journal* reported, the president savaged his proposal, saying it would leave poor people, disabled kids, and the elderly to "fend for themselves."[2]

Without mentioning him by name—not that he needed to—Obama said Ryan's plan "paints a picture of our future that is deeply pessimistic." The congressman was blindsided. He rushed out of the auditorium after the speech, cursing out the president in conversations with his friends back on Capitol Hill. It was a gross breach of decorum by Obama. White House aides worried that he had gone too far; for his part, the president later told journalist Bob Woodward that he didn't know Ryan would be attending, calling the entire episode "a mistake."

The bigger mistake, in the eyes of Republicans, was Obama's own rejection of Bowles-Simpson. It had been a unique opportunity to seize the high ground; Obama could claim that Republicans weren't serious about deficit reduction, that he was willing to make the difficult choices necessary to improve America's fiscal health. "I thought he was going to triangulate us, embrace Bowles-Simpson, make us look like we were right-wingers. But he didn't," Ryan recalls. "At every one of those inflection points in his first term, I thought the guy would go to the middle and kill us and scoop up the center of the country. But he just couldn't help himself. He was a hard-core progressive."

This partisan animosity obscured the conflict smoldering within the new GOP majority.

Boehner had been pleased to see Jim Jordan elected as the new chairman of the Republican Study Committee. Founded in 1973 as a sister organization to the Heritage Foundation, the RSC had become the biggest caucus on Capitol Hill and was home to Congress's most conservative members. Previous chairmen, including Mike Pence, had used the organization to push Republican policy further to the right—often to the chagrin of party leadership. Boehner had reason to be wary of whoever was running the group. But he felt confident that Jordan, whose Ohio district bordered his own, would be more ally than adversary.

That proved naïve. Jordan, a two-time NCAA wrestling champion (collegiate record: 156–28–1), was ready to rumble from the moment he took elected office. Armed with a law degree and a master's in education, Jordan won a seat in the Ohio assembly at age thirty. It wasn't long before he stood out. Squat and rugged, with a rock-chiseled chin and the slightest trace of cauliflower ear, Jordan didn't choose a career in politics for the free cocktail parties. After praying, exercising, and swigging his morning "go-go juice" (half OJ, half Mountain Dew), Jordan would spend the balance of his days seeking out conflict and charging headlong into whatever fray could be found. . In the statehouse, where he wriggled his way to the right of even the legislature's most conservative members, Jordan quickly attained the reputation of a bulldog—or, as Boehner remembers him, "a legislative terrorist."

Interestingly, as a young state lawmaker Jordan had looked up to Boehner, a fellow Buckeye who was then turning Congress upside down. After winning his congressional seat in 2006, Jordan found himself all the more impressed, particularly with Boehner's stewardship of the minority during Obama's first two years. The GOP leader, Jordan felt, had performed brilliantly in holding the conference together and drawing sharp contrasts with the Democrats.

But Republicans now had the majority, thanks to a rowdy freshman class that had arrived in Washington with its fists balled. It was a match made in right-wing heaven: When it became apparent that Boehner had no intention of executing a scorched-earth strategy against Obama, the new Tea Party lawmakers gravitated toward someone who would. "Remember the 'Hell No' speech?" Jordan says with a mischievous grin. "That's the John Boehner we were hoping for."

NINE DAYS AFTER BOEHNER WAS ELECTED SPEAKER, MEMBERS OF THE Republican National Committee gathered in a suburban Washington hotel ballroom to choose their own leader. Reince Priebus's attempted usurping of Chairman Michael Steele lent a melodramatic mood to the proceedings; the onetime allies had not spoken since their fateful meeting at RNC headquarters in October, a fact that did not escape anyone as their loyalists worked the room, twisting arms and trading gossip.

Three other candidates were running, but the main event was Steele versus Priebus. Each had spent recent months dogging the other. Priebus alleged that Steele had been careless with the RNC's finances, spending too liberally and racking up massive debt while doing little to improve the party's technology and ground game. Steele countered that Priebus had been in every meeting with him, advising every decision that was made, and had never once dissented.

Steele was a sympathetic figure in certain respects. He *had* taken over a party decimated by Obama; he *had* raised a respectable amount of money; he *had* overseen one of the most successful election cycles in the party's history. This was all the more impressive considering that his detractors were actively undermining him. Karl Rove, who cofounded the advocacy group American Crossroads in 2010, had urged big donors (former Cowboys quarterback Roger Staubach, for instance) not to give to the RNC. This was representative of a broader, post–*Citizens United* problem for the party: the emergence of super PACs, some run by long-

time insiders and others associated with misfits such as the Koch brothers, served to cannibalize the donor base.

At the same time, Steele was his own worst enemy. He *had* become a
distraction to the party with his frequent gaffes; he *had* alienated some
wealthy patrons; he *had* racked up senseless amounts of debt—more
than $20 million of it.[3] (Steele says he never wanted to take out a line
of credit, and secretly recorded a 2010 meeting to prove that other RNC
officials were in favor of doing so.)

All the lavish spending on hotels and limousines might have been
overlooked, and Steele might have survived as chairman, had it not been
for the *Daily Caller* headline in March 2010: "Michael Steele Dropped
Big Bucks on Bondage Club." A committee staffer had taken a group of
young Republicans to a sex-themed nightclub during a California visit
and dropped two thousand dollars. Steele wasn't in attendance, but the
news handcuffed him for good.

Priebus knew that Steele would be angry at his decision to seek the
chairmanship, though he never fully appreciated how it would be perceived. Still haunted years later by sensationalist talk of his Brutus-like
betrayal, Priebus says Steele's obstinance had left him with no choice.

"For the past six months I was basically pleading with Michael to
understand how seriously bad and sour things were with the members,
and why it would be a mistake to run again," Priebus says. "And when
his ultimate decision was 'I'm running again even though I know I can't
win,' then I felt like I had no obligation to go down with the ship—not
when I could run and fix the problem. If anything, it was disloyal of him
not to support someone who he knew could win, someone who had been
good to him. The loyalty question goes both ways. If anything, he's the
Benedict Arnold, not me."

Steele won 44 votes on the first ballot, second only to Priebus's 45,
but his support dropped in each of the next three rounds. Following the
fourth ballot, Steele took the stage to announce his departure.[4] "I hope
you all appreciate the legacy we leave," he told RNC members. "Despite
the noise—and Lord knows, we've had a lot of noise—despite the difficul-
ties, we won."

Soon thereafter, Reince Priebus was elected chairman of the Republican National Committee, and Michael Steele was exiled. He quickly became a punch line inside the party and persona non grata to its leaders. Once, assuming that he would be chairman for the GOP's 2012 convention in Tampa, Steele had pitched the idea of "the world's biggest block party," an outdoor festival with ethnic food and live music to give Republicans a more relatable flavor. Instead, he would later find himself without an invitation to that very convention—no VIP pass, no floor access, no general admission. He wound up in Tampa anyway, thanks to his new job as a contributor for MSNBC. But the pain of the snub, and of his ouster as chairman, would never recede.

"I have to be honest, the only one that blindsided me was Reince. Because he and I had formed a real good friendship. That hurt more than anything else," Steele says. "He was there for every decision that I made, so it was really painful to sit there and listen to them talk about how much financial ruin I brought on the party. And of course, a part of me couldn't help but think it had to do with my race, because there were chairmen before me who had done a whole lot worse with a whole lot more."

The winding irony is that Priebus, having succeeded the party's first black chairman, would make minority outreach a top priority in the years ahead. He would proclaim that Republicans could not win the White House by targeting only white voters. And he would become chief of staff to a president of the United States who proved him wrong.

"Shakespeare," Steele grins, "has got nothing on this shit."

THE PARTNERSHIP BETWEEN PRESIDENT OBAMA AND SPEAKER BOEHner had the potential to alter the American trajectory for generations to come. There were obvious differences between them. During an early 2011 meeting on the White House patio, Boehner sipped red wine and puffed Camel Lights while Obama drank iced tea and chewed Nicorette gum. (Boehner says Obama, a former smoker, was "scared to death" of his wife and never bummed a cigarette.) But they actually had far more in common: Both men were self-made successes, hailing from humble roots and overcoming long odds to achieve their positions in government. Both carried themselves with a cool charisma that could prove disarming for their opponents. Above all, each believed he owned a certain stature in his party that would allow him to cut a deal with the other.

For all the anti-Obama panic that animated the GOP in 2009 and 2010, Boehner had pushed a more substantive message—"Where are the jobs?"—that revealed a traditional sensibility of how to oppose a president. Boehner wanted, of course, to see Obama defeated in 2012. But he wasn't much for stunts. His mission as the Speaker was to make practical policy gains while demonstrating to the country that Republicans could be trusted as a competent party. Meanwhile, from the White House's perspective, there was a qualitative difference in the way Boehner interacted with Obama compared to the attitudes of McConnell, Eric Cantor, and others.

"When I saw that Boehner was becoming Speaker, I thought that was a positive thing," recalls Obama's vice president, Joe Biden. "I thought there could be actually some work together, some collaboration together, and we could actually get some things done. But I thought, most of all, he was going to treat the president with more respect than some of his colleagues had."

Treating Obama with respect was part of Boehner's problem. Republicans around the country had little regard for the president. Many felt he looked down his nose at their way of life. Some thought he was a serial liar. And more than a few believed he was illegitimate—that he hadn't been born in America and thus wasn't qualified to be president.

Numerous state governments debated newly urgent legislation in 2009 and 2010 requiring presidential candidates to release long-form birth certificates. This paranoia echoed beyond the provinces: Twelve House Republicans cosponsored a similar bill in Congress, lending a higher degree of legitimacy to the conspiracy theorizing. When one of the cosponsors, Texas congressman Louie Gohmert, urged Cantor in a meeting to bring up the bill for a vote, he made his point with the subtlety of a sledgehammer: "*Kenya* hear me? *Kenya* hear me?"

"Louie Gohmert is insane. There's not a functional brain in there," Boehner says, muttering a few expletives for good measure. "I don't know what happened to him."

But Gohmert wasn't an outlier. "I knew people, *smart people*, who were into it," says Karl Rove. "They thought it was this vast conspiracy, that people took this kid who was born in Kenya and faked newspaper clippings from the time of his birth, and documents in the Hawaii state government files, so this Kenyan-born kid could pass for an American citizen and wind up running for president. This was the Manchurian candidate on steroids—not just on steroids. This was the Manchurian candidate on LSD and peyote."

"I dealt with it every day. *Every. Single. Day*," Boehner says. "I went on TV and said, 'Hawaii released the birth certificate; that's good enough for me.' I got my ass chewed out for weeks. You would have thought I was Satan himself."

He adds, "There was at least a couple dozen members who believed it. These are members of Congress, but they aren't in this world by themselves. Once upon a time, they would have belonged to the fringe. They weren't the fringe anymore."

"It wasn't just in the conference. It was at home," says Cantor. "I mean, we would encounter groups of people who absolutely took that to be the truth: [Obama] was not an American, he was a Muslim. . . . There was this rumor mongering, and frankly, I think, a racist play for votes."

Polling throughout Obama's presidency would reveal large numbers of Republican voters harboring stubborn doubts about his citizenship: 41 percent in a 2010 CNN/Opinion Research Corporation survey; 43 percent in 2011, per Gallup; 61 percent in 2015, according to Public Policy Polling; and 72 percent in 2016, as found in an NBC News/SurveyMonkey poll.

Birtherism aside, the reality was that Republicans simply did not like Obama, and for many of them, civility was synonymous with surrender. Boehner learned this the hard way. When he and the president played a round of golf together in June 2011, along with Biden and Ohio governor John Kasich, the blowback was so furious from conservatives—on talk radio, in Boehner's district, even in Congress—that he knew immediately it could never happen again.

"There were actually people who came in, Republican leaders of committees who would sit and say, 'Mr. President, my being here is an act of courage. Do you realize how much damage it does to me to sit with you?'" Biden recalls. His voice pinches with anger. "Can you *imagine* saying that to a president? Well, that was said. That was said by more than two people I can name. And so, John got ripped for doing what the Congress and the president are supposed to do, which is actually see if they can collaborate for the public good. But as the Republican Party became more and more radicalized in the House, John was getting the living devil beaten out of him."

Indeed, this was a strange new world for the Speaker. Boehner had played his share of hardball politics as a lieutenant to Newt Gingrich in the 1990s, an era of vicious political tribalism in its own right. But he had also watched Gingrich strike significant deals with his nemesis, President Bill Clinton. So, when signs mounted in 2011 of a conservative rebellion fermenting inside his conference, Boehner was dismissive. He knew what these hard-charging freshmen wanted, because he'd been in their shoes.

This was a costly misreading of his members. Much had changed since Boehner came to Congress in 1990—the inception of Fox News, the proliferation of super PACs, the decline of trust in government and institutions, the election of a black president—and the Republican Party had evolved accordingly. By the time Boehner came to terms with this transformation, it was too late.

"He thought of himself as someone who was of the Tea Party mentality before the Tea Party was a thing," says Anne Bradbury, who served as Boehner's floor director, one of the top staff positions in Congress. "So, I think there were some assumptions made that he got these people, and that they would see he was one of them. But that really never came together."

OBAMA TRIED TO PUT AN END TO THE INSANITY ON APRIL 27, 2011.

Without advance notice, the White House posted a long-form version of the president's birth certificate online. Suspicion of Obama's citizenship had percolated on the outskirts of the internet since 2008, when his campaign released a copy of the standard "certification of live birth" issued by Hawaii. Imaginations ran wild on the right for the next three years. A new breed of desperado, emboldened by the digital age, took to the internet spreading varying claims of the document's fraudulence. No amount of testimony otherwise—from government officials, genealogical researchers, even conservative pundits—could kill the conspiracy theory. Only by releasing his long-form certificate, the doubters said, could Obama prove himself legitimate.

The president was loath to lend validity to the debate. His sudden decision to share the document caught Washington sleeping, as broadcast networks scrambled to interrupt programming with the news bulletin.[5] "We do not have time for this kind of silliness," Obama said inside the White House Briefing Room. "We've got better stuff to do. *I've* got better stuff to do."

The annoyance in his voice was reserved for one heckler above the rest, the man whose image was now being shown in a split screen next to Obama on the cable channels: Donald J. Trump.

The sixty-four-year-old billionaire, a real estate mogul and star of NBC's hit show *The Apprentice*, had been a fixture of American pop culture for decades. His brand was lent to books and beauty pageants, his name splayed garishly across buildings, commodities, and golf courses. The son of a successful Queens developer, Trump carried a perpetual sense of insecurity in regard to the Manhattan nobility and was determined to make noise with his money, investing in schemes and projects befitting his flamboyant persona. Standing six foot three, with a vulgar New York brogue and his trademark mane of tumid blond, Trump rarely lacked for attention.

He was a latecomer to the birther movement. In fact, he had first commented publicly on Obama's citizenship just a month earlier, during an interview with ABC's *Good Morning America*. Trump called the circumstances surrounding Obama's birthplace "very strange," adding, "The reason I have a little doubt—just a little—is because he grew up and nobody knew him."[6]

This was not true. Obama's upbringing on the big island was thoroughly documented by friends and family members, not to mention verified by journalists and academics. But that didn't stop Trump from peddling falsehoods, with increasing certainty, in the days that followed. On ABC's *The View,* he asked, "Why doesn't he show his birth certificate? There's something on that birth certificate that he doesn't like." On Fox News, he said Obama "spent millions of dollars trying to get away from this issue." On Laura Ingraham's radio show, he said of the certificate, "Somebody told me . . . that where it says 'religion,' it might have 'Muslim.'" And on MSNBC's *Morning Joe,* Trump announced that Obama's "grandmother in Kenya said, 'Oh, no, he was born in Kenya, and I was there, and I witnessed the birth.' Now, she's on tape. I think that tape's going to be produced fairly soon."

In fact, the tape features Obama's grandmother stating repeatedly that she did *not* witness the future president's birth because it occurred in Hawaii and she lived in Kenya. But facts had never stood in the way of conservatives' theorizing about Obama's shadowy past: How he was raised by his radical father (who actually had abandoned the family when his son was two years old); how he inherited an anticolonial bias from living in Kenya (he spent part of his childhood in Indonesia after his mother remarried); how he was a Muslim (despite being baptized in 1998 and writing extensively about accepting Christ after being raised by his nonbelieving grandparents).

Trump would later claim that he never truly believed that Obama was born outside the United States. But Boehner, a frequent golfing buddy, says Trump absolutely did. "Oh yes. Oh yes. He wouldn't have spent the money to send people to Hawaii and do the investigation if he didn't believe it."

Trump's true beliefs, his intentions, his motivations—none of it really mattered. The fact of it was, he could say whatever he felt like, whenever he felt like it, and suffer no consequences. He was a superstar, a brand-name television personality who had spent decades mastering the game of media manipulation. He didn't care if what he said about the president of the United States was unintelligible or factually inaccurate; it would be covered and covered widely.

This was America circa 2011, a nation seduced by celebrity and blissfully unaware of the cancerous effects. That year, another reality television personality, Kim Kardashian, whose career was launched by her role in a leaked sex tape, married basketball player Kris Humphries (not her costar in said tape) in a televised ceremony that drew north of four million viewers.[7] When the couple filed for divorce seventy-two days later, it was reported that they *profited* off their nuptials, having sold their wedding photos to *People* magazine for $1.5 million in addition to receiving the substantial TV royalties. (This, more than any activism on the left, made the case for gay marriage.)

Trump was one of the few people alive who could compete for ratings with the Kardashians. Not coincidentally, the surge of controversy—and publicity—surrounding his birther gambit accompanied the news that Trump was considering a presidential run. Again.

He first flirted with a bid for the White House in 1987, after publishing his best-selling book *The Art of the Deal*. Trump's message over the next three decades would prove fairly consistent—and in certain cases, quite prescient.

"I'd make our allies pay their fair share," he told Oprah Winfrey in 1988. "I think people are tired of seeing the United States ripped off."

Two years later, talking to *Playboy* about the president, Trump said, "I like George [H. W.] Bush very much and support him and always will. But I disagree with him when he talks of a kinder, gentler America. I think if this country gets any kinder or gentler, it's literally going to cease to exist." (In that same interview, he predicted, "The working guy would elect me. He likes me.")

And in 1999, he told Larry King, "I think that nobody's really hitting it right. The Democrats are too far left. . . . The Republicans are too far right. I don't think anybody's hitting the chord."

Trump came closest to pulling the trigger in 2000. Encouraged by former professional wrestler Jesse "The Body" Ventura's winning Minnesota's governorship as a member of the Reform Party, Trump launched an exploratory committee and changed his party registration in October 1999. Testing the waters, he described himself as "very prochoice" and "very liberal when it comes to health care."[8] He also called for tighter immigration restrictions and new trade deals. Perhaps most memorable was his feverish five-month assault on Pat Buchanan, the populist favorite who challenged Bush in the 1992 GOP primary and was now running for the Reform nomination. Trump called the "anti-

Semite" Buchanan a "Hitler-lover" who "doesn't like the blacks" and "doesn't like the gays."[9] Before dropping his candidacy in February 2000, Trump warned of Buchanan's alleged extremism, "We must recognize bigotry and prejudice and defeat it wherever it appears."[10]

Eleven years later, Trump would do something wildly out of character: apologize. Placing a telephone call to Buchanan one day, out of the blue, he told his former rival that he had been wrong to label him a racist. He even asked for forgiveness. Buchanan was stunned.

It was around the time of the Buchanan call, of course, that Trump was kicking off his birther crusade—and pondering once more a campaign for the presidency.

The prospect of Trump running in 2012 jolted the GOP. The expected Republican field was doing little to excite conservatives; Mitt Romney, the right-wing darling of 2008 and the presumed 2012 front-runner, was bleeding support thanks to the attention Obamacare had drawn to the program he had piloted in Massachusetts. Against this backdrop, Trump sent shudders through the party establishment when he accepted an invitation to address the Conservative Political Action Conference in February 2011.

This marked his first appearance at the annual carnival of politics and culture. Entering to the song "Money, Money, Money," Trump fed the roaring masses with a formula that would later become all too familiar: attack, boast, promise. "I like Ron Paul. I think he is a good guy," Trump said of the libertarian icon. "But, honestly, he just has zero chance of getting elected." Trump reveled in the boos from Paul's college-age supporters. "Tactics and strategy are involved in any form of leadership," he went on. "I'm well acquainted with both. I'm also well-acquainted with winning, and that's what this country needs right now: winning." Finally, offering the guarantee of his presidency, Trump pledged, "Our country will be great again."

Less celebrated were the remarks given a night later by Mitch Daniels, the Indiana governor, who was weighing a presidential bid of his own. Having once called for a "truce" on social issues in order to focus on America's fiscal decline—eliciting jeers from the evangelical right—Daniels pleaded with the nation to unify around the sacrifices needed to make America solvent, delivering one of the more compelling speeches by a Republican in the twenty-first century.

"We face an enemy, lethal to liberty and even more implacable than those America has defeated before," Daniels said.[11] "We cannot deter it; there is no countervailing danger we can pose. We cannot negotiate with it, any more than with an iceberg or a Great White. I refer, of course, to the debts our nation has amassed for itself over decades of indulgence. It is the new Red Menace, this time consisting of ink. We can debate its origins endlessly and search for villains on ideological grounds, but the reality is pure arithmetic. No enterprise, small or large, public or private, can remain self-governing, let alone successful, so deeply in hock to others as we are about to be."

Warning his fellow Republicans against pursuing the mutually assured destruction offered by the party's ascendant right flank, Daniels concluded, "Purity in martyrdom is for suicide bombers."

The Indiana governor had offered a vision, one grounded in realism and reasonableness, that elevated common purpose over cultural warfare. But few chose to see it. Trump's alternative, a loud, swaggering, confrontational bravado, was a better fit for the Republican base. It was a clearer diagnosis of the country's condition. And it was a sexier story for reporters to write.

Within a few months, Daniels ended his consideration of a presidential run and his speech was swept into the dustbin of history by the incessant coverage of Trump versus Obama. It came to a crescendo in late April, at the White House Correspondents' Association Dinner, with Trump in attendance as a guest of the *Washington Post*. Obama used the president's traditional stand-up routine to skewer Trump, mockingly acknowledging his "credentials and breadth of experience." As the crowded ballroom turned in his direction, journalists whooping with approval, Trump stared straight ahead. He would have the last laugh.

A PROMISE IS THE MOST DANGEROUS THING IN POLITICS.

George H. W. Bush lost a second term after going back on his famous guarantee, "Read my lips: no new taxes." Lyndon Baines Johnson knew better than to seek reelection after reneging on his assurances that he would not send troops to Vietnam. In the case of House Republicans, their "Pledge to America" of 2010 became a liability the moment they assumed the majority.

The failure to deliver on their single biggest promise, repealing the Affordable Care Act, would come to shape the contemporary party's legacy. But it was, in the early going, the GOP's struggle with some of the smaller objectives that set a foreboding tone. The notion of cutting $100 billion from the budget in year one, for instance, was plainly impossible; the fiscal year was already halfway over by the time the numbers could be crunched, and the Democrats were never going to rubber-stamp such a steep reduction. No asterisk was attached to that particular guarantee when Republicans made it, however, and conservatives were justifiably irate when Boehner and his team attempted to explain the fine print of why such a cut wasn't possible.

That initial fight over the $100 billion cut was a watershed. As the new members pressured leadership to keep the promise, it dawned on them that the promise wasn't *meant* to be kept. This realization is what began sorting Republicans into two distinct camps: one, representing a vast majority, that observed Boehner's message about the realities of governing and resigned themselves to a lemonade-making pragmatism; the other, representing a vocal minority, that dismissed Boehner's call for teamwork and rebelled, convinced that brawling in pursuit of the unattainable was better than accepting half-measures.

It was a catch-22. Republican leaders envisioned using their majority to demonstrate the party's capacity for smart, responsible governance, but they had won their majority by mobilizing the conservative base around patently unrealistic promises. They had set themselves up for failure.

"It was two years' worth of vitriol and venom pointed toward Obama, and once in the majority, they thought we're going to fix it all. And we were the ones who ratcheted that up—'We can set it straight if you just give us the majority,'" Cantor says. "From conservative radio to the blogosphere to cable TV, the expectations rose to a point where it was just unmanageable."

Cantor adds, "For those who wanted to suggest that Republicans weren't fighting hard enough, it was really foolish to think that you were going to beat Obama into submission to abandon everything he stood for—including the bill with his name on it."

Jordan, the conservative ringleader, calls this a cop-out. Using his chairmanship of the RSC to agitate endlessly against Boehner, Cantor, and the perceived passivity of the GOP leadership, Jordan concedes that some of the party's stated objectives were doomed to failure. "But the fact is we made a promise to the voters, and we didn't even try," he complains. "All too often we would do the wimpy thing. We would try to have the debate, it would last twenty-four hours, and then it's like, 'Oh well, we just can't get it done.'"

Cantor, more than anyone, was at the nexus of this divergence. He was younger and more ideological than Boehner, a fact that did not escape anyone in the conference, including the Speaker. Cantor had personally recruited many of the 2010 candidates and sympathized with their desire to fight, to show their constituents they had done everything possible to get results. Yet the second in command had to be cautious. Undermining Boehner would only fuel the perception of a rivalry between them, plunging the party into deeper polarity. Moreover, as their majority took shape, Cantor found himself increasingly sensitive to Boehner's plight and dismayed by the Tea Party's tactics.

"The demands were fine in theory, but put into practice, it just didn't work," he says. "Conservatism was always about trying to effect some progress toward limiting the reach of government. It wasn't being a revolutionary to light it on fire and burn it down to rebuild it. But somehow, that's what the definition of 'conservative' became."

As Congress hurtled toward an August deadline to raise the debt ceiling, a borrowing limit that directs the Treasury Department's payment of costs already incurred, fratricide was beginning to consume the House GOP. It owed to a self-inflicted strategic wound: Republicans had, upon taking the majority, waived the "Gephardt Rule" that had long allowed the debt ceiling to be raised in pro forma fashion. At the time, Boehner's team had encouraged the rank and file to view the debt ceiling vote as a leverage point in dealing with Obama. This advice proved costly.

Boehner started the debt-ceiling talks by publicly demanding one dollar in spending cuts for every new dollar in borrowing authority. Emboldened by the Speaker's objective, Jordan convinced the House conservatives to stake out a position even farther to the right, rallying around a plan called "Cut, Cap and Balance." It would require any debt ceiling hike to accompany a cut in federal spending, a cap on future spending, and a constitutional amendment requiring Congress to balance its budget.

The proposal, which Boehner derided as "Snap, Crackle and Pop," passed the House in mid-July but was rejected out of hand by Obama and the Reid-run Senate. Jordan lobbied Boehner to make a national referendum out of the issue; the Speaker wanted to move on, eyeing August's deadline and hoping to win some concessions beforehand. The tension boiled over in late July, when it was discovered that Jordan's staff had been conspiring with outside conservative groups to pressure RSC members—*Jordan's* members—to vote against Boehner's proposed debt deal. Jordan apologized to the conference, but members shouted, "Fire him!" in reference to the rogue staffer who had undermined the Speaker.

Ultimately, Boehner failed to collect enough Republican votes to present Obama with a unified offer that might have nudged the negotiations rightward. The consequence was twofold. In the short term, it forced the parties to settle on a deal that nobody was happy with, the Budget Control Act, which raised the debt ceiling but introduced automatic "sequestration" cuts to spending if a future agreement were not reached. In the long-term, it stripped Boehner of any negotiating power with Obama. The White House had begun to suspect that Republicans were internally fractured; the debt ceiling implosion confirmed to them that Boehner could not control his members and thus could not be trusted to speak for them.

"He could practically never deliver his votes," says Nancy Pelosi, the House Democratic leader.

Boehner shakes his head. "It's hard to negotiate when you're standing there naked," he says. "It's hard to negotiate with no dick."

The underlying problem bedeviling the GOP was a lack of congru-
ity; that, in turn, could be chalked up to its whiplash-inducing return
to power in 2010. Time in the minority can be enormously beneficial
for a political party—time to reflect, study, question, strategize, change.
Storming back into the House majority just two years removed from
George W. Bush's departure, Republicans, it was clear, were not yet prepared to *be* a majority party.

Of course, history might have reflected something quite different—for
Boehner and Obama, for Republicans and Democrats, for the country—
had the "Gang of Six" not gotten in the way.

On July 19, as the debt-ceiling drama was intensifying, a bipartisan
group of three Republican senators and three Democratic senators had
unveiled the framework of a sweeping fiscal compromise they had been
working toward. What they didn't know—what almost no one in Washington knew—was that Obama and Boehner had secretly been working on their own deal. It was nicknamed "the Grand Bargain," and just
forty-eight hours earlier, the janitor and the community organizer had
shaken hands to seal an agreement that might have altered the arc of
American history.

Their work had begun a month earlier—on the golf course, in fact—
with Boehner's suggestion that the debt ceiling predicament gave them
an opportunity to address America's longer-term crisis: the retirement
of the Baby Boomer generation and the strain it placed on the country's
finances. The Speaker told the president that Republicans could agree to
increased revenue via eliminating tax deductions and loopholes (violating conservative orthodoxy) if Democrats could agree to spending cuts
and entitlement reforms (violating liberal orthodoxy). Obama signaled
his openness to the idea, and for the next five weeks their teams worked
in secret to hammer out a compromise, knowing that fury awaited in
their respective party bases.

On Sunday, July 17, Obama returned from church to meet Boehner
and Cantor at the White House. Some fine-tuning remained, but their
deal was basically done. It included reforms to Medicare, Medicaid, and

Social Security; $1.2 trillion in cuts to discretionary spending; and $800 billion in new revenue. Boehner, wary that Cantor would try to scuttle the deal, had initially kept him out of the negotiations; when Cantor was finally clued in, he warned the Speaker that he shouldn't trust Obama. Now, as they shook hands in the Oval Office, Boehner felt vindicated. "I was one happy son of a bitch," he says.

But two days later, blind to these private dealings, the Gang of Six rolled out its own proposal. The problem was that it included significantly more revenue than Boehner and Obama had agreed to. Watching as several conservative senators endorsed this plan, the president knew there was no way he could sell the Grand Bargain to congressional Democrats.

When the White House reached out to the Speaker's staff, seeking a higher revenue number for their plan, Boehner was aghast. "Are you shitting me?" he shrieked to several of his staffers, a tenor of wrath they had never before heard. "We shook hands!"

The Speaker quickly convened a meeting with Cantor, Paul Ryan, and Kevin McCarthy, apprising them of the circumstances and gauging their opinions about increased revenue. Cantor shook his head in disgust; now *he* was feeling vindicated. The others had largely been in the dark about the deal, but warned Boehner that most Republicans weren't going to go for $800 billion in revenue, much less more. Boehner knew they were right. And he was crestfallen.

As the dazed Speaker walked out to his balcony for a smoke, his chief of staff, Barry Jackson, whispered nervously to friends that a coup was soon to unfold. Not a year had passed since Cantor, Ryan, and McCarthy published a book together, *Young Guns: A New Generation of Conservative Leaders,* that conspicuously excluded Boehner. Now, with word spreading through the conference of the Speaker's shadowy dealings with Obama, he appeared most vulnerable.

Obama phoned Boehner, eager for an update, but the Speaker refused to answer or call back. The Grand Bargain was off.

The two sides would peddle competing versions of what went down. Boehner's team said Obama moved the goal posts; the White House said the Speaker couldn't sell his own members on the deal. Both versions contain truth. Obama *did* renege on their handshake agreement; and Boehner *did* face an uprising among his members. For the Speaker, passing the Grand Bargain through the House would have required leaning on Democratic votes and steamrolling the conservatives—and, in all likelihood, kissing his speakership good-bye.

Looking back, with the nation pushing toward $23 trillion in debt and no mandatory spending reforms on the horizon, Boehner says it would have been worth it. "If I could have pulled this deal off, they could have thrown me out the next day," the former Speaker says. "I would have been the happiest guy in the world."

As their first year in the majority drew to a close, Republicans were forced to reckon with a side effect of their return to power: the rise of the professional right.

Believing they had enabled the GOP's decline by giving George W. Bush's party a free ride, conservative activists were determined to hold the new Republicans accountable. Yet this impulse, however well intentioned, resulted in overreach destructive to their own ends. If politics is the art of the possible, then the influence of outside groups on lawmakers—especially the newest, most susceptible lawmakers—made governing impossible. By going to DEFCON 1 on a weekly basis, threatening reprisals against anyone supporting the leadership on a given legislative issue, the activist warlords locked Boehner into a constant state of lose-lose: If he gave conservatives what they wanted, he would suffer defections from the center-right members and stand no chance of passing a bill; if the conservatives spurned him and he turned to Democrats to make up the margin, he would be accused of selling out the right.

As this dynamic took hold, Republican elected officials began viewing the operational base of the party with suspicion. Conservative political shops in DC were suddenly booming. The email lists and demographic data assembled by consultants and vendors was invaluable; advocacy groups that sprang up with seed money from major conservative donors, most notably the Koch brothers, would use the contact information to build grassroots armies, then issue sky-is-falling warnings of imminent treachery in Congress and ask for money to combat it. That money, in turn, made the groups all the more intimidating to lawmakers. For many Republicans, their "scorecard," a voting record summary graded by Heritage Action, FreedomWorks, or the Club for Growth, became as precious as their pocket-size copies of the Constitution.

People of sincere political conviction are plentiful, both inside the Beltway and beyond. But this purity-for-profit model introduced a new degree of capriciousness to the Republican civil war.

"The grassroots version of the Tea Party [was] earnest, concerned citizens who want limited government and economic liberty," Ryan says. "I think what ended up happening was the conservative industrial complex sort of stood itself up, and you quickly learned you could make money off this stuff. . . . And politicians realized, I don't have to work my way up the committee process. I don't have to pass a bunch of bills and prove my worth. I can just go on Fox News and Rush Limbaugh and be a hero on the Drudge Report."

At that same moment, an ideological insurgency was taking root on the left. It shared with the activist right a fundamental loathing of centralization and a distrust of entrenched institutions. But whereas the Tea Party combated big government and the erosion of traditional culture, the disciples of "Occupy Wall Street" aimed to expose big business and its bottom-line-fueled assault on workers and families.

Over a period of several months in the fall of 2011, tens of thousands of people descended on Manhattan's Financial District to protest the hoarding of wealth by the nation's elite. Some were slackers and hippies with nothing better to do. Others were neo-anarchists inspired by the internet hacking group Anonymous. But many of those who joined the demonstrations were common citizens, laborers justifiably vexed at the rottenness of their fruits while the orchard owners picked the trees clean for themselves.

In a May essay published in *Vanity Fair*, former World Bank chief economist Joseph Stiglitz wrote that 40 percent of America's wealth was controlled by the top 1 percent, and that as a result, "the top 1% have the best houses, the best educations, the best doctors, and the best lifestyles, but there is one thing that money doesn't seem to have bought: an understanding that their fate is bound up with how the other 99% live."[12]

Adopting the identity of "99 percenters," the zealots of Occupy Wall Street were moving the Overton window as it related to prosperity and equity, profits and communities, employers and the employed.

Indeed, for all the cartooning of stoners sleeping inside tents in Zuccotti Park (of which there were many), the Occupy Wall Street movement was peeling back layers of societal self-doubt. The economic recession, coupled with the transformation from an industrial to a tech-based economy, on top of recurring tax cuts for top earners and mounting debt for college graduates, all against the backdrop of trillions spent nation-building abroad, was proving highly combustible. And in the age of social media, organized activism had never been easier.

This revolutionary fervor on the left was just beginning to spark; with a Democratic president in office, it would be years until it breathed the oxygen necessary to grow into a political brush fire. The insurgency on the right, however, had already permeated the governing class.

That said, its aims were proving somewhat abstract. While the Tea Party rose on the strength of its economic argument—that America was spending and borrowing its way to becoming a socialist state—those lawmakers under its banner seemed only selectively animated by fiscal responsibility. Indeed, years later, the right's silence when Trump and his unified Republican government passed massive, deficit-ballooning pieces of legislation suggested that something else had been stirring the party's base.

"I was banging my head against a wall because I just wasn't getting the response from a member that made any sense," Cantor says. "And I think it was because people were in search of something on this menu of fiscally conservative policy positions that didn't quite match with their cultural conservative motivations. That's what's so odd: It was all about the economics at one level, but it didn't quite mesh with this cultural thing that was lighting the house on fire."

Before joining Cantor's team on Capitol Hill, Rory Cooper ran the Heritage Foundation's communications office. He recalls convening a meeting in late 2011 with some of his writers. The reason: A blog post, "Obama Couldn't Wait: His New Christmas Tree Tax," had gone viral.[13]

Once the intellectual engine of the right, Heritage was now trafficking in scandal over a fifteen-cent tax imposed by the Agriculture Department on fresh-cut firs.[14]

"I felt uncomfortable," Cooper says. "But that's what people wanted: cultural clash. A lot of conservatives weren't fighting on policy anymore."

CHAPTER FIVE

JANUARY 2012

"It was like a graduate economics class."

NEWT GINGRICH WAS JUST ABOUT OUT OF IDEAS. THE FORMER HOUSE Speaker, having long since established himself as a visionary at a time when the post-Reagan Republican Party was woefully short on innovative thinking, had come out of retirement to run for president. But he was doing nothing to distinguish himself from the competition and was desperate for a way to break through the stuffy covenants of campaigning.

The primary field was unimposing. Mitt Romney, the front-runner, had expected the stiffest competition to be his old rival Mike Huckabee. But the evangelical-populist favorite of the 2008 campaign had since inked a multimillion-dollar contract with Fox News and was building a mansion in Florida. Prominent conservatives pleaded with Huckabee to run, but he could not be persuaded to give up the comforts of his new life. He ended the suspense by announcing on his Fox News show, following a sublime joint performance of "Cat Scratch Fever" with rocker Ted Nugent ("I make the pussy purr with the stroke of my hand"), that he would stay out of the 2012 race. The show ended with a video message from Donald Trump wishing Huckabee well.

Romney could exhale. Four years removed from being the conservative darling, he was the establishment preference—and given the newfound furor over his health care apostasy in Massachusetts, the right aimed to rally around a viable challenger. Huckabee had been their best bet. The former Arkansas governor's policy record was hardly pristine; conservatives had savaged him over the years on spending and immigration. But campaigns, especially primary campaigns, are an exercise in contrasts, and Huckabee was everything Romney wasn't: authentic, approachable, funny, human. With him out of the way, the road to the Republican nomination looked a whole lot smoother.

There were other obstacles, of course. Michele Bachmann had leveraged her Tea Party stature to raise impressive money. Jon Huntsman, the former Utah governor who had served more recently as Obama's ambassador to China, had presidential looks and royal family resources. Tim Pawlenty, the Minnesota governor, had an understated blue-collar appeal, as did former Pennsylvania senator Rick Santorum. Herman Cain, the former CEO of Godfather's Pizza, had a niche following as the field's only black candidate. And Rick Perry, the Texas governor, had a lush record of job creation amid a barren national economic landscape.

Romney could take no chances: Conservatives were desperate for an alternative to him, and Perry's story—the "Texas miracle" of his state creating nearly half of all new jobs in the United States since 2009[1]—was deeply compelling. So intimidating was Perry's late entry into the race that Romney and the other Republicans ganged up to kneecap him on immigration, specifically the governor's granting of in-state tuition to illegal immigrant students. For Romney, immigration was the only card to play, and he played it ruthlessly, just as he had in 2008.

"Sometimes you have to light a prairie fire to win," recalls Zac Moffatt, Romney's digital guru. "But sometimes it comes back and burns your house down."

In one particularly tense debate, after Perry's rivals took turns lacerating him over offering the in-state tuition to undocumented minors, he told them, "I don't think you have a heart." The audience booed him, then cheered Santorum for calling Perry "soft" on immigration. It was the second-most surreal moment in the primary season, topped only by the time every candidate onstage affirmed that he or she would reject a deficit-reduction deal that proposed ten dollars in spending cuts for every dollar in additional tax revenues, an absurdly absolutist stance given that such an agreement would balance the budget at warp speed without raising taxes on any non-multimillionaires.

One by one, Perry and the others were voted off the island. By the time the Republican primary moved past Iowa and New Hampshire, to the third nominating contest in South Carolina, just four candidates remained: Romney, Gingrich, Santorum, and Ron Paul.

Of this group, Gingrich was in the worst shape. He had finished a distant fourth in Iowa, taking 13 percent of the vote, while Romney, Paul, and Santorum had all finished north of 20 percent. (Santorum technically won, though Romney was announced the winner on caucus night, robbing the underdog of a major momentum boost.) Gingrich had also been blown out in New Hampshire, taking just 9 percent of the vote compared to Romney's 39 percent.

In the modern primary system, a candidate typically needed a strong showing in one of the first two contests to raise the requisite money for his or her campaign to continue into South Carolina. But *Citizens United* had transformed the landscape. One donor could now single-handedly sustain a candidate with millions of dollars in super PAC spending, and Gingrich had the sweetest sugar daddy of them all: Sheldon Adelson, the Las Vegas casino magnate, whose total pro-Gingrich expenditures for the cycle would reach $20 million.[2]

All that money wasn't doing Gingrich any good. His candidacy was on life support, and polls showed Romney up double digits in South Carolina. Gingrich was stumped. He had exhausted every tactic imaginable: staying positive and playing nice with his rivals; eviscerating Obama, even going so far as to call him the "food-stamp president"; and eventually, going nuclear on Romney in response to sustained attacks from the front-runner's camp, alleging that his company, Bain Capital, consisted of "rich people figuring out clever legal ways to loot a company." None of it had vaulted Gingrich into contention. And time was running out.

The two debates in South Carolina offered final gasps of oxygen before the state's January 21 primary. The first forum, hosted by Fox News in Myrtle Beach, got off to a lousy start, as Gingrich stumbled in response to questions about abandoning his positive-campaigning pledge. And then it happened: Juan Williams, the African American moderator, started grilling Gingrich about his recent racially tinged comments, including the "food stamp president" quip. As the crowd hissed at Williams, Gingrich scolded him with a lecture on political correctness that elicited a standing ovation.

It was an uncomfortable snapshot for some in the party: an over-whelmingly white audience booing a black moderator on Martin Luther King Jr. Day, in a state where the Confederate flag still flew on the Capitol grounds. But for Gingrich it was a moment of clarity. Where all his calculated strategies had failed, his off-the-cuff reprimand of Williams had succeeded. Perhaps attacking Romney—or even the president, for that matter—was a waste of time. Maybe there was a greater upside in doing what came naturally to him: tormenting the fourth estate.

"The thing that struck me," Gingrich recalls, "was what conservative audiences reacted to, even more than attacks on Obama, was attacks on the media. You could get a stronger response by taking the media head-on than you could with any other single topic."

Sure enough, three nights later, in North Charleston, Gingrich stole the show with a similar routine. It was all too easy: CNN reporter John King opened the debate with a question about Gingrich's ex-wife's recent claim that he had sought an open marriage. It was like putting a beachball on a tee in front of Babe Ruth. Summoning every fiber of moral outrage, Gingrich tore into King, CNN, and the entire press corps. In a rant heard 'round South Carolina, Gingrich bellowed, "I am tired of the elite media protecting Barack Obama by attacking Republicans!"

"It was an electric moment," recalls Kevin Madden, Romney's long-time senior adviser and communications specialist. "Literally over-night, Newt's favorables and unfavorables flipped in our tracking. We went into those debates ten points up and came out fifteen points down."

Republicans have a rich history of shunning the press. Dwight D. Eisenhower, after leaving office, ripped the "sensation-seeking columnists and commentators" at Barry Goldwater's 1964 convention, saying they "couldn't care less about the good of our party." Vice President Spiro Agnew ratcheted up the rhetoric on behalf of Richard Nixon, giving his famed 1969 speech in Des Moines decrying the "small and unelected elite" who possess a "profound influence over public opinion" without any checks on their "vast power." And, in a less conspicuous fashion, Reagan warred with the White House press corps for most his time in Washington.

Much of this amounted to "working the refs," as a basketball coach does after a tough foul call, in the hope of avoiding the next whistle. There was an age in which the refs were perceived to be impartial: As of 1986, Gallup found that 65 percent of Americans still felt a "great deal" or a "fair amount" of confidence in the press.[3] But over the ensuing three decades, as voters came to view the refs as players looking to dunk on their team, that number plummeted: By the time Trump was elected president, it was 32 percent, and just 14 percent among Republicans.[4]

"Taking on the media, instigating that clash of political civilizations, became Newt's message," Madden says. "I always believed that media bias is a fact, not a message. I don't believe that anymore."

Gingrich won South Carolina in a rout, taking 40 percent of the vote to Romney's 28 percent, a result that would have been unimaginable a week earlier. By using the media as a foil, Gingrich had rallied the conservative base, shattered the establishment's infallibility, and reset the narrative of the campaign. He had also created a blueprint.

That spring, Andrew Breitbart, the combative blogger whose website was gaining cult popularity among the anti-establishment right, died suddenly at age forty-three. This news shook the conservative movement: Breitbart had been pioneering with his vision of an alternative to what he viewed as a biased mainstream media. Upon his death, the leadership of Breitbart's burgeoning empire was assumed by a little-known investment banker with nationalist views and a disdain for the GOP elite. His name was Steve Bannon.

South Carolina was significant less because it was Gingrich's last stand and more because it represented Romney's chance to effectively put away the nomination—and he whiffed.

On paper, there was no reason for a protracted primary. Romney had the biggest bank account, the best campaign infrastructure, the most support from party elders. Sure, he struggled with his aloof image, worsened by his boast that his wife "drives a couple of Cadillacs" and by his telling a crowd in New Hampshire, "I like being able to fire people who provide services to me."[5] Then again, his rivals had the sum emotional connection of a late-night infomercial. Gingrich's ego was surpassed only by his insincerity—he railed against "elites" while being chauffeured between mansions and green rooms. Santorum was so lacking in warmth that his sweater vest became the campaign's spiritual avatar. And Paul, on his best day, was charming in the mold of an aging, disheveled uncle who mutters to himself at holiday gatherings.

For all Romney's flaws, no Republican was equipped to exploit them; they all lacked some combination of money, organization, broad appeal, and political chops. The question was never whether Romney would win the nomination, but how long it would take, and what it would cost him.

Romney recouped his momentum after South Carolina by winning easily in both Florida and Nevada, appearing once more to be the inevitable nominee. But then Santorum, having established himself as the last man standing against the front-runner, swept all three states voting on February 7, Colorado, Minnesota, and Missouri. The result was to stunt Romney's coronation while highlighting his emerging weakness with blue-collar voters in middle America.

Ultimately, Romney's money and manpower were still the difference. As the campaign became a slog, his rivals were exposed as organizationally incompetent. Santorum failed to file complete delegate slates in numerous contests, rendering the vote totals somewhat irrelevant; more embarrassingly, neither he nor Gingrich qualified for the Virginia ballot on Super Tuesday. With competition like this—in the contest to lead the free world, no less—Romney couldn't help but finally pull away. On April 25, after a bruising, four-month-long primary, the RNC declared him the party's presumptive nominee.

Hindsight suggests that however unimpressive on the surface, Romney deserves real credit for his victory. At a time when conservative authenticity had become the currency of a party dominated by evangelical Christians and increasingly populated by working-class whites, Republicans nominated a moderate Mormon with three political reincarnations—one for each vacation home. In the truest sense, Romney was a statesman picked to lead a populist crusade.

For this triumph, he owed a debt of thanks to one man in particular: Donald Trump.

Cable news shows spent much of 2011 speculating, often salivating, over the notion of Trump running for president, and he regularly took advantage of their platforms to critique potential rivals. His favorite target was Romney, whom he derisively labeled "a small businessman," and his favorite venue was Fox News, specifically the *Fox and Friends* morning show, on which he made a weekly Monday appearance. As Trump grew more hostile, Romney's team was split over a possible response. Some wanted him to strike back to prove his toughness, while others thought he should ignore Trump, fearing a pissing match that could not possibly be won.

Trump could sink Romney three different ways: by endorsing someone else, by entering the primary himself, or by running as an independent. None was far-fetched: Trump had, in the summer of 2011, staged a publicity stunt by taking Sarah Palin out for pizza in New York, and was intrigued at the opportunity of tapping into the populist streak she had exposed in 2008. "She's a very good woman, and what they did to her was terrible," Trump recalls. "Working people were angry at McCain, at the McCain campaign, because of what they did to her. She gave that really weak campaign a tremendous, positive jolt of energy, but [they] stripped her of her tremendous assets, her tremendous personality. She had this great vibrancy and they wanted to take it all away."

Romney's brain trust determined that Trump needed to be engaged—but not provoked.

Just as the primary voting was getting under way, Romney's campaign manager, Matt Rhoades, traveled to Manhattan for a meeting with Michael Cohen, Trump's lawyer and "fixer." Inside a conference room at Trump Tower, Cohen probed with concerns about Romney's viability. Rhoades, in turn, made the case for his boss as the surest bet to beat Obama, and pressed for a meeting between Trump and Romney. This dance went on until Cohen was suddenly interrupted by a voice crackling over a speakerphone on the table. It was Trump. He had been listening in the entire time, and was now agreeing to meet with Romney.

The two men soon found themselves standing together on a small stage in Las Vegas—Romney with a nervous smile next to his red-in-the-cheeks wife, Trump at the podium endorsing him. "Mitt is tough, he's smart, he's sharp, he's not going to allow bad things to continue to happen to this country that we all love," Trump said. After a firm handshake, Romney stepped to the podium and announced, "There are some things that you just can't imagine happening in your life. This is one of them."

It was February 2, Groundhog Day, but nobody had seen anything like this from Romney. He was "Massachusetts Mitt," a social liberal, when he ran for Senate in 1994; he was "the conservative's conservative" when he ran for president in 2008; now, having rebranded himself again for 2012, the somber, centrist, establishment-backed Romney was kissing the ring of a carnival barker whose claim to political fame was lying about the president's birthplace.

All these tortuous developments in the Republican primary—Romney's determination to neutralize Trump, his failure to finish off Gingrich in South Carolina, his struggle to put away Santorum thereafter—were indicative of a basic vulnerability: Conservatives did not like or trust the party's nominee.

IN THE FALL OF 2011, A FEW MONTHS BEFORE THE GOP PRESIDENTIAL tournament tipped off, David McIntosh had taken on a new client with a most unusual request.

McIntosh, the former Indiana congressman and cofounder of the Federalist Society, the powerful club of Republican attorneys, had long been a kingpin inside the conservative movement. But nothing had prepared him for this. His client, a major conservative donor, represented a larger group of major conservative donors, and they were prepared to pool their resources to attract a viable challenge to Romney for the GOP nomination. What they needed from McIntosh was a comprehensive analysis of the potential candidates and their records: fund-raising abilities, electoral wins and losses, political strengths and weaknesses, and ideological convictions. Once the dossier was completed, the donors would evaluate the prospects and determine whether any of them was worth recruiting into the race.

"The theory was, even at that late stage, if we could find someone and fund their Iowa and New Hampshire operations on the ground, that would propel them to the nomination," McIntosh says.

The appraisal yielded three distinct conclusions. The first was that the most attractive prospect, Marco Rubio, was too inexperienced—and indeed, he had already ruled out running in 2012. The second was that the most accomplished prospect, Chris Christie, was too moderate for many of the donors. The third was that one man, and one man only, met all the group's criteria: Mike Pence.

They weren't so much drawn to Pence's pious traditionalism. Many of the donors, in fact, were libertarian-minded affiliates of the Koch brothers' network; Romney's vacillations on social issues were the least of their worries. The two major concerns were taxes and entitlements: Romney had hiked tax revenues in Massachusetts (by levying fees and closing loopholes) while implementing a health care system that added significantly to the state's mandatory spending. These were philosophical red flags for the donors, who worried that such heresies reflected a worldview that could lead Romney to decisions even more catastrophic—such as appointing another moderate, à la David Souter, to the Supreme Court.

There were no such doubts about Pence. A fixture of the conservative movement—his soft smile and helmet of prematurely white hair recognizable all about town—the congressman was a consistent voice on every issue, one of the few lonely Republicans who had pushed back against the Bush administration's excesses. With the possible exception of his best friend in the chamber, Arizona congressman Jeff Flake, Pence was the most reliable conservative in the House of Representatives.

McIntosh was struck by the serendipity of the donors' choice. Pence had succeeded him in Congress a decade earlier; the pair went all the way back to Pence's days as a NASCAR-loving, Bill Clinton-bashing radio host in Indianapolis.

There was just one problem: He was already running for governor of Indiana.

Pence had long harbored visions of sitting behind the Resolute desk. After failing in his first two bids for Congress, in 1988 and 1990—the latter loss was defined by his TV advertisement featuring an Arab-dressed actor, speaking in a thick Middle Eastern accent, thanking Pence's opponent for America's dependence on foreign oil—he had stepped back to reassess. He took over as president of a small free-market think tank in

Indiana. He penned an essay, "Confessions of a Negative Campaigner," apologizing for his tactics and vowing never to use them again. When Pence finally succeeded, winning the race to replace McIntosh a decade later, he felt it was the fulfilment of God's plan for his life. He had humbled himself, repented, and was being rewarded. Emerging as a star conservative in the House of Representatives, Pence looked for the next phase of the plan. He suspected it might just include serving as president. But as he surveyed his confidants in early 2011, they all advised against a run. House members don't get elected president, they told him. Go home and be a governor.

Having heeded that advice, Pence was in the driver's seat to be elected Indiana's chief executive. But he was nonetheless captivated by McIntosh's pitch, listening intently to the recruitment effort and chewing it over with his wife, Karen. Ultimately, the timing just wasn't right. He would focus on the governor's race, he told McIntosh, and keep an eye toward 2016 if the Republican nominee failed to defeat Obama in 2012.

With Pence out and the primary season drawing closer, McIntosh's client scrapped the project. It was too much money to risk throwing behind just anyone on such short notice. The ensuing split among donors would reflect the divides across the Republican financial universe: Many went to Romney, eager to curry favor with the likely nominee; others picked a rival horse, determined to deny Romney the nomination; and a few stayed on the sidelines altogether, waiting to see not whether Romney could win the primary, but whether he would choose a running mate who compensated for some of his limitations.

Naturally, those limitations weren't limited to taxes and entitlements.

Even when Romney was Laura Ingraham's pet conservative in 2008, his membership in the Church of Jesus Christ of Latter-Day Saints was the subject of a nasty dog-whistling movement on the right. Socially conservative Americans, particularly those belonging to evangelical churches, had for generations been steeped in the belief that Mormonism is a wicked perversion of traditional Christian doctrine. Even for nonbelievers, the LDS church has been synonymous with an alien community; in 2011 it would become something of a cultural punch line thanks to the Broadway hit *The Book of Mormon*, a foulmouthed tale of two missionaries attempting to proselytize an African village.

In 2008, Romney felt compelled to give a speech on his faith to defuse the antagonism his campaign was eliciting, particularly from voters in southern and midwestern states. But it didn't stop the whisper campaign. And some critics didn't bother whispering.

Robert Jeffress, the prominent pastor of a Dallas megachurch, denounced Romney's religion as a "cult"[6] and implored evangelicals to oppose him. During an organized debate with Christian attorney (and Romney supporter) Jay Sekulow, Jeffress addressed "the hypocrisy" of church leaders who "for the last eight years of the Bush administration have been telling us how important it is to have an evangelical Christian in office who reads his Bible every day. And now suddenly these same leaders are telling us that a candidate's faith really isn't that important." Jeffress added: "My fear is such a sudden U-turn is going to give people a case of voter whiplash. I think people have to decide, and Christian leaders have to decide once and for all, whether a candidate's faith is really important."[7]

Jeffress continued his crusade during the 2012 campaign. A supporter of Perry for president, the pastor used an appearance at the Values Voter Summit in October 2011 to drive a wedge between Romney and evangelicals. "I just do not believe that we as conservative Christians can expect him to stand strong for the issues that are important to us," Jeffress told reporters.[8] "I really am not nearly as concerned about a candidate's fiscal policy or immigration policy as I am about where they stand on biblical issues."

(Four years later, Jeffress would become Candidate Trump's most visible Christian disciple, appearing with the thrice-married, casino-owning candidate onstage in Texas during the heat of the GOP primary race. "I can tell you from experience, if Donald Trump is elected president of the United States, we who are evangelical Christians are going to have a true friend in the White House," he said, according to the *Dallas Morning News*.)

The issue was far from resolved when Romney clinched the GOP nomination in April. A survey released by CBS News and the *New York Times* found that just 27 percent of white evangelical Republicans said they would "enthusiastically" support him against Obama in the fall.

Even as the professional Christian right grudgingly rallied around him—with endorsements from the major evangelical groups and leaders, including, eventually, Jeffress himself—the grass roots remained hesitant. When Romney agreed to give the May commencement address at Liberty University, the Jerry Falwell–founded Christian college in Virginia, the school wound up removing the news from its Facebook page because of the backlash among students and alumni.

"I get it. I'm from a weird religion, too, according to Republicans," says Eric Cantor, who hails from a deeply religious tract of Virginia and heard frequent complaints about Romney's Mormon faith. "My district was sandwiched between the Falwells to the West and Pat Robertson to the East. I'm Jewish, and the district is not even two percent Jewish. We would do polling and one of the most important issues for people was whether the candidate believed in Jesus as their savior. That wasn't good for me."

In retrospect, the distrust of Romney is better understood through a prism of cultural warfare than one of theological creed. At the outset of the primary campaign, the Obama administration mandated that religious institutions must cover contraceptives in employees' insurance plans. In May, on the same day as Romney's commencement address at Liberty, the president announced his support for same-sex marriage. A month later, in the span of two weeks, Obama issued an executive order protecting young immigrants from deportation while the U.S. Supreme Court upheld the constitutionality of the Affordable Care Act.

All the while, America's folk landscape looked like the culmination of conservative jeremiads about decline: *Fifty Shades of Grey,* a novel of true love realized through bondage and sadomasochism, spent more than seven months atop the *New York Times* bestseller list.

"It was a revolution," John Boehner says. "The country was changing right underneath our feet."

The conservative base was on fire. His religion aside, Republicans had reason to worry that Romney—whose adviser had publicly compared the candidate's November strategy to an "Etch A Sketch," shaking off the right-wing positions of the primary and starting over afresh—didn't have the core convictions, much less the stomach, for a fight with Obama.

The president and his allies, on the other hand, couldn't wait to start throwing haymakers.

ROMNEY WAS AN EASY TARGET. THE WAY HE TALKED AND THE WAY HE walked, the haircut and the mannerisms—it was like a black-and-white sitcom character reborn in the age of Technicolor. Once, after a donor meeting in which Romney heard the phrase "No shit, Sherlock" for the first time, he gleefully repeated the quip to his staff—but cleaned it up to say, "No *bleep*, Sherlock."

Even Republicans couldn't help themselves. As he campaigned for House candidates across the country in 2012, Boehner worked humorous digs at Romney into his stump speech, mocking the Republican nominee behind closed doors as someone who had never mowed his own lawn or owned a pair of blue jeans.

This detachment carried significant political risk. In a vacuum, the fact that Romney didn't curse, or didn't drink, or had lots of money, might not have been damaging. But taken with his policy positions, and given how the president aimed to portray him, these characteristics were glaring vulnerabilities. Because strategically, Obama and his allies wanted everything voters learned about Romney to fit into a simple overarching theme: He was not one of them.

The president had little choice but to run a bruising reelection campaign. It was true that Osama bin Laden was dead and the Detroit automakers were alive. But the conditions were categorically ripe for his defeat: There was little tangible progress to show for the stimulus package; the Affordable Care Act was still widely unpopular; unemployment was still north of 8 percent; average gas prices were approaching a record high; and stock markets were barely inching upward.

The bright spot for Obama, politically, was that polling showed him still in decent standing with low- and middle-income voters, most of whom did not blame their hardships on him but, rather, on the previous administration, the broken institutions of government, and what they viewed as a broken economic model that exploited the many for the gain of the few. Against this backdrop, Romney, the billionaire businessman and venture capitalist, represented a perfect foil, someone whose obvious strength could be turned into an insurmountable weakness.

Team Obama spent the late spring and early summer battering Romney with an advertising blitz that defined the Republican nominee in terms he would never recover from: as a cold, bloodthirsty corporate raider who cared more about profits than people. "President Obama was politically wise in characterizing me as some rich Republican that doesn't care about the little guy. There are a lot of reasons why he was able to define me in that way," Romney recalls. "I needed to do a much better job in communicating that the whole reason I was running is for the average American."

The two most memorable spots were so haunting, so brutally effective, that political scientists will dissect them for decades to come. The first, from Obama's campaign, was set to a soundtrack of Romney singing "America the Beautiful," during which headlines flashed across the screen telling of his corporate outsourcing, his Swiss bank account, his tropical tax havens. (The fade to black came as Romney belted out, "And crown thy good with brotherhood . . .") The second, from Priorities USA Action, featured a testimonial from Mike Earnest, an Indiana factory worker who told of building a thirty-foot stage that Bain Capital officials used to announce that they were closing the plant and firing all its employees. "Mitt Romney made over a hundred million dollars by shutting down our plant and devastated our lives," Earnest said. "Turns out that when we built that stage, it was like building my own coffin. And it just made me sick."

The precise impact of the ads themselves would later become the subject of debate, considering how Romney had remained relatively stable at 2 to 4 points behind Obama in the horse race polling. But it was impossible to quantify how this onslaught of negativity—on top of questions about Romney's tax returns and Harry Reid's claim that the GOP nominee hadn't paid any income taxes for a decade—drove media coverage in a way that implanted an asterisk of doubt in the minds of voters whenever they heard Romney tout his economic expertise.

There were critics of Team Obama's strategy. Several of the Priorities USA ads were skewered by fact-checkers; one in particular, which suggested Romney's shuttering of a plant was responsible for an employee's wife dying of cancer, was slammed as egregious and untrue.[9] Meanwhile, several prominent Democrats, including auto bailout czar Steve Rattner and then-Newark mayor Cory Booker, condemned the Bain Capital ads and defended Romney as a practitioner of capitalism. And Reid, for his part, forfeited much of his credibility when it was learned that he'd flat out lied about Romney's tax returns.

But the feathering of Romney was effective largely because Republicans had supplied the tar. Gingrich cronies had flooded South Carolina with ads depicting the predatory ways of Romney and Bain Capital. Perry had called Romney's colleagues "vultures" that were "waiting for a company to get sick, and then they swoop in, they eat the carcass, and they leave the skeleton."[10] Even Palin had gotten in on the act, going on Sean Hannity's television show to question Romney's claim of creating one hundred thousand jobs at Bain and needling him for his failure to release more tax returns.

Meanwhile, Romney had made himself singularly vulnerable when, in 2008, he wrote a *New York Times* op-ed entitled, "Let Detroit Go Bankrupt." The substance was defensible; Romney argued that the automakers, burdened by legacy costs related to lavish union contracts, could be viable long term only by restructuring under rules of bankruptcy. And yet the headline—coupled with Obama's rescue of General Motors and Chrysler—reinforced Romney's negatives in the midwestern states where Obama was most exposed.

The attacks on Romney shouldn't have been surprising. For all the flowery rhetoric, Obama, a student of Chicago-style campaigning, had run a pitilessly negative operation against McCain four years earlier. (And, as Romney observed during the bludgeoning of Gingrich months earlier, "politics ain't bean bag.") What *was* surprising was Romney's failure to defend himself, and to defend capitalism conceptually, from Team Obama's tireless assault.

"We spent the campaign defining his business record. These were legitimate stories and legitimate critiques. What was absent was the other side of the story—they never told it," David Axelrod, Obama's chief strategist, says. "At a time when a lot of people were really jaundiced about Wall Street and financiers, they nominated a guy who spent most of his life in that world. Romney was always going to get the benefit of the doubt as someone who could manage the economy; what was suspect was whether he would manage the economy in a way that was beneficial to the middle class. And he never made that case."

ROMNEY HAD THREE OBVIOUS OPTIONS IN CHOOSING A RUNNING MATE. The first was someone who could help make that economic case, someone with executive or managerial experience who would reinforce his greatest strength. The second was someone who could compensate for a glaring weakness—namely, the lack of enthusiasm in the conservative base, as John McCain had done with Sarah Palin four years prior. The third was someone who could help politically, ideally someone so popular in a swing state that his presence on the ticket could carry it come November.

There was no shortage of choices. Romney could select a successful governor, such as Louisiana's Bobby Jindal or New Jersey's Chris Christie or Minnesota's Tim Pawlenty. He could pick a Tea Party star, such as Marco Rubio. He could choose a geographic complement, such as Ohio's Rob Portman or Virginia's Bob McDonnell.

One name that nobody outside Romney's inner circle took seriously: Paul Ryan.

The congressional GOP was dysfunctional, and Ryan had emerged as perhaps its most polarizing figure. His safety net–slashing budgets were widely viewed as politically toxic: Democrats ran ads depicting Ryan pushing a wheelchair-bound grandmother off a cliff, and Newt Gingrich described the Wisconsin congressman's proposal as "right-wing social engineering."[11]

Furthermore, Ryan had found religion on deficits only once George W. Bush left office, having voted for two wars, enormous tax cuts, and the massive Medicare prescription drug program. Romney had enough problems of his own; why adopt someone else's baggage?

This was the position taken by Stu Stevens, Romney's chief strategist and closest adviser. Determined to make the election a "referendum" on Obama—that is, forcing voters to view their decision through the narrowest possible context of the president's job performance—Stevens was not welcoming of distractions. Anything the GOP did to divert attention away from a limping economy, Stevens told Romney, was a boon to Obama's reelection.

The president wanted a "choice" election, one in which voters judged the incumbent not in isolation, but against the alternative. In this case, that meant framing Obama, the pragmatic protector of the American worker he identified with, against Romney, the ruthless fix-it man who worked off formulas instead of feelings. The problem for Stevens, and for the GOP writ large, was that Romney wanted both a referendum *and* a choice—hence his pick of Ryan.

In truth, Romney never valued a running mate who offered political expediency; he made it known to friends and senior aides that he wanted his selection to signal an emphasis on governing. This was hard to believe given Romney's cautious reputation, but in fact, the Republican nominee had told confidants that he was prepared to be unpopular for the first two years of his presidency because of the cuts he hoped to make—cuts that Ryan, as his junior partner, could help design and pass into law. Obama could barely believe his luck. Ryan, the poster child for policies made synonymous with social Darwinism, was joining the GOP ticket.

When Romney introduced his running mate, on August 11 aboard the retired USS *Wisconsin* in Southern Virginia, he hailed Ryan as "an intellectual leader of the Republican Party" who had been prescient in warning of "the fiscal catastrophe that awaits us if we don't change course." Ryan, who could pass for one of Romney's five sons, came bounding down from the ship wearing a sport coat with no tie.

"President Obama, and too many like him in Washington, have refused to make difficult decisions because they are more worried about their next election than they are about the next generation," Ryan said. "Politicians from both parties have made empty promises which will soon become broken promises—with painful consequences—if we fail to act now."

Although Romney had been pouring time and money into diversifying states such as Virginia and Colorado, the investment was showing little return. In fact, as the campaign progressed, it was becoming apparent that Romney's best chance to beat Obama would be in the Rust Belt, where the president's approval among working-class whites had plateaued over the last several years. Wisconsin was part of the "Blue Wall" of states Democrats had carried in every presidential election since 1992. Romney hoped that Ryan, with his Irish-Catholic roots and midwestern twang, could help put not just his home state into play but Michigan and Pennsylvania as well.

It was an exercise in obliviousness: Romney was confident that certain voters, in a certain part of the country, would respond to a running mate whose governing vision—entitlement cuts, immigration reform, and unfettered free trade—was exactly what they did *not* want.

ROMNEY ALWAYS STRUGGLED TO SELL HIS STRENGTHS. HE WAS A COM-petent, technocrat-minded governor, but he appeared reticent in hyping his record. Reforming the Massachusetts health care system had been a crowning achievement, but he avoided the issue because of the fury over Obamacare. His activity in the LDS church included countless stories of his service to the poor and destitute, but he was hesitant to discuss religion.

This was precisely what Stevens envisioned. *The election should be a referendum on Obama.* Romney wasn't going to win a likeability contest against the president; his advisers wanted a Monster.com election, not a Match.com election. Whenever he saw ads run by the outside super PAC supporting Romney that attempted to humanize him (with stories of how he'd once shut down his business to conduct a search for a missing teenage girl, for instance), Stevens would scream at the television, "Why are you wasting money on this shit?!"

Romney grew more assertive speaking to these themes as the race went on: how he governed pragmatically in a blue state, how he understood the complexities of the health care marketplace, how he served the underprivileged as an elder in his church.

Tellingly, these testimonials had something of an inverse effect on diverging portions of the electorate: It attracted some persuadable voters in the middle, but it did nothing to energize conservatives. Meanwhile, the one trait universally assumed to help Romney connect with the base, his business chops, might have alienated him from it. Pie charts and economic models do little to assuage voter angst. As people watched their jobs disappearing, their communities hollowing out, and their national character changing, they wanted a brawler—not a bookkeeper.

"It became a *Wall Street Journal* campaign," Cantor recalls. "There were rallies in these big airport hangers in rural areas, and he's talking about unfunded liabilities and the entitlement programs and GDP percentage. I mean, it was like a graduate economics class. And it struck me that something was just not clicking. There's no way all these thousands of people that showed up really want to hear this."

What they wanted to hear, many of them, was an echo of their own contempt for Obama.

Four years had provided plenty of ammunition. The president's about-face on gay marriage. His administration's feud with religious groups over contraception and the ensuing talk of a Republican "war on women." His so-called apology tour, in which he traveled the world confessing of America's past arrogance. All of it felt patronizing, disdainful.

But more than anything, it was Obama's perceived exploitation of racially driven identity politics that drove Republicans crazy. Whether it was his election-year legalization of undocumented minors, or his scolding of a Massachusetts police officer for his arrest of a black Harvard professor, or his emotional observation that a murdered black teenager, Trayvon Martin, could pass for his own son, the nation's first black president goaded conservatives in ways that no white Democrat possibly could.

"He took race back to the sixties, as far as I'm concerned. He made everything a race issue, or at least saw it through a racial lens," says Jim

DeMint, the South Carolinian who entered the Senate with Obama in 2004. "The country had moved toward bending over backward to create equality. But then suddenly, with Obama, he just lit the fires. I thought when he was elected that was the big victory, that we had put racism behind us."

Some variation of this strange sentiment, that Obama's presidency ipso facto had ushered in an era of postprejudice America, became a trendy crutch for Republicans as bad actors in the party systematically targeted minority communities with racial gerrymanders and voter-suppression measures.

These institutional labors aside, lesser instances of bald-faced bigotry exploded during Obama's tenure. There was the Speaker of the Kansas House who shared emails calling the First Lady "Mrs. YoMama." There was the Colorado congressman who compared working with the president to "touching a tar baby." There was the Orange County GOP official who circulated an image depicting Obama as a monkey. There was the effort in June 2012, led by Michele Bachmann and Louie Gohmert, to get the State Department to investigate ties between the Muslim Brotherhood and Huma Abedin, the longtime top aide to Hillary Clinton. (Boehner defended Abedin's "sterling character" and called the insinuations "pretty dangerous.")

The contours of the GOP's racial paradox predated Obama. Using the "party of Lincoln" label as protective cover, Republicans could pursue discriminatory policies in one breath while debunking allegations thereof in the next by insisting that their ideological forebears had freed the slaves. Of course, nothing could be farther from the truth: Southern segregationists fled the Democratic Party following Lyndon B. Johnson's signing of the Civil Rights Act in 1964, sparking a decades-long realignment that, with the aid of the "Southern Strategy" employed by Barry Goldwater and Richard Nixon, turned the GOP into the champion of the old Confederacy's states-rights, small-government creed.

As Lee Atwater, the notoriously ruthless political strategist to Ronald Reagan and George H. W. Bush, once observed, "You start out in 1954 by saying, 'Nigger, nigger, nigger.' By 1968, you can't say 'nigger.' That hurts you, backfires, so you say stuff like 'Forced busing, states' rights,' and all that stuff, and you're getting so abstract, now you're talking about cutting taxes, and all these things you're talking about are totally economic. . . . 'We want to cut this' is much more abstract than even the busing thing, and a hell of a lot more abstract than 'Nigger, nigger.'"

Even in the age of a black president—*especially* in the age of a black president—the return on this tactical investment was obvious. Romney was presented with polling on two different occasions that demonstrated the incentive of exploiting the dark side of populism. By making the campaign about identity, personal and national, he could drive a wedge between Obama and blue-collar voters. This was particularly true in the industrial Midwest, where many noncollege-educated whites had soured on the president since voting for him in 2008. But Romney refused. He would not hesitate to take hawkish stances on immigration, and his proposal of "self-deportation" wound up being a gift to Obama's campaign. Racial dog whistling was something else entirely, and Romney made clear that he would not stand for it.

What exasperated Romney, as he endeavored to run a clean campaign, was the perceived double standard inherent to racial politics. That very year, in his home state of Massachusetts, the Democrats' nominee for U.S. Senate, a law professor named Elizabeth Warren, had been exposed for identifying as Native American during her rise in academia. There was no evidence that this had accelerated her career; Warren was regarded as a top legal mind under any circumstance. Yet, as her star rose on the left, Republicans marveled at how a white woman who had claimed "minority law teacher" status at the University of Pennsylvania, and who later allowed Harvard Law School to present her as "Native American," was at best getting away with making a mockery of affirmative-action standards and manipulating them at worst.

The lowlight of perhaps the entire election was when Joe Biden, who had recently impersonated an Indian call center employee and had once called Obama "the first mainstream African American who is articulate and bright and clean,"[12] warned a largely black audience about Romney and Ryan, saying, "They're going to put y'all back in chains."

Contextually, the vice president's comment was part of a broadside against Republicans' deregulation of Wall Street; Democrats said his metaphor was meant to play on Romney's promise to "unshackle" the economy. But the inflection of Biden's voice and the Virginia crowd's reaction told a different story, one from which most reporters quickly moved on.

To the extent Democrats benefited from a double standard, Republicans had only themselves, and their *real* ideological forebears, to blame.

RACIAL POLITICS ASIDE, REPUBLICANS DID HAVE ONE LEGITIMATE BEEF. As the campaign progressed, and Romney exhibited a rare talent for quirk ("The trees are the right height," he said in Michigan), a cruel caricature took hold of the nominee. He was unusual, out of touch, freakishly peculiar.

Romney deserved some ribbing, but the relentless focus on his idiosyncrasies was cheap and callow. One media-made controversy was particularly mind numbing: Romney was widely scored for recalling, during a debate with Obama, how he had assembled "binders full of women" to ensure staff diversity as governor of Massachusetts. Democrats had spent the year cudgeling Republicans for supposedly treating the fairer sex like second-class citizens, yet when Romney offered the insightful story of how he'd once been sent nothing but men's résumés for key job openings, Obama's team, and the press, chose to belabor a single, transient, meaningless sound bite. It was this treatment of Romney that helped numb the electorate to more serious criticisms of the GOP nominee four years later.

Even so, Romney proved to be his own worst enemy.

"There are forty-seven percent of the people who will vote for the president no matter what," said the voice on the recording. "There are forty-seven percent who are with him, who are dependent on government, who believe that, that they are victims, who believe that government has the responsibility to care for them. Who believe that they are *entitled* to health care, to food, to housing, to you name it, it's an entitlement and the government should give it to them. . . . These are people who pay no income tax; forty-seven percent of Americans pay no income tax, so our message of lower taxes doesn't connect. . . . And so, my job is not to worry about those people. I'll never convince them they should take personal responsibility and care for their lives. What I have to do is convince the five to ten percent in the center that are independent, that are thoughtful."

The voice belonged to Romney. He had been taped surreptitiously during a May fund-raiser at the home of a Florida banking executive. Four months later, the tape was leaked to the left-leaning publications *Mother Jones* and the *Huffington Post*. It was an atomic bomb dropped onto the Romney campaign. He had been ruthlessly portrayed as an elitist snob with no feeling for the common man; now he was on tape moaning to fellow nobles about the peasants failing to take responsibility for their lives. There was no spinning the remarks—especially not when Ryan had been *publicly* delivering a similar talk about "makers and takers," dividing the country between those who produced wealth and those who leeched off them.

Some of Romney's allies urged him not to back down, believing that his remarks framed a sharp contrast of capitalistic individualism versus socialized citizenship. Years later, in light of how the 2016 campaign unfolded, some Republicans adopted a revisionist theory that their 2012 nominee would have survived the controversy—and won the electon—if only he hadn't been so apologetic.

But the political problem with Romney's "47 percent" remark wasn't that it offended the delicate sensibilities of the left; it was that it unwittingly marginalized many Americans on the right.

"Most of the tax receipts come in from a certain number of these [wealthy] people, and the redistribution then naturally occurs with entitlement programs. If you're going to take those away, that's not necessarily a position that most of our voters would support," Cantor says, recalling his reaction to Romney's comment. "I'm not so sure the people who were voting for us as Republicans were, on the whole, as ideological as we thought they were."

The country was changing, and so, too, were partisan attitudes. The Republican political class failed to see the ground shifting beneath it, operating as though its voters had a static worldview that aligned with the party's intellectual elite. In fact, evidence had mounted since Bush 43's reelection that on many issues, most notably trade, foreign intervention, and entitlement spending, the party's base had become more populist than conservative.

The hints of this ideological volatility were sufficient to justify the "referendum" strategy, even as it made Romney uncomfortable and Ryan downright angry. The vice-presidential pick joined what he thought was a cause: drawing a bright line between two visions for America. When he realized that wasn't the case, Ryan began griping—to his family, his friends, his fellow congressmen—that Stevens's strategy was making it hard for Republicans to win and, if they did, even harder for them to govern.

"The Bush pivot from a national security campaign to 'I want to do these entitlement reforms' taught me you have to run on that stuff. . . . It convinced me that you have to run campaigns on ideas and you have to make them really clear choices," Ryan says. "Stuart Stevens was the campaign strategist. I came on for the last eighty-eight days, so obviously, it wasn't my campaign. I just think he ran more of an anti-Obama campaign: Obama sucks, therefore vote for Mitt Romney."

THOSE LAST EIGHTY-EIGHT DAYS CONTAINED ENOUGH MELODRAMA TO fill an entire election cycle.

There was Todd Akin and Richard Mourdock, the GOP Senate nominees in Missouri and Indiana, respectively, talking about "legitimate rape" and rape-induced pregnancies as part of God's plan;[13] Obama warning of "a red line" in Syria, raising the specter of American involvement in another war; Clint Eastwood arguing with an empty chair, imaginarily occupied by Obama, during a surreal one-man performance in prime time at the Republican convention; terrorists killing four Americans at the embassy in Benghazi, Libya; three heated presidential debates, including one in which Obama mocked Romney's claim that Russia was America's top geopolitical foe; and Superstorm Sandy devastating the Eastern Seaboard one week before Election Day, killing more than one hundred people and costing $70 billion in damage, capped by Christie, the New Jersey governor, heaping praise on Obama's handling of the disaster.

Despite this roller-coaster final few months of the campaign, the fundamentals of the race remained steady: Obama maintained a modest but meaningful lead over Romney in the key battleground states. Meanwhile, the national polling suggested a dead heat: Surveys from NBC News and the *Wall Street Journal* showed Obama at 48 percent and Romney at 47 percent, while ABC News and the *Washington Post* showed Obama at 49 percent and Romney at 48 percent.

At the president's Chicago headquarters, Axelrod and his team saw no path to victory for their opponent. Romney could conceivably win back the battlegrounds of Florida, Ohio, and North Carolina, but those states would not get him to the requisite 270 Electoral votes. Romney needed something else—Pennsylvania, perhaps, or some combination of Michigan, Wisconsin, and Minnesota—but those states were locked down. The race, they told Obama, was over.

In Boston, the Republican nominee's brain trust had an entirely different outlook. Ohio was in the bag, they told Romney, and both Florida and North Carolina were looking good as well. With Pennsylvania breaking their way late (as a flurry of their internal polling suggested) and Michigan, Wisconsin, and Minnesota all in play, Romney was well positioned. There was none of the backbiting and blame-shifting that defined the final days of the McCain campaign. Everyone in Romney's camp believed he was going to be the president of the United States.

This was a bit jarring for Reince Priebus to hear. The RNC's polling, as well as national surveys he pored over daily, gave no such cause for optimism. "How is it that you guys are the only people in America that have Romney up, while every other public poll shows him down?" the party chairman asked Romney's senior staff on a conference call two days before the election. The session descended into an argument over polling methodology and Priebus hung up, worried that his party's nominee was walking into a buzz saw.

As the candidates ended their campaigning with a final push through the Midwest—Obama in Iowa, Wisconsin, and Ohio; Romney in Ohio and Pennsylvania—both felt certain that victory was at hand.

ELECTION DAY 2012 WAS HARROWING FOR THE REPUBLICAN PARTY.

The party's nominee wasn't the only one convinced he was headed to the White House. His running mate, Ryan, told his wife and children to prepare for a move to Washington—this despite Priebus warning him on Election Day that things didn't look good. But something was in the water. Boehner, McConnell, Cantor—all the party's leaders believed that Romney was going to win, and for the same reason: Their data showed that Obama had bled too much support among working-class and middle-class white voters, especially in the industrial Midwest.

Romney was paralyzed, then, by the returns coming in from Ohio on the night of November 6. He wasn't just going to lose the Buckeye State; he was going to win fewer total votes there than McCain had four years earlier, when the race was barely competitive. The upshot was obvious. Ohio was Romney's strongest state in the Midwest. If he wasn't going to win there, he wasn't going to win anywhere. He wasn't going to win the presidency.

Dazed and devastated, Republicans tried to make sense of what they were seeing. Fewer votes than McCain? In Ohio? How was it possible? The comprehensive answer provided weeks later by Romney's pollster, Neil Newhouse, was that two hundred thousand white voters who turned out in 2008 had stayed home in 2012, the result of disillusionment with Obama and distaste for Romney. But even before that analysis, the exit polling of voters who *did* show up told a simple story:[14]

- 22 percent of Ohio voters said the most important quality they looked for in a candidate is that he "cares about people like me"; Obama won 84 percent of them.
- 56 percent of Ohio voters thought Romney's policies would favor the rich; Obama won 87 percent of them.
- 60 percent of Ohio voters supported the bailout of the Detroit automakers; Obama won 74 percent of them.

It was the same story in exit polling all across the country. Voters perceived Romney to be unsympathetic to the working man, an advocate of the super-affluent, someone who couldn't possibly empathize with the struggles of everyday people. Obama's team spent much of 2012 framing this picture, and with the "47 percent" commentary, Romney had colored it in himself.

"The reason I got involved in politics was to try and help the average American," Romney says. Noting his struggle to connect with those average Americans, he adds, without a hint of irony, "The skill in communicating that is a particular capability that I wish I had in more abundance."

Trump, who attended Romney's Election Night party and recalls with a certain glee watching the candidate's staff agonizing over the results in Ohio, says, "Maybe they focused on the wrong things and in the wrong areas, because they lost Ohio by a fairly substantial amount." When I cite the low turnout of working-class whites, Trump can no longer suppress his grin: "And I brought them out in numbers that they never even knew existed. Because they liked *me*."

Romney (and the party at large) also performed dismally with swing voters. Though he won independents by 5 points, that number was misleading; self-described "moderates" were a much larger chunk of the electorate, and Obama carried that group by 15 points. Meanwhile, Obama won women by 12 points, and Romney won men by 8 points; that combined 20-point "gender gap" was the widest margin seen in a presidential election since 1952, according to Gallup. (Some credit was due Akin and Mourdock, both of whom snatched defeat from the jaws of victory in losing their red state Senate races.)

"The dangerous Mitt Romney, to us, would have been the Mitt Romney appealing to moderate voters and suburban woman. And he never really got there," Axelrod says. "He had to distort himself to win the nomination; he had to present himself as further to the right than he really was. I don't think closing Planned Parenthood was actually a passion project of his. I don't think there was anything in his record in Massachusetts that suggested he would be a fervent anti-immigration foe, and as a businessman he probably felt the opposite way. But he had to paint this portrait of himself that would pass muster in the new Republican Party."

Although Romney failed to turn out white voters in certain states, he did win an impressive majority of those who showed up: 59 percent of whites backed Romney nationwide, compared to just 39 percent for the president. This was 4 points lower than Obama's 43 percent showing against McCain four years earlier, and the worst performance among whites by a Democratic nominee since Walter Mondale during Ronald Reagan's forty-nine-state steamrolling in 1984.

It would have once been unthinkable for a presidential candidate to lose 59 percent of whites and still win the White House. But the acceleration of demographic change in the country made it possible—as did Obama's dominance among minority voters. The president won 93 percent of black voters and 73 percent of Asians. Most alarmingly, he carried 71 percent of Hispanics, the fastest-growing bloc of voters in the country, compared to just 27 percent for Romney, the worst showing for a Republican since Bob Dole in 1996. All told, Romney won just 17 percent of nonwhite voters nationwide.

There were a few bright spots for the Republican Party. Two of its longtime conservative stalwarts in the House, Mike Pence and Jeff Flake, won their statewide races for governor and senator, respectively. The GOP kept the House majority and picked up a Senate seat in Nebraska. And a star was born in Texas, where a conservative firebrand named Ted Cruz scored an upset victory in the primary and was headed to Washington with a full head of steam.

But there was little for the national ticket to celebrate. Romney had held Obama to 39 percent of white voters but still lost. Pushing that number any lower would prove exceptionally difficult—and not necessarily in the party's long-term interest, given how the requisite policy emphases would register with other demographic groups. White voters without a college degree were the fastest-shrinking portion of the electorate, whereas the groups Obama owned (Hispanics, young people, women with college degrees) were booming as a share of the overall vote. Even before Romney's concession speech, the case was being made that Republicans would be competitive in 2016 only by appealing to a broader segment of voters in the diverse states that George W. Bush had carried in 2004: Florida, Colorado, Virginia, Nevada, and New Mexico.

But there was a massive obstacle blocking this approach: the issue of immigration. Romney's hard-line positions and clumsy rhetoric had alienated Hispanics, no doubt, but so, too, had five years' worth of antagonism from Republicans dating back to Bush's failed overhaul. The GOP's perceived hostility toward nonwhites was repelling not just Hispanics and Asians, but also the suburbanites and business-friendly moderates who had anchored the party's coalition for generations. Something had to be done.

"We've got to get rid of the immigration issue altogether," Sean Hannity told listeners on his radio show two days after the election.[15] "It's simple to me to fix it. I think you control the border first. You create a pathway for those people that are here—you don't say you've got to go home. And that is a position that I've evolved on. Because you know what? It's got to be resolved. The majority of people here—if some people have criminal records you can send them home—but if people are here, law-abiding, participating for years, their kids are born here, you know, it's first secure the border, pathway to citizenship, done."

Hell had frozen over. Not only was Hannity of all people publicly endorsing "amnesty," the dirtiest word in the conservative lexicon, but he was placing private calls to Republican leaders, including Cantor and Ryan, urging them to move cursorily in Congress while the issue had momentum.

They were a step ahead of him. The day after the election, Cantor gathered his team in Richmond and announced that he would support offering citizenship to children who had been brought to the United States illegally, a policy Republicans has opposed in the form of the DREAM Act. Boehner went even further. "I think a comprehensive approach is long overdue," he told ABC's Diane Sawyer that same week. "And I'm confident that the president, myself, others, can find the common ground to take care of this issue once and for all."

All the while, inside the headquarters of the Republican National Committee, the chairman's phone never stopped ringing. Donors, elected officials, activists, lobbyists, RNC members—everyone wanted the same thing: a declaration from atop the party that something would be done to prevent another such loss in the future. Priebus had been content to hang back since becoming chairman, toiling behind the scenes to improve the GOP's infrastructure and ground game across the country. He had never believed it was the role of the national party to dictate policy from on high. Now he was prepared to do exactly that.

Gathering five of his closest allies, Priebus instructed them to produce a sweeping report on what had gone wrong in 2012 and how it would be avoided in presidential elections to come. It would lead off with immigration, stressing the need for comprehensive reform, but would also make a host of recommendations about engaging women, minorities, and young people, as well as making smarter investments in technology and data analytics.

Officially christened by RNC staffers as the Growth and Opportunity Project, it quickly earned a more ingenuous moniker: "the autopsy."

CHAPTER SIX

DECEMBER 2012

"There must be atonement!"

THE TIME HAD COME AT THE END OF THE CONGRESS TO CHOOSE A NEW leader of the Republican Study Committee, and Steve Scalise, a Louisiana lawmaker first elected in 2008, wouldn't take no for an answer.

Jim Jordan had spent the past two years relishing the role of John Boehner's personal tormenter, leading one internal charge after another to weaken the Speaker's legislative agenda. Now Jordan was term-limited by the RSC's two-year rotating chairmanship, and his departure was one thing Boehner didn't cry over. The past two years had seen the RSC's membership balloon to record numbers and its relationship with the Republican leadership disintegrate. Boehner and his deputies were desperate to see someone more reasonable take control of the conservative caucus

There had always been tension. Founded back in 1973 alongside the Heritage Foundation by the pre-Reagan luminaries of the right, the RSC's mission was to agitate for legislative outcomes more conservative than the leadership might otherwise permit. The group floundered for decades, counting just a few dozen members, most of whom were considered fringe characters by the party's leadership. Meanwhile, Democrats had controlled the House since 1954, rendering the GOP's far-right wing powerless to dictate policy. That changed in 1994, when Republicans snapped their forty-year streak in the minority and Newt Gingrich became Speaker. Moving quickly to consolidate his power, Gingrich abolished the RSC by rewriting House rules to eliminate its funding. The paranoid new Speaker, already sweating an insurrection from his right flank, thought he had neutralized a potential menace.

It wasn't long, however, before a small crew of House conservatives found a loophole and relaunched the group as the Conservative Action

Team, or CAT. Gingrich watched with angst as David McIntosh, the Indiana congressman appointed its first chairman, handed out lapel pins featuring a roaring mountain lion. As the group grew in size, the members who had resurrected it—known as "the Founders"—established two bylaws: a rotating chairmanship and a private process for choosing each chairman. They worried that moderate Republicans would infiltrate their cabal at the behest of GOP leadership, elevating a company man to lead the group instead of a conservative. Safeguarded by these rules, the Founders restored the name to "Republican Study Committee" and employed a succession of leaders who charged ever harder toward ideological nirvana: Mike Pence, Jeb Hensarling, Tom Price, and Jim Jordan.

The next chairman was hiding in plain sight: Tom Graves, a handsome young Georgian, had been groomed by Jordan to continue the incursion against the GOP leadership, pushing Boehner harder and farther to the right. The Founders, which had since come to include all former chairmen as well as an honorary member, Paul Ryan, were required to interview every candidate interested in the job, including Scalise. But this was a mere formality. The Founders voted unanimously to appoint Graves.

Scalise decided to protest. Past complaints about the Founders' dictatorial process had yielded an asterisk in the bylaws that allowed for any rejected candidate to force a groupwide vote on the chairmanship by collecting signatures from 25 percent of the RSC's membership. Scalise did just that, much to the annoyance of the Founders—and much to the delight of GOP leadership. The Louisianan was campaigning on the promise of a more constructive partnership with party elders, and Boehner and Cantor lobbied furiously behind closed doors in support of him.

Once upon a time, Scalise would have stood no chance. The RSC was too small, its membership too conservative, for someone preaching cease-fire to become its chieftain. But the RSC's numbers had soared in recent years, from fewer than 70 at the turn of the century to upwards of 170 a decade later. More than a sign of the GOP's rightward drift, this explosive growth reflected the necessity for center-right Republicans to identify as anything but. With a Tea Party purge under way, lots of law-makers concluded that belonging to the RSC would enhance their right-wing bona fides. Once the beating heart of conservativism on Capitol Hill, the RSC had become diluted by moderate Republicans who needed street cred to survive. It was this transformation that propelled Scalise to a contentious victory over Graves—and that sparked the first conver-sations about a smaller, spinoff group of House conservatives.

"That Graves race created a rift in our conference that brought this whole Freedom Caucus thing to bear," Ryan says. "Jordan got really up-set about it. Understandably so."

Weeks after the disputed RSC election, another bombshell rocked the conservative movement: Jim DeMint resigned his Senate seat just two years after being reelected to become president of the Heritage Foundation.

The move made sense. He might have been a Moses in the eyes of the base, but to colleagues on Capitol Hill, including many of his fellow con-servatives, DeMint had become a distraction. His crusade against the establishment was unceasing. His talk of purifying the party was ex-hausting. Unlike some of his star pupils, such as Mike Lee, Marco Rubio, and Pat Toomey, DeMint had earned the reputation of a show horse who did more bloviating than legislating.

Heritage, meanwhile, was urgently in need of renewal. Once an intel-lectual giant of the right, crucial to designing the policy achievements of Ronald Reagan and guiding a generation of policymakers thereafter, Heritage suffered greatly from a prolonged stretch of rotten publicity thanks to the ACA's individual mandate being litigated and relitigated. To the extent the venerable institution was still relevant, it was the guer-rilla unit, Heritage Action, that raised the money and earned the head-lines, not the scholarly side of the think tank. Doubling down on that militant approach, Heritage hired DeMint as its decorated new general.

South Carolina's young governor, Nikki Haley, was tasked with pick-ing hes replacement. The speculation centered on two congressmen who were finishing their first terms: Mick Mulvaney and Tim Scott.

Mulvaney had distinguished himself in the freshman class as a mouthy, whip-smart fiscal hawk whose distrust of Boehner was surpassed only by a skepticism of the right's sincerity in avoiding fiscal ruin. (He wanted to cut the defense budget, a nonstarter for most of his colleagues.) Widely seen as a rising star in the party, Mulvaney was known to charm colleagues with an earthy joke one minute and startle them with an expletive-laden rant about corporate subsidies the next.

And then there was Scott. He initially stood out because there were only two black Republicans in Congress and he was one of them. The other, Allen West of Florida, was also elected in 2010, and both were hailed as Tea Party heroes. But the similarities ended there. Whereas Scott was thoughtful and polished, West was impulsive and obnoxious, calling himself "a modern-day Harriet Tubman"[1] while seizing any opportunity to insult Islam, women, liberals, and Obama voters, calling them "a threat to the gene pool."[2] As West became a fixture on Fox News during their first term, Scott's chief of staff received a phone call from a nervous donor. "You guys are falling behind," the donor said. "Allen West is *the* black Republican."

Scott had to laugh. He didn't worry about being *the* black Republican; he worried about being typecast, about being used, about being treated like a prop in a party desperate for outward signs of diversity. Indeed, his first few months in Congress were awful in this regard. The Republican bosses had shoved him in front of the cameras whenever possible, blind to Scott's discomfort at being paraded in front of the press as a rookie congressman still finding his way to the nearest washroom. "Tim was like Elvis Presley. Leadership wanted him all the time to be the party's face on television," says Trey Gowdy, a fellow freshman who would become Scott's closest friend. "It was incredible pressure on someone brand new to Congress."

The pressure had only just begun. When Haley called Scott in December with the news of his appointment, the historical implications were staggering. He would be just the seventh African American to serve in the U.S. Senate—and the first African American *ever* to serve in both chambers of Congress.

The notion of an affirmative-action hire, as grumbled about in certain quarters back home, ignored the fact that Scott had held public office for fifteen years and was easily the most qualified candidate. Still, there was no downplaying the symbolism: South Carolina's Indian American governor, who had overcome a nasty, identity-based whisper campaign in her own election, was bulldozing a major racial barrier on behalf of Scott, a self-made black man from the lethal neighborhoods of North Charleston, just as the national party commenced a public display of hand-wringing over its homogeneity.

"The Republican Party has always been very good at saying, 'We include everyone,' but they've never taken time to show it," Haley said in an interview after Scott's appointment. "When have they ever gone to a minority community and said, 'What do you care about? We're a better country because you're in it.' We can't be this party of old, white men who just say, 'We need diversity' and end it there."

Together, Haley and Scott vowed to each other that they would fight to remake the Republican Party in the image of a diversifying America.

They had no idea what they were in for.

DeMint's Exit in December 2012 was timely for GOP leaders as they faced their trickiest negotiation yet: the fiscal cliff.

January 1 was circled on every congressional calendar. When the ball dropped on 2013, it would trigger a domino effect of economic woe: All the Bush-era tax cuts would expire, raising rates on every American; and the automatic spending cuts crafted during the 2011 debt ceiling crisis would take effect, ripping indiscriminately through the budget and gashing everything from military readiness to safety net programs. With the economy struggling back to its feet after the wallop of recession, going over the cliff was not an option.

The problem, yet again, was ideological disagreement—between the two parties and within the Republican Party.

The president had won reelection campaigning on a proposal to raise tax rates for individuals making more than $200,000 annually. But Republicans would not give an inch. Understanding full well that Obama had the leverage, they argued nonetheless that owners of small businesses would be crushed by such a hike. After weeks of haggling, the president offered a concession: $400,000. It was still unacceptable to Republicans. Most of them had signed a document the "Taxpayer Protection Pledge," sponsored by an outside group, Americans for Tax Reform, that forbade any tax hike for any reason. Even though this circumstance was unique—taxes would *increase* on *everyone* if nothing were done, and Grover Norquist, the group's president, was telling lawmakers that the pledge did not apply—many Republicans didn't care. The nuance would be lost in attack ads from inevitable primary challengers alleging that they had voted to raise taxes. They couldn't take that risk.

So, Boehner made a counteroffer to Obama: $1 million. Anyone making less than that would be spared from a tax increase; anyone making more than that would see their taxes go up. Surely, Boehner thought, after the party's nominee had been bludgeoned as an out-of-touch aristocrat, Republicans would see the value in volunteering a tax hike on millionaires only. The Speaker, having been battered in past negotiations, now thought he had the White House on the ropes. If the House GOP united behind his $1 million proposal, passing it on the floor to demonstrate *their* leverage, Obama's offer would likely go higher—not all the way to $1 million, but higher, protecting more taxpayers along the way. It was quintessential Boehner: He was bluffing with a bad hand, hoping to salvage part of the pot rather than throwing down his cards and walking away from the table with nothing.

But the conservatives didn't see it that way. To them, certain issues were nonnegotiable: guns, abortion, taxes. It didn't matter that Obama had the high ground. It didn't matter that Boehner was trying to make the best of their very bad situation. All that mattered was honoring a commitment, the context and the consequences be damned. "We didn't come to Congress to raise taxes," says Jordan, who led the effort against Boehner's proposal.

As Cantor and Kevin McCarthy ended their vote-whipping effort only to discover that they were well shy of the support needed to pass the $1 million plan, Boehner was devastated. The Speaker was a cool customer, but this defeat nearly broke him. How could they not see? How could they justify opposing a tax hike on millionaires when it would mean a tax hike on everyone making more than $400,000?

Boehner was running out of patience. The day before, Harry Reid had blasted him from the Senate floor, accusing him of running the House like a dictator. "I don't do angry. Nobody on my staff has ever seen me angry," Boehner recalls. "But that little son of a bitch got under my skin." When they arrived at the White House the next morning for a meeting, Boehner spotted Reid talking with McConnell. "I walked right up to him and said, 'Harry, you can go fuck yourself. You ever listen to that shit that comes out of your mouth?'" Boehner imitates a flustered Reid, then adds: "I thought McConnell was going to have a heart attack."

Now, hours later, dejected and teary-eyed, Boehner stepped to the microphone inside a conference room in the House basement. The room was silent. Christmas was less than a week away, the fiscal cliff was looming just beyond, and nobody had a clue as to how this crisis would resolve itself. "Lord," Boehner declared, "grant me the serenity to accept the things I cannot change, the courage to change the things I can, and the wisdom to know the difference."

It was the Serenity Prayer used in twelve-step addiction programs. Republicans had failed to find the necessary votes, Boehner announced, and would have no counter to the president's offer. They were free to go home for the holiday; they would be called back with forty-eight hours' notice to vote on a Senate bill addressing the stalemate.

On New Year's Day, the House passed the Senate's bill—with 85 Republicans joining 172 Democrats—that raised taxes on individuals making more than $400,000, while permanently extending the Bush tax cuts for everyone making less. Boehner and Paul Ryan voted in favor, while Cantor and McCarthy, to the murmurs of their colleagues in the chamber, were opposed.

Boehner made a beeline for his top two lieutenants. "Are you shitting me?" he demanded.

They didn't answer, slipping out of the chamber before things could escalate. Boehner wanted to chase them down, to wring their insubordinate necks. But he couldn't afford to make any more enemies. In two days, the new Congress would convene and members of the House would vote on his reelection as Speaker. As he walked off the floor, Boehner spotted a cluster of young conservatives whispering feverishly to one another. It looked all too familiar. Fifteen years earlier, Boehner had de-

clined to join an attempted coup against Speaker Gingrich. Now he was
the one in the crosshairs.

A JOLTING KNOCK ON THE DOOR SENT THEM SCRAMBLING LIKE TEENAG-
ers at a keg party. Who was it? Were they busted? Should anyone answer?

It was January 2, the night after the tax-hike vote, the night before
the new Congress, and a throng of some twenty House Republicans was
huddled in the Capitol Hill apartment of Tennessee congressman Stephen Fincher. There was no drinking or socializing; the lawmakers carried themselves with an urgency rarely displayed in their day jobs. The
next morning, members of the House of Representatives would elect a
Speaker, and this particular faction had gathered at the eleventh hour
with an extraordinary purpose: to plot a mutiny against Boehner.

Many of the members felt affection for Boehner on a personal level,
his brusque moxie rubbing off on them in ways that were often unconscious. But their tactical disagreements with him were elevated by the
fact that his leadership team did not represent the conference. Boehner,
Cantor, and McCarthy represented states (Ohio, Virginia, and California) that Obama had carried twice, and all three officials identified more
with the party's champagne-sipping managerial wing than its piss-and-
vinegar populist sect.

Boehner suffered the brunt of this frustration and did little to quell
it. The Speaker's approach had begun to grate on members—his sermon-
izing, his secretive negotiations with Obama, and most recently, his retaliation against Republican dissenters. In early December, the Speaker
had authorized the removal of four uncooperative Republicans from key
committees. After two years of brutal infighting, Boehner's punitive
strike was intended to send a message.

It backfired. The members became right-wing martyrs, enlisting outside help to stir outrage against the GOP leadership. "This is not 1995,
when nobody knew what was going on in Washington," Tim Huelskamp,
one of the conservative renegades, told *Roll Call*. "Since then we've got
Fox News. We've got Twitter. We've got Facebook."[3]

Huelskamp, a Kansas congressman representing one of the biggest farming districts in America, had received the harshest sentence of them all: He was kicked off the Agriculture Committee. "He was just a born asshole," Boehner says of his former colleague. "He didn't even have to try."

Incensed, Huelskamp became one of the first to pledge opposition to Boehner and began recruiting others to join a revolt. It was this reputation that earned Huelskamp a phone call from Jim Bridenstine, a young Oklahoman who had just won a congressional seat in November 2012.

Bridenstine had sworn publicly not to support Boehner for Speaker, a promise that energized the base in his conservative district. Once he was elected, however, the arm-twisting began. Tom Cole, the dean of the Oklahoma delegation and a close ally of Boehner's, called Bridenstine repeatedly, urging him to reconsider. When Bridenstine wouldn't budge, the talks got less friendly. Finally, on January 2, as Bridenstine was boarding his flight to Washington, Cole called with a closing threat: If Bridenstine didn't vote for Boehner, he would lose his promised seat on the Armed Services Committee. Bridenstine, a former Navy fighter pilot, was outraged. Hearing his story, Huelskamp invited him to a top-secret meeting that night at a colleague's apartment.

Problem was, Huelskamp hadn't mentioned this to anybody else. When Bridenstine banged on Fincher's door, everyone froze. The room was already rife with tension; some of the attendees, everyone knew, were acting as eyes and ears for the leadership. By relaying updates to Boehner—or, in some instances, to Cantor—the spies would earn eternal goodwill from the men who could dictate everything from committee assignments to campaign contributions.

It was Raúl Labrador, the Tea Party hard-liner from Idaho, who finally answered the door. Standing over six feet tall and weighing every bit of 250 pounds, Labrador decided to moonlight as a bouncer. "We don't know who you are," he told Bridenstine.

"I'm Jim Bridenstine, a new member of Congress. Tim Huelskamp invited me."

"But we don't know *who* you are," Labrador replied. "We don't know *who* you're for."

Bridenstine was bewildered. He had campaigned on a refusal to back

Boehner. And yet these professional politicians, these grown men play-ing Whodunit on a Wednesday night, couldn't identify him. "I don't have time for this," he told Labrador. "Here's my number. Call if you change your mind."

It was barely an hour later when the phone rang. "We need you," Lab-rador told Bridenstine. "And we need other freshmen like you. Bring some buddies."

Bridenstine did as he was told. Before long he was back on Fincher's doorstep, flanked by a pair of fellow newbies, Florida's Ted Yoho and Texas's Steve Stockman, who had also made noise about opposing Boeh-ner. (Stockman had previously served a single congressional term in the 1990s.) Another rookie member, Thomas Massie of Kentucky, who had been sworn in early after winning a special election, was waiting as they stepped inside.

Massie made sure they knew the math: In a Speaker's race, every member of the House is eligible to cast a vote, and Boehner would need an outright majority to win. If all 435 members voted, that meant Boeh-ner needed 218. With 234 Republicans in the chamber, Boehner could lose only 16 of them. If the conservative rebels could collect 17 votes, Boehner would be denied a majority, and another round of balloting would commence. Not in nearly a century had a sitting Speaker been forced to a second ballot; if they could so humiliate Boehner, the think-ing went, he would step aside—or be forced out in a subsequent round of voting.

The incoming freshmen looked around the room with confusion. There had to be two dozen of them in total, more than enough to prevent Boehner from reaching his majority threshold. Why all the fuss?

"You guys don't get it," Labrador told them. "We need thirty."

"That's dumb. Why do we need thirty?"

A hush fell over the mob. It was Bridenstine, the baby-faced door banger, challenging Labrador.

"We need thirty *to get to* seventeen," Labrador growled in response. "Because half of the people in this room are going to cave tomorrow."

Bridenstine glanced from side to side. "Okay. Who's going to cave? Raise your hands."

Nervous laughter. No hands.

"You still don't get it," Labrador said. "There are people in this room working for Boehner. We just don't know who they are."

At this, the chuckling ceased. Bridenstine, the brave novice, glanced all around him, clearly expecting a chorus of vehement denial to Labrador's allegation of espionage. Nothing but suspenseful stares. It was true. They all knew it. Now, so did Bridenstine.

Boehner wasn't the only one with moles. Cantor was keeping close tabs on the meeting, too, which made sense given that some of the rebels were prepared to elevate him to the speakership. At one point, Fincher's phone beeped; he excitedly announced to his colleagues that Cantor was calling and scurried back into his bedroom to speak with the majority leader privately. Adding to the mystery, some Boehner spies were actually *posing* as Cantor spies, pledging fidelity to the number two in order to protect their cover. One of them, Lynn Westmoreland, a Georgia congressman and known ally of the Speaker's, was eventually called out by one of his colleagues. "Why are you here, Lynn? Boehner already put you on good committees."

"Well," Westmoreland said, smiling, "if Cantor's the Speaker, maybe I'll get even better committees."

As eyes rolled throughout the room, Huelskamp whipped out his iPad and tapped out a few words on the screen, showing it silently to the rookies: "Works for Boehner. Don't trust him."

Bridenstine was growing impatient. "Okay," he declared. "Let's just sign our names. That way we're all on the record. A pledge to vote against John Boehner."

This wasn't a new idea; in fact, some of the members had already scribbled their autographs on scraps of paper in an envelope. (Labrador would keep these records for years to come, preserving the sacred text for indebted archivists.) But not everyone was ready to sign. Some were still on the fence about opposing Boehner; others found this ritual of an ink oath a tad ostentatious.

Sensing their reluctance, Fincher, a religious man known to sprinkle his political rhetoric with Scripture, led the group in a rousing prayer.

He then offered a fire-and-brimstone screed condemning Boehner's "sins" against conservatism. "There must be *atonement!*" he cried.

The Republicans exchanged smirks. Even the Speaker's fiercest critics wondered if their daring adventure had turned into a sad sitcom.

NOT EVERYONE AT FINCHER'S APARTMENT SIGNED HIS NAME. BUT THE core conspirators—Labrador, Massie, Huelskamp, among others— awoke the next morning believing they would overthrow Boehner.

By their count, 21 members had either signed the document or sworn their allegiance to the effort, and several more were thought to be considering it. When they huddled inside the Capitol, just before the vote, only one of them announced his defection: South Carolina's Jeff Duncan. Everyone else reiterated their commitment. Whomever they voted for didn't matter, as long as seventeen Republicans rejected Boehner, they would force a second ballot and thrust the House into chaos.

The roll call commenced in alphabetical order. Justin Amash, the first dissident called upon, voted for Labrador. But when the clerk called on Michele Bachmann, a rumored sympathizer, she chose to remain silent. This was the first sign of trouble; members are allowed to skip their turn, but by doing so, they broadcast uncertainty about the outcome. Bachmann wanted to see how the numbers stacked up against Boehner. So, too, did the next rebel called upon, Marsha Blackburn, who suddenly announced, "I have a nosebleed!" and rushed off the House floor.

And so it went: As the clerk worked through the alphabet, only a handful of the sworn anti-Boehner revolutionaries voted against him. One voted "present," hurting the Speaker's vote total without rejecting him explicitly, while many passed on their turn, buying time to decide whether striking at the king was worth incurring his wrath.

When it became clear that the scheme was failing, that too many members had gotten cold feet, Labrador circulated around the chamber advising some of the undecideds that their votes would no longer make a difference. Around this time, Blackburn, a Tennessean who had slyly floated her own name as a possible dark horse candidate for Speaker, reemerged onto the House floor and declared, "I proudly cast my vote for John Boehner." A profile in courage.

When the roll call concluded, 12 House Republicans had refused to back Boehner. Labrador cast no vote at all, nor did Mulvaney, their way of admonishing Boehner without insulting him unduly. Massie voted for Amash. Three of the rebels voted for Cantor, eliciting a rehearsed look of disgust from the majority leader. ("Well," Massie told Cantor afterward, "we threw our support to a Jew, a Puerto Rican, and an Arab, but the white man still won." Cantor did not laugh.)

As for Fincher, the seeker of *atonement!*—he voted for Boehner, explaining to colleagues that he had prayed that morning and felt moved to show mercy. He wasn't alone: Florida congressman Steve Southerland, another pledged mutineer, would later tell friends that he had been reading the Old Testament story of David sparing King Saul's life despite having the chance to kill him. After praying on the House floor, he decided to do the same for the Speaker.

Boehner had survived—bruised, humbled, and fretful. The wounds opened in the previous Congress were bleeding into the new one, and if they weren't bandaged quickly, another uprising was imminent. With the House GOP's annual retreat just weeks away, in Williamsburg, Virginia, the Speaker privately reached out to five of his most respected conservatives: Ryan, Jordan, Hensarling, Price, and Scalise. (The first four had convened a weekly breakfast for years, and invited Scalise to join after his RSC victory.)

Boehner felt a special contempt for Price, the Georgia congressman and medical doctor who carried himself with mannered arrogance. Word had gotten back to the Speaker's office that Price was offering his services to the rebellion, proposing that he become their alternative to Boehner; when they rejected his offer, he voted for the Speaker and slapped his back with a hearty congratulation. "That two-faced prick," Boehner snorts.

Still, this was no time for recriminations. If Republicans were to unite, Boehner needed help from this group (which I dubbed the "Conservative Jedi Council" in *National Journal* magazine, a nickname that somehow stuck). The sequester cuts would soon take effect, and another debt-ceiling deadline was just around the corner. In a series of covert meetings with the Jedi Council, Boehner pledged to champion their priorities for the coming year: rejecting new tax revenues, endorsing a ten-year balanced budget, and upholding the automatic cuts, save for reprioritization to protect the military. In return, Boehner wanted one thing: enough votes to raise the debt ceiling until fall, giving the party some breathing room to notch wins, gain some momentum, and return to the debt fight with a renewed sense of unity.

The Jedi Council agreed, and Boehner presented the agreement in Williamsburg. Most of the members were receptive. Covered by endorsements from the likes of Jordan and Ryan, they felt good about getting everyone back on the same page after two years of dysfunction. But not everyone was sold. Huelskamp and a handful of other malcontents voiced objections, if not to the agreement itself then to the notion of trusting Boehner to follow through. So wary were they of the Speaker's intentions that Huelskamp drafted a document itemizing the precise covenants. ("The Williamsburg Accord" was scrawled across the top in Old English font, a testament to its seriousness and to the social awkwardness of the people we send to Congress.)

Four months later, to the shock of many in the conference, both Boehner and the Jedi Council had delivered. Consistent with the Williamsburg Accord, the sequester cuts went into effect; the short-term "continuing resolution" funding the government was passed with lower spending levels; and the House passed a budget that would balance in ten years. Meanwhile, House conservatives stomached a debt ceiling increase by attaching a provision called "No Budget, No Pay," which forced Senate Democrats to produce their first budget in four years.

There were bumps in the road. Boehner had selectively violated the so-called Hastert Rule, named for the former Speaker, which says a bill can be brought to the floor only if it has majority support from the majority party. The leadership had allowed votes on relief for Superstorm Sandy and a reauthorization of the Violence Against Women Act, both of which passed on the strength of Democratic votes, angering conservatives. But on the whole, Boehner had kept his word to the conference, fostering a newfound sense of cohesion.

"God bless the Speaker," Jordan said that May. "He's done exactly what he promised."

It was a fragile truce. And that spring, as House Republicans watched what was unfolding on the other side of the Capitol, Boehner knew how easily it could shatter.

MARCO RUBIO WALKED INTO THE LION'S DEN WEARING A TENDERLOIN necktie.

It was January 29, 2013. One day earlier, the Florida senator had joined seven of his colleagues (three Republicans and four Democrats) in unveiling the framework of a comprehensive immigration reform bill. The wind was at their backs. Conservative pundits were running out of ink to spill on the GOP's need for a softer image. The national party was writing its "autopsy" as cover for elected officials to take meaningful steps toward reaching new voters. Louisiana governor Bobby Jindal, a rising Republican star, captured the zeitgeist poignantly during a speech to the RNC in January: "We've got to stop being the stupid party!"[4]

The so-called Gang of Eight refused to let this moment go to waste. Their proposal was a compromise. Republicans would get enhanced border security and tougher interior enforcement, including mandatory E-Verify (an internet-based system that checks applicants' eligibility, to prevent businesses from hiring illegals). Democrats would get a long, winding path to citizenship for the estimated eleven million undocumented residents, provided those residents paid back taxes and had committed no crimes.

Many leading conservatives had, in the months since Mitt Romney's defeat, come around to this approach. Rush Limbaugh was not one of them. He recoiled at the Gang's plan. Obama had never been interested in finding common ground with Republicans, Limbaugh told his listeners; having wielded immigration as a political club, the president had beaten them into submission. "I don't know that there's any stopping this," Limbaugh said. "It's up to me and Fox News, and I don't think Fox News is that invested in this."[5]

It was no small act of political courage, then, when Rubio called into Limbaugh's show the day after announcing the Gang's framework, ready to duel with the right's ruling agitator. Limbaugh didn't waste time on niceties. "Why are we doing this?" he asked. Sixteen minutes later, having met with the full force of Rubio's rhetorical prowess, the talk radio bully was blushing with adulation for the freshman senator. "What you are doing is admirable and noteworthy," he told Rubio, wishing him luck with the immigration push. When the interview wrapped, Limbaugh sounded thunderstruck. "Is that guy good or what?"[6]

This was exactly why the Gang had recruited Rubio. John McCain and Lindsey Graham, its two senior Republicans, were known moderates on immigration. And the third Senate Republican, the newly promoted Jeff Flake, was also soft on the issue, even though he had tacked right during his Senate run in 2012, denouncing a comprehensive approach and advocating for border security to be achieved before the undocumented population was dealt with. What the Gang needed was someone with conservative star power to sell the proposal to the base. Rubio, the Tea Party flame of 2010, had it in spades—not to mention policy chops and a straight-from-Hollywood story of living the American dream.

In February, as Rubio spearheaded the immigration push, *Time* magazine featured him on its cover with a suitable headline: "The Republican Savior." People close to Rubio worried that he was sprinting headlong into a legislative quagmire that could derail his promising rise. Then again, he'd just tamed the biggest tiger in conservative media. What could go wrong?

Limbaugh wasn't alone in believing that Obama might tank a compromise bill. The Gang's two senior Democrats, Chuck Schumer of New York and Dick Durbin of Illinois, pleaded with the president to leave the issue alone, fearful that anything with Obama's fingerprints would prove toxic to conservatives. The president reluctantly complied; on the day Rubio spoke with Limbaugh, Obama gave an immigration speech in Las Vegas but choked back the details of his preferred plan, arguing only for "key principles," as the *New Yorker* reported at the time.[7]

With Obama sidelined and the conservative media syndicate on its heels, the Gang of Eight charged ahead, wooing special interest groups and cutting deals with senators in both parties. Momentum is oxygen to the policymaking process. Romney's loss, and the RNC's vituperative autopsy report—released to fanfare in March and calling for sweeping changes to the party's data, technology, and minority outreach programs, all while endorsing comprehensive immigration reform—had set the heaviest of balls in motion. The Gang members needed to capitalize before that ball stopped rolling.

The legislative text arrived in April, thrusting Rubio into a full-scale charm offensive: He visited the offices of *National Review*, dined privately with Fox News personalities, called into talk radio programs, and, on one Sunday, appeared on seven different TV shows to promote the bill. By May it had been debated, marked up, and passed out of the Judiciary Committee, over the objections of two chief opponents, Jeff Sessions of Alabama and Ted Cruz of Texas. And on June 27, the Gang of Eight bill passed the U.S. Senate on a 68–32 vote, with 14 Republicans in favor.

From an aerial view, the ball still looked to be moving forward. On the ground, however, its momentum was arrested. Backlash to the immigration push had built organically throughout the spring, with blogs and local talk radio pummeling the Gang of Eight proposal—even as many prominent national voices remained supportive. As the outcry grew noisier, some advocates on the right got jittery. Having phoned Ryan and Cantor months earlier to lobby for immigration reform, Hannity called back with a sudden warning: "Stay away from the Senate bill," he told them. "It's going to be a career killer."

THE HOUSE AND THE SENATE ARE PROFOUNDLY DIFFERENT INSTITU-tions, not just for their respective traditions and procedures, but because of the inbuilt job descriptions that guide employee behavior.

Senators represent entire states, which affords a broader outlook on policy disputes even in red Wyoming or blue Vermont; they also enjoy the autonomy of serving six-year terms. House members, on the other hand, face reelection every two years. They represent districts that are demographically and ideologically clustered, most of which are locked down by one of the two parties. (The year 2010 saw a massive swing of 64 seats; but that means the other 371 seats, 85 percent of the House, remained loyally partisan.) With 9 of 10 seats safely under one party's control, House members fear primaries more than general elections. And predictably, with primaries drawing just a fraction of the eligible electorate, those voters who participate tend to be the most engaged and the least inclined toward moderation.

Such is the tortured relationship between the two chambers: Senators look down upon the reactionary, hot-blooded House members, while the House members resent the imperious senators. This structural friction will, in some instances, produce sharply diverging approaches from members of the same party. Immigration was one of those instances.

In the wake of Romney's defeat, it was principally the GOP's nationally known leadership—Reince Priebus and his committee members, McConnell and McCain, Boehner and Cantor and Ryan—that made the case for immigration reform. Their argument was the same one made by conservative media figures such as Hannity, Bill O'Reilly, and Charles Krauthammer: Democrats were weaponizing the issue to dominate a changing America over the long haul.

But this reasoning meant little to politicians who live their career ambitions two years at a time, and even less to those whose districts weren't reflective of any such change.

"It's ironic that Reince thought they were helping by issuing their autopsy, because there was this cultural thing going on: 'Here they go again, these out-of-touch people in Washington telling us that we need to let more of the strangers in.' It just poured gasoline on the fire," says Cantor. "Immigration was a problem for the party, but it wasn't a problem in a lot of these districts."

The visual dichotomy of this dilemma is found inside the House chamber. One side of the aisle looks like the country: young and old, man and woman, black and white and brown. The other side looks like the country club: aging white guys. In the 113th Congress, spanning 2013 and 2014, Republicans held 234 seats in the House; 19 members were women, and 9 identified as an ethnic minority. The remaining 205, or 88 percent of the House GOP, were white men.[8] This isn't to say that white men are politically illegitimate or make lousy legislators. But the statistical disparity speaks to the makeup of the districts they represent, and in turn, to those districts' willingness to embrace an America that looks nothing like their microcosm thereof.

Much like President Bush's aides discovered a decade earlier when pushing for prisoner reentry programs, House Republicans wanted nothing to do with immigration reform because they felt it was not relevant to their constituents. "We did a little test whip with our members," Cantor says, "and it went nowhere."

The same was true of legislation affecting another cultural flashpoint in the spring of 2013: guns.

In the wake of the previous December's massacre at Sandy Hook Elementary School in Newtown, Connecticut, which claimed the lives of twenty grade-school children and six teachers—and which came after at least three other mass shootings during Obama's presidency, while predating at least half a dozen others—Congress attempted to act. Addressing the lowest common denominator, West Virginia Democrat Joe Manchin and Pennsylvania Republican Pat Toomey sought to expand background checks on commercial purchases.

But despite the modest aims, and the senators' lifelong "A" ratings from the National Rifle Association, their effort went nowhere. Facing an avalanche of hyperbolic (and often downright false) attacks from the NRA, the bill died in the Senate.[9] Even if it passed, it would have gone nowhere in the House, and for the same reason: The issue was too easily demagogued among Republican voters. The self-preserving instincts of lawmakers were not conducive to any such legislation.

THE HOUSE REPUBLICAN LEADERSHIP WAS BOXED IN.

When it came to immigration, Boehner personally liked the Senate bill and would happily have supported a qualified path to citizenship for both minors and adults. But he had to be careful. The Speaker had survived one attempt on his political life that year; he couldn't afford to invite another by flouting the sentiments of his majority.

Cantor, meanwhile, was no longer in the catbird seat. His recently announced support for the principles of the DREAM Act, which would extend citizenship to illegal youths brought to the country through no fault of their own, had angered many of the Tea Party types who had long preferred him to Boehner. "They were told in our conference, 'If you try this, you're going to be gone," Labrador recalls, referencing a potential vote on comprehensive immigration reform. "And they listened. Boehner was worried about his speakership, and Cantor was worried about not being able to become the Speaker."

As the Senate bill hurtled toward passage, and members' attitudes ranged from uneasy to outright threatening, Boehner and Cantor settled on a passive approach: They would sit back and let the debate unfold freely, not committing to anything one way or the other.

The early returns were actually quite encouraging. On June 5, the RSC and its 170-some members hosted a panel of senators for a huge, bicameral "family meeting" on immigration. The headliners were Rubio and Flake in favor of the Senate bill and Sessions and Cruz opposed. Rubio kicked off the summit by acknowledging the disdain for his bill within the House GOP. But he then pivoted to emphasize the many other areas of agreement across the party's ideological spectrum, urging his brethren not to ignore the major concessions they would receive in exchange doing what was reasonable: pulling people out of the shadows, making them pay taxes, and putting them on a thirteen-year path to citizenship.

As Rubio spoke, intercepting subsequent questions to press his argument, heads in the room nodded. One incident aside—Texas congressman Michael Burgess made a crack about "undocumented Democrats," drawing a glare from Rubio—Republicans left the summit sounding downright bullish on passing an immigration bill. Rubio had done it again. Boehner and Cantor couldn't believe their ears.

And neither could Steve King. The Iowa congressman was known as the House's fiercest immigration hawk—and its most racially polarizing force. King's collection of greatest hits included comparing illegal immigrants to livestock; calling Obama "very, very urban";[10] saying the president "favors the black person";[11] and mentioning that Obama's middle name, Hussein, held a "special meaning"[12] for the Islamic radicals cheering on his presidential run. (To mention King's fixation on Obama's birthplace feels superfluous.)

Leaving the immigration summit that day, King looked deeply un-settled. Comparing himself to Rip Van Winkle, he said, "I went to sleep last year before the election believing that all my colleagues believed in the rule of law and opposed amnesty and understood the impact of amnesty. And then I woke up the morning after the election, and they believed something different." As for Rubio's path to citizenship, and the converts he was attracting, King believed, "There is no upside to it. I can't track their rationale or their logic. I'm flabbergasted that so many otherwise-smart people can come to conclusions that aren't based on any kind of data."

As King walked away shaking his head, Rubio held court with report-ers nearby. Noticing at one point that Sessions had emerged from the room behind him, he moved over, inviting his colleague from Alabama to address the media alongside him. But Sessions declined. Politicians are experts in appearances, and Sessions knew better than to be pho-tographed holding a press conference with the champion for amnesty.

IF VEGAS WERE PLACING ODDS ON FINAL APPROVAL OF COMPREHENSIVE immigration reform, no single event in the twenty-first century would have sparked as much action as that RSC summit with Rubio. For the first time, in the hours thereafter, advocates felt real hope and oppo-nents felt real fear.

Neither lasted long.

That night, Labrador abruptly quit a bipartisan House group that was working on a proposal to mirror the Senate's. Labrador was viewed internally as the "Rubio of the House," a Tea Party favorite whose bi-lingualism and expertise on the issue made him uniquely capable of moving votes. Yet he was farther right than Rubio on the legislative de-tails, most notably arguing for granting legal status instead of citizen-ship. Labrador believed he could convince many conservatives to agree to legalization, but Democrats in both the House and Senate made clear that their support hinged on citizenship. "Without a path to citizenship, there is not going to be a bill," Schumer told reporters that summer. "There can't be a bill."

Labrador also wasn't sold on the security provisions. Senators were promising to double the number of border agents to forty thousand, stationing one along every thousand feet of the Mexican boundary. But this ignored an underlying problem: The American "catch and release" policy, which allowed migrants to be freed upon their apprehension, had become a bad joke in the law enforcement community. Lots of the illegals caught crossing would skip their court date, disappear into the country, and never be heard from again. Unlike most of his fellow conservatives who cowered at the immigration debate, refusing to engage on a federal issue because there was no political upside to straying from the status quo, Labrador was serious about dealing in substance. His exit from the House group, then, signaled a symbolic blow to the reform push just hours after it appeared to have legs.

The next two weeks were instructive for House Republicans. Their office phones shrieked with angry constituent calls. Their consultants warned of dire consequences if they deviated from the simple phrase "rule of law." Meanwhile, their leaders were nowhere to be found. Watching quietly, determined not to get over their skis on the issue, Boehner and Cantor forfeited whatever moment existed to reaffirm the reluctant members.

As his colleagues tiptoed in reverse, Steve King made sure they would never take another step forward. He rented out the East Lawn of the U.S. Capitol for the entire business day of June 19, setting up a stage and attracting thousands of supporters waving signs that read, "The Melting Pot Floweth Over." (Boosting turnout for this event was a neighboring "Audit the IRS" rally, which was sponsored by Tea Party Patriots and held on other side of the Capitol.)

King had boasted for weeks that he had a silent majority of the House GOP on his side. Only a dozen or so joined him at the rally, but everyone else watching from the windows could feel the visceral passion of the activists in attendance. Right after former Fox News host Glenn Beck whipped the crowd of Tea Partiers into a frenzy, King bounded onto the stage. "I can feel it!" he cried. "I can feel we're going to defend the rule of law! We're going to defend the Constitution! We're going to defend our way of life!"

Michele Bachmann received the most boisterous welcome of any speaker that day, calling the event "a beautiful family reunion." It could have been: The crowd was almost exclusively white. At one point, she invited children onto the stage as she argued that bad immigration policy could imperil their future. "Amnesty costs a fortune," she announced, cradling an infant in her arms. "It could also cost us our nation." The crowd responded with booming chants of "U-S-A! U-S-A! U-S-A!"

There was only one senator invited to speak: Cruz. The Texas freshman had made a conspicuous effort to befriend King upon arriving in Washington, and quickly earned an ally in the Iowa lawmaker with considerable clout in the first presidential-nominating state. Cruz, like Rubio, was the son of a Cuban immigrant. But unlike Rubio, he was an immigration hard-liner. Standing before the crowd, Cruz warned that his Republican colleagues—he did not need to mention Rubio, whose name had already been booed countless times that day—were peddling the same proposal that had failed them in 1986: amnesty now, promises of border security later. "If you fool me once, shame on you," Cruz declared. "If you fool me twice, shame on me."

Cruz wasn't wrong. Ronald Reagan had been duped in 1986, offering legal status to nearly three million undocumented immigrants in exchange for enforcement mechanisms that never materialized. Even so, in this case, the Texas senator's motives were suspect.

Much like Obama in the run-up to his own presidential bid, Cruz observed the electoral implications of the problem and saw no upside in seeking out compromise. He sponsored an amendment in the Judiciary Committee that spring to substitute legalization for citizenship, but privately boasted of doing so as a "poison pill" to destroy the overall effort. Cruz, who would deploy the "undocumented Democrats" line while running for president, was concerned more with politics than policy. And the politics were clear: Historical data[13] showed that Hispanics had voted for Republicans in *lower* numbers after Reagan's 1986 amnesty, and present-day feedback showed that the base would impale anyone who repeated the Gipper's mistake.

The Senate bill passed eight days later, but the tide had already turned. Boehner refused to bring it to the floor of the House. He knew that the legislation would pass on the strength of Democratic votes; that only 40 or 50 of his Republican colleagues would support it; and that conservatives would banish him for overruling the popular will of the House GOP. The Speaker was willing to be sacked over a budget deal, but not over immigration.

Boehner had a veneer of plausible deniability in scuttling the immigration bill: Obama, he said, had demonstrated through his unilateral actions that he could not be trusted to enforce immigration laws, old or new. This was a tad disingenuous, but it bought the Speaker time. It also squared with the rhetoric of his most conservative members. They had long accused the administration of being underhanded and deceitful, and in May 2013, just as the immigration debate was reaching its climax, their beliefs were substantiated. The Internal Revenue Service had "used inappropriate criteria that identified for review Tea Party and other organizations applying for tax-exempt status based upon their names or policy positions," according to a blockbuster report by the Treasury inspector general for tax administration.

Since the emergence of the Tea Party in 2010, conservatives had accused the IRS of systematically targeting right-wing groups. Now their allegations had been substantiated. In the long run, the IRS scandal would prove endlessly useful to conservatives in their war on the Obama administration, especially given the dog-ate-my-homework routine from IRS officials claiming to have lost thousands of pertinent emails. In the short term, the inspector general's report handed Republicans the justification they needed to claim they had an untrustworthy partner in the executive branch.

As the August recess approached, Boehner's resolve stiffened. Nothing, not a visit from Ryan or a call from his friend Jeb Bush, could convince him otherwise. The timing wasn't right, Boehner told them. They would revisit immigration in 2014, when things cooled down.

King took a victory lap. Referring to the undocumented minors who remained in limbo, he told Newsmax, "For every one who's a valedictorian, there's another one hundred out there who weigh 130 pounds, and they've got calves the size of cantaloupes because they're hauling

75 pounds of marijuana across the desert." [14] King lauded Boehner for his rejection of the Senate bill—the sort of praise that convinced the Speaker he'd made a big mistake.

Looking back, Boehner says that not solving immigration is his second-biggest regret after the failed Grand Bargain. He blames Obama for "setting the field on fire." But it was the inaction of the House of Representatives—not voting on the Senate bill, not bringing up any conservative alternative, not doing anything of substance to address the issue—that enabled the continued demagoguing of immigration and of immigrants. Ultimately, Boehner's quandary boiled down to a choice between protecting his right flank and doing what he thought was best for the country. He chose the former.

It wouldn't be the last time.

RIGHT AROUND THE MOMENT IMMIGRATION REFORM DIED, SO, TOO, DID one of the longest-standing alliances on Capitol Hill.

Since their inception in 1973, the Heritage Foundation and the Republican Study Committee had worked in tandem. Heritage would supply conservative lawmakers with policy blueprints; conservative lawmakers would keynote Heritage dinners. Heritage would pay for conservative lawmakers to go on retreats; conservative lawmakers would hawk Heritage materials back home and encourage their constituents to donate. For decades, this codependent relationship revolved around the presence of Heritage staffers at the RSC's weekly meeting in the Capitol basement.

But with the creation of Heritage Action in 2010, that bond had begun to fray. Whatever promises of legal separation between the think tank and its lobbying arm proved insincere; the wall crumbled almost immediately and came crashing down entirely once DeMint became president. In raising large sums of money for his Senate Conservatives Fund by picking on "moderate" Republicans, many of whom had solidly conservative voting records, DeMint had created a model for the organization he now led. Heritage Action was increasingly belligerent, baiting Republicans into fights they could not win and then monetizing their failures with fund-raising emails decrying the impotence of the GOP.

The tension boiled over after a June vote on the Farm Bill, a monster piece of annual must-pass legislation that governed both agricultural subsidies and food stamp provisions. Heritage Action argued, as did many conservative Republicans, that the policies should be split into separate bills. It was a reasonable request and a smart fight for Republicans to pick. They had leverage because of the bill's weight; furthermore, legislating these two items together made sense only in Washington, a town that thrives on punting tough decisions and regularly resolves conflicts by larding up legislation with goodies to satisfy both parties.

But Boehner and his team ignored the conservatives, bringing the bill up for a vote that failed in embarrassing fashion. As they scrambled to save face, Scalise promised the party leadership that splitting the farm policy and food stamps into separate bills would deliver the votes. They listened, and he was proved right: Of the 62 Republicans who voted against the first iteration, 48 came around to support the second. ("Incredibly," the *Kansas City Star* reported, "Rep. Tim Huelskamp of KS-01—one of the most farm-centered districts in the United States—was one of just 12 GOP votes against the measure.")

The revised bill passed the House, and conservatives celebrated a major victory. There was just one problem: Heritage Action, which had issued a "key-vote" alert threatening punishment for any members supporting the first bill, had key-voted against the second bill as well. This, despite House leaders doing exactly what the group, and its congressional allies, had called for. Heritage officials explained that *now* their opposition owed to sugar subsidies in the farm-only bill. The goalposts had moved, and the Republican locker room went berserk.

Mick Mulvaney, the hot-tempered South Carolinian, contacted the *Wall Street Journal* about placing an op-ed in the newspaper. He drafted a memo hammering Heritage Action as a bloodsucking enterprise and found a half-dozen willing co-authors. But then, Mulvaney decided to scrap the project. It turned out that his friend Scalise, the RSC chairman, had a different sort of vengeance in mind.

One July morning, Scalise told DeMint that Heritage was no longer welcome at the RSC's weekly meetings. Ed Feulner, the iconic Heritage

Foundation president, had been a founding father to both organizations. Now his successor was responsible for an ugly divorce.

It was a watershed moment inside Washington's conservative movement. Since the dawn of the Tea Party, conservative organizations had pushed Republicans rightward at every turn. Some of it was justified in the pursuit of accountability, but much of it was insincere, empty rage rooted in unreasonable expectations. The revolutionary warfare had succeeded only in amplifying the narrative of a feckless party that needed to be demolished and reconstructed, the result of which was a cyclical, slow-motion collapse of the GOP as a governing entity.

Heritage had pushed too far this time. But it was going to take more than a reprimand from Scalise to turn back the insurgency's advance.

CHAPTER SEVEN

AUGUST 2013

"He led us into box canyon."

THE SENATOR PROWLED THE STAGE LIKE A TELEVANGELIST, A WIRE-less microphone clipped to his crisp, white shirt, sleeves rolled up above the wrist, knifing his hands and arching his cadence to ingest the Dallas hotel ballroom.

It was the third week of the August recess. After a long stretch of high-wire legislative drama surrounding the immigration debate, lawmakers had emptied out of Washington to spend the month back home. What they found, instead of rest and relaxation, was fury among their constituents about something else: Obamacare.

The Affordable Care Act would soon take effect, with federal exchanges opening to the public for enrollment on October 1. House Republicans had voted to repeal it some three dozen times since 2011. But with Harry Reid and the Democrats still controlling the Senate—and the law's eponymous, twice-elected president still in the White House—there was no stopping its implementation. This truth, rooted in the elementary realisms of government, was unacceptable to much of the conservative base. Republican politicians had promised to repeal and replace Obamacare; there had been no caveats, no speed-talking radio voice explaining the mechanical fine print. Their constituents were upset at having been misled. Republicans now faced a choice: Concede the law's irrevocability or wave a white flag and suit up for battle one last time, armed only with the narrative that heretofore the troops had not been fighting hard enough.

Ted Cruz knew a thing or two about narratives. Having watched with childlike fascination in 2010 as the Tea Party movement engulfed the GOP establishment, Cruz, then the solicitor general of Texas, began sketching the contours of an insurgent's campaign for Senate in 2012. It wouldn't be easy. Though he came from modest origins, his father a poor Cuban refugee, Cruz himself had known only success. He had worked in the George W. Bush administration and met his wife, Heidi (who'd since become a Goldman Sachs executive) on the Bush campaign. He had attended Princeton and Harvard Law School, and as Texas's top attorney, he had argued cases before the U.S. Supreme Court. He was, in other words, a career patron of the political class.

When Cruz pitched Bush on his anti-establishment stratagem during a private meeting in 2011, as the journalist Shane Goldmacher reported, the former president replied, "I guess you don't want my support. Ted, what the hell do you think *I* am?"[1]

Cruz was gifted with certain political talents—a keen intellect, linguistic dynamism—that made him instantly formidable. He had a computer-like memory and an uncanny ability for repetition, deploying not just identical phraseology but the same facial expressions and tonal inflections to accompany them. Some of this owed to training: He had spent his teenage years touring the state of Texas delivering the Constitution from memory as part of a free-market troupe and had also been involved in drama club, briefly considering a career as a thespian. (On the Bush campaign, he was known to launch into various recitations of his favorite film, *The Princess Bride*, capturing every line and every character's accent with precision.)

Less appreciated was Cruz's knack for finding a foil. It would serve him well in the years ahead, propelling him to new heights and very nearly to the Republican nomination for president, had he not encountered an opponent with that skill in even greater supply. But in 2012, Cruz had a somewhat smaller target: Texas's lieutenant governor, David Dewhurst, who had a monopoly on the donor class, enjoyed high name identification across the state, and boasted the support of its best-known Republicans, including Governor Rick Perry. Cruz used all this to his benefit. Once polling in the low single digits, he captured the energy of the grass roots and channeled the resentment of the masses, upsetting Dewhurst in the August primary runoff.

Cruz attracted legions of conservative supporters, from Sarah Palin

to the Tea Party Patriots to talk radio host Mark Levin. But no endorsement meant more than the one from Jim DeMint. And now, in Dallas a
year later, the two men were reunited, with Cruz headlining a town hall
event sponsored by Heritage featuring a stage-length banner behind
him: "Defund Obamacare."

The idea was not elaborate: Rather than repeal the health care law,
conservatives argued, Congress could refuse to appropriate the funds
to pay for it. Nor was it new: Tom Graves, the Georgia congressman
who came to Washington in a special election shortly after the law
was passed, had introduced the Defund Obamacare Act in July 2010.
He reintroduced it each of the next two years, though few Republicans
took the effort seriously. It wasn't until 2013, with the specter of the exchanges opening on October 1 and the serendipity of Washington running out of money that same day (the start of the fiscal year), that the
concept gained traction. By holding the rest of government funding hostage, the fantasizing went, Republicans would out-leverage Obama and
compel him to dump his legacy-forging law. Cruz called Graves in July
and asked to sponsor a companion bill in the Senate. A month later, as
DeMint's "Defund Obamacare" tour rolled through Texas, it was Cruz
who stole the show and announced himself as the mission's captain,
daring the party's establishment to stand in his way.

"We have to do something that conservatives haven't done in a long
time: We've got to stand up and win the argument," Cruz declared in
Dallas. "Republicans assume, with any impasse, that President Obama
will never, ever, ever give up his principles—so Republicans have to
give up theirs." Building to a rhetorical crescendo, with the crowd now
chanting the answers to his repeated questions, Cruz asked a final time,
"How do we win this fight? *Don't blink!*"

DeMint was no less dramatic. Calling the Affordable Care Act "probably the most destructive law ever imposed on the American people,"
the Tea Party stalwart declared, "If you're giving up the fight against socialized medicine, you're almost giving up on the country."

Most Republicans didn't feel they were giving up, or blinking, or
abandoning principle. The argument over Obamacare had been lost:

The bill passed the House, passed the Senate, was signed into law, was upheld by the Supreme Court, and was validated by Obama's reelection. Polling that showed the bill's relative unpopularity was meaningless at this point. Unless Republicans believed that the president was willing to abolish the law bearing his name, their threats to defund it could produce only one outcome: a government shutdown.

"I think it's the dumbest idea I've ever heard of," Republican senator Richard Burr, one of Boehner's closest friends in the Congress, told reporters that summer. "Listen, as long as Barack Obama is president, the Affordable Care Act is going to be law."[2]

Burr's remark earned him attack ads on the radio back in North Carolina, courtesy of DeMint's former group, the Senate Conservatives Fund. Meanwhile, DeMint's new plaything, Heritage Action, spent half a million dollars in August on localized ads urging House Republicans to sign a letter that had circulated from an obscure freshman lawmaker, Mark Meadows, urging Boehner to defund Obamacare.

Whatever their tactical preference, the overwhelming majority of Republican voters, activists, and politicians seemed to sincerely believe that the Affordable Care Act represented a threat—if not to their own insurance plans, then to the relationship between the government and its citizenry. That said, the battle over the president's signature law offered a unique window into the shadowy motives and incentives of the leading belligerents in the GOP civil war.

For DeMint, it was an opportunity to rid Heritage of its scarlet letter, the individual mandate.

For the rest of the professional conservative class, it was an opportunity to flex financial muscle while recruiting and mobilizing their armies; Charles and David Koch, through their umbrella group, Freedom Partners Chamber of Commerce, spent more than $200 million on the anti-Obamacare effort, according to the *New York Times*.[3]

For Cruz, it was an opportunity to establish ideological supremacy among the nascent 2016 Republican field, capitalizing on Rubio's immigration stumble. (During a meeting that summer in Mike Lee's office, as Cruz and top conservative activists plotted the defunding plan, Rubio arrived late. "The prodigal son is here," he said, smiling.)

And for Meadows, the little-known freshman congressman, it was an opportunity to make a name for himself.

AMERICAN POP CULTURE WAS ROCKED IN 2013 BY THE RELEASE OF THE Netflix series *House of Cards*, an adapted version of the British drama that follows one exceptionally cunning and ruthlessly ambitious politician's rise to power. Kevin Spacey portrays Francis "Frank" Underwood, a Democratic congressman who lies, betrays, swindles, and murders his way to the top of American government. The show was a commercial dynamo at the height of the Republican drama inside of the real Congress. And if there was one person on Capitol Hill who looked in the mirror and saw Frank Underwood, it was Meadows.

The freshman lawmaker from North Carolina wasn't a bad person, and he certainly wasn't a killer—not in the literal sense, anyway. But there was something about the way he worked a room, the way he perched his glasses low over his nose for effect, the way he would feed a group of reporters one thing and then walk away texting a favored reporter something contradictory.

There was also something cryptic about his past: A self-described "fat kid" and social misfit from Florida, Meadows lost weight, married at age twenty, and, after randomly choosing the mountains of North Carolina for a honeymoon, fell in love with the area, so much that he and his wife eventually moved there.[4] First opening a sandwich shop, then selling it to become a real estate broker, Meadows made enough money to loan his congressional campaign $250,000, essentially buying both the GOP nomination and the general election in his freshly gerrymandered western North Carolina district.

We first interacted over several breakfasts in the middle of 2013, consistent with my efforts in covering Congress to build relationships with new members. Meadows wasn't like any of the others—or like any other politician I'd come across. He was disarming, with an easy smile and a sluggish southern drawl. He was engaging on policy matters. But what set him apart was the questions he asked—about the media, the coverage of Capitol Hill, how reporters' sourcing worked, what he needed to do to get his name in the paper. It was obvious that Meadows wanted to be a player.

Cue the release of his Obamacare letter.

It took serious gumption for a freshman lawmaker eight months on the job, but Meadows clearly saw a vacuum waiting to be filled. Cruz and Lee were leading the fight on the Senate side; nobody had yet orchestrated a real pressure campaign in the House. McConnell could only do so much: Despite a primary challenge from his right in 2014 that he was monitoring obsessively, the Senate GOP leader had the cover of a Democratic majority to deflect blame for Obamacare's implementation. Boehner had no such luxury. As House Republicans returned from the August recess emboldened by the anger on display in their districts and itching for a showdown with Obama, the Speaker knew there would be no talking them down.

There had been a cooling-off period for both parties after the president's reelection and his second inaugural. That period was long gone. Events that summer, including CIA contractor Edward Snowden's leaks showing illegal mass surveillance and Syria killing nearly fifteen hundred of its citizens in a chemical attack on the one-year anniversary of Obama's "red line" remark, exacerbated partisan tensions and fueled the declining trust in government.

The acceleration of cultural conflicts throughout the year—Obama's push for gun control, his unilateral action on climate change, the Supreme Court's rulings striking down California's gay marriage ban *and* the federal Defense of Marriage Act—had pushed traditionalists to the edge. The broader societal landscape did little to soothe the sense that things were spiraling. The Oxford dictionary shortlisted *twerk* as the word of the year, but opted instead for *selfie*, newly popular among not just Hollywood celebrities but politicians as well. Miley Cyrus was Google's most-searched person. Even in the Vatican, a redoubt of orthodox thinking, newly elected Pope Francis was sounding squishy, doing little to pull conservatives back from the brink.

Obamacare's approaching silhouette sent them over it.

BOEHNER AND CANTOR COULD SEE IT COMING. OBSERVING THE BREWing storm over the August recess, they prepared various trial balloons to float, hoping to prevent the zero-sum warfare their members wanted.

First, on September 9, Cantor outlined the leadership's preferred plan to the conference: They would force both the House and Senate to vote on defunding Obamacare but would not tie those votes to the rest of the government's funding, as a way of avoiding a shutdown. Conservatives booed Boehner and Cantor out of the room.

At this point, Boehner ditched large-scale diplomacy and began calling small cliques of members into his office for a reality check. "Don't do this. It's crazy," the Speaker told them. "The president, the vice president, Reid, Pelosi—they're all sitting there with the biggest shit-eating grins on their faces that you've ever seen, because they can't believe we're this fucking stupid." Not only would Democrats never abandon the president's bill, Boehner warned them, but a shutdown would overshadow the rollout of the Obamacare exchanges October 1, which members in both parties privately expected to be a logistical nightmare.

It made an impression. But so, too, did the countervailing influences from outside Congress. When Texas congressman Pete Sessions announced his opposition to the defund plan, the Senate Conservatives Fund labeled him a "RINO" (Republican in Name Only) and threatened to recruit a primary opponent. That same week, a Tea Party activist launched her campaign against Sessions and promptly received the endorsement of FreedomWorks. For the sake of context, at that time, Sessions had an 85 percent lifetime score with the Club for Growth, a 97 percent lifetime score with the American Conservative Union, a 100 percent lifetime score with National Right to Life, and an A+ rating from the National Rifle Association.

Republican in Name Only?

Aggravated by these developments, Hensarling stood up in the weekly RSC meeting and delivered a fiery rebuke to the outside groups, as *Roll Call* reported at the time.[5] A fellow Texan and revererd archconservative, Hensarling took out his voting card, which members use on the House floor, and held it up. He reminded his colleagues that nobody else—especially not Heritage Action—controlled their voting cards.

But the conservatives were unmoved. And Boehner was beginning to understand why: Whenever the rowdy elements of his conference had pushed too far over the past couple of years, the Speaker had pulled them

back, preached patience, told them, "Live to fight another day." That day
had arrived. If they didn't go to war against Obamacare now, they never
would. With renewed whispers of his weakness gusting through the
GOP, and his members girding for a game of chicken with the president,
Boehner jumped into the driver's seat and throttled up.

On September 18, less than two weeks before a potential shutdown,
Boehner gathered his troops in the House basement and delivered the
news. Conservatives would get a vote on exactly what they wanted: a
short-term continuing resolution funding the government through December 15; an extension of the lower, post-sequester spending levels;
and a permanent ban on funding for the Affordable Care Act. The room
erupted in applause. "I think our leadership has got it just right," Jim
Jordan said afterward. Was that the sound of conservatives cheering
Boehner? "Oh, yeah," he said, grinning. "Heck yeah."

Thus began one of the more futile negotiating periods in congressional history.

Boehner's version passed the House but was rejected in the Senate,
where Harry Reid stripped out the anti-Obamacare language and sent a
"clean" funding bill back to the lower chamber. (The Democratic leader
was not moved by Cruz's twenty-one-hour speech in opposition, during
which he promoted a new hashtag, #makeDClisten, and read Dr. Seuss's
Green Eggs and Ham for his daughters watching at home.)

At that point, Boehner and Cantor had a decision to make. There
was no use volleying an identical bill back to the Senate; the shutdown
was now less than a week away, and Democrats, even those who had
voiced concerns about the health care law's readiness, were not going
to defund it. After surveying their options and twisting some arms, the
GOP leadership rallied its members around a new bill, this one delay-
ing Obamacare's implementation by one year, repealing the medical
device tax, and designating pay for military members in the event of a
shutdown. Conservatives on the floor celebrated when it passed just after midnight on September 29. Boehner had given a concession, but the
House had held its ground. By lowering their asking price, Republicans
hoped, maybe Senate Democrats would come to the table.

Then again, maybe not. "Today's vote by House Republicans is pointless," Reid said shortly after the House bill passed. "Republicans must decide whether to pass the Senate's clean [bill], or force a Republican government shutdown."[6]

When the Senate officially rejected that version on September 30, with hours to go before the shutdown, House Republicans made a third concession. This time they passed a bill that would delay implementation of Obamacare by a year but keep the medical device tax, offering instead a populist amendment to strip health care subsidies for federal politicians and their staffs.

Once again, Harry Reid refused to flinch. It was like watching a speeding car negotiate with a brick wall.

When the clock struck midnight on October 1, Republicans scurried around the Capitol, many of them sporting mischievous grins. Democrats marched as though they were in a funeral precession, wearing rehearsed looks of melancholy. These optics were jarring and spoke to the national divide in public opinion. The vast majority of the country was assigning blame to Republicans. But many of them didn't care. They weren't elected by the vast majority of the country; they were elected by their districts, most of which were safely red and rewarding of any last-ditch effort to defeat Obamacare.

Just after midnight, David Schweikert, a Tea Party congressman from Arizona who had been kicked off his top committee a year earlier for his frequent votes against the leadership, told me a story. Hours earlier, he had participated in a telephone town hall with constituents back in his district. Rubbing his hands with glee, Schweikert relayed that nearly all of them were supportive of the shutdown and blamed Senate Democrats for their unwillingness to negotiate over Obamacare. "They get it," he said, practically squealing.

But did the Republicans get it?

The policy implications aside, the politics made sense for many in the House GOP: Their voters, by and large, weren't going to punish them for a government shutdown under these circumstances. But it wasn't just the shutdown Congress had to deal with. The Treasury Department had already announced that the country would run out of borrowing authority on October 17. This was the debt-ceiling deadline that House

Republicans had agreed to punt back at Williamsburg when Boehner explained how a series of wins would give them momentum heading into negotiations. Instead, with a possible default looming in sixteen days, they had now backed themselves into a government shutdown with no apparent exit strategy.

"We have to get something out of this," Marlin Stutzman, a conservative Indiana congressman, told the *Washington Examiner* in the wee hours of October 1. "And I don't know what that even is."

Fight. It became the defining word of the modern Republican era. As feelings of desertion took root during this period of dizzying cultural and economic transition, voters came to crave one quality above all others in their elected officials: a willingness to scrap, claw, kick, and bite on their behalf, demonstrating an understanding of their frustrations and their fears.

It's why Donald Trump, despite innumerable manifest flaws, won the presidency in 2016.

It's why Ted Cruz, despite obvious political defects, was the Republican runner-up.

And it's why John Boehner, despite their prior threats on his political life, won conservatives' trust in October 2013.

The Speaker had not ordered his troops into combat and watched their slaughter from horseback far removed from the front lines. Instead, Boehner had gone into the trenches, displaying an uncommon defiance of political convention at the outset of the shutdown and demonstrating to the world that he stood shoulder to shoulder with his Tea Party platoon. When an anonymous White House official said that Obama was "winning" the shutdown, Boehner responded by whacking a newspaper against the lectern at his press conference. "This isn't some damn game," he scolded.[7]

But it was a game. And Boehner's indignation was a charade. He privately believed the shutdown was idiotic, accomplishing nothing but to steal the public's attention away from the ugly rollout of the Obamacare exchanges—as he'd predicted. And yet, as with the immigration dilemma months earlier, the Speaker knew there was no way to exert his will without alienating much of the conference. Once again, Boehner chose to protect his right flank by feeding the party's most self-destructive appetites. And this time, it made him a hero on the right.

"It's easier to follow somebody who you know is willing to fight," Labrador, who orchestrated the coup against Boehner nine months earlier, said three days into the shutdown. With his bold subversion of Obamacare, Labrador said, Boehner was suddenly revealing himself as "the leader we always wanted him to be."

"He's leading," added Ted Yoho, the Florida freshman who had been part of the anti-Boehner mutiny. "He listened to membership, and he's put himself out there, and he's standing strong. We're all so proud of him right now."

But all of Boehner's leading—all his *fighting*—wasn't getting the Republican Party anywhere.

The purpose of the shutdown had been to highlight a bloody last stand over Obamacare, demonstrating to the Republican base how its leaders were willing to go down swinging. Several days into October, talk of health care had largely dissipated. The exchanges had opened. Democrats had not given an inch. And Republicans were no longer asking them to. Instead, lawmakers in both parties had turned their focus to the October 17 deadline for a debt limit increase.

Whatever internal GOP schisms had existed prior to the shutdown were swelling by the minute. Roughly a third of the House Republicans wanted to reopen the government immediately, with or without concessions from the Democrats. Another third was similarly eager to end the crisis but wanted to save face, needing something to show for their efforts. The final third seemed not to have a care in the world. They faced little blowback in their deep-red districts; moreover, many of them believed they were beginning to wear down the White House. Obama hadn't budged on his health care law, they told each other, but he wouldn't allow the country to default on its debt, giving them more leverage the nearer they got to October 17.

"As it starts getting closer to the debt ceiling date, the president feels more and more pressure," Labrador said at the time. "I think there's a good chance we can both get things we want, because he understands that we've never gone past that debt line."

Specifically, House conservatives believed they could hold out for entitlement reforms. If Obama wanted any lengthy extension of the debt limit, they argued to their leadership, Republicans should get proportional tweaks to either Medicare or Social Security in return.

Boehner was mystified by this but decided to play along. Why not? His conference had felt like a pressure cooker for the past two and a half years, and the shutdown was letting out steam. There was no way Obama would reward their tactics with changes to entitlements. But now that he had assumed the fighting position, it wouldn't hurt to hold it just a few days longer—allowing more steam to escape in the process.

On the Senate side, Cruz's Republican colleagues were less sanguine.

Nearly all the GOP senators, even those who initially vouched for Cruz's plan, had turned against him. Only Lee, the Utah conservative who had co-led the defund strategy, remained in the Texas freshman's corner. This angered many of the influencers in the conservative movement; they felt a special disdain for Jeff Flake. The Arizona senator had been a conservative stalwart in the House, regularly making life miserable for the GOP leadership. He promised to continue those ways when seeking a promotion to the Senate in 2012. Yet his first year in the upper chamber had been defined by membership in the Gang of Eight—and now by a refusal to follow Cruz and Lee into battle during the shutdown.

"It wasn't a gradual change with him," says Erick Erickson, a prominent activist and blogger who endorsed Flake in his Senate race. "It was a radical shift from where he was in the House, within one year, to where he was in the Senate."

"I didn't want anything to do with it," Flake says, looking back on the shutdown fight. "I said when I came to the Senate, I wasn't going to force somebody else to carry my water like I had on the TARP vote. . . . This notion that you could defund Obamacare on an appropriations bill—you really couldn't. And it wasn't responsible governing."

Cruz proved masterful at weaponizing such critiques. Whether it was the "Beltway insiders" or the "Republican establishment" or the "elite media" or the "Washington cartel," the senator found ways to cast himself as a lonely light flickering in defiance of the darkness. Republicans could win the shutdown, Cruz argued, if they didn't surrender; if they united to filibuster the Senate's clean funding bill and forced Obama to the negotiating table.

Some of his conservative allies outside Congress bought this idea, as did the Tea Party congressmen who huddled with him frequently in the early days of October. But Cruz's colleagues, who probed him behind closed doors for an explanation of his endgame, did not.

"He was making it up as he went along. He led us into box canyon," says Bob Corker, the Republican senator from Tennessee. "It was clear as a bell what was happening. He had no plan, but a base of people actually believed that Ted Cruz was doing the Lord's work."

Predictably, Boehner's effort to seek concessions from the White House went nowhere. America doesn't negotiate with terrorists, and Obama wasn't going to negotiate entitlement reforms with Republicans who had shut down the government over an entirely separate issue. As this became clear, a small clutch of hard-liners in the House retreated to a most unexpected position: that the Treasury Department's debt limit deadline was toothless.

"Nobody thinks we're going to default on October seventeenth," Tim Huelskamp, the Kansan, said.

"There's always revenue coming into the Treasury, certainly enough revenue to pay interest," said Justin Amash of Michigan.

"We're not going to default. There is no default," said Mick Mulvaney, the South Carolina congressman and future director of the Office of Management and Budget. "There's an OMB directive from the 1980s— the last time we got fairly close to not raising the debt ceiling—that clearly lays out the process by which the treasury secretary prioritizes interest payments."

The October 15 announcement from Fitch, one of the big three credit rating agencies, that it was reviewing the country's AAA credit rating for a possible downgrade, did nothing to change these assessments.

"I remember that day, sitting there with Ted Yoho," Cantor recalls, "and I said to him, 'Are you ready? If we default, are you ready to accept whatever consequences there are? You can't tell me what they're going to be because I can't tell you, either. We've never been there.'"

The Florida congressman's response: "Absolutely."

House Republicans had started their morning conference meeting by singing "Amazing Grace." But a day that began with hymns of resurrection ended with bagpipes of burial. With Harry Reid and Mitch McConnell close to a deal that would raise the debt ceiling and reopen the government, Boehner had thrown out a last-ditch plan to extract some small concessions from the White House. But his conservatives, and Heritage Action, rejected it as insufficient. After sixteen days of incremental retreat with nothing to show for it, Boehner told his team that the House would follow the Senate's lead and end the crisis.

When Boehner shared this news, telling his House Republican colleagues, "We fought the good fight, but now the fight's over," they responded with a standing ovation. The Speaker was slack-jawed. A month earlier, in this very room, the conservatives had booed him for proposing a short-term funding bill and a temporary extension of the debt limit. Now, after getting their teeth kicked in for sixteen days, they were going to pass exactly that language. And the room was cheering.

"I'm thinking to myself," Boehner recalls, "*this place is irrational.*"

PAUL RYAN WAS NEARLY UNRECOGNIZABLE WHEN HE RETURNED TO Congress after the 2012 election.

Once upbeat and outgoing, he was now irritable and introverted. Convinced that Mitt Romney would be president, Ryan had prepared himself for the role of the vice presidency, working closely with the transition team to craft a plan for the new administration's first two hundred days. Instead, the Wisconsin congressman found himself back on Capitol Hill.

He didn't want to be there. Ryan had told his wife, Janna, that he might retire, and it took a long phone conversation with Boehner to persuade him to stay. But he had no patience for the pep talks, the funny looks, the constant inquiries from his colleagues. In a meeting one day, after being asked for the umpteenth time what he'd learned while running for vice president, Ryan snapped, "The Electoral College matters. That's what I learned."

In fact, he had learned a whole lot more. Raised comfortably as a fifth-generation resident of Janesville, Wisconsin, a town where the extended Ryan clan is royalty, the congressman had lived much of his life in a bubble. While studying economics and political science at Miami University in Ohio, he interned on Capitol Hill and caught the bug, moving to Washington full time after college and working his way up the Republican food chain. His formative experience was apprenticing at a think tank led by the former New York congressman Jack Kemp, a proponent of inclusive, sympathetic conservatism. Ryan was more philosophically doctrinaire, an Ayn Rand devotee who ate free markets and breathed rugged individualism. Even as Ryan matured politically, winning his House seat at age twenty-eight and becoming a specialist in budgetary matters, the ideological blinders never came off: He believed wholeheartedly in the concept of "makers and takers," a line he deployed to cite the divide between America's productive, hardworking citizens and those who snacked indolently on the fruits of that labor.

The 2012 election changed him. Traveling the country, exposing himself to new audiences and different ideas and worlds that resembled neither Washington, DC, nor Janesville, Wisconsin, Ryan grew ashamed of his insularity. A summit with civic leaders, learning of their rehabilitation programs, hit Ryan particularly hard. His mentor, Kemp, had made poverty and social mobility the causes of his career. Yet Ryan, for all Kemp's tutelage, was too focused on budget line items to appreciate what they represented. Accentuating the realization of his own failure was that of Romney's campaign. Ryan had choked back his criticisms as Romney ignored nontraditional Republican voters—and as the GOP itself alienated women, minorities, and young people with its policies and rhetoric.

Processing the gut-wrenching loss in November 2012, Ryan committed himself to reinvention. He would be wiser, more amenable, less reactionary. He stopped saying "makers and takers" and apologized to people back home he might have offended. He also undertook a quiet journey to better grasp the country's problems. Channeling the ghost of Kemp, Ryan took a special interest in poverty. Contacting Bob Woodson, a longtime civil rights advocate and leader in the black community, Ryan asked for a tour of facilities around the country that helped the poor and addicted. It struck Woodson as a publicity stunt, but Ryan said he wanted no media present. Woodson was still skeptical. "And then every month, for about the next four years, we went to a different city, we met different groups, and he deepened his understanding of these people," Woodson recalls. "I witnessed a transformation in him. He's traveled to more low-income black neighborhoods than any member of the Black Caucus that I know of."

The most outward sign of Ryan's change could be seen in Congress. He supported raising the debt ceiling. He voted to reauthorize the Violence Against Women Act. He refused to sign Meadows's letter demanding that Obamacare be defunded, warning colleagues that Cruz's approach amounted to "a suicide mission." Ryan's actions prompted whispers in the conservative movement and puzzled looks from longtime comrades. They worried that the presidential campaign had broken him, that he had lost his nerve.

The tipping point was Ryan's deal with Patty Murray.

When Congress voted on October 17 to reopen the government on a short-term basis, the condition was the convening of a bicameral budget panel to address the country's long-term fiscal challenges. This was thought to be an empty gesture, given that neither party had shown any real willingness to cede ground on the issues of spending and taxation.

To the extent that conservatives worried about the budget talks, it was because some Republicans had advocated getting rid of the sequester cuts—the first spending reductions the GOP had achieved in years. When Ryan was appointed the chief Republican negotiator, those concerns melted away. He wasn't just their fiscal Goliath; he had brokered the Williamsburg Accord, which stated that the sequester cuts could be traded only in exchange for dramatic entitlement reforms. Conservatives needn't fear that Ryan would break that promise, and certainly not in a negotiation with Murray, a liberal senator from Washington State.

But Ryan himself had a different view. With the automatic spending cuts growing more severe, and another government shutdown looming because of Washington's inability to govern on annual budgets, the Wisconsin congressman approached the negotiations with a dealmaker's mentality. He would be willing to give away some of the automatic cuts if Murray were willing to make specific, offsetting cuts elsewhere that would reduce the deficit—without any new tax revenue.

This seemed preposterously unlikely. It wasn't exactly Washington's golden age of bipartisanship: On November 21, as Ryan and Murray hammered away in negotiations that proved surprisingly leak-proof, Harry Reid invoked the "nuclear option" in the Senate, changing the body's rules to prevent the minority party from filibustering presidential nominees (Supreme Court justices not included). The vote, which passed mostly along party lines, with 3 Democrats joining the GOP in opposition, represented rock bottom for the Senate. Once a fraternity-like society that thrived on relationships and decorum, the "world's greatest deliberative body" had devolved into a bad faith blood feud between parties catering to pugnacious bases. Reid was goaded into the convention-shattering decision by McConnell's policy of blocking a historic number of Obama's judicial appointees, but whatever short-term gain awaited the Democrats was offset by questions of damage to the institution—and fear of unintended consequences. "You may regret it," McConnell said on the Senate floor, "a lot sooner than you think."[8]

Official Washington was dumbfounded, then, when Ryan and Murray announced a budget deal on December 10. The toplines were straightforward: Their plan would fund the government for two full years at new, slightly higher spending levels, but would reduce the deficit and save $28 billion over a decade—all without raising taxes.

Ryan was thrilled with the deal. Yes, he had broken the terms of the Williamsburg arrangement, but there was no question that the budget compromise moved the country in a fiscally conservative direction. "I deal with the way things are, not necessarily the way I want things to be," Ryan said after the agreement was unveiled. "I have passed three budgets in a row that reflect my priorities and my principles and everything I want to accomplish. We're in divided government. I realize I'm not going to get all of that."

The budget deal revealed a new schism on the right—between some, like Ryan, who had come around to the concept of incrementalism, and others who rejected the notion of half a loaf. Tom Price, Ryan's friend and vice chairman of the Budget Committee, attempted to rally conservatives around the plan. "It is increasingly obvious that success, particularly in divided government, has to be measured in positive steps, not leaps and bounds," Price said.

But that sentiment was drowned out by loud opposition. The Club for Growth blasted the deal as "budgetary smoke and mirrors."[9] Talk radio host Mark Levin told Ryan the agreement was "Mickey Mouse."[10] And Heritage Action, which also key-voted in opposition, called Ryan's work "a step backward" for conservatism.[11]

Boehner had always ignored such external criticisms, but he could no longer bite his tongue. Ryan had volunteered on behalf of Boehner's congressional campaign as a college kid, and the Speaker felt a fatherly bond with the Wisconsin congressman. He could not stomach watching these professional purists eviscerate his young budget chairman.

"I think they're misleading their followers. I think they're pushing our members in places where they don't want to be," Boehner said of the critics. "And frankly, I just think that they've lost all credibility."

Boehner's battering of the outside groups was an assault on the conservative movement itself. And it wasn't happening in isolation. On December 11, in another Mafioso-style move, RSC chairman Steve Scalise fired the group's longtime executive director, Paul Teller, an integral player in Washington's conservative scene, for leaking member-level conversations to the outside groups in the hope that they could turn on-the-fence members against bad legislation.

"We are saddened and outraged that an organization that purports to represent conservatives in Congress would dismiss a staff member for advancing conservatism and working with conservatives outside Congress," the leaders of Heritage Action, FreedomWorks, and other activist outfits said in a statement responding to Teller's firing. "Given this action . . . it is clear that the conservative movement has come under attack on Capitol Hill."[12]

In the middle of all this stood Ryan, once the golden child of conservatism, who seemed more bemused than beleaguered by the right's turning against him. "It's a strange new normal, isn't it?" he said.

Ryan always knew the compromise would draw opposition from Tea Party lawmakers. But the toughest disagreement was with his fellow Jedi Council member, Jim Jordan. "Eleven months ago, our conference made a decision . . . that we will not get rid of the sequester unless and until we get the kind of big savings in mandatory programs that put our nation on a path to balance in ten years," Jordan explained. He called Ryan's deal with Murray a "marked departure" from their Williamsburg agreement and mobilized his allies to defeat it.

Other conservatives piled on—Mulvaney said it wasn't "hard-core" enough, and Labrador called it "really a terrible plan"—but Jordan's dissent was the most consequential. He and Ryan had shared breakfasts together for years, bonding over talk of sports, families, a common philosophy. Jordan had defended Ryan earlier in the year against accusations that his friend had gone soft. But he could no longer ignore the evidence of his own eyes. Ryan's breach of the Williamsburg arrangement wasn't just a disagreement; it was an act of duplicity. And Jordan would never let it go.

Thirteen days before Christmas, the House passed Ryan and Murray's bill, the Bipartisan Budget Act of 2013, in lopsided fashion: The tally was 332–94, with 169 Republicans supporting the legislation and only 62 opposed. It then passed the Senate and was signed into law by Obama.

The deal's success marked Ryan's promotion to an essential player in Washington—no longer an ideologue, but a seasoned and accomplished policymaker who had secured real progress in a divided government and had faced down his own base to sell it.

Ryan's triumph was just as meaningful for Boehner. The Speaker had rung in the New Year amid swirling rumors that a mob of conservative malcontents was orchestrating a coup d'état aimed at overthrowing him in humiliating fashion. He had not only survived but thrived, uniting the conference around a plan, sidestepping the land mine of immigration, and earning newfound respect by giving his trigger-happy hardliners the shutdown shootout they craved. Topping it all off, Boehner had outmaneuvered his enemies by putting Ryan in charge of budget negotiations, baiting the right into criticisms of a bill that passed with enormous support. After three years of the Tea Party dictating terms to the GOP, its influence was on the decline.

As Boehner walked off the House floor, shaking hands and patting backs and looking forward to the bottle of cabernet waiting inside the Speaker's suite, he knew this Christmas would be merrier than the last.

CHAPTER EIGHT

APRIL 2014

"We called them 'the Caveman Caucus,'
and we needed to crush them."

JOHN BOEHNER SIPPED HIS BLACK COFFEE WHILE STARING INTO THE
soul of Roger Ailes.

It was a sunny Monday morning in New York, and the renewing sights
of springtime felt fitting to the Speaker of the House. Ever since the government shutdown of the previous fall, the worm had turned inside
the Republican Party. The civil war raged on. But it was the rebels who
were now on the run—and the establishment was striking back. Having
exposed the strategic clumsiness of the Tea Party delegation, and tri-
umphed over the right with Ryan's budget compromise, congressional
leaders and their establishment allies looked ahead to the 2014 elections
as a chance to seize back control of the GOP.

This depended in large part on neutralizing the conservative news
media—or at least, defusing its explosive predispositions.

The proliferation of right-wing reporting and punditry in the
late 1990s had once been a blessing for the GOP. The impeachment
of Bill Clinton, the election of George W. Bush, the wars in Iraq and
Afghanistan—wherever there was controversy, conservatives had been
able to depend on friendly voices covering it.

But the disruption of recent years—the implosion of Bush, the election of Obama, the arrival of the Tea Party—had upended that business
model. Politics was no longer symmetrical. To channel the populist fury
of its audience, conservative media began targeting the GOP elites with
the same mendacity that it displayed in attacking Democrats. The irony
was inescapable: Republicans had spurned legacy journalism outlets for
their perceived bias and dishonesty only to receive heaping portions of

both from the likes of Fox News, talk radio, and the ever-expanding constellation of conservative blogs, websites, and social media feeds.

"When we won the majority in 1994, we barely had talk radio. The only people using the Internet were a couple geeks in Palo Alto. There was no Facebook, Twitter. There was no Breitbart.com. And there was no Fox News," Boehner says. "It's hard to calculate how much more information people have about their government than they did back then. A hundred times? A thousand times? They've been buried under all this information, and much of is either untrue or misleading, and it has pushed people farther and farther away from the middle and into their echo chambers."

Boehner had given up on getting fair treatment from talk radio. Once upon a time he spoke regularly with Sean Hannity, and he would play golf with Rush Limbaugh during frequent trips to Palm Beach. But those relationships had frayed since his becoming Speaker. Nuance and pragmatism don't play well on the airwaves; there was little audience to be gained by realistically assessing the expectations for Republicans in a divided government under a Democratic president. In Boehner's view, it was the sudden popularity of fanatical radio host Mark Levin in the years after 9/11 that influenced the others. "Levin went really crazy right and got a big audience, and he dragged Hannity to the dark side; he dragged Rush to the dark side," Boehner said. "I used to talk to them all the time, and suddenly they're beating the living shit out of me."

But there was still hope for Ailes. Boehner had known the Fox News chairman and CEO since the early 1990s, when he was a rookie congressman and Ailes was a powerful media consultant to then-President George H. W. Bush. Their friendship grew over the years, and even when Fox News was bludgeoning his speakership, Boehner would always find himself breaking bread with Ailes during swings through Manhattan.

This meeting would have a different tone. Boehner was at his wit's end with cable news and its insatiable appetite for conflict-driven coverage. Ailes was giving a platform to people who had no reasonable claim to one; his network was incentivizing lawmakers to do more wrecking than building, more gossiping than governing. This wasn't merely a question of ideology. There was a difference, to Boehner, between Jim

Jordan and Louie Gohmert: Both congressmen made life miserable for their party's leadership, but only one could offer a lucid rationale for why.

Boehner could deal with fringe characters in his conference: Gohmert, Steve King, Michele Bachmann. What he couldn't deal with anymore was seeing them on national television, broadcasting their bat-shittiness to tens of millions of people in prime time.

The Speaker had come to ask Ailes a favor: *Stop putting these people on your network.*

Ailes was not inclined to agree to this. His on-air talent, and the guests they booked, were part of a well-oiled ratings machine. Dictating a blacklist to Hannity at the behest of the Speaker would not go over well. But Ailes was not unsympathetic to Boehner's plight. He, too, had observed that the GOP was becoming anarchic; Ailes had even agreed to give the Gang of Eight some breathing room at the outset of their immigration push. (It didn't last.) Moreover, his boss, News Corp executive chairman Rupert Murdoch, was a known moderate on certain issues and had voiced discomfort with the GOP's absolutist wing and its allied hosts on his channel.

"What happened to immigration reform? Why not pass that bill?" Murdoch had asked Eric Cantor during a dinner in the fall of 2013.

"Rupert," Cantor replied, "Have you *watched* your network?"

Now Ailes was giving Boehner the same answer that Murdoch had given Cantor months earlier. "Don't worry about them," Ailes said, referring to his resident provocateurs. "They're just getting ratings."

But the Speaker wouldn't be dismissed that easily. He had come equipped with a sweetener for his request, something that the Fox boss could sink his teeth into. Ailes owned a special loathing for the Clinton family and particularly for its matriarch, whom he found manipulative and unfit for office. Boehner knew this. And although he did not share the right's loathing of Hillary, the Speaker was actively building an in-house opposition research firm to damage her presidential prospects in 2016. Unbeknownst to the public, Boehner was about to launch a select committee in Congress to investigate her handling of the Benghazi attacks that killed four Americans while she was secretary of state. He had come to give Ailes a heads-up, hoping it would persuade the cable king-pin to pull the "crazies" off his airwaves.

Boehner's plan backfired. Rather than rejoice at the Speaker's news, the word *Benghazi* tripped a switch. Suddenly high-strung and wary of his surroundings, Ailes proceeded to unpack for Boehner the outlines of an elaborate, interconnected plot to take him down. It started with Ailes's belief that Obama really *was* a Muslim who really *had* been born outside the United States. He described how the White House was mon-itoring him around the clock because of these views. He concluded by assuring Boehner that his house had been fortified with combat-trained security personnel and "safe rooms" where he couldn't be observed.

"It was the most bizarre meeting I'd ever had in my life. He had black helicopters flying all around his head that morning," Boehner recalls. "It was every conspiracy theory you've ever heard, and I'm throwing cold water on all this bullshit. Ratings were ratings to Murdoch, but I began to realize that Ailes believed in all this crazy stuff."

The Speaker had come with hopes of quieting the furor on Fox News. He left more concerned than ever about the threat it posed to the country.

REGROUPING AT THE DAWN OF 2014, THE REBELS LICKED THEIR WOUNDS and pondered two principal lessons learned from the past year.

First, they recognized, their problem was less with any one person—even Boehner—and more with the top-down processes of the House, where legislative influence is derived from seniority, fund-raising abil-ity, and proximity to power. The incentive structures of Congress were beginning to shift; many Republicans now feared the ire of their base more than a rebuke from their party's leadership. But this transforma-tion wasn't happening quickly enough for the insurgents in the House GOP. Only by reforming the process, and breaking into the inner sanc-tum of power, could they transform Congress.

Second, they agreed, no one outside their small circle could be trusted anymore. Cantor had undermined them by offering citizenship to illegal minors. Steve Scalise had fired the RSC's popular executive director and booted the Heritage Foundation from its weekly meetings. Even Paul

Ryan had sold them out, trading away the sequester cuts and basking in the afterglow of a bipartisan compromise they had vigorously opposed.

The result of these twin realizations—a need to disturb the procedural status quo and an imperative to distinguish themselves from the rest of the conference—was a series of embryonic talks about a smaller, purer group of conservatives. It would be invitation-only. It would have strict rules governing how members could vote. It would, if properly organized and executed, empower the rebels to serve as a veto on their party's leadership, denying them the numbers needed to pass ligislation.

None of this would be easy. Standing up a new caucus takes time, money, and logistical savvy. The House hard-liners who wanted freedom from the RSC and its nominal "conservative" membership faced a series of hurdles before they could spin off as their own autonomous group. In the meantime, they had the House Liberty Caucus.

Justin Amash, the libertarian Republican from Michigan, had established the group as a loosely organized luncheon for a couple of his fellow Ron Paul acolytes. They gathered every few months, debating the Fourth Amendment while munching on deli platters and sipping cans of Cherry Coke. The caucus was an afterthought on Capitol Hill. Yet, in early 2014, as the insurgents wandered in the party's wilderness, seeking a territory of their own, Amash allowed his friends (Jim Jordan, Mark Meadows, Raúl Labrador, Mick Mulvaney, and Thomas Massie, among others) to adopt his House Liberty Caucus as their home base.

The group began meeting every other week. Two dozen members arrived in secret, swearing to safeguard the discussions held within. There were no leaks, no spies working for Boehner or Cantor. After three years spent battling their leadership within the House majority, worrying all the while about sabotage and betrayal, House conservatives finally had a safe haven.

It was a breakthrough for the rebels, and an inflection point for the party. After four decades as the tip of the conservative movement's spear, the Republican Study Committee was losing relevance, its ideological intensity mitigated by its swelling membership. With three out of every four House Republicans now belonging to the RSC, many of them viewing the $5,000 membership dues as an investment in preempting a primary challenge from the right, the organization could not possibly play the intraparty hardball its founders had envisioned.

"The RSC today covers a fairly broad philosophical swath of the party. It's no longer just the hard-core right-wingers," Mulvaney said upon joining the House Liberty Caucus.

"When working with like-minded people," Amash added, "you need something a little more nimble that doesn't dilute its positions because of the size of the group."

Nimbler, smaller, and more secretive, the breakaway faction of right-wingers got busy mapping out its strategy for 2014. Their first target was institutional apathy. Congress has a rich tradition of doing as little as possible in even-numbered years, and the rebels hoped to change that. They wanted the party to pursue major legislation—on taxation, welfare, privacy, and health care, for starters—instead of simply running out the clock until Election Day.

Boehner and McConnell had different ideas. The president's approval ratings were middling. His signature law was proving increasingly unpopular. Republicans were poised to expand their House majority *and* win back the Senate—as long as they didn't overplay their hand. "If your opponent is committing suicide," Boehner warned his troops, "Why shoot him?"

Yet again, in the spring of 2014, a fundamental schism was being laid bare—this one about power and its inherent purposes. The conservatives insisted that the election should be waged around ideas, even if those ideas might cost the party votes; the leadership argued that the party's best chance for implementing those ideas was by winning elections first. The conservatives thought the leadership cowardly; the leadership thought the conservatives reckless.

By the time Boehner visited Ailes in New York City, the rebels' interest in procedural changes was taking a backseat to their contempt for the party's establishment. They could not begin to fix Congress without replacing its most powerful figure. Once again, they started scheming to oust Boehner. This time it would be different: By organizing early in the year, they told themselves, they would lock down the votes needed to prevent Boehner from winning another term as Speaker. They would tell him as much in private after the November elections, preventing an ugly scene from unfolding on the House floor in January 2015.

Little did they know, Boehner was already plotting his exit strategy.

A COMMON EXPLANATION FOR THE TEA PARTY'S ELECTORAL SUCCESSES of 2010 and 2012 was that the Republican establishment had over-reached. By endorsing the likes of Charlie Crist over Marco Rubio and David Dewhurst over Ted Cruz, this thinking went, GOP insiders had unwittingly aided the opposition by stoking antagonism toward the paternalistic party elite.

This was not exactly wrong: Amid a groundswell of resentment toward Washington, the self-important endorsements from politicians, party leaders, and committees had backfired.

Yet it missed the bigger picture. The establishment's mistake wasn't in going too far, but in not going *far enough*. Party officials had spent the past two election cycles pretending that the old rules still applied, that voters would fall meekly in line, that candidates without traditional support would wither and die. In a political climate defined by the extremes, freezing cold or scorching hot, the Republican establishment had been lukewarm, offering respectable support but nothing in the way of overwhelming force.

That could no longer be the case. With a host of vulnerable Senate Democrats facing reelection, Republicans could flip the chamber in 2014—but only if they nominated the right candidates. That meant playing aggressively in primaries. That meant counteracting the right's energy and money. And that meant marginalizing fringe conservative candidates who could not win in November. If Republicans were going to take back the Senate, they couldn't afford any more Christine O'Donnells.

"We had taken a passive view of involvement in primaries. In 2014, I said the business model has got to change," McConnell recalls. "It wasn't so much a philosophical thing; it was getting quality candidates who can actually appeal to the general electorate. I wasn't offended by the Tea Party. We were glad to have their support. But in order to win in most states you have to have somebody who can be presentable to a larger electorate, and [the Tea Party] produced some people who simply couldn't win."

With the aid of their burliest outside allies, the U.S. Chamber of Commerce as well as Karl Rove's group, American Crossroads, McConnell and the Republican establishment set about smothering the Tea Party.

"We called them 'the Caveman Caucus,' and we needed to crush them," recalls Scott Reed, the Chamber's senior political strategist, who coordinated with state and local affiliates to raise and spend nearly $20 million in Republican primary fights that year. "It was a turning point for us. We felt like we were taking back control of the party in 2014."

Nobody had entered 2014 wearing a brighter bull's-eye on his back than McConnell. The bespectacled, gray-haired Senate leader, perpetually poker-faced and soft-spoken in a manner that belied his barbarous instincts, was a political institution unto his own. He had spent the past three decades building the Kentucky GOP from the ground up, earning priceless goodwill and collecting favors across the state. But his DC deal-making and bring-home-the-bacon politics were poorly suited to the Tea Party era. With his numbers sinking in Kentucky, a chorus of conservative outside groups—FreedomWorks, Tea Party Patriots, the Senate Conservatives Fund—made a show of rallying around McConnell's challenger, a veteran and manufacturing executive named Matt Bevin.

But nobody knew McConnell's flaws better than McConnell. Having worked tirelessly to forge an alliance with Rand Paul, the Tea Party favorite, McConnell won the junior senator's endorsement in 2014. He also hired the Paul family's political consigliere, Jesse Benton, as his campaign manager. (Benton would later be recorded saying he was "holding my nose"[1] working for McConnell, citing the advantage it could lend Rand Paul's 2016 presidential bid.)

Meanwhile, McConnell's team built an encyclopedia-thick opposition research dossier on Bevin, blanketing the airwaves with attack ads the week his rival entered the race. They branded him "Bailout Bevin" for state funds he'd accepted to rebuild a factory, undermining his conservative bona fides and neutralizing attacks on McConnell's TARP vote. Buried under millions of dollars in negative ads from McConnell and his outside partners, Bevin's campaign never got off the ground. Mc-Connell creamed him by 25 points.

A similar pattern played out across the country, as establishment-favored candidates used massive war chests to beat back primary opponents from the right.

In Louisiana, where Democratic senator Mary Landrieu was deeply vulnerable after voting for the Affordable Care Act, a retired Air Force colonel (and self-professed alligator wrestler) named Rob Maness won the support of Sarah Palin, Phyllis Schlafly, numerous talk radio hosts, and more than a dozen Tea Party groups. But GOP leaders weren't taking any chances. They drowned the state in financial and structural support on behalf of a centrist congressman, Bill Cassidy, whose allies roasted Maness for suggesting that he would have opposed the Hurricane Katrina relief package if he had been in Congress. Maness wound up taking just 14 percent in Louisiana's all-party primary, and Cassidy easily topped Landrieu in the general election runoff.

In North Carolina, where another incumbent Democratic senator was on the ropes, a spirited conflict broke out between the GOP's warring factions. On one side, a multitude of conservative leaders and organizations endorsed Tea Party activist Greg Brannon; on the other, party elders such as Mitt Romney and Jeb Bush, as well as McConnell and his allied groups, threw their weight behind Thom Tillis, the Speaker of the state House. Tillis out-raised Brannon by a nearly a four-to-one ratio, pulling away to win the primary and sparing Republicans "the kookiness of a candidate who thought *Marbury v. Madison* was wrongly decided and the U.N. was trying to destroy our suburbs," as *Slate*'s David Weigel put it. In the year's most expensive race, Tillis edged Democrat Kay Hagan by roughly 45,000 votes.

And on it went: In every contested Senate primary of the 2014 cycle, and in nearly every contested House primary, the forces of the establishment suffocated the right-wing insurgency.

Even more heartening to Republican leaders was the apparent lack of enthusiasm, or fresh thinking, on the left. President Obama's party had become stale, as is customary six years into most any administration. Democrats across the country were struggling to find a coherent message to run on. Gas prices had dropped well below three dollars per gallon, and there were growing signs of economic recovery, but it still felt too sluggish for voters to reward at the ballot box.

In lieu of any powerful economic argument, many Democrats settled on cultural warfare, painting Republicans as extremists who would subjugate women and starve the poor. Abortion, once a break-glass-in-case-of-emergency issue for Democrats, was becoming a thematic cornerstone to the party's campaigns, sometimes with disastrous results. In Colorado, incumbent Democratic senator Mark Udall spent so much time talking about "reproductive rights," in television ads, on the campaign trail, and during debates, that he was nicknamed "Mark Uterus."[2] Rather than focusing on ISIS or climate change, Udall talked incessantly about birth control and abortion, prompting the liberal *Denver Post* editorial page to endorse his Republican opponent, Cory Gardner, because of Udall's "obnoxious one-issue campaign." Gardner, a top recruit of McConnell's, would go on to flip the seat in November.

Throughout the year, however, there were hints of long-term trouble for the GOP's shot-callers. Even as they flooded the competition with dollar signs in 2014, a new class of donor was emerging—not the archetype patron of the elite, the fiscally conservative and socially liberal elbow-rubber looking for an ambassadorship down the road, but the true-believing types with millions to burn and ideological firepower to spare.

In the spring of 2014, at the Club for Growth's donor conference in Palm Beach, former congressman and MSNBC host Joe Scarborough was one of the headliners. He pandered to the audience by calling himself a "Club conservative" based on his old voting record, praising the group's work to hold the GOP accountable. When he finished, a hand shot up. It belonged to a young woman many of the attendees did not recognize. She stood up and began dressing Scarborough down. "How dare you call yourself a conservative?" she asked him. "I've watched your show. I've watched you calling Ted Cruz a phony. You're nothing but a pompous sellout."

The donors murmured to one another. Who *is* that?

"Rebekah Mercer," one Club staffer whispered to another.

THE ONE-SIDED RESULTS OF THE 2014 PRIMARY SCORECARD DID NOT always reflect a show of strength by the Republican elite. Rather, in certain races, the intraparty feuding exposed the fatal deficiencies of both teams.

The Mississippi Senate primary was one such instance. The contest signified rock bottom for Republicans in 2014, featuring race-baiting advertisements, dirty tricks aimed at unseating the party's most endangered senator, and a last-second Hail Mary to save him.

Thad Cochran, a septuagenarian incumbent who faced serious and legitimate doubts about his capacity for executing the duties of a U.S. senator, was staring down defeat. Despite substantial assistance from GOP heavyweights in both Washington and Mississippi, Cochran was losing ground to his young primary challenger, Chris McDaniel. An attorney and bomb-throwing state senator, McDaniel had channeled the anti-Washington zeitgeist as well as anyone in 2014, winning the support of myriad Tea Party politicians and conservative outside groups and even an endorsement from Donald Trump.

The contrast couldn't have been starker: While the animated young McDaniel enlisted the likes of Sarah Palin to draw enormous crowds, Cochran's team kept the incumbent in a bunker, avoiding the public (and the media) almost altogether, certain that the decrepit senator would do or say something disqualifying to his reelection.

And then, six weeks before the primary, Cochran received a gift from the political gods. Sneaking into a local nursing home where the senator's infirm wife lived, a local Tea Party activist and pro-McDaniel blogger snapped a photo of a bedridden Rose Cochran and posted it online, part of a smear campaign aimed at stoking speculation that Senator Cochran was cheating on his sickly wife with a younger woman.

The perp was arrested less than two weeks before the primary. Shortly thereafter, several other McDaniel allies were arrested as part of a police investigation into charges of exploitation of a vulnerable adult. The avalanche of negative publicity in the contest's closing days resulted in McDaniel finishing with 49.5 percent of the vote—more than Cochran but just shy of the 50 percent threshold needed to avoid a runoff.

The ensuing three weeks offered a political soap opera for the ages.

Recognizing that McDaniel's support among conservatives ran far deeper than did Cochran's, allies of the incumbent senator schemed to turn out Democrats—specifically, *black* Democrats—in the runoff. This was perfectly legal. But the tactics toward that end were most unsavory: The political machine of longtime Republican governor Haley Barbour made covert payments to an African American activist group that in turn produced television ads and mailers accusing the McDaniel campaign of preying on racial divisions and attempting to suppress the black vote.[3] McDaniel cried foul, justifiably so. Yet he was hardly a sympathetic figure; with the nursing home scandal and his history of racially incendiary remarks on talk radio, few Republicans felt inclined to jump to his defense. (Furthermore, it wasn't until after the runoff had concluded that the truth behind the ad campaign fully materialized.)

Meanwhile, Cochran's other allies looked for a cleaner but equally effective way to hit McDaniel. At the U.S. Chamber of Commerce, Reed, its top strategist, decided that Cochran needed the jolt of a celebrity endorsement. His first choice would be a famous football player to vouch for the senator's toughness in the pigskin-crazed state. The Chamber reached out to Archie Manning, the legendary New Orleans Saints quarterback, to star in a pro-Cochran ad. But Manning was recovering from surgery and had to pass. Reed then reached out to Manning's son, Peyton, a future Hall of Famer in his own right. Peyton declined but recommended his younger brother, Eli. As it turned out, the Giants quarterback was indeed interested—until his agent stepped in and put the kibosh on any political involvement.

Reed was ready to give up. Then a colleague suggested Brett Favre. It was perfect: The legendary Green Bay Packers quarterback was a Mississippi boy, born and bred, who had since returned to the state after retiring from the NFL. Favre immediately agreed to Reed's request. (What Reed didn't learn until later: Favre's agent was from the same

hometown as McDaniel and despised the Senate hopeful.) The Chamber
sent a video crew to Favre's house the next day, and the old gunslinger
came speeding down the driveway on an all-terrain vehicle, shirtless
and bearded, to meet them. The video crew exchanged bewildered
looks. Favre's advertisement, which promptly began airing in the southern Mississippi market, blew up overnight—thanks in part to the NFL
retweeting it—and breathed life into Cochran's moribund campaign.

It was desperately needed. Cochran looked like a dead man walking,
politically and otherwise. He lived around the corner from his campaign headquarters but needed to be walked home. He never wanted to
campaign. On infrequent bus rides around the state, he would do little
but sleep. John McCain, who traveled to Mississippi to campaign on his
colleague's behalf, called Reed one day from the road. "You don't understand," McCain said, "It's like fucking *Weekend at Bernie's* down here."

Cochran won the runoff by some 7,700 votes, with black participants
figuring decisively in the margin thanks to noticeably higher turnout
in urban precincts. McDaniel refused to concede, threatening legal action that he parlayed into a subsequent failed run for federal office a few
years later.

The Mississippi campaign, for all its melodrama, had come to symbolize the ugliness of the party's internecine struggle and the fundamental
weakness of both sides. Cochran represented an aging, compromised
Republican establishment and its willingness to do anything to cling to
power. McDaniel embodied the fringe portrayal of the right, someone
who gave little thought to being thoughtful, who specialized in picking
fights instead of offering policy ideas.

Meanwhile, one of the men charged with conspiring to photograph
the senator's sickly wife, Mark Mayfield, vice chairman of the Mississippi Tea Party, committed suicide while awaiting a grand jury trial.

Cochran may have won, but everyone in the GOP was losing.

ON THE FIRST SUNDAY IN JUNE, CRUISING THROUGH THE SOUTHERN
Virginia countryside, Paul Ryan felt compelled to make a phone call.

Things were going remarkably well in the Republican primary season, and Ryan, riding back from a weekend retreat with major party

donors, was relieved. He had always considered himself a movement conservative, but the purist outlook put forth by the Tea Party, crystallized in the opposition to his budget deal, had forced him to reconsider his affiliations.

Once encouraged by the anti-establishment uprising, believing the GOP needed an injection of fresh blood and daring ideas, Ryan had come to view it as a threat to conservatism. With the Tea Party having thus far failed to score a major victory in 2014, GOP leaders were feeling emboldened. Some even went so far as to suggest—as Boehner did to donors that spring, and as Ryan had done that Sunday afternoon—that immigration reform would be back on the table following the primary season.

"Hey, I'm seeing *Cantor* signs everywhere," Ryan told his friend, the majority leader, over the Bluetooth speaker. "You're going to be fine on Tuesday. And once you're through your primary, we're going to get to work on immigration."

Cantor agreed that they would. His primary election was two days away, but the result seemed like a foregone conclusion. Having compiled eight figures in his campaign accounts over the past two years, Cantor faced token opposition in the form of Dave Brat, a small-school economics professor who had raised roughly $200,000 for his bid to take down the next Speaker of the House.

Boehner had made nothing public; nor had he confirmed his plans to Cantor. But it was increasingly clear that the top House Republican would retire at year's end, and that Cantor, the undisputed heir apparent, would become the first-ever Jewish Speaker of the House, fulfilling a lifelong dream and marking a momentous feat for his community.

Caught up in his historic career arc, Cantor spent much of 2014 doing member maintenance: tending to fragile relationships, comforting aggrieved parties, soothing frail egos. All the while, Brat was portraying him as a career politician, an elitist who'd fallen out of touch with the district and was dangerously soft on immigration. At first, Cantor responded with a shrug. His consultant Ray Allen continually reassured him, "As long as you're on the right side of two issues with the base, guns and abortion, you can't lose."

But the red flags were visible to anyone interested in seeing them.

Cantor's district had been redrawn to include more of the sprawling, rural, noncollege-educated areas outside Richmond. Many of the voters in these areas didn't know him, and many of those who *did* know Cantor didn't like him. Meanwhile, as the majority leader flew around the country headlining $10,000-per-plate dinners, Brat was running a dogged shoe-leather campaign in the district, canvassing neighborhoods and talking up constituents Cantor had long taken for granted.

By late April, Cantor decided that he wouldn't take any chances. His campaign went up with a negative ad picturing Brat's face on a cartoon body, depicting him as a "liberal college professor" who was shaky on tax increases. But it backfired: Brat, who had no money to run ads of his own, felt a sudden surge of publicity surrounding his campaign. His rallies grew in size. Local grassroots groups began volunteering on his behalf. Around that time, Cantor had a series of worrisome run-ins with the Tea Party. Activists rejected his attempt to install a longtime friend as a local GOP chairman. They blocked his slate of delegates from being seated at the district party's convention. And at the convention itself, some of them booed Cantor in front of his family, startling the congressman and stirring a sudden disquiet within his team.

The most vexing scene played out at the State Capitol thirteen days before the primary election. As Brat held a press conference accusing Cantor of peddling "amnesty," Luis Gutierrez, the House Democrats' most flamboyant immigration reform advocate, held a competing press conference nearby accusing Cantor of blocking the Gang of Eight bill.

The majority leader, it appeared, was a man without a country.

Gutierrez's visit to the district sparked suspicion in the conservative blogosphere that Cantor, who had traveled the country with the Illinois congressman talking up immigration reform, had orchestrated a bit of Kabuki theater aimed at shoring up his right flank, using Gutierrez as a foil to prove his own conservative bona fides.

This theory probably lent to Cantor's political team far more cunning than it deserved. Yet such an effort would be consistent with the congressman's lurch to the right in the campaign's final six weeks. He blocked an amendment from coming to the House floor (one he had previously supported) offering citizenship to illegal immigrants who

enlisted in the military. He ran advertisements that touted his harsh
dealings with Obama over the comprehensive reform. He distributed
flyers that credited him with thwarting a plan "to give illegal aliens
amnesty."[4] After spending the past year touting his plan to provide citizenship to illegal minors, Cantor's sudden shift in tone was whiplash-
inducing.

And not terribly convincing.

The headline on Breitbart.com said it all: "DAYS FROM PRIMARY,
ERIC CANTOR POSES AS ANTI-AMNESTY WARRIOR."

Breitbart's wall-to-wall coverage of the primary race in its final days
was not coincidental. The right-wing website had pummeled the Gang of
Eight proposal throughout 2013, seemingly hell-bent on dashing Marco
Rubio's future presidential prospects. Now another handmaiden of the
elite had strayed into the crosshairs, and Breitbart's executive chairman, a native Virginian named Steve Bannon, was insistent on his reporters flooding the zone with bruising coverage of Cantor during the
primary's home stretch.

Breitbart wasn't alone in hammering Cantor. Two of conservative
talk radio's finest flamethrowers, Mark Levin and Laura Ingraham,
used the majority leader as a political piñata throughout 2014. Ingraham was particularly harsh, even paying a visit to the district to headline a Brat rally less than a week before the primary.

"We are slowly losing our country," she told an overflow crowd, according to one of several stories published by a Breitbart reporter on
the scene. "Who do you think Barack Obama and Nancy Pelosi want to
win this primary? They want Eric Cantor to win because Eric Cantor
is an ally in the biggest fight that will occur in the next six months in
Washington . . . and that is the fight over immigration amnesty."[5]

When Brat took the stage, he added, "A vote for Eric Cantor is a vote
for open borders. A vote for Eric Cantor is a vote for amnesty. If your
neighbor votes for Eric Cantor, they're voting for amnesty."

Cantor's allies were rattled but hardly resigned. The campaign's
pollster, John McLaughlin, had conducted a survey in late May showing Cantor leading Brat by 34 points. The noise on the ground, and on
conservative talk radio, and in the right-wing corners of the internet,

was just that—noise. Cantor couldn't lose. Not to *Dave Brat*. So confident was the majority leader's team that on the morning of the election, rather than pounding the pavement in search of every last vote, Cantor was speaking at a fund-raiser in Washington—on behalf of a colleague.

It was symbolic of how the majority leader had prioritized his career. Staffers would later think back to a decade's worth of postponed or canceled meetings with constituents, almost always due to fund-raisers or lobbyist receptions or member-driven events. Cantor was a master of the inside game, collecting chits and building the brand he would need to secure the most powerful office in Congress. But it came at a price. As the returns came in on June 10, it was evident that residents of Virginia's Seventh District assigned greater importance to the title of representative than that of Speaker of the House.

For the biggest political upset modern Washington had ever seen, the results were anticlimactic. Brat won the primary by 11 points, beating Cantor even in his own home county. The victory was so lopsided that the majority leader's complaint of Democrats crossing over to defeat him, which turnout patterns showed had contributed to his demise, fell on deaf ears. Cantor didn't just lose; he got destroyed. And so, too, did the prospects for immigration reform.

"I don't think the split in the Republican Party is going to be made up with new Latino voters or new black voters or new Asian voters," Ingraham said on Fox News that night, savoring her victory lap. "In fact, I think somebody who runs on immigration reform—or amnesty or whatever you want to call it—in 2016 would probably do worse than Mitt Romney did in 2012."

THE SPEAKER'S CELL PHONE BUZZED. HIS CHIEF OF STAFF WAS CALLING, and Boehner, midway through a meal at Trattoria Alberto, his favorite Capitol Hill *ristorante*, nearly didn't answer. When he did, the news practically knocked him off his chair: Cantor had lost. "I was pissed," Boehner says. "Because in my mind, I was done."

In a private memo written in November 2013, titled "The End," Boehner's top staffers had presented him with three choices for retirement. Option one meant announcing in January 2014 his plan to leave at year's end. Option two meant announcing that plan in August. Option three meant announcing it in November, after the midterm elections. Boehner had ruled out option one, refusing to make himself a yearlong lame duck and invite further discord inside the House GOP. But he had never decided between options two and three. And now, with his replacement sidelined, it was not clear he could choose either.

Boehner hung up and dialed Ryan. Explaining that he had long planned to retire after 2014, and that Kevin McCarthy, the third-ranking House Republican, was nowhere near ready to become Speaker, he asked Ryan to replace Cantor as majority leader, effective immediately, and slide into the speakership in 2015.

"You've got to do this job," Boehner told him.

"There's no way I'm doing this job," Ryan replied. "You've got to stay.'"

Boehner dragged on a Camel Light and cursed Cantor's name into the summer night's air. For all the chariness between them early on, the Speaker and the majority leader had developed a solid working relationship and a level of trust that could be understood only inside the leadership's foxhole. Boehner was well prepared for Cantor to take over, confident that he was leaving the party—and more important, the institution—in capable hands. So much for that.

Then the Speaker received a call from an old tutor, Newt Gingrich.

"Hey, Boehner, try this on for size," Gingrich said. "What do you think happens if Cantor doesn't spend a dime on that race, doesn't mention his opponent's name once, just ignores him completely and pretends there's no primary at all?"

"I think he wins by forty points," Boehner replied. "That was the worst campaign ever run."

But the Speaker couldn't afford to dwell on the disappointment of Cantor's improbable defeat. The forces that had crushed his heir apparent threatened to swallow up Washington itself. Congress was consumed that summer by the issue of—what else?—illegal immigration, with a crisis unfolding at the southern border. In fiscal year 2014, a total of 68,541 unaccompanied children had been apprehended at the U.S.-Mexico border, a 77 percent increase over the previous year, according to Vox.[6] Conservatives accused Obama of providing a magnet for the illegal minors with his executive actions to provide amnesty; the White House blamed Republicans for railroading a bipartisan bill that would have secured the border.

If it wasn't obvious after the summer of 2013, it was after Cantor's loss: Immigration reform was dead. Sean Hannity had been right. It was a career killer.

A LEADERSHIP SHUFFLE IN THE REPUBLICAN RANKS THREATENED TO expose the intraparty schisms anew. Cantor announced that he would be leaving Congress early, resigning both his seat and his position below Boehner. That meant a special election to name a new majority leader and, if McCarthy succeeded in moving up, another special election to replace him.

McCarthy was a curious figure. Universally viewed as a pitiful whip, someone with neither the legislative guile nor the meat-grinder maliciousness required to steer the membership, he was also generally well liked, an easygoing Californian who was more a buddy than a boss. McCarthy also benefited from the same unspoken realization that buoyed Boehner and Cantor: The GOP leadership had been dealt an exceptionally tough hand after the 2010 election, charged with supervising a rowdy bunch of revolutionaries who refused to play by the customary rules.

One of those revolutionaries, Raúl Labrador, took exception to McCarthy's likely promotion. "What I found most objectionable was not Kevin, but the process: You're next in line and you get to move up without even being challenged," Labrador says. "It was everything that's wrong with Congress."

For all the beefs with GOP leadership, and the conservative qualms with McCarthy, nobody was stepping forward to thwart his coronation. So, Labrador took it upon himself. He waged a sacrificial lamb campaign, arguing for sweeping structural changes that would make Congress a bottom-up institution by empowering individual members to drive a wide-open policymaking process free of meddling from the party's leadership. His pleas fell on deaf ears. The truth was, most rank-and-file members of Congress had come to appreciate the heavy hand of leadership, recognizing the fine line between inclusivity and anarchy. Even among Labrador's fellow 2010 classmates, many had come around to view Boehner's iron fist as necessary—reductive, certainly, but effective in corralling an unruly conference.

Take Tom Graves, for example. Once the handpicked conservative to lead the Republican Study Committee, and a co-architect of the Defund Obamacare strategy, Graves had retreated from the front lines in 2014. He worried that the party was inflicting too much damage on itself in the name of ideological rigidity. Inside the room, as the House Republicans prepared to vote, Graves watched his former mentor Jim Jordan stand to nominate Labrador for majority leader. Then Graves, to the shock of conservatives in the crowd, rose to nominate McCarthy.

Labrador was disgusted. "I have never seen a person change so much over a period of time," he said of Graves. "He's totally different than he was when he first came here."

But Labrador had changed, too. "I'm no longer mad at the leadership. It's not their fault. It's really the membership that has failed, not the leadership," he says. "The membership wants leadership to exercise a strong hand because they want this game to continue. It protects them from making tough decisions. . . . It's much easier to go along and get along with leadership, to do what the special interest groups want you to do, because they're all going to give you money for your campaign and help you get reelected."

McCarthy won the internal election in a rout. And to replace him as majority whip, Republicans chose Steve Scalise, who, while alienating some of the more vocal conservatives with his chairmanship of the RSC, had endeared himself to a much broader swath of the conference.

Boehner had a new leadership team but an old set of problems. Obama was threatening further unilateral action on immigration and health care. The Speaker's members were demanding a more forceful response; Boehner obliged them in the form of a lawsuit against the president alleging abuses of executive power in implementing Obamacare, strategically filed one day after the administration expanded the DACA program to shield another four million illegal immigrants from deportation.

But the Speaker could not satiate the bloodthirst of his base. The conservative insurgency, kept at bay for much of the year, had regained its strength thanks to Cantor's defeat and Obama's brazen defiance of the coequal legislative branch. The time and money spent by Republican elites trouncing far-right challengers in 2014 would be an asterisk in future political science textbooks. Far from being tamed, the GOP's fratricidal tendencies had been further emboldened.

Jon Runyan, a former all-pro tackle in the National Football League, won a congressional seat in 2010 before abruptly quitting in 2014. Days before his departure, I asked him what the biggest difference was between playing in the NFL and serving in Congress as a Republican. "When you're on the football field, you only hit the guy wearing the other jersey," Runyan said. "Up here, the jerseys don't matter. You have no idea who's going to hit you."

THE PARTY WAS AT WAR WITH ITSELF, AND SO WAS THE COUNTRY.

In the summer of 2014, a pair of high-profile killings of unarmed black men by white policemen revived the national argument around race and equality—and predictably, fractured it along tribal boundaries.

The deaths of Eric Garner in Staten Island and Michael Brown in Ferguson, Missouri, just outside St. Louis, and the subsequent decisions not to indict the officers involved, set off a national furor. The fatal incidents became intertwined and practically synonymous. Politically, the proximity of timing in the Garner and Brown cases, and their conflation in the national subconscious, amounted to a choose-your-own-adventure experience for Americans already living in silos.

Liberals angered by generations of unchecked police misconduct in minority communities (enabled by systemic inequalities in the judicial system) saw white cops getting off scot-free for murdering unarmed black men.

Conservatives riled by an ethos of disrespect toward law enforcement (one perpetuated by popular culture, rap music in particular) saw black rioters ravaging their own communities in response to the killing of known criminals.

Many whites scoffed at the sight of black celebrities striking the "Hands up, don't shoot!" pose, considering that the slogan, inspired by

Brown's killing, was rooted in an account[7] that was proved false. (As the *Washington Post*'s Fact Checker determined, "'Hands up, don't shoot' did not happen in Ferguson."[8]) Many blacks, meanwhile, fumed at how their appropriation of such a symbol was construed through the lens of a single incident rather than a vast body of racial injustice.

A Pew Research poll released late in the year confirmed this experiential and identity-driven disconnect. A full 80 percent of black respondents said that the grand jury had erred in failing to indict the officer who shot Brown, and 90 percent said the same of the officer who killed Garner. Among white respondents, those numbers were 23 percent and 47 percent, respectively.[9]

Obama made a game attempt to split the baby, sensitive to the warring perceptions that he was either stoking racial divisions by empathizing with the struggles of the black community or turning his back on his heritage by touting the difficult, admirable work of law enforcement.

It did little good. A *Politico* poll taken in the aftermath of the Ferguson unrest found that just 6 percent of voters in battleground states and congressional districts thought race relations had improved under the first black president, while 46 percent said they had gotten worse.[10]

The Democratic Party was already bracing for a bruising midterm election. A polarizing president and a summer framed by racial and cultural friction weren't helping their cause.

EVEN BEFORE ELECTION DAY 2014, CONSERVATIVES HAD BEGUN LOOKing ahead to the 2016 presidential campaign.

It was an article of faith on the right that the past two general elections had been lost not because of Obama's popularity, but because Republicans had nominated moderate opponents who failed to mobilize the party's base. This was the result of divided loyalties: In 2008, conservatives split their votes between Mitt Romney and Mike Huckabee, allowing John McCain to win with plurality support; a similar dynamic played out in 2012, when Romney won the nomination thanks to prolonged divisions between Rick Santorum and Newt Gingrich. In truth, it seemed there was no pleasing the base; many conservatives had not been happy with the GOP's presidential nominee since 1984.

With 2016 approaching, and Jeb Bush making noise about raising unprecedented sums of campaign cash to clear the Republican primary field of potential challengers, leaders of the conservative movement agreed on an urgent priority: to coalesce behind a single candidate, as early as possible, to stand a chance of defeating the establishment.

Spearheading this effort was Tony Perkins. A former Marine, police officer, and state lawmaker in Louisiana, Perkins was best known as president of the Family Research Council. But his more covert, more consequential title was as president of the Council for National Policy. Founded in 1981 as a rallying point for politically active Christian conservatives, CNP had evolved into a forward operating base for the entire conservative movement, an umbrella organization that housed the leaders of the biggest national and state-based activist groups on the right. Meeting in private three times each year, CNP's invitation-only membership would host prominent guests to give lectures, legislative updates, and insights from Washington's smoke-filled rooms.

Perkins had long planned to use CNP to advance his pet cause. Setting out in 2013 to build a coalition of activist leaders who accepted his premise that defeating the establishment depended on the movement uniting behind one candidate before the primary season began, Perkins assembled a roster of heavyweights. His secretive sect, known simply as "The GROUP" on email chains, convened for the first time in New York City in August 2014. Previewing their objective, Perkins told his comrades that they would get a good look at their choices at the following month's CNP summit in Atlanta.

Sure enough, on back-to-back nights in September 2014, the titans of the conservative movement auditioned two leading men to be their champion in 2016: Cruz and Huckabee.

Both delivered dynamo performances (speeches, followed by Q&A sessions) and both had deep support in the room. Huckabee had spent the past two years listening to friends, many of them CNP members, telling him that he would have won the nomination, and the White House, had he run in 2012. Hearing their pleas for him to enter the 2016 race, Huckabee was laying the groundwork for his second presidential bid.

Cruz, meanwhile, had spent every waking moment since arriving in the Senate working methodically to forge alliances across the conservative movement—none more intimate than with Perkins himself. The senator, who lived a block from the posh Capital Grille in Washington, had a back-corner booth reserved almost every night of the week. He packed his dinner schedule with useful meetings: donors, think-tankers, lobbyists, House members, and activists. Perkins was his most frequent companion. Even though the CNP president had been one of the voices whispering in Huckabee's ear to run, he was beginning to fall for the new kid in class.

With all the bubbling excitement over the upcoming presidential season, the 2014 midterm elections went somewhat overlooked. And indeed, the results did not strike many Republicans as exceptionally consequential. Yes, Republicans expanded their majority in the House of Representatives. And yes, the GOP picked up 9 Senate seats and regained control of the upper chamber, making McConnell the new majority leader.

Yet the core realities of divided government remained. Republicans could now pass bills, including the repeal of the Affordable Care Act, through both houses of Congress and send them to the president's desk. But Obama still sat behind that desk. And there was no way to force his compliance with any of the GOP's legislative priorities.

As it turned out—shockingly—Republicans couldn't agree on what those priorities should be.

Conservative advocacy groups, for instance, pumped millions of dollars into Republican general election campaigns with the hope of abolishing the Export-Import Bank, a government agency that provides loan guarantees to foreign entities to ensure the purchase of U.S. goods. It was a niche cause that sought to make a symbolic point about the dangers of "crony capitalism." But GOP leaders scoffed at the argument after Election Day.

This infuriated right-wing activists and certain donors, particularly the Koch brothers, who spent more than $300 million on the 2014 election cycle and viewed the Export-Import fight as the least Republicans could do to bare their ideological fangs with total control of Congress.

As the intraparty recriminations began anew, and the familiar grumbling grew louder, Republicans quickly put 2014 in the rearview and began focusing on the wide-open presidential campaign ahead. After eight years of being told what they couldn't do because Obama occupied the Oval Office—and with the party establishment quickly rallying around its favorite dynastic son, Jeb Bush—conservatives viewed 2016 as their make-or-break year.

It was impossible to understand, at that time, how the GOP Senate takeover of 2014 would alter the trajectory of American politics forever.

CHAPTER NINE

JANUARY 2015

"I thought, 'Well, that's it. He's finished.'"

STUBBORN ISLANDS OF SNOW DOTTED THE LANDSCAPE OF DOWNTOWN Des Moines, a melting remnant of winter as balmy temperatures topping fifty degrees welcomed the presidential circus to Iowa.

Saturday, January 24, marked the unofficial kickoff to the 2016 Republican primary: Congressman Steve King had partnered with Citizens United to host the "Iowa Freedom Summit," a nine-hour buffet of rhetorical red meat flung to the caucus-going masses.

Ten future presidential candidates would deliver speeches at the cattle call. Marco Rubio was not among them. Just as Jeff Sessions had known better than to pose for the cameras with Rubio during the immigration fight, Rubio could not stomach being photographed behind a lectern plastered with the name of King, the federal government's most notorious race-baiter.

Another potential heavyweight in the GOP field, Jeb Bush, skipped the shindig for similar reasons. Bush was married to a Mexican immigrant and felt personally offended by the party's tone toward foreigners. Having gone further than perhaps any Republican in recent memory to destigmatize the issue—even saying that immigrants who entered U.S. illegally were committing "an act of love"[1] for their families—Bush would not have found a welcome audience at King's event.

There was no such dilemma for Donald Trump.

"We have to build a fence. And it's got to be a beauty. Who can build better than Trump?" he told a delighted audience, a steady blend of laughter and applause filling the auditorium. "We don't have the best coming in. We have people that are criminals, we have people that are crooks. You can certainly have terrorists. You can certainly have *Islamic* terrorists. You can have anything coming across the border. We don't do anything about it. So, I would say that if I run and if I win, I would certainly start by building a very, very powerful border."[2]

It was a preview of what would become vintage Trump. There were digs at the GOP's impotent governing class. ("Everything about Obamacare was a lie. It was a filthy lie. . . . And what are the Republican politicians doing about it?") There were strange boasts. ("Our president is either grossly incompetent, a word that more and more people are using—and I think I was the first to use it—or he has a completely different agenda than you want to know about.") There were paroxysms of populism. ("I'll probably be the only Republican that does not want to cut Social Security. . . . Get rid of the waste, get rid of the fraud, but you deserve your Social Security.")

And of course, there were unsolicited attacks. "It can't be Mitt because Mitt ran and failed," Trump said of the 2012 nominee, who was mulling a third campaign. "You can't have Romney. He choked."

As for the other establishment favorite, Trump cautioned, "The last thing we need is another Bush. . . . His brother gave us Obama. Abraham Lincoln coming home back from the dead could not have won the election because it was going so badly and the economy was just absolutely in shambles that last couple of months."

Trump's bombast was not off-putting in the least to his audience; it was precisely what they expected. He was a larger-than-life character, someone with whom Americans of all ages had become familiar after decades of his manipulating the media-entertainment complex. At any political venue, in any state, even his best-known rivals needed to introduce themselves—if not by name, then by deed. Trump faced no such barrier to entry. Even though *The Apprentice* was declining in viewership, its early seasons had been a blockbuster breakthrough, reestablishing Trump's household name and bolstering his image as a successful executive. He was universally recognized and, increasingly on the right, seen as a kindred spirit, his rants against political correctness resonating more with each passing day.

To conservatives, the nation's self-portrait was becoming unrecognizable. Having only just lost the battle over same-sex marriage, merely their latest defeat in the broader culture war, they were fighting on new terrain: Transgenderism, the *T* in "LGBT," although it had been broadened to "LGBTQ," the final letter standing for "Queer" or "Questioning." (The question gnawing at conservatives was how they'd allowed the left to annex one-fifth of the alphabet in pursuit of social justice.)

Anyone attempting to wish away the issue couldn't do so for long. In June 2015, the month Trump would launch his presidential campaign, *Vanity Fair* revealed its forthcoming magazine cover featuring the Cold War–era Olympic hero formerly known as Bruce Jenner. Having elected to transition and live as a woman, Jenner, once an exemplar of American masculinity, appeared on the cover wearing a revealing corset. The headline: "Call Me Caitlyn."

Many conservatives said they were less irked by anyone's sex change per se than by the rapidly evolving set of rules imposed by polite society. Not only would it be wholly unacceptable to question Jenner's new look, but the pronoun police stood ready to detain anyone using *he* instead of *she*, even accidentally. More bothersome, throughout 2014 and 2015, numerous state legislatures were debating bills relating to transgender bathroom usage, including in K-12 public schools, an issue most Americans could not have imagined reading about just a few years before.

Trump was better equipped than anyone to tap into this unease. Even as a cosmopolitan moderate on social matters, he possessed an innate understanding of the cultural disquiet gripping middle America and proved remarkably effective at exploiting it. By the time he finished in Des Moines, speaking to a crowd clad in flannel jackets and John Deere caps, the billionaire businessman had earned a standing ovation. "I know what needs to be done to make America great again. We can make this country great again," he declared. "And I am seriously thinking of running for president because I can do the job."

Not that anyone took this proposition seriously.

"The chances of him running for president are roughly equal to the chances that Earth will be overrun by Ewoks by Memorial Day," Mark Sappenfield wrote in the *Christian Science Monitor* after Trump's speech. "He was there for microphone and the money shots of his legendary hair."[3]

"Donald Trump is doing his tease with the public and media," Will Rogers, the Republican chairman of Polk County, the state's largest, told McClatchy at the event.[4]

The skeptics were on solid ground. As BuzzFeed reporter McKay Coppins had written in a 2014 piece, "36 Hours on the Fake Campaign Trail with Donald Trump," the real estate tycoon had been giving presidential head fakes for a quarter century, using the specter of a campaign to keep his name and company brand in the news with no intention of ever following through.

Trump had revved up the old routine in time for 2016, visiting some of the early nominating states and making noise about an actual, all-joking-aside campaign. But nobody was buying it. For most of those in attendance, me included, Trump belonged in the same category as Sarah Palin: an "entertainer," as King said when introducing the reality television personality.

In fact, this was an insult to the future president. Whereas Trump actually spoke of policy, however fleetingly and unintelligibly, the former Alaska governor delivered a speech that was incoherent bordering on clinically insane.

"GOP leaders, by the way, y'know the man can only ride ya when your back is bent," Palin said. "So strengthen it. Then the man can't ride ya. America won't be taken for a ride."[5]

At another point, Palin remarked, "When will they let us control our own care? When will they do to stop causing our pain, and start feeling it again? Well, in other words, um . . . is Hillary a new Democrat or an old one? Now, the press asks, the press asks, 'Can anyone stop Hillary?' Again, this is to forego a conclusion, right, it's to scare us off, to convince that—a pantsuit can crush patriots?"

It was an appalling display from the person whom John McCain had proposed to place one heartbeat away from the presidency. With much of the national media and GOP consultant class assembled in Des Moines for the event, Palin's crackpot appearance became the talk of the town. Even her staunchest defenders on the right wondered aloud whether something had gone wrong, seriously wrong, for the vice-presidential nominee turned carnival barker.

CELEBRITY ENTERTAINMENT NOTWITHSTANDING, THE STARS OF THE day were Scott Walker and Ted Cruz.

Walker, the Wisconsin governor who had crushed the state's unions (only to survive a recall election in 2012 and then win a full second term in 2014) put to rest doubts about his political skills. He reminded the Republican faithful of a salient fact: While the national spotlight fixated on the GOP's dysfunction in Washington, its governors and state legislatures were acting as highly effective and fiercely partisan laboratories of democracy, churning out tax cuts, balanced budgets, deregulated state economies, and expanded school-choice programs.

(Left unmentioned were the party's more distressing ignominies out in the provinces. This included Michigan governor Rick Snyder's negligent management of Flint, a once-thriving city placed under state supervision after decades of industrial rot doomed the economy, only for its drinking water to be inadvertently poisoned as the result of a cost-cutting maneuver.)

Walker gave a spirited and commanding talk, weaving mention of his Iowa upbringing into a tale of his clashes with the left in Wisconsin and the blueprint they provided for reclaiming the White House in 2016. Once viewed by the national political class as too bland to be considered a serious contender, with a single speech Walker had vaulted into top-tier status.

And then there was Cruz. Nobody arrived in Iowa with higher expectations. Not only had his campaign-in-waiting been working the state hard, recruiting volunteers and hunting for high-profile endorsements, but Cruz's celebrity had soared during his brief time in Congress. Two years into his freshmen term, the Texas senator had replaced Jim DeMint as the right's favorite street fighter in Washington, a man whose take-no-prisoners approach made him a hero to the grass roots nationwide. And Cruz had a unique Iowa advantage: Steve King.

He had worked the congressman hard from the day he arrived on Capitol Hill: long dinners, spontaneous coffees, countless bottles of red wine, even a pheasant-hunting expedition. ("Both of us popped our guns to our shoulders and shot simultaneously as if it were one bang," King recounted afterward. "That pheasant folded in a cloud of feathers.") King couldn't announce his intentions just yet; remaining neutral would allow him to influence the race and build suspense, making his eventual endorsement all the more impactful. But the congressman was all-in on Cruz. Not since Ronald Reagan, King would whisper to his friends and allies, had conservatives seen someone of this talent.

Cruz was already operating under a strategic theory of the race: It would boil down to an establishment favorite and a conservative challenger. By merging two distinct lanes of the Republican electorate, evangelical Christians and Tea Party populists, behind his candidacy, he could emerge as the consensus anti-establishment candidate.

Previewing this pursuit, Cruz built his introductory speech to Iowans around paeans to his spirituality (at least a dozen of them) and denunciations of empty rhetoric from his fellow Republicans. "Talk is cheap," Cruz warned the crowd. "The Word tells us, 'You will know them by their fruit.'"

One conservative favorite and possible 2016 contender wasn't in attendance: Mike Pence.

Like Rubio, he wasn't keen to share a stage with King. Pence had, while in Congress, compiled a record of unquestioned conservatism on almost every issue—except that of immigration. Along with his dear friend Jeff Flake, Pence had pushed for a path to citizenship for illegal immigrants. He acknowledged the principled opposition from within his ideological and partisan tribes; what he could not suffer was the nakedly nativist instincts of some on the right who called themselves Christians while showing no compassion for some of the most vulnerable among us.

Watching the debate unfold over the Gang of Eight's bill in 2013, Pence felt fortunate to be hundreds of miles away. Congress was ugly, messy, perforated with career potholes. The governorship offered a cooler environment for doing the people's business, and a cleaner path to the presidency. The problem was, Pence had struggled to find his footing in Indianapolis. He enjoyed the perks of being home, attending frequent NASCAR events and becoming friendly with Colts quarterback Andrew Luck over chats at their shared downtown barbershop. But after more than a decade in Congress, his 2013 reimmersion into Hoosier politics had been rocky.

Mitch Daniels, Pence's predecessor, had been arguably the most effective governor in the country. Adjusting to a new job was hard; securing policy wins that would distinguish him from Daniels and raise his profile ahead of a possible presidential run was even harder. Pence cut taxes to the extent possible; Daniels had already slashed them to historic lows. He invested heavily in K–12 education. And he worked out a compromise with the Obama administration to accept additional Medicaid funding, under the Affordable Care Act, on the condition that his plan include some conservative strictures.

Yet these were not parade-inducing feats in the eyes of the GOP base. In fact, during a trip to Washington, Pence was frog-marched into the Heritage Foundation, where movement leaders demanded an explanation for why he had accepted Obamacare money at all. Meanwhile, Pence's proposed formation of a state-run media service, JustIN, which would use taxpayer money to hire reporters and editors to publish news articles about government deeds, was met with so much ructious mockery that the idea was quickly scrapped.

The governor very much wanted to run for president. His longtime pollster and adviser, Kellyanne Conway, stood ready to move her family to Indianapolis to help guide his campaign. But he needed a signature win. So, in early 2015, when Indiana Republicans began pushing legislation aimed at protecting religious liberties, an issue of urgency to the evangelical wing of the GOP base, Pence saw it as a no-brainer.

The month after King's shindig in Iowa, as the Indiana State Senate began considering the Religious Freedom Restoration Act, Pence joined a rally at the statehouse in support. The bill was rushed through the legislature and arrived on his desk by late March. Despite impassioned objections from Democrats, who claimed it would permit discrimination against gay Hoosiers, the governor signed it.

All hell broke loose. Facing a sudden national uproar and an all-out revolt from the state's business community, Pence agreed that the legislation needed fixing. But the changes that he agreed to didn't satisfy either side: Liberals thought he hadn't gone far enough to address the state-sanctioned discrimination, and conservatives thought the governor had caved on a fight of immense cultural and political importance.

His closest allies marveled at how badly he'd mishandled the crisis. Pence had always been shrewd as a tactician and velvety smooth as a messenger, sidestepping the land mines that claimed so many politicians' careers. But the religious liberty fight had been an unmitigated fiasco. Having kept a low profile for the first two years of his governorship, Pence found that his maiden venture into the national news ahead of a potential White House campaign in 2016 couldn't have been more damaging. He looked weak, lubberly, indecisive, unprepared.

He could forget about running for president. He would have a tough enough time running for reelection.

JIM BRIDENSTINE HAD NO PATIENCE FOR TOUGH TALK. TWO YEARS EARlier, in January 2013, he had stumbled upon a Keystone Cops mutiny of House Republicans pledging to remove John Boehner as the Speaker of the House. But nearly half of them got cold feet. It convinced Bridenstine and his fellow freshman agitator, Thomas Massie, that private assurances were no longer enough. If the rebels were going to oust Boehner, they told one another in 2014, the mutiny would have to come out of the shadows. By pledging their opposition to the Speaker in tweets or press releases or Facebook posts, the conspirators could be held accountable.

Whatever traction this idea gained early in 2014, as House conservatives continued to sulk over their defeats the previous fall, vanished in the wake of Cantor's stunning primary defeat. In one sense, Dave Brat's win could have opened the revolutionary floodgates, prompting a total overhaul of the House GOP leadership. But that prospect was too daunting even for the unruliest renegades. They still didn't trust Boehner, but with no obvious alternative, they weren't prepared to thrust the House into chaos—at least, not yet.

It was two days after Christmas Day 2014 when Massie, a curly-haired, MIT-educated inventor and robotics engineer who comported himself with a quirky impishness, pulled into a McDonald's drive-through. He was stopping to buy breakfast for his sons on their way to a gun range. A sign hung on the talk box: "Next Speaker Please." Massie tweeted a picture with an implicit reference to Boehner.[6] It gained little attention at first—there had been no talk in recent months of any organized attempt to overthrow the Speaker—but a little while later, Massie's phone chirped. "So," Bridenstine asked him, "Are we gonna do this?"

With the Speaker's election just ten days away, Massie and Bridenstine flew into recruitment mode. They made the criteria clear: Unlike with the 2013 rebellion, this time anyone joining had to publicly declare his or her opposition to Boehner. The pledges started trickling in—a few and then a dozen, with the figure suddenly climbing toward twenty. There was another difference this time around: After consulting with the House Parliamentarian, Massie and Bridenstine decided they would put other members' names into nomination for the Speakership. This was highly unusual; typically only the two party leaders are nominated at the beginning of the Speaker vote, and anyone dissenting is free to vote for whomever else they identify. By nominating Republicans not named "Boehner," the rebels would rob weak-kneed conservatives of the excuse they had given to their constituents in 2013: that nobody was running against Boehner, so they had to back him.

Louie Gohmert volunteered his services to Massie and Bridenstine, arguing that he could represent the symbolic alternative. Not wanting to hurt the overeager Texan's feelings, but not wanting someone of his temperament as the face of the anti-Boehner movement, they promised to nominate Gohmert on the floor only if other Republicans stepped forward as well. This prompted Ted Yoho to volunteer—which didn't exactly solve the problem. Finally, Massie and Bridenstine convinced Daniel Webster, a respected former Speaker of the Florida House, to allow himself to be nominated, lending fresh legitimacy to the cause.

The January 2015 rebellion was most notable for whom it did not include: Raúl Labrador and Mick Mulvaney, two chief organizers of the 2013 effort, as well as Jim Jordan, the recognized leader of the House conservatives. At the beginning of the roll call vote, the most bullish expectations were that 20 members might oppose Boehner. But even as the nays surpassed that, Labrador and Mulvaney and Jordan didn't budge.

"You're not organized enough. It's too late," Labrador told Massie. The Kentucky lawmaker was incensed. Other nervous conservatives, he believed, were holding back because Jordan and his friends were.

Ultimately, it was another politician's death that might have resurrected Boehner's career: More than a dozen Democrats were out of
town attending the funeral of former New York governor Mario Cuomo,
lowering the total number of votes cast—thereby also lowering Boehner's threshold for reaching a majority. Whereas the rebels would have
needed 29 votes to force a second ballot in a fully populated House, that
number now stretched high into the 30s. When the gavel fell, Boehner,
watching on his office television while huffing cigarettes at a pace his
friends had never seen, survived 25 defections to remain Speaker of the
House.

After four years of making life miserable for Boehner, some of the
House's most problematic members had decided to lay off—and in doing
so, they incurred the fury of their constituents back home. Jordan, Labrador, and Mulvaney were inundated with angry phone calls and emails;
Labrador alone received more than seven thousand negative comments
on his Facebook page, he told friends. The animus they had stirred toward the Speaker, which was turbo-charged through the filters of talk
radio and social media, had come back to haunt them.

Luckily, they had a plan in the works that would satiate their voters'
bloodthirst and put Boehner back in the crosshairs.

Months earlier, in a final attempt to reclaim control of the Republican Study Committee, with vows to rewrite its rules and restore its sedi-
tious reputation, Mulvaney had run for chairman. But he was defeated.
Once again, the RSC's bloated numbers had worked against the "hard-
core" base, and once again, the leadership had played a role, whipping
support for Bill Flores, a Texan and former oil-and-gas executive.

It was a breaking point for the conservatives. Justin Amash's group,
the House Liberty Caucus, had served its purpose as a temporary bunker. But now they needed a new outfit—one committed to a certain ideology, yes, but even more so dedicated to tactics that would make them
enemies of the Republican state. The group would need bylaws codify-
ing their strategy of strength in numbers. They would need 29 members,
enough so that if they voted as a bloc, they could defeat the leadership on
any given vote—whether on a "rule" that dictated floor procedures or on
the legislation itself.

Weeks after the Speaker election, as House Republicans gathered in Hershey, Pennsylvania, for their annual retreat, a group of nine conservatives put the finishing touches on their new vehicle. All it needed was a name. After debating a host of dreadful suggestions, they settled on House Freedom Caucus because, as Mulvaney later told the *New Yorker*, "It was so generic and so universally awful that we had no reason to be against it."

Another name they jokingly considered was the Reasonable Nutjob Caucus. It was good for some laughs; members such as Mulvaney and Labrador had long defended themselves as more pragmatic than the party's leaders gave them credit for. But the name also carried an implication: Not just anyone would be allowed in. The architects of the cabal, Jordan, Mulvaney, Labrador, Amash, and Mark Meadows, didn't want the group defined by some of the louder and less thoughtful Republicans in their conference. That meant, at least initially, no Massie, no Michele Bachmann, no Steve King, and no matter how many times he asked, no Louie Gohmert.

"They felt the conservatives needed a sensible effort—not a Thomas Massie/Louie Gohmert effort," Massie says. "They told each other, 'We're not gonna let the crazy ones in.'"

Massie had big plans of his own. A few days after the failed coup against Boehner, he hosted an academic from the Congressional Research Service in his office. He wanted to know about an obscure parliamentary maneuver known as "the motion to vacate the chair."

BUSH WAS THE FIRST HORSE OUT OF THE GATE—SORT OF.

In December 2014, the consubstantial son and brother, respectively, of the last two Republican presidents announced the formation of Right to Rise PAC, which would serve as an exploratory committee and fundraising vehicle for his own White House run. Campaign finance laws forbade coordination between candidates and their affiliated PACs; by withholding his official candidacy, Bush was able to work in concert with his new super PAC to raise unlimited sums of money from the country's biggest donors. It was a post–*Citizens United* loophole that no presidential candidate had ever exploited, and Bush took full advantage, raising $100 million in a period of six months.[7]

It was a breathtaking amount of coin to throw at someone who had yet to shake a hand or kiss a baby. Bush's team took to dubbing their financial conquest "shock and awe," a preemptive show of force meant to clear the primary field of potential foes. (Unfortunately, given its more recent applications, the term foreshadowed Bush's woeful quagmire of a campaign.)

The strategy worked at first: Romney, who had weighed a third campaign, saw much of his donor base defecting to Bush and announced that he would stay on the sidelines.

But not everyone was so deterred. In fact, dollar signs notwithstanding, there wasn't much to be daunted by. Bush had been an imposing figure in Florida, widely viewed as one of the most ruthlessly effective governors in America and a paragon of conservative policymaking. But he had left office nearly a decade ago—with his brother still in office, social media in its infancy, and the Tea Party's emergence still several years off. The game had changed. There were always going to be concerns about fatigue with the family brand—hence "Jeb!" as his logo—but the more existential predicament for Bush was communicating with a GOP electorate that had been speaking a different language since he left office.

Nobody understood this better than Rubio. The onetime Florida lawmaker had learned at Bush's knee in Tallahassee, and the governor had helped him ascend to the most powerful office in the statehouse. When he became Speaker, Rubio was gifted a large, golden sword (that of Chang, "a great conservative warrior") by Bush, who choked up in the House chamber during Rubio's swearing-in ceremony, "I can't think back on a time where I've ever been prouder to be a Republican, Marco."[8]

Despite those ties, Rubio saw Bush's blind spots—his support for Common Core education standards, his moderation on certain social issues, his support for immigration reform that made his own efforts look tame by comparison—and knew that his old mentor would struggle to connect with the contemporary Republican base.

Bush never saw him coming. Having locked up virtually all of Florida's major donors and political colossi, not to mention having helped Rubio win his Senate race four years earlier, Bush spent the early months of 2015 dismissing speculation of a challenge from his apprentice. "Listen," he told a group of Florida Republicans during a meeting in Washington, just after the New Year, "I really believe in my heart that Marco will not run against me."

It was a fundamental miscalculation—of the climate, of the party, and of Rubio himself. If the 2010 Senate campaign had taught Rubio anything, it was that old rules no longer applied. He had embarrassed Charlie Crist despite being told to wait his turn. Now friends in Florida were telling him the same thing. Rubio wasn't hearing it. The senator believed himself to be a figure of Obamaesque proportions, someone uniquely suited to a new era of American politics, one where experience mattered less than raw talent.

"I loved watching Michael Jordan play basketball, because he could do things with the basketball that were not teachable," Whit Ayres, Rubio's highly regarded pollster, told a group of reporters two weeks before his client announced for president. "Marco Rubio is the Michael Jordan of American politics."[9]

On April 13, inside the Freedom Tower in Miami, the Cuban American equivalent of Ellis Island, Rubio launched his candidacy by throwing thinly veiled haymakers at both Bush and Hillary Clinton, who had formally entered the race one day earlier.

"Just yesterday, a leader from yesterday, began a campaign for president by promising to take us back to yesterday," Rubio said, savoring the punch line as his crowd booed. "Yesterday is over. And we are never going back." Cheers filled the building. "We Americans are proud of our history, but our country has always been about the future. Before us now is the opportunity to author the greatest chapter yet in the amazing story of America. But we can't do that by going back to the leaders and ideas of the past. We must change the decisions we are making by changing the people who are making them."

It was the third launch, in as many weeks, by the Senate's trifecta of talented freshmen.

On April 7, Kentucky's Rand Paul had announced his own campaign in Louisville, standing on the shoulders of his father's efforts in 2008 and 2012. Promising to break up the stale intellectual duopoly in Washington, the younger Paul was less doctrinaire and more calculating than his dad while peddling a comparably nonconformist platform. Central to it was a renunciation of the GOP's muscular foreign policy and a pledge to restrict America's military adventurism abroad. This had been appealing to a war-weary nation (and party) in the years after Bush left office: With 9/11 fading in the rearview mirror, two wars dragging on in the Middle East, and the appetite for intervention continuing to diminish even among majorities of Republican voters, Paul had reason to feel confident that his candidacy would meet the moment.

In July 2014, the RealClearPolitics average of national surveys showed Paul atop the Republican field.[10] That same month, NBC News polls showed him leading in New Hampshire and tied for first place in Iowa.[11] His presidential stock peaked with an August 2014 *New York Times Magazine* feature headlined, "Has the 'Libertarian Moment' Finally Arrived?"

It had indeed—and it departed just as quickly. In the weeks after that piece was published, the Islamic State, or ISIS, which had announced the formation of a caliphate to govern Muslims worldwide, released videos depicting the beheading of two American journalists. With the spectacular savagery piercing Western consciousness—the executioner was dubbed "Jihadi John" by media outlets—Obama delivered a prime-time address in September pledging to "destroy" ISIS.

Time magazine featured Paul on its cover the next month, naming him "The Most Interesting Man in Politics." Intended to capture his rise, the story instead marked the onset of his decline. Paul had already dropped to 12 percent in the RCP national poll average, from 14 percent in July; by Christmas, he was at 9 percent. The crash continued throughout 2015, interrupted by only a fleeting bounce after his April campaign launch. In late July, he was below 6 percent, and by October, one year after *Time*'s cover, he hovered at just over 2 percent. Once considered a front-runner for the GOP nomination, Paul was a nonfactor before the first votes were cast.

"Two people were Senator Paul's undoing in the presidential race," says Chip Englander, his campaign manager. "Donald Trump and Jihadi John."

CRUZ WAS THE FIRST OF THE 2016 REPUBLICANS TO FORMALLY DECLARE his candidacy. Having spent the previous six months recruiting what would become regarded as the sharpest, most data-savvy team in the Republican field, he and his top lieutenants were stumped on the question of where to launch the campaign. Iowa was too obvious, campaign manager Jeff Roe warned on a final conference call. They discussed other options: Cruz's hometown of Houston; historical sites around Texas; even the Reagan Library in California. As the call dragged on, one of Roe's employees at Axiom Strategies in Missouri sent his boss an email: "You should do it at Liberty."

The line lit up with opinions. Liberty University, founded in Lynchburg, Virginia, by the late fundamentalist preacher Jerry Falwell in 1971, required students to attend convocation. This would provide Cruz a built-in audience of enthusiastic young people as the backdrop for his big announcement. But there was risk involved in launching at a university so associated with the Moral Majority. Some staffers argued that it would narrow Cruz's appeal, backing him into a corner and forcing him to go all-in on Iowa.

But Iowa was going to be a must-win anyway, and evangelicals were going to be his base. The campaign's only choice was to embrace these realities. Weeks later, once more pacing the stage in his best Joel Osteen impersonation, Cruz declared on the campus of Liberty, "God has blessed America from the very beginning of this nation, and I believe God isn't done with America yet."

In short order, a small constellation of super PACs was set up in support of his candidacy, and promptly brought in tens of millions of dollars, more than any Republican save for Bush.[12] This was a shock to the Republican system. Cruz was the disruptor in chief on Capitol Hill, a reputation he clung to in campaigning for the presidency. In the summer of 2015, the Texas senator called McConnell a liar on the floor of the Senate, a spectacular breach of etiquette (even though McConnell had, in fact, misled Cruz and other senators about the reauthorization of the Export-Import Bank). In the not-distant past, a freshman would have been cast out into political purgatory for such an assault on the majority leader. Instead, Cruz was raising more money for his White House bid than his team knew what to do with.

If all went according to plan, Cruz believed, he would meet Bush in the middle of the bracket, each of them having advanced through the preliminary rounds of the Republican tournament.

But in the summer of 2015, only one candidate was shocking and aweing the primary competition—and it wasn't Bush.

ON, JUNE 15, INSIDE AN AUDITORIUM AT MIAMI DADE COLLEGE, BUSH jumped into the race by downplaying his dynastic connections—"It's nobody's turn," he declared—and reinforcing his image as someone who could appeal to a cross-section of the electorate, just as the RNC's autopsy had prescribed. "As a candidate, I intend to let everyone hear my message, including the many who can express their love of country in a different language," Bush declared to what was, for a Republican event, a strikingly diverse audience.

And then, he added in his fluent *español*, "Ayúdenos en tener una campaña que les da la bienvenida. Trabajen con nosotros por los valores que compartimos y para un gran futuro que es nuestro para construir para nosotros y nuestros hijos."

Translation: "Help us to have a campaign that welcomes you. Work with us for the values we share and for a great future that is ours to build for us and our children."

It wasn't clear, however, which values Bush was referring to—or how widely they were shared in the new Republican Party.

The next afternoon, thirteen hundred miles to the north, Trump descended into the lobby of his Fifth Avenue skyscraper on a gilded escalator, Neil Young's "Rockin' in the Free World" blaring in the background. The atmosphere could not have been more dissimilar to that at Bush's rally. (Some of the supporters wearing "TRUMP" shirts were paid tourists brought in off the streets of Manhattan.) Also somewhat different was the material offered by the candidate.

He made fun of Obama for playing so much golf. He noted how the recent Republicans to launch campaigns had "sweated like dogs" during their events, questioning how such people could defeat ISIS. He mocked Secretary of State John Kerry for breaking his leg in a bicycle accident.

The speech was not without substance. "I will build a great, great wall on our southern border," Trump declared, "and I'll have Mexico pay for that wall."

It was the type of audacious promise that no elected official would dream of making. But Trump turned it into a staple of his campaign. Unlike most Republican politicians who took hardline positions on immigration as a matter of economic policy, arguing that the cheap labor depressed American wages, Trump primarily framed the influx of people as a threat to the nation's security—and its identity.

"The U.S. has become a dumping ground for everybody else's problems," Trump argued. "When Mexico sends its people, they're not sending their best. They're not sending you. They're not sending you. They're sending people that have lots of problems, and they're bringing those problems with us. They're bringing drugs. They're bringing crime. They're rapists. And some, I assume, are good people."[13]

His presidential run was three minutes old, and already Trump was undermining the last three years' worth of Republican outreach to the Hispanic community.

He didn't think much of the RNC autopsy. "Big waste of money," he scoffs. "I didn't waste a lot of time reading it." Instead, he was doing a different type of outreach.

"Remember the first time you went to an ATM to withdraw cash and it asked whether you wanted it in English or Spanish? You had no idea how many Americans were outraged," Boehner says. "People want to look at people who look like them. They want to live with people who talk like them. When Trump's making all this noise, you can see who it was appealing to."

In the ensuing weeks, the newly declared candidate would draw tens of thousands of people to his campaign stops. He promised to turn the country around. He served up paeans to its past glories. He offered himself as the vanguard of a movement. "The silent majority is back!"

Trump told an uproarious Phoenix crowd in July. "We're going to take our country back!"[14]

To many on the American right, this was the din of deliverance. Although the economy was continuing its steady reclamation, with unemployment dropping to 5.3 percent in June 2015, feelings of cultural unrest were growing more viscerally tangible. The European refugee crisis, which saw tens of thousands of Middle Easterners fleeing war-ravaged nations and spilling over the porous European borders, put conservatives on high alert. Obama's stated goal of admitting ten thousand Syrian refugees into the United States was met with fierce resistance from House Republicans, fueled in part by a series of Islamic terrorist attacks in France.

Conspicuously, there was no comparable sense of urgency around that time in response to teenager Freddie Gray dying at the hands of the Baltimore police department after suffering a nearly severed spinal cord in the back of a police van; or to Walter Scott being shot in the back by a South Carolina policeman who was caught on video planting a weapon on Scott's body.

Ten weeks after Scott's murder, a twenty-one-year-old white man named Dylann Roof walked into the Emanuel AME Church in Charleston and joined its black parishioners for a Bible study. After nearly an hour with the group, Roof reached into his backpack, pulled out a .45-caliber pistol, and executed nine people, including the church's pastor, state senator Clementa Pinckney. ("He was a giant of a man and a prince of a fella," recalls his friend, Senator Tim Scott, who still saves the final text message he received from Pinckney.)

South Carolina governor Nikki Haley promptly called for, and signed a bill mandating, the Confederate flag's removal from the statehouse grounds. National Republicans lauded Haley's leadership, but a scab had been opened over America's oldest wounds: A CNN poll taken before the flag's removal found 75 percent of southern whites described it as "a symbol of pride" while just 18 percent called it "a symbol of racism."[15]

The summer of conservative discontent climaxed on June 26, 2015, when the Supreme Court ruled 5–4 in the case of *Obergefell v. Hodges* to legalize same-sex marriage nationwide. Public opinion had shifted with such neck-breaking haste that the verdict was no surprise: In 2004, 60 percent of Americans opposed same-sex marriage, with just 31 percent approving, according to Pew. By 2014, the numbers were 52 percent approving and 40 percent opposed, an astonishing net swing of 41 points in the span of a decade.[16]

The ruling lent a deeper conviction to the right's sentiment of being under siege—from culture, from a changing country, and from government itself.

MARK MEADOWS HAD NEVER QUITE FIT THE MOLD OF A FREEDOM CAUcus radical. Unfailingly polite and winsome, with the faintest trace of a sweet-tea accent and his hand always on someone's shoulder, the North Carolina congressman was as threatening as a sweater-clad kitten.

He was conservative, sure, but nobody's idea of a firebrand. When word leaked to the GOP leadership that Meadows had been involved in the plotting against Boehner in 2013—even though he ultimately did not oppose him—the brand-new lawmaker requested a meeting with the Speaker. "He's on the couch, sitting across from me in my chair, and suddenly he slides off the couch, down onto his knees, and puts his hands together in front of his chest," Boehner recalls. He says, 'Mr. Speaker, will you please forgive me?'"

Boehner's chief of staff, Mike Sommers, who witnessed the encounter, said it was "the strangest behavior I had ever seen in Congress."

The Speaker, for his part, chalked it up to a "nervous new member who wanted to be liked." Boehner told Meadows not to worry, that they were moving on.

For a few months, House Republicans recall, Meadows kept his head down, doing little to draw attention to himself. But then came his letter that summer calling for the defunding of Obamacare. After opposing the Ryan-Murray budget compromise and joining the crew of restive conservatives who broke away from the Republican Study Committee the following year, Meadows decided to vote against Boehner's reelection as speaker in 2015.

"And then he sends me the most gracious note you'll ever read, saying what an admirable job I've done as Speaker," Boehner says, shaking his head in bewilderment. "I just figured he's a schizophrenic."

There were other indications that Meadows was less politically genteel than he let on. In early 2015, Mark Walker, a former minister and a new Republican in the North Carolina delegation, was pulled aside by Meadows on the House floor. "You voted the wrong way," Meadows told him. Walker was confused. On the bill before them, he had sided with Heritage Action, the default decision for an ambitious conservative. And yet Meadows, a Freedom Caucus cofounder, was voting the other way. After some persuasion, Walker changed his vote, believing Meadows knew something he didn't. Instead, as it turned out, Meadows simply didn't want his new colleague earning a better grade on the annual Heritage scorecard, a bit of subterfuge that did lasting damage to the relationship between the two lawmakers.

Meadows wasn't done making moves. After helping to establish the House Freedom Caucus, he and other conservatives threatened to block funding for the Department of Homeland Security unless GOP leaders defunded Obama's programs protecting certain classes of illegal immigrants from deportation. Their stance was not unreasonable: Boehner had promised a forceful response if the White House continued making immigration law from the executive branch. "When you play with matches, you take the risk of burning yourself," Boehner warned Obama during a press conference after the midterm elections. "He's going to burn himself if he continues to go down this path."[17]

But months later, Boehner urged his conference to avoid a fight over DHS funding. When the conservatives cried foul and vowed to go their own way, the American Action Network, a powerful outside group staffed by Boehner loyalists, began targeting the likes of Meadows and Mulvaney with ads in their congressional districts accusing them of being "willing to put our security at risk by jeopardizing critical security funding."[18]

The Freedom Caucus went ballistic. Members began railing against big, bad General Boehner firing at his own foot soldiers. Their outrage was somewhat amusing: House conservatives had used Boehner as a punching bag for the past four years, taking shots at the Speaker for sport, yet were incensed whenever his leadership allies dared return fire.

Meadows made a statement of protest by refusing to pay his dues to the National Republican Congressional Committee. He also began whispering to friends about orchestrating a new, and operationally different, coup against Boehner. Rather than wait for the official Speaker vote at the beginning of the next Congress, Meadows explained, they could use a parliamentary device to force a vote on Boehner whenever they wanted. The tactic, known as vacating the chair, was rarely used and unfamiliar to many of the conservatives. But it was simultaneously being studied by Massie, the Freedom Caucus outcast who had been researching the procedure for months. As spring turned to summer, Meadows and Massie began comparing notes.

In Congress, the smallest legislative splashes can create the biggest waves. Such was the case in June, when Boehner moved to hold a vote on Trade Promotion Authority, or TPA, granting the president leeway to negotiate trade deals that would come to Congress for up-or-down approval. The issue was not thought to be controversial. Conservatives were free-traders, after all. Even Ted Cruz, whose right-wing antennae were better tuned than those of any Republican alive, had written a joint op-ed with Paul Ryan in April calling for speedy passage of TPA.

But the House rebels wouldn't fall in line: 34 of them voted to block a rule allowing for a vote on the legislation, forcing the leadership to rely on Democratic votes for passage—and prompting Boehner to strip Meadows of his subcommittee chairmanship.

A long view of the policy issue shows that it marked an inflection point in the GOP's relationship with trade. Trump had spent the last two years explicitly threatening to slap tariffs on China and Mexico and other commercial partners, something unheard of from any presidential hopeful in recent memory, much less a Republican. It moved the needle: When TPA came to the Senate for approval, Cruz voted no, shocking everyone in the chamber, including his closest allies.

The backstory was simple enough: Cruz's campaign had been poll-testing the issue and realized that he might suffer crippling blowback from blue-collar voters if he continued to support the agreement. This would be a far steeper price to pay, he decided, than a few weeks of headlines about flip-flopping and the inevitable furious phone call from Ryan, who wondered how Cruz could hang him out to dry on the issue. "It wasn't good for me, it wasn't good for the party," Ryan recalls. "But it was good for Ted."

The Freedom Caucus's defiance of Boehner, and the Speaker's retaliation, represented a point of no return. Boehner tried to keep one step ahead of the conservatives. In July, for example, he called for a congressional investigation into Planned Parenthood after undercover videos showed a top-ranking official with the organization talking callously about the supply of aborted baby parts. (The videos implied that Planned Parenthood was illegally profiting from selling them; this was not the case, though the group's leadership apologized for the tone taken by their employee.) Even in this instance, Boehner failed to capture the mood of his right flank: Conservatives wanted to promote the videos by showing them in House hearings, but the Speaker held back, worried that they would be sensationalizing the issue, especially given the legal challenges contending that the videos had been shot illegally.

The relationship between the leadership and the conference's right wing was no longer salvageable. The only question was what conservatives planned to do about it. Meadows wanted to go after Boehner immediately, forcing a vote that summer on the Speaker's future. Having dissected the parliamentary maneuver with Massie, each lawmaker had prepared a separate draft to file with the House clerk. They both had also sent copies to the outside groups by mid-July, needing air cover in the event of a guerrilla-style attack on the warlord himself.

But the Freedom Caucus leadership was opposed. Calling an emergency meeting of the board, Jordan, the group's chairman, told Meadows that the timing wasn't right, that there was no organized opposition, that no alternative speaker had stepped forward, that the Freedom Caucus would suffer irreparable damage to its credibility by lunging at Boehner but failing to take him down. Labrador agreed. So did Mulvaney and Amash and the other board members present. Meadows was on an island. "We didn't think it was the right timing," Labrador said. "And we were trying to give Boehner an opportunity to change."

But Meadows would not budge. Whipping out his cell phone, he played a voice mail for the group. It had been left that morning by his son, Blake, who told his father how proud he was of him for standing on conviction, and quoted Theodore Roosevelt: "The credit belongs to the man who is actually in the arena . . . if he fails, at least he fails while daring greatly, so that his place shall never be with those cold and timid souls who neither know victory nor defeat."

Meadows's comrades were puzzled. They had never known him to act on rash emotion. But there was no stopping him. He was determined to file paperwork with the House clerk that would amount to an attempt on Boehner's political life. What some of them didn't realize was that Massie was standing at the ready to file his own version of the motion if Meadows didn't follow through. That was a chance Meadows couldn't take.

Of course, what *nobody* knew—outside of Boehner's three most trusted staffers—was that the Speaker had already settled on his exit date. After the midterm elections, Boehner would announce his retirement on his birthday, November 17.

He never got that chance.

On the afternoon of July 28—*Meadows's* birthday, and the last day before Congress adjourned for its August recess—the North Carolina congressman strolled up to the House clerk and handed over a piece of paper. The document claimed that Boehner had endeavored to "consolidate power and centralize decision-making," while "diminishing the voice of the American People" and using his office to "punish Members who vote according to their conscience instead of the will of the Speaker."[19]

As Jordan and Labrador pulled Meadows aside, "raking him over the coals," according to Massie, the Kentucky congressman leapt in with his congratulations. Contrary to what Jordan and Labrador thought, Massie and Meadows believed the timing of this ambush was perfect.

"The leadership's job was to keep 218 frogs in a wheelbarrow, but we were going into the August recess, and the wheelbarrow would be unguarded," Massie explains. "Jim and Raúl were telling Mark, 'Don't put this on the Freedom Caucus.' They didn't want to be responsible. But now they take credit for it."

Boehner's allies were out for blood. Ryan raced to the Speaker's office, where he was joined by several like-minded members, all of them imploring Boehner to call up the motion and hold a vote immediately, that same day. It would be a show of strength, a middle finger to the Freedom Caucus, putting the right-wing absolutists in their place once and for all. But Boehner waved them off. He wanted to think. None of them realized that he was planning to leave in less than four months anyway. "All these Republicans were going to get crap at home for supporting me, only to have me leave soon after that," Boehner recalls.

Calling a private meeting the next day with some of the Freedom Caucus and Jedi Council members, Boehner and Jordan agreed that it made no sense to hold the vote before August, since the Republicans who supported him would spend the month getting pummeled in their districts. "He's like, 'I don't want to make members take that vote,'" Ryan recalls Boehner saying. "Totally selfless. Always thinking about protecting the membership."

By going it alone, disregarding the wishes of his closest friends and confidants, Meadows had plunged the House of Representatives into turmoil and earned himself more media coverage than any second-term congressman in recent memory.

"I don't like being in the limelight," he told a mob of reporters awaiting him off the House floor.

THE FOLLOWING WEEK, THOUSANDS OF PEOPLE PACKED INTO THE Quicken Loans Arena in Cleveland for quite possibly the most anticipated presidential debate in modern American history. Reince Priebus, the chairman of the Republican National Committee, was thrilled to show off the historically diverse presidential field. No longer could the GOP be mocked as the party of old white men. There were two Cuban Americans, sons of immigrants both. There was a woman who'd run a Fortune 500 company. There was a black man. And there was an Indian American (although Bobby Jindal would reject that label, having made the rallying cry of his campaign a rejection of "hyphenated Americans").

When Fox News anchor Bret Baier opened the debate with a gimmick question, asking the candidates to raise their hand if they would *not* commit to supporting the eventual nominee of the party, Trump shot his arm into the air. "If I'm the nominee, I will pledge I will not run as an independent," he said, drawing a robust round of boos from the Republican faithful.

Suddenly, Priebus was reminded of his nightmare scenario. Ever since Romney's loss to Obama, he had labored to get the Republican Party out of its own way—not just on policy, but on process. The 2012 primary had stretched on nearly five months and featured upwards of twenty debates and forums, an atmosphere of anarchy that took a brutal toll on the party's general election readiness. Priebus had effected sweeping changes to the primary structure, most notably a condensed nominating calendar and half the number of debates. It was all in the service of producing a quality nominee as quickly as possible with minimal intraparty damage done.

And then along came Trump.

Priebus had initially laughed off the billionaire playboy's candidacy. When young staffers at the RNC approached the chairman with concerns earlier that summer, Priebus rolled his eyes. "Donald Trump is never going to be our nominee." But there was nothing funny about the situation now. Priebus still didn't think Trump was going to win, but he was increasingly fretful of the damage he could do. The candidate's insulting of immigrants, for instance, was negating hard-won public relations victories for the national party.

In fairness, Trump was an equal-opportunity offender. A month into his campaign, he mocked John McCain for being shot down over Vietnam. "He's not a war hero. He's a war hero because he was captured," Trump said during a forum in Iowa. "I like people who weren't captured." (Trump took five draft deferments, including one for bone spurs in his heel, and later boasted of spending those years avoiding sexually transmitted diseases, calling it "my personal Vietnam."[20]) The POW comment was met with such anger from within the political class, including a rare on-record rebuke from Priebus, that Trump appeared mortally wounded.

"When that happened, I thought, 'Well, that's it. He's finished,'"

Mitt Romney recalls of Trump's remark. "I thought his campaign was over." Romney pauses, then shakes his head: "There were *several times* I thought his campaign was over."

Instead, while his candidacy took shape in the summer and fall of 2015, Trump proved astonishingly resilient. As he made an art of scandalizing the Republican establishment with his ad hominem vilifications and general affronts to decency, Trump's poll numbers climbed steadily upward, the traditional consequences for such behavior nowhere to be found.

Meanwhile, for all the concerns about Trump sullying the GOP's brand from within, Priebus worried more about the prospect of Trump destroying the party from without: If he ran as an independent, he would siphon away millions of conservative votes and hand Hillary Clinton the presidency. It would be Ross Perot all over again. Trump had already been making such noise, rightly suspecting that the GOP's top officials might plot to sabotage him. Now, not sixty seconds into the first televised debate, with a record-shattering audience of twenty-four million viewers, he had threatened to leave the Republican Party.

As the debate progressed, a different thought occurred to Republicans in Cleveland. Trump might not leave the GOP; Trump might *take over* the GOP.

The booing of his very first response aside, the crowd seemed to be delighting in Trump—his shameless bragging, his sparring with the moderators, his unrepentant earthiness. He could do no wrong. When Megyn Kelly, then the reigning queen of Fox News, pressed Trump on his calling women "fat pigs, dogs, slobs, and disgusting animals," and asked him about his history of degrading them with sexual innuendos, Trump interrupted, "Only Rosie O'Donnell." The debate hall shook with laughter. A minute later, he added, "I think the big problem this country has is being politically correct." The audience cheered.

After the debate, Trump went on CNN and insinuated that Kelly's hostile questioning was due to her menstrual cycle. "You could see there was blood coming out of her eyes, blood coming out of her—*wherever.*" As with the McCain controversy, Trump didn't flinch in the face of criticism. And as with the McCain controversy, he didn't suffer from it one bit. The Republican front-runner was a lot of things, but he wasn't apologetic. And the American political industry loved him for it.

"Reince made some great reforms to the party and to the nominating process," says Hugh Hewitt, the conservative radio host, who co-moderated several of the GOP debates. "But what he didn't know was [that] of the seventeen candidates, one of them was an honest-to-goodness television star, who knew how to turn the primary into a reality TV show."

After the first debate, the U.S. Chamber of Commerce held a donor retreat in which Scott Reed, its senior political strategist, presented an impossible data set. Having crunched the statistics on social media to determine which of the GOP candidates was getting the most attention, Reed's group of data geeks determined that 82 percent of the online conversation revolved around Trump. The remainder amounted to digital bread crumbs for a few of his starving opponents.

This was unnerving to party leaders. Trump had spent the opening months of his campaign mired in one firestorm after another, the severity of which would have been career-crushing for any other politician. Yet his poll numbers kept rising. Registering at just 6.5 percent in the RealClearPolitics national average one month before the debate, Trump had soared to 24.3 percent by the time Fox News hosted the clash in Cleveland. And nothing—certainly not his vulgar insult of Kelly—appeared capable of bending his trajectory downward.

With a sprawling, historically large field of seventeen Republican contenders, Priebus and Fox News had agreed to limit the prime-time debate to the ten top-polling candidates, relegating the other seven to a "kids' table" debate earlier in the day. Trump hated the even number of participants—"Because that meant two people were at the center," he says—and personally lobbied Roger Ailes to change the number of people onstage to either nine or eleven.

It was one favor that Fox News couldn't do for him. When the survey averages were tabulated, Trump was in the pole position and Bush was positioned beside him, at center stage. They were the tallest, the best known, the highest-polling. But the similarities ended there.

Trump was made for these moments, having spent decades mastering camera angles and production quality, distorting his expressions and gestures for maximum dramatic impact. He was having the time of his life.

Bush was not. Awkward and reticent, with his six-foot-four frame coiling into itself due to poor posture, the former governor was already sore about having to compete with the Judas known as Rubio. Now he was forced to endure the indignity of sharing top billing with a man who had spent the last year mocking his family.

Trump could read the repulsion on his rival's face. At one commercial break, he turned to Bush. "Jeb, how you doing?" he asked.

"I'm fine, Donald."

"So, where are you going after this?"

"Headed to New York for some fund-raising events tomorrow."

Trump beamed. "You want a ride? I've got my plane here. We're heading back tonight."

Bush stared blankly. "No. I'm good. We've got a ride."

"You sure?"

Bush nodded briskly.

"Okay. Let me know if you change your mind."

Trump, feet still positioned perfectly over his stage mark for the television cameras, turned toward his family in the front row and winked. It was a down payment on the space he would occupy inside Bush's head for the duration of the campaign.

CHAPTER TEN

SEPTEMBER 2015

"We fed the beast that ate us."

For John Boehner and his team, there had been a million lo-gistical hurdles to clear in preparing for the visit from Pope Francis: coordinating security logistics, arranging meetings by protocol, allot-ting space on the Capitol lawn for spectators, securing tickets for rela-tives and friends, including John Calipari, the University of Kentucky men's basketball coach. Reared in Catholic pews, instructed in Catholic schools, guided by his Catholic faith, the Speaker had spent twenty-five years daydreaming of bringing the pope to address a joint session of Congress. It was finally happening.

The Speaker also had a personal wish. He had asked if the supreme pontiff would baptize his one-year-old grandson. Vatican officials had gently denied the request, citing limitations in their scheduling.

Yet now, standing just a few feet away inside the Speaker's suite, with Boehner's family assembled in front of him, Pope Francis was smil-ing down at the baby and asking his assistant for some water. Boeh-ner began to sob uncontrollably. He looked to make sure a camera was trained toward the baby. He could already see the photo resting on the mantelpiece in his Ohio living room. Pope Francis received the glass of water . . . and then tipped it back, drinking every last drop. Boy, the Speaker thought to himself, you really *are* a Boner.

But nothing, not even the Holy Father's unwitting head fake, could ruin this day for Boehner.

In an institution thriving on cynicism and spite, Francis's visit of-fered an oasis. His speech, delivered in heavily accented English, had included ideological catnip for both parties. "We, the people of this con-tinent, are not fearful of foreigners, because most of us were once for-eigners," Francis said, drawing booming applause from the Democrats.

He then earned a rousing ovation from the Republicans: "The Golden Rule also reminds us of our responsibility to protect and defend human life at every stage of its development."

Still, partisanship did not rule the day. Boehner and Pelosi beamed at each other during the speech. Lawmakers of warring tribes snapped photos with one another and their families. Catholics of every gender, ethnicity, sexual orientation, and political affiliation gathered on the lawn outside Boehner's balcony to catch a glimpse of Pope Francis and share with him a moment of prayer.

"I never saw members happier than they were the day the pope was there. Democrats, Republicans, House, Senate—everybody was *happy*," Boehner recalls of that sunny day in late September. "I looked up and thought, you know, it's not going to get any better than this."

After the pope's caravan had departed the Capitol, Boehner called Jim Jordan and three other Freedom Caucus members into his office. The Speaker had not yet brought up a vote on Mark Meadows's motion to vacate the chair. He saw no reason to—not with his planned retirement in November. But now, with the sweetness of the day strumming his reflections on life and legacy, Boehner wanted to know: Did the rebels *really* still want a vote to throw him out?

"We tried talking Mark out of it before the recess. We didn't think the timing was appropriate," Jordan explained. "But after everyone went home and got an earful about it, now everyone's all fired up. There are more people ready to vote against you now than there were before."

Boehner nodded and let out a chuckle. He knew Jordan was telling the truth: With the outside groups, blogs, and talk radio shows firing on all anti-establishment cylinders since the moment Meadows filed his motion, August had not been kind to any Boehner-allied House Republicans. A swelling group of members would feel obligated to vote against the Speaker; it would represent a choice between saving their career or saving his. Boehner understood.

What Jordan and the others didn't know was that his job was never going to be in jeopardy. Earlier in the month, Boehner's chief of staff, Mike Sommers, had written him a memo, titled "Save the Institution," explaining that his survival would be ensured if Pelosi had Democratic members vote "present" when the motion came up, lowering his threshold of needed support. If Democrats cooperated, Boehner could win with a simple majority of Republican votes cast, which was never in doubt. The Speaker had broached the idea with Pelosi, and she agreed.

"You can't have thirty people in your caucus decide they're going to vacate the chair," Pelosi says. "He knew I had—not his back, but the institution's back."

And yet this scheme never sat well with Boehner. He wanted to leave on his own terms, not hang around for a few extra months on the strength of Democratic votes, an outcome that would only exacerbate the party's internecine tensions.

The night of the pope's address, the Speaker hosted dozens of friends and family members for a wine-soaked celebratory dinner at Trattoria Alberto, on Capitol Hill. Boehner then withdrew to his nearby apartment and told his wife, Debbie, that he was thinking of announcing his retirement in the morning. "And then I went to bed and slept eight hours. Like a baby. It was unbelievable."

After his customary breakfast at Pete's Diner, the greasiest spoon on Capitol Hill, the next morning, "I looked at that statue of the Virgin Mary next to St. Peter's Church and I decided, All right, today's the day," Boehner recalls.

He was at peace. Strolling into the morning's House GOP conference meeting, the Speaker told his members that he would retire from Congress at the end of October, leaving a month for Republicans to choose his successor. Jaws hit the floor. Boehner had tipped off McCarthy just moments before stepping to the microphone—"Get ready," he told him, grinning, "I'm out of here"—and the majority leader now wore the look of a defendant unexpectedly sentenced to death.

But one thing nagged at Boehner: the perception that the Freedom Caucus, and particularly Jordan, whom he calls "a legislative terrorist," and Meadows, whom he considered "Jordan's puppet" and "a perfect fucking idiot," had forced him out of Congress. So, he decided, when addressing an overflow press conference just after informing colleagues of his decision, to walk into the room singing, "Zip-a-dee-doo-dah, zip-a-dee-day, my oh my, what a wonderful day."

Boehner was not forced out—at least, not in any technical or parliamentary sense. And there is no question he felt unburdened by the decision finally to throw in the towel. But the singing routine masked the hurt he felt at the circumstances of his departure. "He was just kind of emotionally done," recalls Anne Bradbury, Boehner's floor director. "The fact that he felt like he'd given and given to the conference and the country, and this is how he was rewarded, when he didn't want to be there anyway—" She stops herself. "It was very disheartening for him."

The president was stunned at the news. When the Speaker called him that morning, Obama pleaded with him to stay on the job. "Boehner, you can't do this, man," the president said.

Boehner told him that there was no turning back.

"I'm gonna miss you," Obama said.

"Mr. President," Boehner replied, "yes, you are."

Conservatives were less gracious. That morning, in a speech to the Values Voter Summit in Washington, an activist confab organized by the Family Research Council, Marco Rubio broke the news of Boehner's retirement during his speech. The crowd leapt to its feet, roaring with a brutishness straight from the ancient Roman Colosseum.

Seated at a conference table in Coral Gables, Florida, that same morning, Jeb Bush saw yet the latest sign of an anti-establishment revolution—hardly the sort of environment conducive to his winning the presidency. But that wasn't his first reaction to Boehner's decision. Glancing across the table at Eric Cantor, the co-chairman of his Virginia campaign, Bush remarked that in a normal world, Cantor would be preparing to assume the Speakership.

As if he needed reminding.

TRUMP HAD EVERY REASON TO RAISE HIS HAND IN CLEVELAND.

The truth was, from the moment he stepped onto his golden escalator in the middle of June, Republican Party leaders had privately plotted against him. Senior members of Congress, governors, major donors, influential lobbyists, and many top conservative activists—all of them wanted to take Trump out. Nowhere was the cunning more concentrated than inside the Republican National Committee.

Having spent the past three years working to fortify the GOP's electoral vulnerabilities and safeguard it from another humiliating November defeat, Reince Priebus's members urged him to move swiftly to distance the party from Trump. Their trepidation was understandable: Trump was not a Republican. He had held positions for decades that ran counter to party orthodoxy: pro-choice, antiwar, pro-universal health care. He had also donated considerable sums of money to Democratic politicians, an apostasy that would have spelled doom for most any aspiring Republican in most any race, all the way down to city council.

But getting rid of Trump wasn't so simple. Parties are inherently open entities; there were no formalized rules governing the qualities, characteristics, or policy stances required to run for office as a Republican or a Democrat. No practical mechanism existed for shunning Trump from the GOP or forbidding him from identifying as a member thereof. Certainly, there were ways to stack the deck against him. But what made this problematic was Trump's knowledge of the plotting going on inside the party—and his repeated public warnings that he would run as an independent in 2016 and bury the Republicans' chances if, in fact, he felt they were sabotaging him.

All this put Priebus in an impossible position. He didn't want Trump to be the Republican Party's nominee for president. But he was convinced it could never happen anyway—"Not in a million years," he told friends—and saw no wisdom in alienating someone with the pop celebrity of Kim Kardashian and the political etiquette of Joseph McCarthy.

Two of Priebus's lieutenants, chief of staff Katie Walsh and communications director Sean Spicer, were appalled by Trump and urged the chairman to undermine his candidacy before it gained more steam. A number of senior RNC members pushed the idea of banning Trump from the debates altogether, with the justification that he had no history of identifying with the party. Spicer, a longtime GOP flack who was running point on the debate arrangements with the TV networks, endorsed this argument. But Priebus refused. What kept the chairman up at night wasn't the prospect of Trump winning the primary, but of him demolishing the party—in 2016 and perhaps for years to come—with an independent bid.

After months of delicate discussions with party elders across the country, Priebus settled on the safest solution he could think of: a loyalty pledge. All the candidates running for president would be asked to sign their names to a piece of paper stating their promise to support the eventual Republican nominee. Several state parties already required presidential candidates to sign a similar document when filing paperwork to qualify for the ballot,[1] which provided Priebus with plausible deniability that he was singling out Trump.

He was, of course, and everyone knew it. But instead of backing Trump into a corner, the move increased his leverage exponentially. Here he was, a first-time candidate for office, and the RNC chairman was improvising new guidelines in the hope of mitigating the risk he posed to the party.

Trump could hardly believe it. His running as an independent was as realistic as his promise to make Mexico pay for a border wall. Assembling the manpower to clear procedural hurdles and qualify for enough state ballots to be relevant to the general election would cost upward of $10 million. And time was of the essence; deadlines loomed early the following year, while many states had "sore loser" laws that banned a losing primary candidate from running as an independent. Trump was aware of these restrictions, having been briefed on the nightmarish logistics of running a third-party race, and he had zero intention of actually following through.

But Priebus couldn't take any chances.

Spooked by the opening act of the Cleveland debate, the chairman spent the next several weeks back-channeling with Trump about his willingness to sign a loyalty pledge. The document, Priebus told him, was as much about protecting Trump as protecting the party: Many of his Republican rivals would be irate if Trump won the nomination, Priebus pointed out, but they would be duty-bound to support him by signing their names.

Trump, who had earned a fortune negotiating deals far more intricate than this, strung Priebus along, savoring the sight of the chairman squirming in angst over the indecision of the party's unwelcome guest. At last, with the second GOP debate approaching in California, Trump told Priebus he would sign the pledge, on one condition. Priebus needed to come to New York. Trump was too busy, he told the chairman, to come to Washington.

Priebus was thrilled. His allies were not. It struck them as a power play. Trump believed he had cornered the GOP, and now he wanted to prove it to the political world. "Don't do this, Reince," Matt Moore, the South Carolina GOP chairman, told his friend. (South Carolina's loyalty pledge was a template for the RNC's certificate.) "You're the party boss. Make *him* come to DC and sign it. Then go out to the cameras and declare victory."

But Priebus did not share this concern. It was a distinction without a difference, he argued. Whether the pledge was signed on Fifth Avenue or on Capitol Hill, all that mattered was Trump agreeing not to run as an independent.

Priebus raced to New York on September 3 and watched as Trump autographed a document with the RNC's insignia and handed it back. The candidate then shooed Priebus out a back door of Trump Tower and went down to address the media by himself. "The RNC has treated me with great respect," Trump told a packed press conference.[2]

The irony was nothing short of sublime. For the past several years, conservative malcontents in the activist and media classes had branded anyone they disagreed with a RINO, Republican in Name Only. Now they were falling for a presidential candidate who had spent decades as a Democrat, who had donated generously to liberal causes, who had hosted Bill and Hillary Clinton at his wedding, and whose *only* connection to the *Republican* Party was his *name* on a piece of paper.

As Trump danced in the end zone and Priebus sipped a frosty Miller Lite on his train ride back to Washington, I spoke to John Ryder, the RNC's general counsel, to ask whether the pledge was legally binding. "Uhhh, legally binding?" he responded. "No. No. I think it's politically binding."

Ryder wouldn't elaborate on what he meant by that. But two things were obvious. First, the pledge wasn't worth the paper it was printed on. Second, and more consequentially, Trump had outmaneuvered the

Republican Party. By drumming up a month's worth of reality TV–style suspense over his empty threat to flee the GOP, he had starved his opponents of oxygen in the press, elevated his own brand above that of the party's, and scared the RNC chairman into making accommodations that no candidate could rightly expect.

TRUMP WAS STILL SMILING LATER THAT MONTH WHEN THE REPUBLIcan candidates met for their second debate, at the Ronald Reagan Presidential Library in Simi Valley. Once again, Trump's deficiencies and offenses were manifest. And once again, they did nothing to damage his standing. His poll numbers continued to climb after an evening spent dominating the cameras and the conversation. He called Rand Paul ugly. He called Carly Fiorina beautiful—by way of defending himself for having previously called her ugly. He called Jeb Bush "low energy." And, as would become custom, he refused to back down. Having recently suggested that Bush would have had different immigration policies had he not married a Mexican woman, Trump scoffed at Bush's attempt to force an apology.

It was a clarifying moment. In the run-up to the debate, Bush's senior advisers, worried that the candidate's Charmin-soft caricature was killing him, wondered aloud whether Bush should physically confront Trump; not necessarily punching the bully in the nose, but intimating a threat with his body language, mustering outrage over Trump's insult of his wife. Instead, Bush's approach had all the confidence of a puny kid whose father—or older brother—had just trained him to throw his first punch. (Trump would later recall his surprise at how Bush had barely raised his voice over the affront to his wife.)

After his two feeble attempts to force an apology failed—"I won't do that, because I did nothing wrong," Trump said, adding that he'd heard "phenomenal things" about Columba Bush—the scion of America's premier political dynasty turned almost helplessly to the cameras and framed the discussion in more transcendent terms.

"We're at a crossroads right now: Are we going to take the Reagan approach? The hopeful, optimistic approach?" Bush asked the audience.

"Or the Donald Trump approach? The approach that says that every-thing is bad, that everything is coming to an end?"

THIS QUESTION ANSWERED ITSELF. THE COUNTRY WAS NOT FEELING terribly hopeful or optimistic, and truth be told, the sour mood owed as much to one candidate's demonizing as to another candidate's ser-monizing. The reason Trump was able to get away with calling his rivals ugly, with insulting prisoners of war, with belittling women and using vulgar language, was that Americans, particularly conservatives, were becoming numb to the outrage culture.

On that very same night, just after the California debate concluded, the season premiere of *South Park* harvested this zeitgeist with flawless hilarity. The animated show, which follows the lives of a group of foul-mouthed elementary school kids, opened its nineteenth season with the introduction of a new villain, "PC Principal."

Militant and overbearing, with a puffed-out chest and a brimming list of grievances, PC Principal bullies the children who possess any-thing other than fully enlightened views of the world. In the first epi-sode,[3] PC Principal punishes anyone in South Park who dares describe Caitlyn Jenner as anything less than "stunning and brave."

Recruiting a like-minded army of young, white social-justice war-riors, PC Principal sets out to reeducate South Park. What the show cap-tured brilliantly was how the paroxysm of virtue-signaling had choked our capacity for engaging those with whom we disagree; how the fear of offending had diminished our ability to talk honestly and laugh openly. (Months earlier, Jerry Seinfeld made headlines by announcing that he no longer performed stand-up comedy shows on college campuses be-cause of the students' sensitivities.)

Against this backdrop, Trump's talent for afflicting offense, and his aversion to apologizing, made him a demigod to portions of the popu-lation.

What *South Park* fans might have missed was the show's subtler crit-icism of what had yielded the social-justice mentality in the first place: institutional racism and economic inequality, compassionless individu-ality and consequence-free bigotry. Indeed, the show lampoons the supercilious nature of the left and the reactionary nature of the right with equal effectiveness: In the season's second episode, as a teacher at South Park Elementary launches a presidential campaign based on building a wall to keep out Canadian immigrants, PC Principal forces the school's faculty to take "Canadian-language" classes to better serve their vulnerable migrant population.

The political guile of Trump was in reducing these nuanced and necessarily complex debates to their lowest common denominator. Taking the blanket complaint of "political correctness" and weaponizing it, he discovered that there was everything to gain from challenging the pearl-clutching ethos of the progressive base—even when he went too far.

That fall, Trump surprised exactly no one by affirming his support for the nickname of Washington's professional football team. For the past several years, the left had waged an unrelenting assault on the name "Redskins," calling it insensitive to Native Americans. In the spring of 2014, some months after Obama used his bully pulpit to call for the name to be changed,[4] fifty senators (none of them Republican) sent a letter to NFL commissioner Roger Goodell urging him to take action.[5] They called the team's name "a racial slur" and asserted that "Indian Country has spoken clearly on this issue."

In fact, to the extent that Native Americans had weighed in, a body of polling, research, and interviews suggested that most of them fell somewhere between indifferent and supportive. The previous fall, when the Redskins hosted a group of Navajo code talkers at a home game, honoring their World War II service with a ceremony on the field, the group's vice president, Roy Hawthorne, told the Associated Press, "My opinion is that's a name that not only the team should keep, but that's a name that's American."[6]

As with all things Trump-related, however, this proved to be a slippery slope. Emboldened by being on the winning side of this issue, the GOP front-runner saw no downside to pushing the envelope.

Within a few months, he was targeting Elizabeth Warren, the Massachusetts senator who, despite her obvious whiteness, had claimed Native American identity in her rise through academia. Trump settled on a sobriquet for Warren, one that might have been mildly offensive to the right had it not elicited such disproportionate wrath from the left: "Pocahontas."

PAUL RYAN WAS HOLED UP IN HIS OFFICE ON A CRISP THURSDAY MORN-ing in early October, tapping final revisions into a document on his computer, when the phone rang. It was the majority leader.

"Hey, just finishing up the speech," Ryan told Kevin McCarthy. In a few hours, Republicans were expected to choose McCarthy, in an internal vote, to succeed Boehner as Speaker of the House. Ryan had volunteered to deliver remarks and formally place his friend's name into nomination. "What's up?"

Around that same time, Boehner pounded the gavel inside the mostly empty lower chamber, opening the House for business. Then he walked off the podium, through a swinging door, and into the Speaker's lobby, a hallway decorated with monarchic portraits of his predecessors. Waiting for him, whispering furiously, were his chief of staff and McCarthy's chief of staff. "Uh-oh," Boehner responded. "He doesn't have the votes?"

It was a surprise only in the sense that McCarthy was running unopposed. Yet no one could claim to be shocked at his sudden collapse: McCarthy was hardly an inspired choice for the most powerful position in Congress.

A gregarious Californian with an easy laugh and a perfectly coifed swatch of silver hair, he enjoyed strong personal bonds across the Republican Conference. And he was universally respected as an electoral sage with a mental Rolodex of districts, voting histories, and demographic trends. But serving as Speaker of the House requires more than relationships and political knowledge; the job demands intuition and temperance, unwaveringly sound judgment and coolness under fire. McCarthy's possession of these attributes was shaky at best.

Just recently, he had boasted on Fox News that the House GOP's probe into the Benghazi attacks had damaged Hillary Clinton's presidential prospects—after two years of Republicans denying any partisan motivation behind the committee's work. It was precisely this sort of unmoored loudmouthery McCarthy's associates worried about. And it wasn't their only concern.

For several years, rumors had percolated inside the House about an extramarital adventure involving McCarthy and a colleague, Renee Ellmers of North Carolina.[7] McCarthy denied the affair, as did Ellmers, though somewhat less vigorously. Their colleagues weren't sure what to believe. "I never bought it. I thought she was nutty," Boehner says. "She had this fixation on Kevin."

But McCarthy's biggest problem was the Freedom Caucus. After more than four years of living under the thumb of Boehner, conservatives weren't going to robotically promote the next in line. To win their support, they told McCarthy, they would need concessions—ideally, a seat at the leadership table for one of their own.

This was an impossible ask. Jordan was loathed by much of the House GOP for his seek-and destroy tactics; he would never receive the votes for majority leader or even majority whip. Meadows was despised for his treatment of Boehner. None of their Freedom Caucus comrades was well known or well liked enough to stand a chance, either.

Hoping for an unlikely assist, McCarthy placed a call to Ted Cruz, explaining his untenable position with House conservatives and wondering if the Tea Party favorite might weigh in on his behalf or at least not do something to derail his unsteady candidacy. Cruz vowed neutrality, nothing more or less. It was a window into the dizzying, upside-down world of GOP politics: The man poised to become Speaker of the House believed his fate could be determined by a freshman senator.

When McCarthy suggested a compromise to Jordan, offering to put forth Trey Gowdy, the popular South Carolinian who chaired the Benghazi committee, as his majority leader, the Freedom Caucus balked. If not a leadership position, Jordan told McCarthy, conservatives might settle for an infusion of their members onto the Steering Committee, an influential panel that appoints chairmen and hands down committee assignments. When McCarthy told him that he could not deliver on this, Jordan made it clear that a sufficient number of Freedom Caucus members would block his promotion to the speakership.

"It's not going to happen, Paul," McCarthy told his friend over the phone.

Ryan knew what was coming next. McCarthy made the case that

Ryan should step up and become Speaker, arguing that he was the only Republican capable of uniting the conference, a sentiment echoed throughout his conversations with colleagues over the following week. Ryan was not interested in the job, and everyone knew it. He had insisted to Boehner and others, after Cantor's loss in the summer of 2014, that he would "never" be Speaker. Now he was repeating himself to McCarthy.

Pacing briskly through Statuary Hall en route to his office suite, Boehner placed his first call to Jo-Marie St. Martin, his general counsel. Boehner was worried that if word leaked of McCarthy's withdrawal, and if they knew how to manipulate the bylaws, conservatives could seize control of the meeting and nominate anyone they wished. Under *Robert's Rules of Order*, the conference could be forced into an interminable number of voting rounds until someone emerged with a majority. "I was *not* gonna let that happen," Boehner says.

The House GOP meeting began with a prayer and the Pledge of Allegiance. Then, the conference chair, Cathy McMorris Rogers, recognized McCarthy, whom everyone in the room, save for Boehner and Ryan, was expecting to make his closing pitch for the job. Instead, McCarthy offered a tearful exit from the race.

Boehner immediately asked for recognition from the chair. "I move to adjourn," he shouted, nodding to McMorris Rogers to bang the gavel. Before anyone knew what had hit them, the meeting was over. McCarthy was out, and Ryan was making a beeline out of the room and toward his office, avoiding the media throngs sure to descend on him after learning the news.

Boehner, lighting a cigarette as he returned to his office, was preparing to pull out the stops. McCarthy had been right about one thing: Ryan was the only House Republican who could unify the party. Now, Boehner felt, he had a responsibility to help Ryan see that.

For the next twelve days, Ryan's phone did not stop ringing. First, it was Boehner, explaining the situation and impressing upon his old friend that he had no choice in the matter—the party needed him. Then it was Mitt Romney, his former running mate, saying much the same. Then it was Priebus, his longtime pal and fellow Wisconsinite. Then it was a chorus of senators, lobbyists, donors, think-tankers, all the allies he'd compiled over two decades in Washington.

At one point, Ryan turned his phone off, disappeared into the woods outside Janesville by himself, wielding a bow and arrow, crouching in a cramped tree stand for nearly an entire day. When he switched the phone back on, it buzzed once more. The voice on the other end belonged to Cardinal Timothy Dolan, archbishop of New York and president of the U.S. Conference of Catholic Bishops, telling Ryan of his obligation to serve at a moment of national uncertainty. Ryan, an observant Catholic, was vexed and bemused.

He called Boehner. "You son of a bitch," he told the Speaker. "You sicced Dolan on me?"

"Yes, I did," Boehner replied. "And I'll have the pope calling your sorry ass next if you don't smarten up."

Ryan sighed. His opposition to serving in a leadership position was rooted in family: The congressman had lost his father at fifty-five, and neither his grandfather nor his great-grandfather had lived to see sixty. Ryan also carried tremendous guilt about living away from his wife and three young children, not to mention his brother and wife and their kids, who lived a few blocks away, on top of a bevy of cousins and family friends in Janesville. The town had always been his refuge; he already spent Mondays through Fridays in Washington and had no appetite for jet-setting the country in his spare time, giving speeches and shaking down donors.

But what if he didn't have to? Ryan convened a call with his top advisers and told them he would consider the job with two conditions: that he be given weekends to spend with his family and that the whole of the House GOP support his candidacy. Having watched the Freedom Caucus heap abuse on Boehner, Ryan would not step into the job unless its members were on record with their support of him.

This ultimatum irked some members in the Freedom Caucus. Seeing their leverage slip away, Raúl Labrador and Justin Amash implored their colleagues to keep a poker face. They would demand a summit with Ryan, airing their gripes pertaining to Boehner's closed-off legislative process and his punitive tendencies.

Some of their comrades agreed, but others arched an eyebrow. The group had effectively driven Boehner out of town; now they were going to play hardball with Ryan, the closest thing to an ideological soul mate any of them could imagine holding the Speaker's gavel?

The summit was anticlimactic. Ryan shared the conservatives' frustration with how the House had been run and pledged to restore balance between the chamber's powerful chairmen and its back-bench members. A small clique of hard-liners decided to hold out, but a supermajority of the Freedom Caucus, roughly 30 of its three dozen members, swore their support to Ryan.[8]

Boehner did his best to provide a smooth transition for his successor. A web of thorny legislation awaited action in late 2015: a debt ceiling deadline, an unresolved budget to keep the government open, and a fight to strip funding from Planned Parenthood, among other items. Boehner promised Ryan he would "clear the barn," legislative-speak for passing a sweeping set of bills to provide a clean slate for the new Speaker. It was cathartic for Boehner. Not only would he be pissing off the conservatives for a final time, but he would also have a chance to unload on Obama, who had different ideas for handling the fall's agenda.

On a conference call with the president and Mitch McConnell one week before his departure, Boehner recalls Obama voicing his displeasure with the Speaker's preferred path before venturing into one of his patented homilies. Boehner held the phone away from his ear for several moments, gesturing to his staff to bring him an ashtray. When he listened back in, Obama was still going.

"I waited three, four, five minutes, and finally I said, 'Mr. President, I didn't get on this goddamn phone call to listen to you lecture me one more time!'" Boehner recalls. "Then I hung up. I'm sure McConnell was shitting in his pants."

An hour later, Boehner's staff informed him that an agreement was in place with the Senate and the White House.

Watching from across the Capitol, McConnell mourned his counterpart's departure while rejoicing at his opportunity to lead a more civilized institution. "I don't have the luxury of having some kind of philosophical purity test, and I've never tried to have one. But it is noteworthy, we don't have a Freedom Caucus in the Senate," McConnell says. "I think we play well with others—almost all of us. Our members are more pragmatic because they just have broader constituencies, whereas, in the House, obviously that's sometimes quite different."

On October 29, Ryan was elected Speaker of the House, and Boehner was sent into the sunset with a prolonged standing ovation, tears streaming down his face as he walked to the back of the chamber to witness the coronation of his successor.

Boehner was leaving a complex legacy. As a young House member, he had been instrumental in cleaning up Congress. As a committee chairman, he had written and ushered through one of the premier policies of the Bush administration. And as Speaker, Boehner had accomplished more than the conservatives would ever give him credit for: securing meaningful spending cuts under a Democratic president; protecting the overwhelming majority of Americans from a tax hike; banning earmarks and keeping them banned despite the negotiating leverage it robbed him of; and his proudest accomplishment, fixing a nagging problem with the Medicare payment formula that could produce nearly $3 trillion in savings over the ensuing three decades.[9]

Yet these will be overshadowed by posterity's more existential observations: That Boehner's twenty-five years in Washington saw the dissolution of a party, the vandalizing of a government, the splintering of a nation. That Boehner watched as the GOP transformed from the party of George H. W. Bush into the party of Donald Trump. That Boehner funded and helped recruit a class of majority-makers who drove him from office and destabilized the Congress he cares deeply about.

The triumph of John Boehner was that he achieved reform and ascended to the speakership while rising above the uncompromising dogma of both parties. The tragedy is that he came to Congress an insurgent only to be swallowed by the insurgency, and that he wasted momentous opportunities, as with the shutdown and immigration battles of 2013, to lead in a way that might have quelled it.

Mike Sommers, the Speaker's chief of staff, says it best: "We fed the beast that ate us."

CHAPTER ELEVEN

OCTOBER 2015

"When I listen to Donald Trump,
I hear the America I grew up in."

WHEN PAUL RYAN ACCEPTED HIS PROMOTION TO SPEAKER, A JOB HE did not want, leading a party and an institution that were increasingly ungovernable, a principal justification was the chance he saw to spearhead an intellectual renaissance in the GOP.

Republicans had once prided themselves on belonging to the "party of ideas," as Obama had himself described the Reagan-era GOP while running for president in 2008.[1] But the buzz of Reaganism had long since turned into a hangover. And the new Speaker, a politician whose values were alchemized in a conservative think tank, sensed an opening.

Ryan had never gotten over the Republican ticket's loss in 2012. The race was eminently winnable, what with the fundamentals of slow economic growth, lagging public confidence, and mediocre approval ratings for the president. Ryan felt certain that the reason Mitt Romney had failed to turn Barack Obama into Jimmy Carter 2.0 was that, unlike Reagan, Romney (and the party) lacked a sharp, contrasting vision to offer the country. "It taught me that you have to run campaigns on ideas, and you have to make them really clear choices," Ryan recalls.

At the outset of Obama's second term, some on the right undertook a serious effort at reinvention. Against a blank canvas of introspection, a bloc of thoughtful reform conservatives emerged with a new agenda, earnest and cerebral and prescient in identifying the blind spots of the modern GOP. They were dubbed "Reformicons," and their quarterback was Yuval Levin, a former Bush 43 adviser who in his thirties launched a quarterly journal called *National Affairs* that became the handbook of the brainiac right.

Levin and a loosely affiliated squadron of academics, think-tankers, journalists, and political strategists designed a fleet of forward-looking free-market solutions that shared a simple premise: that the post-Reagan GOP had become reflexively servile to corporations and the wealthy and no longer offered much to the middle- and working-class Americans left behind by the forces of globalization, deindustrialization, and an uneven recovery from the Great Recession. It was, at its core, the same critique that would drive Trump to see political gold in the "American carnage" of hardscrabble towns battered by decades of economic dislocation.

"Reaganism arose to deal with barriers to prosperity being put up by an overly aggressive, interventionist government, and obviously there are still such barriers in the way," Levin says. "But what we have now more obviously is the breakdown of fundamental institutions, from the family and community, to the very nature of the workplace for a lot of Americans."

His crew's ideas were provocative and compelling: tax reform centered on child tax credits to benefit working families and earned-income credits to incentivize work; eliminating subsidies across the board to level the playing field for little guys competing with Big Business; overhauling the immigration system to prioritize high-skilled labor; and limiting, perhaps temporarily halting, the inflow of low-skilled workers.

They gained a critical mass of media attention with op-eds, speeches, and policy conferences in 2013 and 2014. For the first time in two decades, there was authentic energy penetrating the party's political class, if not its blue-collar base, that could be traced to new intellectual experimentation rather than old ideological rhetoric.

In Congress, the Reformicons found natural allies in the GOP's swelling crowd of Gen X legislators who felt a certain detachment from establishment orthodoxy. Chief among them was Ryan, now the highest-ranking official in the Republican Party. He knew he might not stay on top for long; he certainly hoped that the next president would be a Republican. But more specifically, he hoped that the next president would be a Republican willing to challenge the status quo.

Surveying the GOP presidential field and seeing several like-minded individuals—Bush, Rubio, and even Wisconsin governor Scott Walker, less a conservative visionary than an accomplished agitator—Ryan knew that his speakership, in partnership with one of them as president, could result in a policy revolution for a party stuck in the 1980s.

So, he seized the Speaker's gavel and got to work, crafting a comprehensive set of proposals on poverty, health care, and taxation that Republicans could run on in 2016, and that could serve as a ready-made agenda for whichever kindred spirit won the White House.

And then, as Ryan so delicately puts it, "Donald Trump sort of overtook things."

IT WASN'T MERELY THAT TRUMP HAD ESTABLISHED A COMFORTABLE lead in virtually every national poll of the Republican primary. It was that he was driving the conversation and dominating the media coverage like no presidential candidate America had ever seen. Every interview, every press conference, every early morning tweet or late-night leak from his campaign blocked out the sun.

Trump had spent decades manipulating the New York City tabloids like a puppeteer. Now the candidate was doing likewise to the political press corps. Nightly newscasts worked him into every show. Cable networks carried nearly all his campaign events live. Even the Sunday shows, hallowed for their self-important equanimity, got in on the act, allowing Trump to call into the programs rather than appear on set—something unheard of for any other politician.

When the primary contest had concluded, independent estimates suggested that Trump had received more than $3 billion in free media coverage. And he did not let it go to waste.

It has often been said that Trump has no core ideology, that he is a man without conviction. This is dangerously false. Any casual examination of Trump's writings and remarks going back three decades reveals an opportunist who, while fluid in partisan affiliation and most of his policy positions, cleaves to a few bedrock beliefs. They revolve around the notion that globalization is irredeemably injurious to American society; more specifically, that unrestricted levels of immigration, uneven trade deals, and unchecked foreign cheating have undermined the American business and the American worker.

None of these arguments, in isolation, is necessarily wrong or even wrongheaded. Indeed, Trump's ascent in 2015 was a confirmation of the novel, systemic problems plaguing much of the electorate and the failure of both parties to advance relevant solutions for addressing them.

Yet his policies, rather than leaning forward into the challenges posed in a hyperconnected new century, suggested turning back the clock, looking inward in the hope of returning America to familiar terrain rather than daring to discover the uncharted. Trump spoke like the CEO of an aging conglomerate bereft of new ideas, one that recycles vintage labeling to inspire nostalgia instead of creating new products to attract the next generation of consumers.

The marketing campaign was called "Make America Great Again." And it sold like hotcakes—particularly when printed on his iconic red baseball cap.

"When I listen to Donald Trump, I hear the America I grew up in. He wants to make things like they used to be," Pam McKinney said outside a Trump rally in Arizona in 2016. She and her husband, Lee Stauffacher, had recently moved there to escape the "welfare state" of California.

"Where I grew up, in the San Joaquin Valley, it was a good, solid community, but it fell apart when the government started pandering to all of these immigrants who don't understand our culture and don't want to assimilate," she said. McKinney stiffened. "I'm okay with immigrants as long as they're legal. But they need to assimilate to our culture. They can have their culture at home. In public, you're an American. They're celebrating their own holidays instead of ours."

She continued: "I was born in the fifties, when women stayed at home and men went to work and houses and cars were affordable. We had manufacturing jobs, good jobs. We used to farm in the San Joaquin Valley. It was called the Bread Bowl of America. Now we get our fruits and vegetables from South America. I remember praying in school, but then that got stopped, too. Trump gives us a chance to take things back."

America during the rise of the forty-fifth president was witnessing a sweeping and unprecedented demographic transformation, becoming younger, better educated, more diverse, more urban, more secular, and more dependent on a globalized economy. These trends showed no sign of reversal, hence the RNC project attempting to recalibrate a party that had long depended on older, white, rural, working-class, religious voters. The biggest driver of America's change was the ethnic diversification of the electorate and its political implications.

California became a majority-minority state at the turn of the century. By 2016, whites were 38 percent of its population and dwindling;[2] in turn, the GOP became extinct. McKinney and Stauffacher fled to Arizona, only to feel a sense of déjà vu: Over the past twenty-five years, the state's Hispanic population had nearly tripled, and whites had gone from 74 percent of the population to 56 percent. Minorities would be the majority by 2022, and Democrats planned to end the GOP's monopoly on the state. (Clinton's campaign would spend millions in Arizona while all but ignoring the traditional Democratic stronghold of Wisconsin.)

"The good people like us are leaving California because of all that—the influx of immigrants, many of them illegal, who are getting state ID cards, welfare benefits, and other government programs, and not even assimilating," Stauffacher, a Navy veteran, said. "And now it's happening here. This state is up for grabs. The entire country is changing because they're letting people in who will only vote for Democrats."

This is what "Make America Great Again" conveyed to many voters. Others heard a message that was altogether different—not an identity-based message, but an anti-elitist screed, or a populist call for government reform. The genius of the catchphrase, and what made Trump's candidacy so effective, was its seamless weaving of the personal and cultural into the political and socioeconomic. His was a canopy of discontent under which the grudging masses could congregate to air their grievances about a nation they no longer recognized and a government they no longer trusted.

"The country's morals have changed," Helen Best, a retired cardiac technician, said outside a Republican campaign event in her native

North Carolina a few weeks before Election Day 2016. "People say it's just a changing of the times. But why do we need to change at all?"

AS UNEASY AS REINCE PRIEBUS FELT WITNESSING TRUMP'S ASCENT, the party chairman couldn't help but marvel at the way voters responded to him. It was unlike anything he had seen in a quarter century of Republican politics. Priebus had consulted the best pollsters and strategists about broadening the GOP's socioeconomic appeal; in his native Wisconsin, he was always vexed at seeing rural, religious, blue-collar voters side with an increasingly urban, coastal Democratic Party. Trump offered a revelation: These voters were far less likely to respond to policy arguments than they were to emotional appeals aimed at their long-simmering sense of grievance, displacement, and marginalization.

"Everyone told them that they needed to shut up, that their views weren't culturally proper anymore, that society is moving in a direction that they don't fit into," Priebus says. "And then, Donald Trump comes along and starts saying the same things they've been thinking, and suddenly it's okay again. There's just this feeling among people, among classes, that have felt left behind, not heard, ridiculed, pushed down upon. And he became their vehicle."

During a speech in Burlington, Iowa, in late October 2015, just before Ryan assumed the speakership, Trump drew thunderous ovations from a capacity crowd when promising to punish illegal immigrants, confront China, shred existing trade deals, and pummel the elites funding the campaigns of Bush and Hillary Clinton. But the biggest applause line of the event came when Trump pledged, extemporaneously, to end the so-called "war on Christmas" waged by Obama and his cabal of secularists.

"I'm a good Christian, Okay? Remember that," Trump said, smirking. "And I told you about Christmas—I guarantee if I become president we're going to be saying 'Merry Christmas' at every store!"[3] (The merits of his anti-"Happy Holidays" shtick aside, Trump's assertion of spiritual aptitude was puzzling. Months earlier, he boasted that he had never asked God for forgiveness, the central tenet of the Christian faith.[4])

Trump would soon accelerate the cultural-political warfare to levels yet unseen. After terrorists in Paris killed more than 130 people and injured another 400 during coordinated attacks in November, Trump called for a government database tracking Muslims in the United States while monitoring activities inside mosques. In the past, Trump had repeatedly spread the false story that "thousands" of New Jersey Muslims were celebrating after the 9/11 attacks. (Mocking a disabled *New York Times* reporter who corrected his version of events, Trump curled his hand and jerked his arm around, saying, "The poor guy, you've gotta see this guy."[5])

The following month, in the aftermath of an Islamic-inspired terrorist attack that killed fourteen people in San Bernardino, California, the Republican front-runner's campaign issued a statement that read, "Donald J. Trump is calling for a total and complete shutdown of Muslims entering the United States until our country's representatives can figure out what is going on."

Bush said Trump had become "unhinged." Rubio described the idea as "outlandish."[6] Mike Pence, the Indiana governor, called the proposed ban "offensive and unconstitutional."[7]

That same month, Trump told *Fox and Friends* that he would order the American military to kill terrorists' family members in order to defeat ISIS. Dismissing the implications of violating international law, Trump blamed the United States for "fighting a very politically correct war."[8]

The Pentagon warned that Trump's rhetoric would boost the recruiting efforts of ISIS and other terrorist groups, fueling the group's narrative of a zero-sum clash between followers of Islam and their Crusader enemies in America.

As if intentionally pushing the limits to see what else he could get away with, Trump in December joined the *InfoWars* program hosted by Alex Jones, the country's most prominent conspiracy theorist. Jones existed so far outside the mainstream of American political thought that party officials thought it was an elaborate prank when they heard that Trump was going on the program. But it wasn't. "Your reputation is amazing," Trump told Jones. "I will not let you down."[9]

The "reputation" to which Trump referred? Jones had built a cultlike following on the fringes of the internet by proclaiming that September 11 was an inside job, as was the Oklahoma City bombing, both of them "false flags" choreographed by the U.S. government to expand its tyrannical powers. Jones had also insisted that nobody was killed at Sandy Hook Elementary School in 2012; that the first-graders slaughtered in their classroom and buried by their parents were child actors.

None of this seemed to bother Trump; he was told by Roger Stone, his longtime henchman and a veteran tinfoil hat wearer, that he had a huge following among the *InfoWars* audience. (Conspiracy theorizing wasn't limited to the fringes of the right, as proved by the birther movement, and later, the Fox News–fueled nonsense that a DNC staffer named Seth Rich had been murdered because he'd been the source of a massive email leak.) To most anyone with a brain and a soul, Jones was a demonic influence. To Trump, he was someone who could help him win the presidency, what with his "amazing" reputation.

Yet none of this—the buddy act with Jones, the proposed Muslim ban, the idea of committing war crimes—damaged Trump's candidacy. From the middle of November though the end of December, he jumped from 24 percent to 37 percent in the RealClearPolitics average.

RYAN WAS STRUGGLING TO MAKE SENSE OF HIS NEW JOB. EVERYTHING had been so much easier when he was in the policy business: numbers, legislative text, committee hearings. Now he was in the personnel business. And while he had some appreciation for the challenges his predecessor had faced, the scope of instability inside the conference, and the party, didn't fully dawn on him until he'd moved into the Speaker's suite.

Ryan was miserable almost from the very start. He had given up his dream job chairing the Ways and Means Committee. He was refereeing near-daily disputes among various factions inside the conference. He likened himself to a prisoner some days, and other days to a teacher at a day care center. "Just getting people to agree on how to do things that are in their own interest is hard to do. Getting people to agree, getting to consensus, on things that are basic and axiomatic, is really hard to do," Ryan said. "You need more of a degree in psychology than you need in economics."

One day, Boehner was back in town for a meeting and decided to pop into Ryan's office unannounced. Ryan was cheered momentarily—only to wag a finger in Boehner's face, warning him not to dare light up a cigarette, explaining that it had taken months to get the smell out of the office. Then Ryan looked at him wearily. "This job is a lot harder than I thought," he said, sighing.

Boehner laughed so hard that he spent the rest of the day coughing.

Recalling the conversation later, Ryan added, "And I wanted to say, 'You ass, you stuck me with this sh—'" He swallows the rest of the sentence.

The new Speaker did find a way to exact revenge. Having inherited his predecessor's security detail, Ryan let the agents grow unruly, Navy SEAL–style beards and texted photos of them to Boehner. This was a serious affront to the Rat Pack sensibilities of the former Speaker, code name "Tan Man," who had demanded that his detail be freshly shaven every day. Boehner was not amused.

As Ryan wrestled with his new role, he struggled also with the trajectory of the GOP race. Trump's crowds and poll numbers were growing by the day, while the Speaker's preferred horses were falling hopelessly far behind.

Walker had abruptly dropped out after the September debate in California. Things weren't going much better for Bush, whose "joyful" candidacy offered all the pleasure of a root canal. Having once led the field, registering at 18 percent in the RealClearPolitics average as of mid-July, Bush had dropped below 4 percent by December. The "low energy" tag from Trump had proved debilitating, capturing the caricature portrayed on *Saturday Night Live* of Bush as docile and disinterested. Strangely, nothing could have been further from the truth; Bush was known to barely sleep, to answer hundreds of emails per day, and to work with a metabolism that exhausted staffers half his age.

But it wasn't just the nickname that hurt Bush; nor was it just Trump's bullying. Coming off two poor debate performances, Bush's campaign telegraphed a coming attack on Rubio, his old protégé, in the October 28 debate in Colorado. When Bush began by criticizing Rubio's missed votes in the Senate, Rubio flipped the script. "Jeb, I don't remember you ever complaining about John McCain's vote record," he said, recalling Bush's support for the 2008 nominee. "The only reason why you're doing it now is because we're running for the same position, and someone has convinced you that attacking me is going to help you."

The audience cheered, and someone whistled loudly. Bush folded his hands together and smiled timidly. He began to respond, but Rubio wasn't done, and the senator again overpowered his old friend. "Here's the bottom line," Rubio said. "My campaign is going to be about the future of America. It's not going to be about attacking anyone else on this stage. I will continue to have tremendous admiration and respect for Governor Bush. I'm not running against Governor Bush. I'm not running against anyone on this stage. I am running for president, because there is no way we can elect Hillary Clinton to continue the policies of Barack Obama."

The applause grew louder yet. Once more Bush attempted to respond, and once more he was drowned out—this time by a combination of the audience, the moderators, and a smirking Trump proclaiming to the masses, "I told you that they did not like each other!"

Moments are the currency of a presidential campaign: the acts, the exchanges, the gaffes that break through the clutter of the news cycle and inform voters' view of candidates. This was Bush's weakest moment to date, one from which he could never fully recover. He spoke the least of all ten candidates onstage that night, according to a *New York Times* tally,[10] and would continue to see his airtime fade in future debates. Bush had been castrated on national television—and not by Trump, whose harrying had become expected, but by Rubio, whom he had targeted with a premeditated, unsolicited attack.

The learner had slain the master.

RUBIO REPRESENTED THE LAST, BEST HOPE FOR RYAN AND THE REFOR-micons. With a platform heavy on vocational training, higher-ed reform, and answers to automation, Rubio urged voters to peer around the corner at the challenges of the twenty-first century.

He constructed his candidacy around the notion of an inverted economic landscape. Illustrating the scale of change Americans were living through, Rubio noted how the biggest retailer in the country, Amazon, didn't own a single store; the biggest transportation company, Uber, didn't own a single vehicle; and the biggest lodging provider, Airbnb, didn't own a single hotel. This would require, Rubio argued, a foundational reimagining of the relationship between business, the government, and its citizens.

For all the talk of a historically crowded race, it was down to three horses: Rubio was in third place, at 11 percent in the RCP average; Cruz, who had surged on the strength of a behemoth grassroots operation, sat in second place, at 18 percent; and Trump had double his support, registering at 36 percent.

The wild card was Rubio's courtship of evangelical voters. Once widely assumed to be angling for the support of centrist, business-friendly Republicans, the Florida senator had managed to thread the needle, running an everything-to-everyone campaign. However unsound strategically, this approach kept him in play for the support of social conservatives who did not trust, or did not like, Cruz.

And there were plenty. Some leading activists found Cruz inauthentic to the point of fraudulent; others complained of his social awkwardness, his struggle to make small talk or laugh in a way that wasn't contrived. (Cruz's aides, at the outset of his campaign, had to stress to him the importance of making eye contact with strangers in elevators.) One influential woman in the conservative movement told Cruz's staff that she was simply creeped out by his inhuman disposition.

But these were minority views. Having spent the better part of three years tirelessly pursuing the support of activist leaders and their grassroots followings, Cruz had established himself as the clear favorite to land their support. Now it was just a matter of Tony Perkins, Cruz's chief ally, closing the deal.

Perkins and his group of conservative movement heavyweights had met for the past sixteen months with the narrow purpose of consolidating the right's support around a single challenger to the establishment's favored candidate. They were closer than ever on December 7. Huddled in a boardroom inside a Sheraton Hotel just outside Washington, the group seemed to be closing in on a decision. A supermajority of the group, 75 percent, was required to bind its membership in support of a candidate, and Perkins was working like mad to line up the votes.

After four intense rounds of balloting, with lengthy prayer sessions in between, the participants were physically and emotionally drained. It looked like an impasse was at hand. Cruz continued to hold a lead but was short of the 75 percent supermajority threshold. As several groups split off into side meetings, Perkins dropped in on each of them, pleading his case. Conservatives have worked toward unity for two years, he told them. We are *this* close.

And then, on the fifth ballot, Cruz hit 75 percent.

The impact was felt immediately. Three prominent participants, direct-mail pioneer Richard Viguerie; the National Organization for Marriage's Brian Brown; and the Family Leader's Bob Vander Plaats, a social conservative kingmaker in Iowa, announced their support of Cruz within seventy-two hours of the Sheraton meeting.[11]

This barely scratched the surface. An avalanche of endorsements was forthcoming from conservative leaders, including James Dobson, founder and chairman emeritus of Focus on the Family; Ken Cuccinelli of the Senate Conservatives Fund, and from Perkins himself, among a chorus of other right-wing rainmakers.

The conservative movement, in its official capacity, had unified. Now, if only there were an "establishment" champion for them to face off against.

FOR ALL THE LAWLESSNESS THAT GOVERNED THE 2016 REPUBLICAN campaign, two rules were constant: Trump was the front-runner, and nothing could be done about it.

A telling example came during the December 15 debate in Las Vegas, between Trump and Hugh Hewitt, the conservative radio host who was co-moderating. The pair had a complicated history: Trump had appeared often on Hewitt's show, going back to the spring of 2015, but Hewitt always seemed to stump him with policy questions.

A few months earlier, when Hewitt had asked Trump about the Quds Force, Iran's guerrilla military unit, Trump responded by talking about the mistreatment of the Kurds.[12] He later claimed he'd misheard

Hewitt's question. But this made no sense: Hewitt had begun by mentioning the Quds's leader, General Qasem Soleimani, a name frequently
in the news at that time. There had been no mix-up. Trump was simply
unschooled.

The candidate had blamed Hewitt for the blunder, brushing him off
as a "third-rate radio host" on MSNBC's *Morning Joe*. (The treadmill-
viewing choice of official Washington, *Morning Joe* offered comforting quarter to the GOP front-runner for a long time before serving as a
group therapy session during his presidency.)

Privately, Trump was seething. He could deal with garden-variety
indignities; the man owed much of his fame to assessing the business
acumen of Gary Busey and Meat Loaf on *The Apprentice*. But being made
to look stupid was intolerable. Trump had dialed Hewitt the next day in
a rage. "Don't be fucking around with me like that!" he screamed.

Meanwhile, Trump's campaign manager, Corey Lewandowski, pe-
titioned the RNC to have Hewitt removed as a co-moderator of future
debates. No such luck.

When the Vegas debate rolled around three months later, Hewitt
decided to test Trump in a different way. He would ask the candidate
about a subject they had covered previously on his radio show: the nuclear triad, America's ability to launch atomic attacks from the air, land,
and sea. Trump hadn't been familiar with the terminology the first time
Hewitt asked; the radio host wondered whether he would be now.

In response to Hewitt's question, Trump produced ninety seconds'
worth of word salad about the importance of nuclear weapons. When
Hewitt pressed him, asking which leg of the triad he considered the
most crucial, Trump flailed. "To me, nuclear is just, the power, the devastation, is very important to me," he replied.

Since the conclusion of World War II, global order has been administered via the threat of nuclear warfare. But Trump, in applying for the
job of controlling the largest stockpile on the planet, was blatantly illiterate as to its usage. Making this all the more unforgivable to Hewitt
was the fact that he'd asked Trump about the subject months earlier.
"He wasn't motivated by what he didn't know," the radio host recalls.

He threw the follow-up to Rubio, expecting him to savage the GOP front-runner for his witlessness. Instead, Rubio offered viewers a gentle tutorial on America's nuclear capabilities.

"Marco treated it like a Sunday school class instead of looking at [Trump] and saying, 'You're running for president. How do you *not* know what the nuclear triad is?'" Hewitt says. "He could have embarrassed him, but Trump bluffed his way through it. He bluffed his way through the entire campaign."

On the sidelines of the debate, during an intermission, Priebus walked over to a friend. "Now if only someone would ask him the difference between Sunni and Shia," the chairman whispered.

IF TRUMP'S INADVERTENT IGNORANCE OF POLICY BASICS WASN'T GOING to hurt him, then why would his deliberate contravening of political norms?

This was the question Republicans were forced to grapple with as 2015 came to a close. It wasn't just that his rivals' attacks on him had backfired, or that voters didn't seem to care that he lacked a basic understanding of certain issues. What made Trump's enemies most nervous, what exasperated them and kept them up at night, was how he could get away with saying whatever he wanted.

Examples of this in the first six months of his campaign already numbered too many to count. But few were as audacious as his praise for Vladimir Putin.

Trump had hinted in the past at his respect for the Russian strongman, having felt a bond with him over their shared disdain for Obama and Hillary Clinton. In November, the GOP front-runner told *Face the Nation* of Putin, "I think that I would probably get along with him very well. And I don't think you'd be having the kind of problems that you're having right now."[13]

These comments seemed harmless enough at the time. A month later, however, the long-distance brotherhood was in full bloom.

"He is a bright and talented person without any doubt," Putin said during a year-end press conference, according to Russian state media. He called Trump "an outstanding and talented personality" and described him as "the absolute leader of the presidential race."[14]

Trump, ever a sucker for a compliment, responded in kind. "It is always a great honor to be so nicely complimented by a man so highly respected within his own country and beyond," he said in a statement.

The next day, on MSNBC's *Morning Joe*, Trump shrugged off the Kremlin's brutal reputation: suppressing homosexuals, torturing prisoners, murdering journalists and political dissidents. "He's running his country. And at least he's a leader, unlike what we have in this country," the candidate said. "I think our country does plenty of killing also."[15]

This constituted a radical break with traditional American foreign policy and its emphasis on denouncing autocrats and promoting democracy. It was also a sharp departure from recent Republican dogma. Echoing the tough-on-Russia rhetoric of Mitt Romney four years earlier, the other GOP hopefuls took turns calling Putin a "gangster" and a "KGB thug." Outraged at Trump's remarks, Romney tweeted, "Important distinction: thug Putin kills journalists and opponents; our presidents kill terrorists and enemy combatants."

Thinking, wishing, hoping that this time, finally, his insufferable nemesis had jumped the shark, Bush told CNN, "To get praise from Vladimir Putin is not going to help Donald Trump."[16]

He was wrong.

What Bush and his Republican peers failed to understand was the degree to which Putin had become an appealing figure for many on the American right—not for the particulars of his government's cruelty, necessarily, but rather, for the masculinity he radiated in such sharp contrast to his U.S. counterpart.

This was happening long before Trump began singing the Russian leader's praises. Back in September 2013, Marin Cogan wrote in *National Journal* magazine[17] about the cult following Putin was amassing on the American right with his macho exploits: tranquilizing a tiger, hunting a gray whale with a crossbow, riding war horses, catching gigantic fish. He was always shirtless and never afraid, Rooseveltian testosterone oozing out of every pore.

Cogan noted how, on popular websites such as Cracked and theChive, slideshows of Putin's legend were labeled "The World's Craziest Badass" and "The Real Life Most Interesting Man in the World," drawing millions of eyeballs and enhancing the reputation of Russian's tyrant among a certain segment of red-blooded American males.

Observing this very phenomenon in early 2014, the conservative author Victor Davis Hanson wrote, "Obama's subordinates violate the law by going after the communications of a Fox reporter's parents; Putin himself threatens to cut off the testicles of a rude journalist."[18]

Still, Trump's insistence on toadying to the virile Russian tyrant was strange, even for Trump. The GOP nominee had freely shunned Euro pean allies, talked tough on China and Japan, and emasculated America's two intracontinental allies. Yet he was going out of his way to avoid any utterance of negativity about Vladimir Putin. Nobody knew quite what to make of it.

On ABC's *This Week*, when host George Stephanopoulos raised the allegations of Putin murdering his opponents, the GOP front-runner grew defensive.

"Have you been able to prove that? Do you know the names of the reporters he has killed?" Trump responded.[19] "He's always denied it. It's never been proven that he's killed anybody. So, you know, you're supposed to be innocent until proven guilty. At least, in our country."

CHAPTER TWELVE

JANUARY 2016

"My party is committing suicide on national television."

UPPING HIS HAND, PALM FACING DOWNWARD, ROTATING IT FROM nine o'clock to three o'clock, Ted Cruz would reassure them, "There's a natural arc to Donald Trump's candidacy." It was the same speech he had been giving—to friends, staffers, donors, anyone who would listen, really—since June 16, 2015, the day Trump descended his gilded escalator in Manhattan.

Cruz had come to Washington intent on making enemies, using his first two years in Congress to compile an unrivaled record of aggression toward the political class and its conventions. Yet he could no longer be considered the preeminent instigator in the GOP field. Trump offered the same arsenious approach as Cruz but without the professional constraints. Cruz was still a politician, after all, one who had to worry about long-term career prospects and constituents back in Texas even while chasing the presidency. His opponent was free of such concerns. Anything Cruz said, Trump could say with triple the bombast; anything Cruz did, Trump could do more aggressively, more emphatically, more audaciously. If Cruz was bringing a knife to the Republican primary fight, Trump was packing a nuclear-tipped bazooka.

This distressed the senator's allies. Believing that his capacity for winning the nomination stemmed from his distinction as *the* insurgent in the race, some urged Cruz to confront Trump head-on, calling attention to his decades of commentary that strayed from conservative orthodoxy. The senator had accumulated ample credibility on the right as an equal-opportunity truth teller, someone unafraid to reveal the ideological doublespeak practiced by the most powerful members of his own party. Why not, having turned the right against the likes of Marco Rubio and Mitch McConnell, do the same to Donald Trump?

Because, Cruz believed, Trump was not built to last. His poll numbers would rise with the tide of free media, but once political gravity took hold and the news coverage exploited his obvious lack of preparation for the job, Trump would suffer a mass defection of supporters. They would be looking for the next-best wrecking ball to swing at Washington, and Cruz would be their obvious choice.

Trump's candidacy was fanning the flames of the very anti-establishment mood Cruz needed to win the nomination and ultimately the White House. There was nothing to be gained by attacking someone who was "renting" his supporters, Cruz argued, especially when Trump had shown a proficiency for emasculating whichever rival (Rand Paul, Jeb Bush, Rick Perry) had dared to engage him.

"I like Donald Trump. I'm glad he's in the race. I think he is having many beneficial effects on the race," Cruz had said that summer. "He is attracting significant crowds and significant passion of people who are ticked off at Washington, fed up with politicians who say one thing and do another. The last thing I want to do is have a bunch of Washington politicians insulting and condescending to these hardworking Americans who are rightly and understandably frustrated with the direction this country is going."

Cruz added, "Many of the Republican candidates have gone out of their way to take a two-by-four to Donald Trump. I think that's a mistake. I have deliberately declined to do so, and indeed have bent over backward to sing his praises."

That strategy made sense—for a time.

With seventeen candidates in the field, and weaker prey such as Mike Huckabee and Jeb Bush to target, Cruz could afford to be patient. Many Republicans believed, in the summer and fall of 2015, that Trump's campaign was a publicity stunt; that he was generating huge ratings to promote his hotels, particularly his new project in Washington, and feed his business ego; that he would never actually compete in Iowa or New Hampshire, much less go the distance and win the party's nomination. It was defensible, then, for Cruz to focus his fire elsewhere. "There are seasons to a campaign," he told his allies. "We will deal with Trump if and when necessary."

The problem with Cruz's approach was that it undermined his own brand. He was a brawler, but also a dispenser of brutal honesty, someone whose word could be taken to the bank. ("TrusTed," his campaign banners read.) Cruz didn't just lay off Trump; he spent the months of June through December lavishing praise on his opponent in the hope of seducing the supporters whom Cruz believed would not, *could not*, ultimately pull the lever for such a man. But many conservative voters didn't see the Trump whom Cruz saw, a soulless, philosophically hollow showman. They saw a brash renegade taking on a broken system on their behalf, someone whose credibility had been vouched for by leading figures on the right, including Cruz himself.

In mid-December, Cruz spoke to me at length about his team's considered outlook of the race: He believed he would win the Iowa caucuses, and predicted that Rubio, who had become the favorite to emerge from the "moderate lane," would need to win the New Hampshire primary to keep pace. Never once did Cruz mention Trump's name; it was as though the front-runner could be wished out of existence.

This was telling. Despite a race that was unconventional in every sense, Cruz was clinging to his original, most conventional view of it. He took to reminding anxious donors and friends that America had never elected someone with neither political experience nor military experience. Despite the media's obsession with Trump, Cruz insisted, the primary contest would still come down to a collision between an establishment favorite and a conservative favorite, with Trump's supporters abandoning him long before that day arrived.

Trump knew better. "I could stand in the middle of Fifth Avenue and shoot somebody and wouldn't lose any voters, okay?" he said in Iowa the following month. "It's like, incredible."

NOBODY KNEW WHAT TO MAKE OF RUBIO'S CAMPAIGN.

As a candidate, he was the total package: intelligence, personal magnetism, stirring oratory, policy chops, a gripping biography, and a message that could unite the GOP's disparate factions. But his campaign did little to accentuate these strengths. For much of 2015, while most of his Republican rivals stumped their way across the early states, Rubio was nowhere to be found. He had enjoyed a solid bounce from his mid-April launch, but he was doing nothing to build on it. Rubio spent much of the summer avoiding voters so conspicuously that opposing campaign officials would ask reporters of his whereabouts. Officially, Rubio aides claimed that he was on fund-raising swings. And yet, for the entire third quarter of 2015, he raised less than $6 million,[1] a haul dwarfed by those of Bush, Cruz, and Ben Carson. Even Carly Fiorina had raised more.

At the same time, news clips piled up detailing Rubio's poor attendance record in the Senate. Stories alleged that he'd missed half his committee meetings; others claimed he had the Senate's worst voting record. The South Florida *Sun-Sentinel* demanded his resignation in an editorial.[2] The "truant senator" narrative, combined with his absences from the trail and his lackluster fund-raising numbers, flummoxed friend and foe alike. If he wasn't barnstorming across the early states, and he wasn't collecting campaign dough, and he wasn't voting in the Senate, what exactly *was* he doing?

Rubio's absenteeism was especially baffling in Iowa, where GOP officials wondered why he wasn't making a play for the swaths of center-right voters desperate for an alternative to Trump and Cruz. But Rubio's team, reluctant to raise expectations in any given state, kept playing hard-to-get. By the time Thanksgiving arrived, frustrations were boiling over. Prominent Republicans scolded Rubio in private for his failure to organize in the Hawkeye State, as stories abounded of his team missing easy opportunities to reach voters: The time a line of people waited for him after an event, while his field staffers ate pizza backstage; the appearance he canceled at a major evangelical gathering for no apparent reason; the Saturday he spent in Iowa watching football with his state chairman, Jack Whitver, rather than holding public events.

In response to the uproar, Rubio's campaign manager, Terry Sullivan, told the *New York Times,* "More people in Iowa see Marco on *Fox and Friends* than see Marco when he is in Iowa."

This only made things worse. Rubio had been dogged by criticisms that he was running his campaign from a television studio; now his campaign manager was confirming it. However tone-deaf Sullivan's remark may have been, it contained no small kernel of truth. The unwritten rules of presidential campaigning were being rewritten in real time. In the four decades since the modern nominating system was canonized—Iowa, New Hampshire, then South Carolina—candidates had endeavored to achieve as much face time with voters as possible: coffee shops, high school gymnasiums, church sanctuaries, and, when in Iowa, Pizza Ranch restaurants.

But Trump, equipped with universal name identification and unceasing media coverage, was atop the polls despite shaking fewer hands—"I'm a total germaphobe," he explained—than anyone in the three early states could remember. On top of this, candidates who *did* spend lots of time on the ground and who *did* boast booming field organizations, such as Bush, had nothing to show for it. Studying their surroundings, Rubio's team made the calculation that their time and money would be better spent on television than on ground operations.

It appeared good enough for third place in Iowa. That would be good enough to earn a ticket to New Hampshire. But then what? As the voting season neared, Rubio's team continued to blanch publicly and privately at the basic question of what its path to victory looked like. "To win the nomination, you have to win states," Stuart Stevens, Romney's 2012 chief strategist, says. "I love Rubio and I love his guys . . . but my question for them was always: Where are you going to win?"

The dam broke in January. The Rubio brain trust had outlined an unconventional sequence in which Rubio would place third in Iowa, second in New Hampshire, and first in South Carolina. Nicknamed "3-2-1," the strategy banked on a rapid winnowing of the field. Rubio's team felt that a third-place finish in Iowa, ahead of establishment-friendly competitors such as Bush, Chris Christie, and John Kasich, would vault him ahead of the pack in New Hampshire. If he finished second to Trump there, he could consolidate the center-right vote, which, added to his share of conservatives, would give him a winning plurality in the three-man race with Trump and Cruz in South Carolina.

By designating South Carolina as their must-win state, Rubio's team was investing in a home game. Much of the candidate's high command either hailed from or had deep connections to the Palmetto State; Jim DeMint, the former senator, had brought Rubio there with such frequency during his 2010 campaign that the underdog candidate became an adopted son. Rubio also had trip aces up his sleeve, three waiting high-profile endorsements in the state, national figures all: Governor Nikki Haley, Senator Tim Scott, and Congressman Trey Gowdy.

Rubio and Scott had developed a particularly close kinship in the Senate, joining each other for Bible studies and pickup basketball games. Rubio had brought his family to Scott's church in Charleston; Scott had backed Rubio during the darkest moments of the Gang of Eight affair. Now, as Rubio's endgame finally came into focus, he found himself sharing a stage with Scott at a feel-good Republican event that only Reince Priebus could love.

It was January 9, and Scott was teaming with Paul Ryan to cohost a "Poverty Summit" in Columbia, South Carolina. Staged as a nontraditional conversation about conservative solutions to socioeconomic immobility, Ryan and Scott strained to display a GOP that cared about building trust with minority communities more than walls along the southern border. One by one, the presidential hopefuls took the stage to discuss the plight of poor Americans, offering commentary rarely if ever heard in contemporary Republican politics.

Rubio was thoroughly in his element, peddling his up-by-the-bootstraps biography and pitching a forward-looking vision of conservatism that made the regular suspects swoon. Arthur Brooks, the brilliant president of the American Enterprise Institute, called the event "a new day for the Republican Party and the conservative movement." Mika Brzezinski, the cohost of MSNBC's *Morning Joe*, who took part in a panel discussion, remarked, "*This* is a Republican Party that can win the White House."

Two candidates declined to attend: Trump and Cruz.

The symmetry was inescapable. A year earlier, Congressman Steve King had unofficially kicked off the 2016 Republican primary by hosting an event dominated by red-meat rhetoric that showed the hardline im pulses of the party. Rubio had refused to go, not wanting to identify w King, and Trump and Cruz had won rave reviews. Now, a year later, the voting soon to begin, Rubio had distinguished himself at an event designed to showcase the GOP's softer side. But Trump and Cruz, not wanting to identify with Ryan, had refused to attend.

It would soon become clear which version of the party Republican voters preferred.

THE TRUCE BETWEEN TRUMP AND CRUZ COULD NOT HOLD FOREVER. The physics of a presidential campaign would not allow it: Two candidates, occupying the same space, are bound to collide. When they finally did, it was unlike anything the Republican Party had ever witnessed.

Their rivalry began in earnest with a *New York Times* story in December that reported that Cruz had questioned the "judgment" of both Trump and Ben Carson during a private fund-raiser in Manhattan.[3] Cruz objected to the reporting, prompting the *Times* to release audio of his remarks. This was validation for Trump, who had been predicting that Cruz would soon start attacking him. Cruz quickly tried to pull back. On December 11, as Trump began needling him with tweets, Cruz sent a tweet of his own: "The Establishment's only hope: Trump & me in a cage match. Sorry to disappoint—@realDonaldTrump is terrific."

But the genie could not fit back into the bottle. Over the ensuing seventy-two hours, Trump began hurling a hodgepodge of insults. He questioned the authenticity of Cruz's faith, telling a rally in Iowa that "not a lot of evangelicals come out of Cuba, in all fairness."[4] He accused Cruz of being in the pocket of big oil and slammed his flip-flopping on ethanol subsidies. And he told Fox News, "I don't think he's qualified to be president," saying that Cruz carried himself "like a little bit of a maniac" in the Senate.[5] (Apropos of nothing, it was around this time that Trump said that Hillary Clinton "got schlonged" by Barack Obama in the 2008 primary.[6])

Cruz held back. Having moved to the top of the polls in Iowa, a confrontation with Trump seemed less than ideal. But it soon became apparent that unilateral disarmament was not practical. As the calendar turned to 2016, Trump intensified his attacks, questioning whether ʌz, born in Canada to an American mother, was qualified for the pres-ʈy.

Cruz replied to the Birther 2.0 routine by tweeting a clip from *Happy Days*, suggesting that Trump had jumped the shark. When it became clear that he hadn't—that the front-runner was gaining ground in Iowa with his insinuations—Cruz finally decided to return fire.

"Donald comes from New York and he embodies New York values," Cruz told a New Hampshire radio host on January 12, finally snapping a seven-month streak of playing nice with the GOP front-runner.[7] "The Donald seems to be a little bit rattled." Cruz repeated these rehearsed lines that night on Megyn Kelly's Fox News program.

By the time the remaining candidates gathered in South Carolina for a debate on January 14, all other story lines were considered side dishes to the delicious, long-awaited main course of Trump versus Cruz. And it did not disappoint.

"Back in September, my friend Donald said he had his lawyer look at this from every which way and there was no issue there. There was nothing to this 'birther' issue," Cruz announced on the stage. "Since September, the Constitution hasn't changed. But the poll numbers have."[8]

The crowd roared with approval. When the moderators pressed Trump on why he was now raising the issue against Cruz, he could only shrug, "Because he's doing a little bit better."

Trump would soon have his revenge. When the moderators asked Cruz to explain his cryptic references to "New York values," he responded, "Most people know exactly what New York values are," adding that "not a lot of conservatives come out of Manhattan." Trump was visibly offended by the remark and delivered a devastating counterattack. Standing at an adjacent lectern, he recalled the spirit of New Yorkers in the aftermath of September 11 as a point of national pride. Cruz himself was forced to applaud, and when Trump charged that Cruz's comments were "insulting," the Texas senator did not raise an objection. It was Trump's finest debate showing to date, and Cruz's worst.

It wasn't the only tough moment for Cruz. After a prolonged back-and-forth with Rubio over the 2013 immigration bill, a dispute that had become increasingly heated between the two senators, Rubio unloaded on Cruz. He accused the Texan of sticking a finger in the wind not only on immigration, but also on ethanol subsidies and national security, saying he had joined with Rand Paul and Bernie Sanders in voting for a defense budget that slashes military spending. "That is not consistent conservatism," Rubio said. "That is political calculation."

Cruz smirked. "I appreciate you dumping your oppo research file on the debate stage," he said.

The sparring between Cruz and Rubio would make for compelling melodrama in the months ahead. But more than eleven million people had tuned in on this occasion to watch the clash of the titans: Trump versus Cruz. It was clear now, with three weeks remaining until the Iowa caucuses, that the gloves were off. "I guess," Trump told CNN following the debate, "the bromance is over."

WHAT WOULD THE 2016 REPUBLICAN PRESIDENTIAL RACE HAVE looked like without Trump?

The January 28 debate in Iowa, the final gathering of the candidates before votes were cast, offered a glimpse into this alternative history. Because Fox News was hosting, and due to the bad "blood" still lingering with Megyn Kelly, Trump skipped the event, choosing instead to hold a rally for veterans just down the road, in Des Moines.

The debate was undeniably duller without him: fewer outbursts, fewer eyeballs, fewer clicks. For the journalistic establishment's eternal virtue-signaling about all things Trump, in truth, it had grown reliant on him. Its most trusted properties and personalities spent the campaign milking him like a cash cow, starving the other candidates of oxygen at pivotal junctures in the race. "I didn't anticipate that Trump would receive over three billion dollars in free media. There is no precedent for that in the history of the United States of America," Cruz says. "Our campaign raised over ninety-one million, which is the most any Republican primary candidate for president has ever raised. Ninety-one million is a ton of money, unless you're facing three billion of free media on the other side."

As then-CBS executive chairman and CEO Les Moonves observed during Trump's ascent, "It may not be good for America, but it's damn good for CBS."[9]

Trump may have been wise to stay away that night. Even as his support swelled among a base of populist supporters—including Sarah Palin, who flew to Iowa and gave a memorably strange speech endorsing him—a general panic over his viability was beginning to blanket portions of the American right. Days before the debate, *National Review*, the esteemed publication of conservative opinion, had announced a special issue of its magazine, "AGAINST TRUMP," featuring some two dozen essays from leading conservatives voicing their resistance to the Republican front-runner. It was the first real showing of organized opposition to Trump, and it came as a surprise even to some employees of the publication. (At the time, I was *National Review*'s chief political correspondent, reporting on the straight news of the race, and I was informed of the issue just hours before it published online.)

Cruz assumed the role of archvillain on stage in Trump's absence, reasserting his preeminence as a provocateur and reminding voters of his legend as the *original* outsider. His audition as leading man, while beneficial in certain respects, also invited an unprecedented amount of dogpiling. Seemingly everyone on stage took a turn swinging at the Texas senator. It was a favorite pastime in Iowa of late.

For months, Cruz's bus had been shadowed through the state by an RV owned by America's Renewable Future, a group targeting his opposition to ethanol subsidies. Then, in mid-January, the *New York Times* reported that Cruz had not disclosed loans from Citibank and Goldman Sachs (where his wife, Heidi, worked) that helped fund his 2012 Senate race.[10] His opponents pounced on the chance to expose Cruz, the self-styled populist hero, as a privileged insider. Less than a week later Palin flew into Iowa to endorse Trump; that same day, January 19, legendary Iowa governor Terry Branstad said of Cruz, "I think it would be a big mistake for Iowa to support him."

Worse yet, he was also coming under attack from evangelicals. After *Politico* reported that Cruz told a New York fund-raiser that opposing same-sex marriage would not be a top priority,[11] Rick Santorum's campaign said Cruz "makes Mitt Romney and John Kerry look consistent." When BuzzFeed reported that he tithed nowhere near the biblical 10 percent rate, a pro-Huckabee group ran a TV ad in Iowa labeling Cruz a "phony" Christian.

And then there was Trump.

The front-runner had continued to stage attacks on Cruz: his citizenship, his poor relationships on Capitol Hill, his sweetheart loans for the 2012 campaign. Yet none of this seemed terribly personal—at least, not by Trump's standards—until late January. In preparation for a major address to Liberty University, the nation's largest Christian college, Trump asked Tony Perkins for some pointers. Perkins provided a few suggestions, including a verse he jotted down as "2 Corinthians 3:17." ("Now the Lord is the Spirit, and where the Spirit of the Lord is, there is freedom.")

When Trump pronounced the book as "Two Corinthians," drawing laughter from the audience and spawning coverage of his manifest lack of scriptural intimacy, he was furious at Perkins, calling a day later to chew him out for the lack of clarity. But when Perkins endorsed Cruz less than a week later, Trump's ire turned toward the candidate himself, believing the entire affair had been an orchestrated act of sabotage.

Trump grew more certain of this when Cruz's allies began using the "Two Corinthians" mishap to mock him in the final days before the Iowa caucuses. One of the offenders, evangelical figurehead Bob Vander Plaats, president of an Iowa group called the Family Leader, used an appearance with Cruz to excoriate Trump's lack of Christian virtue.

"You know," Trump told one Iowa official, "these so-called Christians hanging around with Ted are some real pieces of shit."

Cruz was similarly at his wit's end with Trump. The night before the Des Moines debate, Cruz scalded his "narcissistic, self-involved" rival during a local pro-life rally. At dinner with friends afterward, the Texas senator vented his frustrations in uncommonly blunt fashion. "If you're a faithful person, if you believe that Jesus Christ died for your sins, emerged from the grave three days later, and gives eternal life, and you're supporting Donald Trump," Cruz told his friends, "I think there's something fundamentally wrong with you."

These tensions built to a crescendo in the campaign's final days. Trump's decision to skip the debate placed Cruz at center stage, subject to two hours of attacks. The next day's *Des Moines Register* led with a headline that summed up the weeks of persecution: "ROUGH NIGHT FOR CRUZ."

Which made it all the more impressive when Cruz won the Iowa caucuses on February 1.

It was a testament to the campaign's stellar organization and top-notch data analytics program, which mined the state's GOP electorate for its most receptive voters and then swamped them with microtargeted ads, mailers, and phone calls. That said, even Cruz's data gurus had low-balled voter turnout. Four years prior, a record-breaking number of Iowans (121,503) had voted in the Republican caucuses. All the campaigns were banking on a sharp uptick this time around: 135,000, or even 150,000, perhaps. The craziest, most bullish estimates reached 175,000.

When all the votes were tallied, nearly 187,000 Iowans participated in the GOP caucuses.[12]

Cruz captured 27.6 percent of the vote; Trump finished second with 24.3 percent; and Rubio took third with 23.1 percent. No other candidate reached the double digits.

For Cruz, Iowa represented more vindication than victory. Once dismissed as a quixotic candidate, only later to be told that his financial and organizational strength would be wasted because of Trump's all-eclipsing presence, Cruz believed his caucus triumph represented a breakthrough: a win for the most hated politician in the party and a loss for the front-runner who talked of nothing but winning. On a stage inside the state fairgrounds that night, Cruz looked beyond Iowa, *way* beyond Iowa, going so far as to preview portions of a speech he intended to give later that year when accepting the Republican nomination in Cleveland.

But his success did not come without controversy.

Seventeen minutes before the caucuses were called to order at locations all around Iowa, a CNN reporter tweeted the news that Carson was headed home to Florida after the caucuses instead of traveling on to New Hampshire and South Carolina. The cable network immediately picked up the story and ran with it, suggesting that Carson was suspending his campaign. Having set up a sophisticated instant-alert system with their volunteers and precinct captains across the state, Cruz's team blasted out a message informing them that Carson was quitting the race and urging them to "inform any Carson caucus-goers" to vote for Cruz instead.

Carson finished with 9.3 percent of the vote, roughly equivalent to his recent polling in Iowa, but he blamed Cruz for his defeat. On a phone call the next day, Carson asked for a public apology; Cruz issued one immediately. Carson wasn't satisfied. Over the next week he tortured Cruz, portraying his opponent as conniving and untrustworthy. Carson knew he was not going to win the nomination. But he felt a newfound resolve to prevent Cruz from winning it.

In this, he made a powerful new ally: Trump.

The front-runner had long suspected Cruz of playing dirty tricks, and now he had solid proof. After boarding his plane at the Des Moines airport, Trump placed a phone call to Jeff Kaufmann, the Iowa GOP chairman who had just declared Cruz the winner.

"You know what the Cruz people did. They threw the vote," Trump told Kaufmann. "I think you need to publicly disavow the result."

Kaufmann told Trump he couldn't do that. It would be another black eye for Iowa, four years after the party mistakenly declared Mitt Romney the winner over Rick Santorum.

A long silence. "You should disavow the result," Trump said. "Think about it, will you?"

RUBIO WAS ROUNDLY RIDICULED FOR DELIVERING WHAT SOUNDED LIKE a victory speech after his third-place finish in Iowa. But in some ways, he *had* won: Presidential politics are all about narratives and expectations, and Rubio captured 23 percent of the vote, just 1 point behind Trump, in a state where polls had projected him in the mid-teens. More important, his next-closest competitor was Carson, at 9 percent. Huckabee and Santorum quit the race after Iowa, freeing up more voters, and Rubio's rivals in the establishment lane had become afterthoughts.

The polling in New Hampshire reflected this new reality. Rubio, who for weeks had been stuck in the low teens in a five-way cluster with

Bush, Christie, Cruz, and Ohio governor John Kasich, suddenly broke out. Several reputable surveys showed Rubio jumping to 17, 18, and 19 percent in the immediate aftermath of Iowa's caucuses, establishing clear separation from the non-Trump pack.

Heading into the February 6 debate in Manchester, on a Saturday evening three days before the state's primary, Rubio was positioned to complete step two of the process: a second-place finish that would send his centrist rivals packing and set up the three-way contest in South Carolina that Rubio's team craved.

The governor of New Jersey had other ideas.

Christie had once been the hottest commodity in Republican politics. His upset victory in 2009 had injected vitality and personality into a party woefully short on both. His truculent style and larger-than-life aura were a perfect fit for the state; when a group of top GOP donors pleaded with him to run for president in 2012, Christie refused, saying there was more work to be done in New Jersey. He did it well, reforming the state's pension structure and winning multiple fights with the teachers' unions, earning himself approval ratings that topped 70 percent. After his deft handling of Superstorm Sandy, Christie coasted to reelection in 2013 by 22 points—in one of America's bluest states—and was positioned as a top-tier contender for the presidency in 2016.

And then came "Bridgegate." Many local Democratic officials endorsed Christie in his 2013 reelection bid; one who did not was Mark Sokolich, the mayor of Fort Lee. In retaliation, a top Christie aide emailed one of the governor's allies at the Port Authority: "Time for some traffic problems in Fort Lee." On the first day of school that September, the Port Authority unexpectedly shut down multiple road lanes on the New Jersey side of the George Washington Bridge, causing mass delays and prompting an investigation that exposed the administration's plans for political retribution. Christie was never proved to have had knowledge of the scheme, but the scandal engulfed his second term, sinking his approval ratings and his presidential prospects.

Once considered a favorite in New Hampshire, leading some of the earliest surveys taken in 2013 and 2014, Christie had been reduced to also-ran status by February 2016. He did not break single digits in any poll of the state in the final four weeks before primary day, and in part, he blamed Rubio, whose super PAC had dropped millions of dollars slamming his record.

Christie was not going to win New Hampshire—or the Republican nomination—but he could still take Rubio down with him.

Rubio's greatest vulnerability was his protective casing. Despite the observable political gifts, his candidacy was carefully stage-managed. Not only did his campaign keep him under wraps, but everything he said and did seemed carefully rehearsed. His remarks about biography, policy matters, and political disputes were often streamlined down to the syllable. Being "on message" is vital to campaigns, but Rubio grew disciplined to the point of absurdity. His insularity and highly mechanical messaging had become a subject of fascination in the political world, not just for reporters but also for rival campaigns.

Christie telegraphed his coming attacks on Rubio in the February 6 debate, and when the lights went on he wasted no time prosecuting his case that the forty-four-year-old first-term senator was not prepared for the presidency. Responding to Christie's charge that he shared Obama's meager qualifications, Rubio offered a practiced rebuttal, arguing that Obama's inexperience had not kept him from effecting a calculated makeover of American government. "Let's dispel once and for all with this fiction that Barack Obama doesn't know what he's doing," Rubio warned. "He knows exactly what he's doing. Barack Obama is undertaking a systematic effort to change this country, to make America more like the rest of the world."

When Christie responded by pressing the Obama comparison, warning voters "not to make the same mistake we made eight years ago," Rubio returned fire by highlighting New Jersey's credit downgrades. Then, curiously, he repeated his earlier remark almost verbatim. "Let's dispel with this fiction that Barack Obama doesn't know what he's doing. He knows exactly what he's doing," Rubio said. "He is trying to change this country. He wants America to become more like the rest of the world."

Christie turned to the audience. "That's what Washington, DC, does," he announced. "The drive-by shot at the beginning with incorrect and incomplete information and then the memorized twenty-five-second speech that is exactly what his advisers gave him."

The crowd, having noticed Rubio's rhetorical repeat, laughed and cheered.

It could have ended there. Instead, Rubio continued to engage him. After criticizing Christie's handling of a recent snowstorm, Rubio, sounding like a malfunctioning robot, repeated himself a third time. "Here's the bottom line: This notion that Barack Obama doesn't know what he's doing is just not true," Rubio said. "He knows exactly what he's doing."

"There it is!" Christie blurted out, to the delight of the crowd. "The memorized twenty-five-second speech!"

Standing two lecterns away from Rubio, with only Trump between them, Cruz thought to himself, *Ho-lee crap.*

It was a truly unforgettable exchange; pundits dubbed it a "murder-suicide" likely to bury both campaigns. Rubio acquitted himself well for the remainder of the event. Yet he knew, walking offstage, that nobody would remember anything but his verbal glitch.

He had arrived in New Hampshire with the wind at his back, a second-place finish looking certain. Instead, Rubio's debate perfor-mance doomed him. Over the ensuing seventy-two hours, his standing in the state collapsed. He placed an embarrassing fifth in the primary, taking just 11 percent of the vote.

Trump was dominant, taking 35 percent and leaving just 16 percent for John Kasich, the why-can't-we-all-just-get-along Ohio governor, who capitalized on Rubio's implosion to finish as the runner-up. Cruz, who was not thought to be competitive in New Hampshire, had a sur-prisingly strong third-place showing. Even Bush, whose campaign was on life support, finished narrowly ahead of Rubio.

It was little consolation that Rubio bested Christie, who promptly exited the race following his sixth-place showing. By finishing behind Kasich and Bush, both of whom claimed justification to carry on with their campaigns, Rubio saw the three-way contest in South Carolina slip through his fingers.

"Those thirty seconds or sixty seconds in New Hampshire," Rubio says, shaking his head. "That was a big moment, because of that tactical mistake. Had we performed better in New Hampshire, the race could have gone on a different trajectory."

A WEEK LATER, AS THE CAMPAIGN CONTINUED ON IN SUNNY SOUTH Carolina, a GOP operative named Marc Short arrived in bitter-cold Kansas for a meeting he hoped would turn the tide of the race.

Trump's demolition of the field in New Hampshire had set off alarms across the right. What began as a joke—the prospect of The Donald as The Nominee—was suddenly a very real possibility. Rubio was mortally wounded. Kasich had no money or organizational muscle. Bush looked like the biggest flop in recent presidential memory. Of the remaining candidates, only Cruz appeared capable of thwarting Trump's advance. This was cold comfort to the graybeards of the GOP establishment.

But it wasn't just the party's elite who were panicking. Short, the longtime consigliere to Mike Pence and now the president of Freedom Partners Chamber of Commerce, the umbrella group in charge of political activities for the donor network led by Charles and David Koch, believed Trump embodied an existential threat to conservatism. Worried that he would soon be unstoppable, Short had led a small team to Koch Industries' headquarters in Wichita to present a detailed plan for subduing the front-runner. What he needed was approval from Charles Koch to organize an eight-figure spending blitz against Trump on Super Tuesday, March 1, when eleven states would vote. Short hoped to hammer Trump in the states where he was most vulnerable, depriving him of delegates and undermining the narrative of his inevitability.

But there was an unwelcome surprise awaiting Short's crew inside a conference room: A number of top executives and advisers from across the Koch enterprise had been invited to attend the meeting. They represented the so-called corporate side of Koch world, which had long warred with the "political side" of the empire, particularly over the consequences of the brothers' campaign-related activities. One of America's most valuable companies, Koch Industries, a producer of everything from toilet paper to jet fuel, was increasingly synonymous with the Koch brothers, a fact that worried their bean counters and stockholders.

Facing protests, boycotts, and attacks on them by name from some of the country's top Democratic officials, the brothers' business associates grew antsier by the day.

There was another unexpected development in store: After Short presented his proposal, Charles, whose greenbacks and green light the political side was soliciting, asked his corporate lieutenants to cast an up-or-down vote. One by one, they voiced their opposition to going to war against Trump. Everyone figured that Charles would still have the last word, weighing the pros and cons of meddling in the race. Instead, he shrugged. The majority, he said, had spoken.

The verdict was indicative of what the Koch brothers' allies described as a long-term "realignment" of resources, with their money and focus steered away from elections and toward a slew of the more intellectual, policy-oriented projects on which they had historically lavished their fortune. Charles, in particular, had grown exasperated with the lack of return on their mammoth investments in recent years, and was not keen to throw bad money after good.

In the short term, however, it meant losing their top political staffer. Short quit Freedom Partners within the week, going to work for Rubio's campaign and gushing to Koch world staffers about "being in the fight" to stop Trump.

Meanwhile, the party's leaders were also striking out. In the twenty-four hours immediately following the New Hampshire primary, a senior official with the Republican National Committee reached out to several prominent DC consultants, veterans of past presidential campaigns, to ask for help. The RNC official warned that Trump was becoming a runaway train. Something needed to be done, they said, and fast—without the national party's fingerprints. Could they organize against him on the fly? Build a coalition of household names? Run TV ads? Raise money for an anti-Trump effort?

The consultants were aghast. It was the middle of February, two primary contests into the nominating season. The time to organize against Trump was months ago. No last-minute opposition campaign, led by Beltway political fixers, was going to derail him.

Stymied, party leaders took matters into their own hands. In just a few days, the candidates would convene for yet another debate, this one in Greenville, South Carolina, on February 13, one week before the state's primary. Desperate to halt Trump's momentum, the chairman of the South Carolina GOP, Matt Moore, colluded with top RNC brass, including communications director Sean Spicer and chief of staff Katie Walsh, to stack the debate hall in Greenville with Rubio supporters.

The resulting two-hour melee, moderated by CBS News, was even uglier than usual.

Trump was booed lustily and repeatedly, so much so that Google reported a 1,400 percent spike in searches for "Why are people booing?"[13] The answers varied, but the loudest objections to the front-runner were voiced during his criticisms of George W. Bush. With the former president set to visit the state days later to campaign for his brother, the moderator, John Dickerson, asked about Trump's comment in 2008 that Bush should have been impeached.

"George Bush made a mistake," Trump said, fighting through waves of booing. "We all make mistakes. But that one was a beauty. We should have never been in Iraq. We have destabilized the Middle East. . . . They lied. They said there were weapons of mass destruction. There were none. And they knew there were none."[14]

It was an unthinkable statement for a Republican to make about the party's last president, particularly in the veteran-heavy, pro-military Palmetto State. But Trump wasn't done yet. After Jeb Bush responded— "While Donald Trump was building a reality TV show, my brother was building a security apparatus to keep us safe"—Trump went nuclear.

"The World Trade Center came down during your brother's reign. Remember that," Trump said.

Gasps could be heard in the auditorium.

It wasn't the last confrontation of the night. When Cruz slammed Trump's onetime advocacy for partial-birth abortion and his recent statement of support for the organization Planned Parenthood, Trump called him "the single biggest liar," citing the Carson incident in Iowa. Cruz, in turn, reminded voters that Trump had compared Carson to a "child molester" earlier in the campaign. (On a positive note, Trump vowed to refrain from further vulgarities on the campaign trail, days after referring to Cruz as a "pussy."[15])

And then, at long last, Cruz and Rubio dropped the gloves for good.

After a heated back-and-forth regarding the 2013 immigration fight, with Cruz alleging that Rubio supported amnesty (he did) and Rubio accusing Cruz of supporting a dramatic increase in legal immigration (he did), the Texas senator seemed intent on scoring a personal point against his colleague from Florida. "Marco went on Univision in Spanish and said he would not rescind President Obama's illegal executive amnesty on his first day in office," Cruz said.

"I don't know how he knows what I said on Univision," Rubio shot back, "because he doesn't speak Spanish."

Red in the face, Cruz blurted out several lines of sloppy *español* as the crowd buzzed. The mutual contempt between them, simmering below the surface for years, was now on full display. Cruz believed Rubio was a phony conservative with few core convictions; Rubio saw Cruz as a craven opportunist who had forsaken the Hispanic community and his own heritage. Rubio's insult was the political equivalent of the rap battle diss in the film *8 Mile*, when the freestyler played by Eminem mocks his suburbanite opponent: "This guy's a gangster? His real name's Clarence!"

Despite the ad hominem pettiness that defined much of the debate, the stakes, for the party and the country, were suddenly even higher. Earlier in the day, Supreme Court justice Antonin Scalia was found dead inside his room at a Texas hunting ranch. (Trump couldn't help but wade into a new theater of conspiracy theorizing, telling one talk radio host of Scalia's death, "They say they found a pillow on his face, which is a pretty unusual place to find a pillow."[16])

The Senate's new majority leader, Mitch McConnell, had immediately announced that Republicans would not hold a vote to confirm a new justice in an election year. If he held to his word—a big *if*, in the minds of many conservatives—it wouldn't just be the presidency hanging in the balance on November 8. It would be a Supreme Court seat, plus the likelihood of one or two more appointments.

The friendly debate had given Rubio a much-needed boost. Trump, on the other hand, walked off the stage in a fury. When the event ended, his campaign manager, Corey Lewandowski, cursed out multiple RNC officials and threatened a boycott of future debates. With his post–New Hampshire autopsies now being put on pause, Rubio pulled out the stops, calling in the endorsements of Nikki Haley and Tim Scott.

It was no small decision for Haley. Widely considered prime vice-presidential material for whichever Republican won the nomination, she also harbored future White House ambitions of her own and was reluctant to pick sides in such an ugly primary. She initially demurred, promising both Jeb and George W. Bush that she would remain neutral. Ultimately, however, Haley felt an obligation, if not to her friend Rubio, then to the cause of defeating Trump by any means necessary. The governor had become outspoken in disavowing the GOP front-runner, telling friends that she feared he might destroy the party and do irreparable harm the the country.

"I wanted somebody," Haley said when endorsing Rubio, "that was going to go and show my parents that the best decision they ever made was coming to America."

The endorsements from Haley and Scott—as well as of Trey Gowdy, the popular congressman who chaired the Benghazi committee and was making life miserable for Hillary Clinton—breathed life into Rubio's candidacy. For several days the crew of telegenic, next-generation conservatives barnstormed the state together, just the image Priebus had dreamed of when drafting the RNC's autopsy.

But it wasn't enough. Rubio finished second in South Carolina, taking 22 percent of the vote, but in a virtual tie with Cruz, whom he topped by a thousand votes. Trump beat them both by 10 percentage points.

Bush finished a distant fourth and promptly departed the race. It was an ignominious end for a candidate with dynastic riches but scant organic support; a candidate who suffered humiliations at the hands of both Trump and GOP voters, some of whom he asked to "please clap" in New Hampshire; a candidate whose super PAC had spent nearly $35 million attacking Rubio, according to *ProPublica*, and just $25,000 attacking Trump.

No Republican did more to criticize Trump during the primary than

Bush. But no candidacy did more to symbolize the decline of the GOP and the ascent of its unlikely new torchbearer.

IT WAS DESPERATION TIME FOR RUBIO. HE HAD DECLARED HIS INTEN-tion to remain in the race until Florida's primary on March 15. But a victory there was already looking unlikely, and it would be downright impossible if he didn't score some points on Super Tuesday, March 1.

In their final chance to make a national impression before Super Tuesday, Rubio and Cruz formed a tag team during the February 25 debate in Houston, emptying a dump truck of opposition research against the front-runner.[17] It might have come in handy, say, six months earlier: How Trump had defrauded students at a university bearing his name; how he had hired foreign workers, oftentimes illegals, ahead of Americans; how his ties and suits were made in Mexico and China; how he repeatedly mismanaged his companies into bankruptcy.

But the verbal drubbing of Trump in Houston was too little, too late. He was gaining altitude and growing more emboldened by the day.

At a rally in Nevada after his South Carolina win, Trump dialed up the rhetoric in reference to a protester being escorted out of the audience. "I love the old days. You know what they used to do to guys like that when they were in a place like this? They'd be carried out on a stretcher, folks," Trump said. He added: "I'd like to punch him in the face."[18] (Weeks earlier, he'd offered to "pay the legal fees" of anyone willing to "knock the crap out of" a protester.[19]) Trump knocked the crap out of his opponents in Nevada, taking 46 percent of the vote.

Three days later, on February 26, Chris Christie, the New Jersey governor and former contender for the nomination, became the first prominent Republican official to endorse Trump. "New lesson kids," Bush chief strategist David Kochel tweeted. "Sometimes, the best option for the fat kid is to just hand his lunch money over to the bully!"

Then, two days later, at a massive rally in Alabama ahead of the state's Super Tuesday primary, Jeff Sessions became the first Republican senator to endorse Trump. "I told Donald Trump this isn't a campaign, this is a movement," Sessions declared from the stage. "Look at what's happening. The American people are not happy with their government."

That same week, Trump added another ally, albeit one with a much lower profile. Having recently finished managing her father's presidential campaign, Sarah Huckabee Sanders was a seasoned strategist, a good communicator, and had a deep knowledge of the upcoming southern states. Despite the early wins, Trump's campaign was still a fly-by-night operation. Sanders, after meeting with Trump aboard his campaign plane and agreeing to come on as a senior adviser, offered an injection of veteran savvy.

With the polls in the eleven Super Tuesday states showing little movement, and Trump on a glide path to the nomination, his rivals emptied out their ammunition lockers.

Cruz began hitting him on a topic that, strangely, had gone largely unmentioned during the GOP primary: Trump's refusal to release his tax returns. Rubio, for his part, decided to break character in a fateful moment of attempted levity. During a February 28 rally in Virginia, Rubio (dubbed "Little Marco" by the GOP front-runner) decided to fight Dumpster fire with Dumpster fire. "I'll admit he's taller than me. He's like 6'2", which is why I don't understand why his hands are the size of someone who's 5'2". Have you seen his hands?" Rubio asked, the audience delighting in his new routine.[20]

"And you know what they say about men with small hands?" Rubio continued, grinning. As the crowd hooted and hollered, the senator hedged, "You can't trust 'em! You can't trust 'em!"

Rubio also observed, noting how Trump often teased him for sweating, "He doesn't sweat because his pores are clogged from the spray-tan he uses. Donald is not gonna make America great, he's gonna make America orange!"

The senator's friends were horrified. He had spent the past decade-plus distinguishing himself as a serious, sober-minded policymaker with an inspiring life story to boot. Now he was getting into the mud with Trump, cracking jokes about the size of a rival candidate's penis.

Super Tuesday offered no validation of Rubio's newfound approach. Of the eleven states voting, Trump won seven and Cruz carried three while Rubio's lone victory came in Minnesota, a race that was called so late in the night that it barely registered. The results did nothing to alter the broader trajectory of the race: Trump was on his way to becoming the GOP nominee, save for a dramatic intervening event.

Mitt Romney had one in mind.

He had spent months biting his tongue as it pertained to Trump. This was in part because he believed the party would rally around a strong alternative, and in part because he knew he wasn't an ideal messenger, having accepted Trump's endorsement in 2012, only to lose to the GOP's bête noire in a race many thought winnable.

But Romney could no longer stay silent. In a speech at the University of Utah on March 3, he urged voters to act strategically in the months ahead by backing whichever candidate had the best chance to win their state—rather than voting their preference—in the hope of denying Trump the delegates needed to be nominated outright in Cleveland. The Republican nominee of 2012 was calling for a political conspiracy to facilitate a brokered convention in 2016.

He also denounced the GOP front-runner in the harshest terms imaginable. "Here's what I know: Donald Trump is a phony, a fraud," Romney said. "His promises are as worthless as a degree from Trump University. He's playing members of the American public for suckers: He gets a free ride to the White House, and all we get is a lousy hat."[21]

It was more fight than Romney had ever shown against Obama, a fact that Trump and his acolytes used to paint the 2012 nominee as a weak-kneed traitor to the party.

But to Romney there was a qualitative difference: Trump was doing and saying things that Obama had never done or said. This wasn't about the party, Romney told his friends. It was about the country. Some warned him against it nonetheless; they worried that he would look duplicitous, especially when he declined to make any mention in his remarks of having accepted Trump's endorsement four years earlier.

Romney didn't much care. "I felt he was taking advantage of those who are racially insensitive or worse. And that led me to say, I've got to speak out now. I can't just be on the sidelines," he says. "I know not a lot of people pay attention to the former nominee, who doesn't even have a political office right now. But for me, for my family, for my grandkids, I didn't want them to say, 'Hey, where were you when this was going on?'"

The verdict among senior Republicans was unanimous: Romney had strengthened Trump. Having never channeled the cultural and economic frustrations of the party's base, Romney did nothing but demonstrate a familiar tone-deafness by attacking the man who was succeeding where he had failed.

Trump seemed to sense as much. The evening of Romney's speech, the GOP front-runner was in a jovial mood upon arriving in downtown Detroit for a Fox News–sponsored debate. It would be Trump's first face-to-face with Megyn Kelly since their dust-up in August, and his first encounter with Rubio since the comments about his spray tan and his "small hands." Walking up to Rubio backstage before the event, Trump spread his fingers and thrust his palm toward his rival's face. "Look at these," he grinned. "What are you talking about? My hands are not small!"

When the lights went on, the entire opening sequence was a fever dream. Trump began by calling Romney "an embarrassment" who was trying to remain "relevant."[22] He was then asked to renounce, after initially failing to do so in a recent CNN interview, the Ku Klux Klan. (He obliged: "I totally disavow the Ku Klux Klan.") Finally, after Rubio was asked to explain his recent detour to Gutterville, Trump jumped in. "He referred to my hands—'If they're small, something else must be small,'" Trump said, mimicking Rubio. "I guarantee you there's no problem. I *guarantee*."

That was just the first five minutes.

Jamie Johnson, an Iowa GOP official and activist, tweeted during the debate, "My party is committing suicide on national television."

RUBIO'S CAMPAIGN WAS BUILT ON NARRATIVE AND MOMENTUM, A house of straw that was pitifully staved by the gale-force winds of Trumpism. Unlike Cruz, who was anchored by a sprawling, disciplined field organization, Rubio had nothing to fall back on during rough patches in the campaign. He was, by every metric, more likeable, more relatable, more personally popular than his rival senator. But Cruz was running a tactically superior campaign.

The Kansas caucuses, held on Super Saturday, crystallized the shortcomings of Rubio's candidacy. He secured the endorsements of the state's major players, ranging from Senator Pat Roberts, an establishment mainstay; to Governor Sam Brownback, a Baptist-turned-Catholic social conservative stalwart; to Congressman Mike Pompeo, a Tea Party favorite whom conservatives had tried to recruit to run against John Boehner for Speaker. Yet Rubio had virtually no ground game in Kansas.

For Cruz, the opposite was true; he boasted no name-brand backers but owned the best organization in the state, an especially important advantage in caucus contests where participation is lower. The result was predictable: Cruz won Kansas with 48 percent of the vote, followed by Trump at 23 percent and Rubio at 17 percent.

It was not for a lack of trying on Pompeo's part. On caucus day, inside a convention center adjacent to a voting site, the congressman took the stage in front of thousands of caucus-goers and ripped into both Trump and Cruz. He accused Trump of being immoral and possessing dictator-like qualities that were dangerous to the country; he dismissed Cruz as a legislative thespian, someone more interested in stealing the spotlight than governing the country. Listening backstage, Trump turned to Jeff Roe, Cruz's campaign manager, and asked, "Who the hell is that?"

Roe replied that it was a congressman named Mike Pompeo. "Should we go put a scare into him?" Trump asked.

The two men walked to the opposite wings of the stage. There they stood—Trump, well over six feet tall with a world-famous scowl; Roe, every bit of three hundred pounds, with a hitman's goatee—shooting daggers at the congressman as he spoke, hoping to throw him off his game. (Pompeo, a former Army officer who graduated first in his class at West Point, was not rattled.)

Rubio went scoreless on Super Saturday, with Trump and Cruz splitting the four contests. As the losses mounted, and the life drained from his campaign, Rubio made an impassioned final plea. "Every movement in human history that has been built on a foundation of anger and fear has been cataclysmic in the end," he warned voters in Kansas. In Idaho, he added, "Don't give into the fear. Do not allow the conservative movement to be defined as anger."

And yet, anger was the currency of the campaign. It was a reality that Rubio himself had acknowledged and attempted, in his own milder manner, to harness. "Every traditional institution in America is failing you," he had told voters in Kansas, according to the AP, naming "the media . . . higher education . . . big business . . . and, by the way . . . your politicians and your political parties."

What Rubio took pains to avoid discussing during the 2016 campaign was how immigration played into that anger, and how his role in the Gang of Eight may have doomed his candidacy from the jump. Having retreated from his support for comprehensive immigration reform, Rubio chose during the primary to promote incremental efforts that would achieve the same result but in piecemeal fashion. This seemed, to him, a decent way of modulating his brand without actually changing his position. He still supported a path to citizenship, even though he wouldn't have uttered that phrase for all the Cuban coffee in Miami.

Try as he might, Rubio could never escape the scarlet letter of the Gang of Eight. Interestingly, though, his theory that Republican voters were broadly sympathetic to his views on immigration—more so than to those of the Steve King/Jeff Sessions wing of the party—was validated by the election data.

In twenty-five of the twenty-six states with exit polling, Republican primary voters ranked immigration dead last among the list of concerns, behind jobs and the economy, government spending, and terrorism. And in eighteen of the twenty states where the question was asked, a majority of GOP voters preferred legalization for undocumented immigrants as opposed to mass deportation.

Rubio had long argued in private that the loudest voices in the party were driving the argument on immigration and that those voices were not representative of the party's electorate as a whole. The data collected during the GOP primary, in the midst of Trump's rise, suggested he was right.

If Trump miscalibrated slightly on immigration, he was directly over the target on another issue of national métier and economic identity: trade.

The GOP had spent the past half century assuming that its voters viewed free trade as a positive, both for themselves and for the country in aggregate. With a few notable exceptions (Pat Buchanan, Ross Perot), the party had embraced the onset of globalism as an economic boon to the world's largest economy. But the twin phenomena of outsourcing and automation, on top of the broader transition to a tech-based economy and the lack of retraining programs for a generation of workers who had never earned college degrees, created a political powder keg. In 2016, it exploded.

The most visible evidence came from the Michigan primary on March 8. Trump won the GOP contest easily, taking 37 percent to Cruz's 25 percent with scraps for Rubio and Kasich. According to exit polls, half of GOP voters there were whites without a college degree; Trump dominated, winning 46 percent of them. On a key question for that demographic, a majority of all Michigan GOP voters (55 percent) said trade with other nations "takes away U.S. jobs." Trump won 45 percent of those respondents, compared to Cruz's 22 percent.[23]

Even more fascinating was the result on the Democratic side: Bernie Sanders, who trailed in Michigan polls by more than 20 points, stunned Clinton in the biggest upset of their primary battle, edging her by some 20,000 votes. A plurality of Michigan's Democratic electorate, 36 percent, were whites without a college degree; Clinton lost those voters badly to Sanders, 58 percent to 41 percent. An even bigger majority than in the GOP primary said that trading with other countries "takes away U.S. jobs," and Sanders won those voters by double digits, once again claiming 58 percent to Clinton's 41 percent.[24]

Taken together, these outcomes revealed not just the shifting political landscape but the unique opportunity for Trump to win the presidency. As the primary season progressed, voters in nearly every state across industrial middle America echoed their brethren in Michigan, telling exit pollsters that trade was doing more harm than good to the American economy.

These states—Michigan, Pennsylvania, Wisconsin—were thought to be Democratic locks, having not voted for a Republican presidential candidate since 1988. Yet, with Trump's Michigan triumph, and the polling on trade and other issues of economic nationalism, a once-faint sketch was coming into focus: The GOP, with Trump as its nominee, could redraw the electoral map in 2016.

CRUZ WAS RUNNING OUT OF TIME. HE WAS FALLING FURTHER BEHIND IN the delegate count, and despite his public insistence to the contrary, there was little evidence that he would win a one-on-one duel with Trump. Having once envisioned a "natural arc" to Trump's campaign, Cruz later adjusted his prognosis to set a ceiling of 20 percent on his rival's support. And then 25 percent. And then 30. And then 35.

By early March, with Trump clearing 40 percent in multiple nominating contests, it was clear that something radical needed to happen, and quickly, to prevent him from running away with the nomination. That's when an idea took root among Cruz's staff: a ticket with Rubio. The Florida senator was treading water and heading for a certain exit after losing Florida; what would happen if he teamed up with Cruz, running as his vice-presidential-pick-in-waiting?

Cruz was lukewarm to the idea. The two senators had a strained relationship, and the last several months, including Rubio's jab about not speaking Spanish, had been especially spiteful. But his outlook brightened upon seeing the polling. According to numbers compiled by Cruz's gold-standard data analytics team, a Cruz-Rubio ticket would demolish Trump in head-to-head competition in the remaining primaries, often winning more than 60 percent of the vote.

Cruz's pollster, Chris Wilson, called Utah senator Mike Lee to share the campaign's findings. Lee had not endorsed in the primary; he was close friends with both Cruz and Rubio. Reviewing the data, Lee sprang into action. He called dozens of hotels in the Miami area, needing one with an underground parking garage and an elevator that could ferry guests directly up to a private suite. Upon securing such an arrangement, at the Hilton Miami Downtown, Lee called Rubio to set up a meeting for March 9, one day before the Republican debate in nearby Coral Gables. He then informed Cruz that Rubio had agreed to a secretive sit-down at five o'clock that afternoon. Cruz cleared his campaign schedule and held his breath.

When the day arrived, several hours before the scheduled meeting, Lee received a text from Rubio. He did not feel right about the situation. The meeting was off.

The Utah senator, sitting in the hotel suite with his wife, was dejected. He viewed Trump as a threat to every principle he had entered public life to protect, and believed that the only way to defeat him, at this point, was if his two friends joined forces. With Rubio no longer interested, Lee grabbed hotel stationery and began sketching out his endorsement of Cruz, which he would announce the next morning. He was the first senator to endorse the Texan, an indication of Cruz's popularity among his colleagues. It was a fatal blow to Rubio—endorsing a nemesis on his home turf less than a week before the Florida primary—but Lee felt his friend had left him no choice.

Rubio says it "wasn't a serious consideration" to form a ticket with Cruz on top. "Mike, at the end, saw two friends running," Rubio says. "If I wasn't going to be president, there comes a point at the end of the campaign when you've invested almost two years of your life into it. The last thing you're looking forward to is now joining up on the ticket with somebody else you were just competing with. It was a zero percent chance that that was going to happen."

Cruz, for his part, still wonders what might have been.

"I believe it would have broken the race open. It would have been decisive in terms of winning, and for the life of me, I will never understand why Marco didn't come to that meeting," he says.

"When I lay awake at night frustrated about 2016, that night is the moment I go back to most often. I actually kick myself that I didn't go literally beat on his hotel room and talk to him," Cruz says. "I never had the chance to even talk to him."

The next evening, as the candidates and their teams arrived at the South Florida venue for what would be the final primary debate of 2016, a physical tension filled the air. Lee, having greeted Rubio, stuck close to Cruz's entourage. Rubio's crew gave him the cold shoulder. And Trump, taking it all in with childlike glee, eventually made his way over to Lee. "Hey," he said, grabbing the Utah senator by the arms. "Good luck with that endorsement."

The debate did nothing to alter the inevitable. On March 15, Trump trounced Rubio in the Florida primary, winning 46 percent of the vote to the home-state senator's 27 percent and driving him from the presidential race.

"It is clear that while we're on the right side, this year we will not be on the winning side," Rubio told a crowd of several hundred supporters at Florida International University. Rubio used his closing remarks to condemn the GOP front-runner (though not by name) for using "fear" to "prey upon" the insecurities of voters.

It was the end of a brief and underwhelming era. Rubio, the brightest star in the GOP galaxy on whom the hopes of so many in the party's establishment rested, was done. The "Republican Savior" had fallen short of messianic. "Michael Jordan" had missed his shot.

Having promised not to run for reelection to the Senate in 2016, Rubio appeared to be on his way out of politics—maybe even permanently. It would have made for an anticlimactic ending to a career that had held such promise just a few years earlier. Eventually, however, at the prodding of Mitch McConnell, and after a shooter killed forty-nine people at a gay nightclub in Orlando that June, Rubio changed his mind. He reentered the race to keep his Senate seat, all but clearing the Republican primary field and cruising to a second term that November.

As for the *presidential* primary field—and with due respect to Kasich, who won Ohio's primary on March 15, gaining token justification to keep campaigning—it was now down to two: Trump and Cruz.

CHAPTER THIRTEEN

MARCH 2016

"I think it was Roger's dying wish to elect Donald Trump president."

SCOTT REED DIDN'T RECOGNIZE THE NUMBER BLINKING ON HIS CELL phone screen. But there was no mistaking the gruff, gravelly voice on the other end of the line: It belonged to Paul Manafort.

A native of New Britain, Connecticut, where his extended clan was best known for its sprawling construction enterprise, the young Manafort grew up obsessed with the other family business: politics. He helped his father win three terms as the town's mayor, earned his undergraduate and law degrees at Georgetown University, and landed a job in the Gerald Ford administration soon thereafter.

Manafort caught his break in 1976, when the former California governor Ronald Reagan challenged Ford in the GOP presidential primary. Enlisted to help protect Ford's delegates from defecting to Reagan at the contested convention, he proved so effective that he was tasked with corralling the entire Northeast delegation. His role in sealing the nomination for Ford turned Manafort into a major player. In 1980, he partnered with veteran GOP operatives Charlie Black and Roger Stone to found the lobbying giant Black, Manafort and Stone, a firm described by *Time* magazine as "the ultimate supermarket of influence peddling."[1]

It was Stone, the Nixon-era hatchet man, who in the mid-1980s introduced Manafort to his friend, the New York real estate scion Donald J. Trump. At that time, Manafort's star was rising rapidly inside the GOP: He helped to elect Reagan twice, then George H. W. Bush in 1988. When Kansas senator Bob Dole won the Republican nomination in 1996, Manafort was charged with running the convention.

It was in this capacity that Manafort worked side by side with Reed,

Dole's campaign manager, a fellow northeasterner with a similar taste
for fine suits and expensive cocktails. They had known each other since
Reagan's reelection campaign in 1984, but Reed eyed his old friend
warily. He had heard all the stories, including the one about Manafort's
lifestyle turning lavish after millions of dollars in Filipino government
money, illegally earmarked for Reagan's reelection campaign and allegedly funneled through Manafort's consulting firm, never surfaced in
the United States.[2]

Twenty years after their work for Dole, Manafort was calling Reed,
the senior political strategist at the U.S. Chamber of Commerce, to
break some news: He was taking a job with Trump.

Reed was dumbfounded. Manafort certainly didn't need the money:
Between the legal revenues and the offshore bank accounts stuffed with
proceeds from his underhanded work on behalf of dodgy despots, including the pro-Russian leader of Ukraine, he was easily worth many
millions of dollars. And though Manafort had always been a gambler,
teaming with Trump struck Reed as a dicey bet. All the candidate's un-
foreseen successes in 2016 had come *in spite of* a functional campaign.

Trump had no serious organization to speak of, no overarching strategy guiding his efforts. There was the raw passion of his supporters;
the input of a few friends and unofficial advisers; the cloak-and-dagger
counsel of Stone; the guidance and unfailing loyalty of his children; and
there was Corey Lewandowski, the campaign manager, more street
fighter than savant, who fed Trump's belligerent instincts but lacked
any reasoned vision for reaching 270 electoral votes.

"Trump doesn't have a real campaign—it's just a bunch of guys
lighting everything on fire," Reed warned Manafort. "There's no organization, there's no infrastructure. If you join Trump, you'll wind up
running the campaign."

Manafort insisted he would not. Trump, he explained to Reed, was
growing ever-more suspicious of the party's efforts to defeat him at the
convention. There was an emerging, noisy "Never Trump" movement—
comprising activists, consultants, even some party officials—that aimed
to deny him the nomination by whatever means necessary. The immediate goal was to prevent him from collecting the 1,237 delegates needed

to clinch the nomination. But there was also talk of amending the rules in Cleveland to allow for "bound" delegates, those rightfully belonging to Trump, to vote against him on the convention floor.

"He just wants me to run the delegate operation," Manafort told Reed. "I'm going to make sure he secures the necessary delegates and secures the nomination. Nothing more."

"Look, Paul," Reed said. "I don't know exactly what you've been doing. But I know you've been in the Ukraine, with the penthouses and the vodka martinis and the caviar and the women on each arm. You had better be very careful. Remember the golden rule of politics: Nothing stays a secret. And believe me, with Trump, *everything* will come out eventually."

Manafort assured Reed that he would be aboveboard. He swore that, for whatever roguish work he'd taken on since the Dole days, he would not be getting into any trouble with Trump.

"Everything we'll be doing is legal," Manafort said.

TRUMP HAD SPENT SEVERAL DECADES BUILDING HIS OWN ASSOCIA-tions with the louche and depraved. As his campaign for the presidency gained surprising credibility, few of these allies proved as valuable as David Pecker.

As the chairman and CEO of American Media Inc., the country's largest tabloid publisher, Pecker had enjoyed a longtime symbiotic relationship with Trump, whose celebrity owed in large part to his engagement of the New York City gossip rags. The two men became close friends, sharing dinners at Mar-a-Lago and rides on Trump's private plane.

In August 2015, two months after Trump announced his bid for the presidency, they came to an understanding. In a meeting first reported by the *Wall Street Journal*, Pecker offered to protect Trump from women who came forward alleging sexual escapades. He would use AMI and its biggest brand, the *National Enquirer*, to "catch and kill" on behalf of the candidate: purchasing testimonies that could be damaging to Trump, having the women sign exclusivity and nondisclosure agreements, and then burying the stories for good. Trump loved the idea, and instructed Michael Cohen, his lawyer and fixer, to work in concert with Pecker.[3]

The arrangement would prove extraordinarily beneficial—at least, in the short run. Over the ensuing year, Pecker and Cohen defused two bombshells that might have blown up Trump's campaign. The first deal was with a former *Playboy* model, Karen McDougal, who approached AMI with details of her extramarital romance with Trump. Pecker bought the rights to her story for $150,000. Cohen, meanwhile, brokered an agreement with adult-film star Stormy Daniels, paying $130,000 in hush money to conceal her past sexual relationship with Trump.

All the while, Pecker was playing another role in Trump's run for the White House: that of lead blocker.

In September 2015, as Carly Fiorina gained steam in the GOP primary, rising all the way to third place in the RealClearPolitics polling average, the *National Enquirer* ran a piece calling her a "homewrecker" who had lied about her "druggie daughter."[4] (It was a reference to Fiorina's sharing the story of her stepdaughter who had died of an overdose.) The next month, as Ben Carson nipped at Trump's heels, the *Enquirer* reported that the "bungling surgeon" had ruined several patients' lives and had even left a sponge inside one woman's brain.[5] In December, as Marco Rubio moved into third place, the *Enquirer* published a story on the Florida senator's "cocaine connection," detailing his brother-in-law's incarceration for drug dealing.[6]

These were mere appetizers for Pecker and the *National Enquirer*. The entrée would be Ted Cruz.

In early March, the *Enquirer* formally endorsed on its front page: "TRUMP *MUST* BE PREZ." As it became apparent that the front-runner's path to the nomination had one remaining obstacle, Pecker and his minions turned their attention to Cruz.

The *National Enquirer* had run one piece in February, "Ted Cruz Shamed by Porn Star," about the senator's unwitting casting choice of a softcore adult actress in a campaign ad. But the story fell flat. Pecker's team dug deeper. Over the next month they turned over every rock of Cruz's personal and political life looking for dirt. At one point, AMI reporters visited the Capital Grille in Washington, Cruz's neighborhood haunt, offering cash to restaurant employees in exchange for compromising information on the senator. The waitstaff, having befriended Cruz (despite, in many cases, their wildly diverging political views), re-fused to cooperate.

On March 28, the same day Manafort's hiring was reported by the *New York Times*, the *National Enquirer* went nuclear. The tabloid published four stories pertaining to Cruz that day. But the biggest, its "Special Report," suggested that Cruz had carried on numerous extramarital affairs. Having been tipped off that this bombardment was on its way, Cruz chose to call his wife, Heidi, so that she wouldn't be blindsided. She laughed so hard, so hysterically, that her husband was mildly offended.

But whatever humor they found in the situation soon dissipated. Two days later, amid a flurry of other hit pieces on Cruz, the *Enquirer* piled onto its original report by printing the images, eyes blurred out, of five women the Texas senator had allegedly cheated with.[7] Three of them, it reported, were former staffers; one was a "sexy" schoolteacher; and the fifth was a DC prostitute.

The tabloid had finally broken through. Mainstream media outlets were forced to cover the allegations and the candidate's reaction. Google searches for "Ted Cruz affair" spiked. The hashtag #CruzSexScandal went gangbusters on Twitter. Cruz blamed Trump for the onslaught. "I want to be crystal clear: these attacks are garbage," the candidate wrote on his Facebook page. "For Donald J. Trump to enlist his friends at the National Enquirer and his political henchmen to do his bidding shows you that there is no low Donald won't go."

Trump responded, true to form, on his own Facebook page. "I have nothing to do with the National Enquirer and unlike Lyin' Ted Cruz I do not surround myself with political hacks and henchman and then pretend total innocence," he wrote. "Ted Cruz's problem with the National Enquirer is his and his alone, and while they were right about O.J. Simpson, John Edwards, and many others, I certainly hope they are not right about Lyin' Ted Cruz."

Amazingly, this was not the low point of the Trump-Cruz rivalry.

The week prior, an anti-Trump super PAC published a Facebook ad featuring a 2000 photo, taken for British *GQ*, that showed Melania

Trump nude. The ad, which targeted Mormon voters ahead of Utah's March 22 caucuses, read, "Meet Melania Trump. Your Next First Lady. Or, You Could Support Ted Cruz on Tuesday."[8]

Infuriated, Trump warned the world via Twitter that he might have to "spill the beans" on Heidi Cruz—whatever that meant. Trump later retweeted an unflattering photo of his opponent's wife that was posted in juxtaposition to a flawless-looking Melania Trump. The caption read, "No need to 'spill the beans.' The images are worth a thousand words."

Cruz finally lost his cool. "Donald, you're a sniveling coward," he said during a campaign stop in Wisconsin, looking straight into the camera.[9] "Leave Heidi the hell alone."

The mainstream media couldn't help but cover the story as the professional wrestling melee that it was—schoolyard taunts, nude women, the "cage match" Cruz had once scoffed at. The *National Enquirer* had sparked the fracas, a real-time embarrassment for the world's leading liberal democracy, but its more sophisticated counterparts in the Fourth Estate had fanned the flames, dedicating hours of breathless blow-by-blow coverage. Trump, a master manipulator of the media for so much of his adult life, had done it again.

Ultimately, it wasn't David Pecker and the *National Enquirer* that thwarted Cruz's candidacy. It was Roger Ailes and Fox News.

THE CRUZ CAMPAIGN HAD BEEN NEGOTIATING A SIT-DOWN INTERVIEW with Sean Hannity one day before the *National Enquirer* story broke alleging the senator's extramarital adventures. Wanting a forceful response but needing to move on from the story, Cruz's spokeswoman, Catherine Frazier, negotiated a deal with Hannity's producers. He would ask the candidate a single question, at the top, about the *Enquirer* report. Then they would turn to substantive matters.

But Hannity had other ideas. When Cruz dismissed the story as nonsense, attempting to pivot to discuss other topics, the Fox News host would not let him. Hannity continued to raise questions about the *Enquirer* story. Cruz grew angry. Finally, he blew up at Hannity, telling him the story had been planted by one Trump hack, Stone, the fabled "dirty trickster," in the publication of another Trump hack, Pecker.

Hannity's response? Stone had assured him, personally, that he had had nothing to do with the *Enquirer* report.

"Sean," Cruz exclaimed, "you're too damn smart to believe that."

The exchange was hot—so hot that it never aired. When Cruz's campaign staff tuned in for the segment, the tense back-and-forth wasn't included. They were mystified. As they discussed the reasoning, they decided that Fox had cut that portion not just because it made Hannity look bad, but because it made Trump look bad.

This was a recurring theme of the campaign, much to the chagrin and bewilderment of Cruz. He had been a mainstay on Fox News for the past three years, earning copious amounts of coverage for his crusade against the party establishment. But now he was being shoved aside. The network had a favorite new iconoclast, someone brasher and even more swashbuckling than he. This was, in its broadest sense, a reflection of the core dynamic between the two candidates.

"What Donald Trump did," observes Jim DeMint, "is out-Cruz Ted Cruz."

What Trump also did was out-*hustle* Cruz. The senator was a demon on the campaign trail, frequently making five or six stops on a bus each day, shaking hundreds of hands and taking more questions—from voters and reporters—than any other canditate. But those long days often turned into late nights. To wind down his brain, Cruz would ask a staffer to go buy a bottle of pinot noir and host the traveling team in his hotel suite, sipping wine and debriefing on the day's activities. This meant, at the instruction of Cruz himself, no campaign events before ten in the morning and, sometimes, no morning events at all.

By contrast, Trump (who does not drink) was always up before six, and typically dictating the day's news cycle with his Twitter feed. He met a fraction of the voters Cruz did, but knew, somehow, that it didn't matter. For a first-time candidate with no real consultants guiding him, Trump's instincts as a campaigner were phenomenal. And for a septuagenarian who would subsist on fast food and as many as twelve Diet Cokes a day, Trump's stamina was almost supernatural. He was game to go anywhere, engage anyone, and stay on offense at all hours of the day—an insurgency-style campaign that proved impossible to keep up with.

As the field winnowed down to what was essentially a *mano a mano* showdown, Fox's attitude toward Cruz became more pugnacious. In March, two paid contributors to Fox phoned the candidate with an ominous warning. "We're not allowed to say anything positive about you on air," they told Cruz. He thought it was a joke; they assured him it was not. "You've got to talk to Roger," one of them said, referring to Ailes. Cruz had already been trying. The senator once enjoyed a friendly relationship with the Fox News chairman, joining him for private breakfasts when he visited New York. But since the end of 2015, Ailes had not been responding to Cruz's calls.

A year later, when Ailes passed away, Cruz would tell friends, "I think it was Roger's dying wish to elect Donald Trump president."

The most galling expression of this, in the eyes of Cruz, came on the evening of April 5. The results of the Wisconsin primary were coming in. It was a huge prize for both candidates, with 42 delegates up for grabs, and an absolute must-win for Cruz. The campaign would move later that month into the northeastern states, Trump's backyard, and his rival's only chance was to arrive with a head of steam.

Everything had gone right for Cruz in the state. In populous southeastern Wisconsin, where conservative talk radio was renowned both for its influence and its pragmatic streak, Trump's negatives had soared sky-high in the polls. Two outside groups, the Club for Growth and Our Principles PAC, blanketed Wisconsin's airwaves with anti-Trump ads. The state's GOP establishment, led by Governor Scott Walker, rallied around Cruz as the party's last, best hope for toppling Trump. (Speaker Paul Ryan, who had spent the last four months ripping the front-runner behind closed doors, remained publicly neutral.) And swarms of pro-Cruz volunteers and super PAC workers descended on the state, seizing upon the lull in the primary schedule to out-organize the competition as they had done in neighboring Iowa. As the primary neared, polls showed Cruz opening up a double-digit lead in Wisconsin.

Cruz was ecstatic. He viewed Wisconsin as a watershed in the race, proving his capacity for beating Trump one on one and laying a blueprint for how to stop him in other contests. This was willfully naive; the stars had aligned in the Badger State in ways Cruz's team could not

hope to replicate elsewhere. Still, taking the stage in Milwaukee to celebrate his victory, Cruz called the Wisconsin result "a turning point,"
and a "rallying cry" for Republicans to defeat Trump. He touted his
consecutive delegate conquests in four states—Utah, Colorado, North
Dakota, and Wisconsin—before declaring, "We've got the full spectrum of the Republican Party coming together and uniting behind this
campaign."

Once Cruz had shaken hands and posed for pictures to commemo-
rate his triumph, he climbed onto his campaign bus and dialed into Fox
News. The initial signs were positive; Hannity's program was showing
images of him, not Trump, on the screen. (Cruz had become accustomed
to seeing all three cable networks, CNN, MSNBC, and Fox News, showing his opponent simultaneously, a situation his advisers referred to as
"The Full Trump.") Relieved, Cruz settled in to hear the analysis. Just
then, however, Hannity began discussing the night's developments with
Laura Ingraham.

"If Ted Cruz can keep beating Donald Trump in state after state after
state—" she began.

"Can he?" Hannity interrupted.

"I don't see that happening in a place like New York and especially the
New England states," Ingraham replied.

"Yes, New York's got to be Trump's firewall," Hannity said. "He's going to win New York. He's up by thirty-four points."

Cruz's grin turned into a grimace.

When Hannity returned from commercial, he was joined by news anchor Bill Hemmer, who ran down the slate of upcoming state contests.

"New York—winner take all. Right now, Trump looks pretty good in
New York," Hemmer said. "End of the month here, you've got five states
in the Northeast. Trump looks pretty good in all five. And then we clear
the month of April and move to May. And on May third is Indiana. We've
looked at the numbers so far. We're crunching them. Looks pretty good
for Trump. Go a week later, West Virginia looks good for Trump. That's
winner-take-all, by the way . . ."

Cruz leapt from his seat. "What the *fuck*?" he screamed at the television. His staffers were startled and more than a bit surprised. Their boss

was not the emotional sort. But months of building antagonism toward Trump, and frustration with Fox News, could no longer be suppressed.

Cruz flopped back into his seat. He had just secured his biggest victory to date, yet he felt deeply defeated.

IN THE TWO WEEKS BETWEEN WISCONSIN ON APRIL 5 AND NEW YORK on April 19, the Cruz campaign laid the groundwork for its last stand.

Despite the candidate's public projections of confidence, everyone knew Trump was poised to steamroll through the northeastern primaries and crush Cruz in late April. If that happened, the campaign would need an abrupt, high-profile victory to stop the bleeding. Their best shot: Indiana on May 3.

The state offered 57 delegates, an electoral jackpot that, if hit, could make Trump's delegate math unworkable. Cruz's team began throwing everything they had into Indiana, hoping to reapply the formula that had worked in Wisconsin. But despite some demographic similarities, Indiana bore little electoral resemblance. There was no multimillion-dollar assault from outside groups on Trump. The conservative talk radio army was nowhere to be found. And unlike in Wisconsin, where Cruz was backed by much of the party establishment, Indiana's top officials showed no signs of support. Trump was far too popular in the state for Republican leaders to risk disaffecting their base by denouncing him.

Mike Pence was Exhibit A.

The governor loathed Trump, his longtime friends and allies whispered at the time, viewing his personal indiscretions and campaign rhetoric as destructive to the cause of conservatism. But Pence was in no position to do battle with the GOP front-runner. He had been damaged goods since early 2015, when the religious liberty dispute blew up in his face. The governor's actions had alienated almost every constituency imaginable—the left, the socially moderate center, the business community—when he first signed the legislation, and then, for good measure, the conservative base and evangelical right when he backtracked. By the spring of 2016, things looked grim. Pence's popularity had tanked, his approval rating was underwater in public and private polls, and he was running even in his race against Democrat John Gregg.

Almost uniformly, Pence's friends believed his political career was slipping away. The last thing he needed was a war with Trump.

But Cruz wouldn't go away. For several weeks in April, he put a full-court press on Pence: phone calls, text messages, emails from mutual friends. He finally secured a lengthy private meeting, and later, a formal invite to the Indianapolis GOP spring dinner, where Cruz gave a speech and sat at the governor's table. The senator implored Pence to do what was right, not just for his candidacy but for the conservative movement.

The governor began to wear down. For all his political ambition and keen sense of self-preservation, Pence was a true believer. All the way back to his earliest days as a think tank president and talk radio host, Pence had approached politics with a zealot's sincerity. In 1999 he wrote an opinion piece trashing the Disney film *Mulan*, the story of a Japanese girl who disguises herself as a man to join the military. "I suspect that some mischievous liberal at Disney assumes that Mulan's story will cause a quiet change in the next generation's attitude about women in combat and they just might be right," Pence warned.[10]

With his party's nomination potentially hanging in the balance, and a like-minded conservative pleading for support to stop someone they both viewed as unfit for the office of president, Pence thrilled Cruz by informing him of his endorsement. There was one condition: Pence said he would not, *could not*, disparage Trump the way Walker had in Wisconsin. He would endorse Cruz but say nothing negative about Trump.

As the final arrangements were made for Pence's endorsement, Cruz offered another surprise to the voters of Indiana. On April 27, six days before the state's primary, Cruz introduced Carly Fiorina, the onetime Hewlett-Packard CEO and a former rival in the Republican race, as his running mate.

Cruz was desperate for a shift in momentum. One day earlier, Trump had swept him in the April 26 primaries in Connecticut, Delaware, Maryland, Pennsylvania, and Rhode Island, with Cruz failing to break 25 percent in any of the contests. Fiorina made perfect sense as a vice-presidential pick: She had endorsed him sometime ago and proved herself to be an effective surrogate, especially when it came to connecting with conservative women who liked Cruz's policies but found him personally unpleasant.

The *National Enquirer* was not impressed. "Carly Fiorina Plastic Surgery—Fake Face of Ambition," its headline screamed.

Two days after the Fiorina announcement, on April 29, Pence announced his support for Cruz—in a fashion even more lukewarm than anyone in Cruz's camp had anticipated.

"I particularly want to commend Donald Trump, who I think has given voice to the frustration of millions of working Americans with a lack of progress in Washington, DC," the governor said on a local radio program. "And I'm also particularly grateful that Donald Trump has taken a strong stance for Hoosier jobs when we saw jobs in the Carrier company abruptly announce leaving Indiana not for another state but for Mexico."

Only then did Pence transition to his endorsement. "I'm not against anybody, but I will be voting for Ted Cruz in the upcoming Republican primary," he said.[11]

Pence later noted his admiration for John Kasich as well, and encouraged Indiana voters to make up their own minds.

Cruz had Fiorina and Pence in his corner; Trump had Bobby Knight. The legendary University of Indiana men's basketball coach, more famous for his chair-chucking antics than for his 763 career wins on the Hoosier bench, campaigned with Trump around the state in the days before the primary.[12] "That son of a bitch can play for me!" Knight cried at one campaign event.

Cruz, for his part, tried to reenact a scene from the film *Hoosiers*, staging a rally inside a local gym and measuring the ten feet between the floor and the bucket. He called it a "basketball ring," a jarring malapropism in the hoops-mad state that did little to quell talk of his weirdness.

As they scratched their heads over Pence's tepid show of support, Cruz's staff discovered that the governor was of little use anyway. Their polling revealed that Pence was more unpopular than originally thought. Only in two areas of the state were the governor's numbers right side up *among Republicans*. Having once envisioned a four-day sprint across the state with Pence in tow, Cruz's team now worried that it might do more harm than good. Ultimately, the governor did not join Cruz on the stump until May 2, one day prior to the primary. Meanwhile, Trump missed no opportunity to mock Pence's endorsement, calling it "a very weak one" that came in response to pressure from big donors.

"I think what he said about me was nicer than what he said about Cruz," Trump said the day before the primary.[13] "All the pundits said, 'You know what, I think that was maybe the weakest endorsement in the history of endorsements.' In the end, they had to re-run the tape just to find out who he was endorsing."

A game-changer in the Republican primary Mike Pence was not.

CRUZ KNEW THE END WAS NEAR. ON SATURDAY, APRIL 30, AS HIS WIFE campaigned in Indiana on his behalf, he flew to California for the state's Republican convention. With its June 7 primary marking the grand finale of the GOP primary schedule, and a whopping 172 delegates at stake, California had become an object of obsession inside the party. Trump led Cruz by more than 400 delegates heading into Indiana, but the question remained whether he could reach the "magic number" of 1,237 needed to clinch the nomination.

Cruz had long taken a defiant stance, insisting to his top aides and biggest donors—and to himself—that he would remain in the race all the way through California. Very recently, however, he had begun to reconsider. He was only four years into his national political career; at forty-five years old, his future in the Republican Party was limitless. While Cruz was painstakingly close to the biggest prize in party politics, Trump's lead appeared increasingly insurmountable. And for as much as he had come to despise the GOP front-runner, Cruz had also come to recognize the transcendent connection Trump had with the party's base. Would it be worth making so many enemies, and tarnishing his strong second-place showing, in the pursuit of a victory that seemed unattainable?

Complicating this question was the continued presence of John Kasich. The Ohio governor had won exactly one nominating contest—*in Ohio*—yet remained an active candidate. He had no money and no campaign infrastructure across the country, but the media coverage of his centrist messaging was effective enough to peel off chunks of delegates in any number states. If the nominating fight was going to result in a brokered convention, every single delegate would count. And if Cruz was going to pursue the long-shot strategy of winning under such a scenario, he needed Kasich out of the race.

In the bowels of a Hyatt Regency near the San Francisco airport, not far from where the state's GOP convention was unfolding, the Texas senator stepped into a top-secret meeting with the Ohio governor.

"We can't beat Trump two on one," Cruz told Kasich. "One of us has to drop out. That's the only chance we have for a Republican to win the nomination."

"Do what you need to do, Ted," Kasich replied. "But you need to understand under no circumstances am I getting out of this race. I'm going all the way to the convention in Ohio. Nothing can change that."

Cruz frowned. "John, do you realize the consequences of that? You are making it certain that Donald Trump will be the nominee."

"Ted," he replied, "I am not leaving this race."

Dismayed, Cruz flew back to Indiana and informed his senior staff that preparations should be made for his withdrawal from the primary. He had employees all over the country, most especially at the headquarters in Houston, who had never been out on stump with him. He and his campaign manager, Jeff Roe, wanted them flown to Indiana on Tuesday. It would be their last chance to feel the heat of the campaign trail.

Cruz was nursing open wounds as the final hours of his campaign wound down. Naturally, Trump found a way to fill them with salt and lemon juice.

"His father was with Lee Harvey Oswald prior to Oswald's, you know, being shot. I mean, the whole thing is ridiculous," Trump said about Cruz's father, Rafael, during a Fox News interview on the morning of the Indiana primary.[14] "What was he doing with Lee Harvey Oswald shortly before the death? Before the shooting? It's horrible."

Trump's remark was in reference to a *National Enquirer* "World Exclusive!" published on April 20 that implicated Rafael Cruz in the assassination of President John F. Kennedy. Shockingly, it could not be confirmed by other news organizations or corroborated by law enforcement sources.

Cruz had tried to discover a peace about his pending departure from the campaign. But Trump's provocation triggered something he had buried deep inside: a gush of pure, unrestrained hatred for the man Republicans were choosing as their standard-bearer.

"I'm going to do something I haven't done for the entire campaign.... I'm going to tell you what I really think of Donald Trump," Cruz told reporters shortly after Trump's Fox News appearance.[15] "This man is a pathological liar. He doesn't know the difference between truth and lies. He lies practically every word that comes out of his mouth, and in a pattern that I think is straight out of a psychology textbook, his response is to accuse everybody else of lying. The man cannot tell the truth, but he combines it with being a narcissist—a narcissist at a level I don't think this country's ever seen. Donald Trump is such a narcissist that Barack Obama looks at him and goes, 'Dude, what's your problem?'"

Calling his archnemesis "a serial philanderer" who is "utterly amoral," Cruz concluded, "Donald is a bully . . . Bullies come from a deep, yawning cavern of insecurity. There is a reason Donald builds giant buildings and puts his name on them everywhere he goes."

Trump's response was vintage: "Today's ridiculous outburst only proves what I have been saying for a long time, that Ted Cruz does not have the temperament to be president of the United States."

Hours later, Trump trounced Cruz in the Indiana primary, winning by 16 points and capturing all of the state's 57 delegates.

Cruz promptly quit the race. "From the beginning I've said that I would continue on as long as there was a viable path to victory," he said, his wife, Heidi, standing by his side. "Tonight, I'm sorry to say, it appears that path has been foreclosed."

The next morning, Kasich headed for the Columbus airport. He had back-to-back fund-raisers scheduled in Washington. Sitting on the runway, however, he experienced an abrupt change of heart. "Screw it," he told his traveling companions. He wanted to drop out of the race, too.

When Cruz learned of Kasich's decision, the color went out of his face.

He looked gravely ill for the day's remainder. Two friends who were with the senator worried for his health.

Reflecting on the campaign in its final hours, Cruz believed he had been done in by two incidents he would give anything to have back: the perceived cheating against Carson in Iowa and Rubio's refusal to form a ticket in early March. Now there was a third: Kasich's bluff in California. The trilogy of regrets would haunt Cruz in the months, and years, to come.

As Kasich walked off his plane in Columbus, and Cruz rued the hand of providence back home in Houston, their opponent celebrated with friends and family in New York City.

Reince Priebus called to offer congratulations. Donald J. Trump would be the Republican Party's nominee for president in 2016.

CHAPTER FOURTEEN

MAY 2016

"Now that you've gone this far, there's no going back."

HUNDREDS OF PROTESTERS, REPORTERS, AND UNAFFILIATED GAWK-
ers swarmed outside the offices of the Republican National Committee on First Street Southeast, a few short blocks from the Capitol.
The circus had come to town. As Donald Trump's entourage pulled up,
sneaking him into a side entrance of the building, the gawkers gawked.
The reporters shouted questions. And the protesters hoisted signs:
"R.I.P. G.O.P."

Inside the party headquarters, Paul Ryan stewed. This wasn't what
he had signed up for. Trump had looked increasingly viable when the
new Speaker took over for John Boehner the previous October, but Ryan
never, ever, took seriously the prospect of the reality TV star winning his
party's nomination. Everything Ryan knew about politics told him that
it couldn't happen. Nervous nonetheless, he checked in often with his
old pal from Wisconsin, Reince Priebus, to make sure. Priebus's answer
was steady throughout the summer and fall: "Not gonna happen." Yet, as
the calendar turned to 2016, the chairman's certitude softened. When
they talked just before Christmas, Priebus broke the news. Trump, he
told Ryan, might just win the nomination after all.

This sent the Speaker into a panic. Having been on the GOP ticket
four years prior, having seen the devastation wreaked by Mitt Romney's
insularity, Ryan had returned to Congress a changed man. Everything
he had done, including accepting the promotion to Speaker, had been in
service of softening the GOP's brand to reach a broader swath of a diversifying nation. This would allow Republicans to win elections and sub-
sequently pass meaningful policy reforms.

Trump was dashing those dreams. Ryan had to remain neutral in the
race; as Speaker, he would be chairing the party's convention later that

summer. But as Trump's momentum built, so, too, did Ryan's naysaying. He denounced Trump's proposed Muslim ban, saying it's "not what this party stands for, and more importantly, it's not what this country stands for."[1] He slammed him for his strange hesitation in disavowing David Duke and the KKK. He blasted him for suggesting there would be "riots" in Cleveland if he were denied the nomination.[2]

As Ryan worked himself into a lather, whispering to Republican allies about Trump's instability and immorality, the GOP front-runner was busy steamrolling the competition. By late April, Trump was already turning his attention to Hillary Clinton. "I think the only card she has is the woman's card. She has nothing else going. Frankly, if Hillary Clinton were a man, I don't think she would get 5 percent of the vote," Trump said. "The beautiful thing is, women don't like her."[3]

Ryan's warnings about Trump—that he was exploiting voters' fears; that he was using "identity politics" to turn working-class whites against brown and black Americans; that he was ethically bankrupt and dangerously divisive—were shared by his peers in the governing class. But the Republican primary voters felt differently. They had elevated the brash political neophyte over a primary field that many party elders felt was their deepest, strongest, and most diverse in at least a century.

The Speaker was not ready to follow the voters' lead.

"I'm not there right now," Ryan told CNN on May 5, two days after Trump became the GOP's de facto nominee. "I think what is required is that we unify this party. And I think the bulk of the burden on unifying the party will have to come from our presumptive nominee."

Trump responded in a statement that read, "I am not ready to support Speaker Ryan's agenda." Trump also suggested that Ryan ought not to serve as the convention's chairman.

Ryan, in turn, offered to step down if Trump so requested. The Speaker's performance was that of a political Hamlet, pondering the existential ramifications of subjugating himself to the evil new king.

It was against this backdrop, on May 12, that Trump arrived at RNC headquarters. On the itinerary was a roundtable discussion with all the GOP congressional leaders. But first, privately, Trump would meet with Ryan and Priebus.

The party chairman was desperate to broker a truce. Sitting them down in his office, Priebus tried to clear the air, talking of "party unity" that could only come from the two men setting aside their differences. Trump and Ryan, like a pair of high-schoolers called into the principal's office after fisticuffs, listened silently, recalcitrance written across their faces. When Priebus finished, Ryan told Trump he wanted to show him something. It was a PowerPoint presentation. The country was drowning in red ink, Ryan explained, and could be saved from a debt tsunami only by a reforming of the tax code and a restructuring of Social Security and Medicaid. Flashing the first slide onto a monitor, Ryan prefaced his remarks by clarifying the basic distinction between mandatory spending and discretionary spending.

After Ryan popped the second slide onto the monitor, Trump interrupted him. "Okay, Paul, I get the point," he said. "What's next?"

Ryan was astonished. He shot a look at Priebus. The party chairman avoided eye contact.

"The meeting was great," Priebus tweeted a short while later, after Trump convened with the larger group of congressional officials. "It was a very positive step toward party unity."

The Speaker played along. He told reporters that Trump had been "warm and genuine" in their interactions. But Ryan, the last holdout among the GOP's elected leadership, remained cold to the idea of endorsing the party's presumptive nominee. Indeed, he still couldn't get his head around the fact that Trump *was the party's presumptive nominee*. With all that baggage, after all those years of all those controversies, how had no opposition research surfaced to sink his candidacy? And what would happen if it finally did, just in time for the general election?

Trump didn't like Ryan. He found the Speaker dull and supercilious, "a fucking Boy Scout," as he told friends after the meeting. But the party's new standard-bearer was not averse to being schooled by the GOP establishment. Trump did not suffer from a lack of teachability; he simply preferred to dictate the flow of information, rather than be dictated to. Lengthy briefings and conference calls were never a staple of his executive style. He favored an aggressive, inquisitive approach, learning about issues, and about people, with rapid-fire questioning, consuming what he needed from the answers and discarding the rest.

After eliminating his final competitors in early May, Trump knew that he needed a crash course on what lay ahead. This was how he came to sit down with Karl Rove.

Trump didn't particularly like Rove, either. He found the "architect" of George W. Bush's winning campaigns to be haughty and condescending. For much of the past year, Trump had raged against Rove when reading his columns in the *Wall Street Journal*, many of which were pitilessly critical of the GOP front-runner. On numerous occasions, Trump reached out to a mutual friend, the casino magnate and GOP megadonor Steve Wynn, asking him to relay his displeasure to Rove.

In early May, Rove's phone rang. "Karl, kiddo, I talked to Donald and he wants you to write something nice about him," Wynn said. "He won the Indiana primary. Can you write something nice about him?"

"As a matter of fact, I just got done writing a column, and I said some nice things about him," Rove replied. "Would you like to hear it?"

Rove read portions aloud. He said that Trump had "bludgeoned 16 opponents into submission" and "rewrote the rule book," beginning the column with a blunt declaration: "No one has seen anything like this."

Wynn approved. But the next morning, he called Rove back. Trump hated the column. Rove had castigated the candidate for his endless string of insults, called the JFK–Rafael Cruz talk "nuts," and written, "Trump's scorched-earth tactics have left deep wounds that make victory more uncertain."[4]

Wynn read Rove the riot act on behalf of his friend. But then he added something surprising: Trump wanted to sit down to talk strategy. "He says he wants to meet with you and get your advice," Wynn told Rove. "He knows you did this twice."

A few weeks later, on May 23, Rove surveyed the nine-hundred-square-foot living room of Wynn's apartment in New York City. The setting was fabulous: Situated on the thirtieth and thirty-first floors of the Ritz-Carlton Hotel, the ballroom turned domicile featured, among other things, fifteen-foot cathedral ceilings, a library, a media room, and a private terrace overlooking Central Park South. Rove had arrived two hours early, wanting to keep the meeting private and avoid the media scrum surely accompanying Trump. Yet the candidate arrived by himself, right on time, without any entourage or fanfare. He, too, seemed intent on secrecy.

Trump and Rove had met before: In 2010, Rove traveled to Trump Tower to solicit funds for his super PAC, American Crossroads. He walked out with a $50,000 check. The small victory earned Rove some ribbing from Steven Law, a former Mitch McConnell aide and American Crossroads' president. "Congratulations," Law told Rove. "I think you're the first Republican I've ever known to get a check from Donald Trump."

There wasn't much foreplay when they sat down across from one another inside Wynn's opulent living room. Trump asked Rove what he needed to know. Rove, in firehose fashion, launched into his lecture on the contours of the Electoral College.

"You have to have a strategy to get to 270. We had several paths to get there," Rove began. "We had the traditional battleground states, which were Florida, Ohio, New Hampshire, Iowa, New Mexico, Colorado, Nevada, Michigan, Pennsylvania, Wisconsin, Minnesota. And we had four battleground states that had traditionally been carried by Democrats: Arkansas, Tennessee, Kentucky, and West Virginia."

"West Virginia?" Trump interrupted. "I did really good in the primary there. I can win West Virginia—that's a big Republican state."

"Well, in 2000 it wasn't," Rove explained. "Bob Dole had lost it by fifteen points four years earlier. The last time it had gone for Republicans in an open-seat presidential race was 1928, and it took nominating a New York Catholic to bring all the Methodists and Presbyterians and Baptists out of the hills and hollows of West Virginia to vote Republican."

Rove worked his way around the map. When he reached the West, he focused on four states, Montana, Nevada, Arizona, and Oregon, explaining that Bill Clinton and Al Gore had each carried at least one of them. Trump, Rove said, would need to win at least two—and probably three— to stand a chance in 2016.

"Oregon? I can win Oregon," Trump said excitedly. "I did really good in the primary there."

"No, you can't," Rove cautioned. "In 2000, we had Ralph Nader on the ballot there, and he had a real following in Portland and Eugene; the state had just elected a Republican U.S. senator; they had Republican constitutional officers; they had a Republican majority in the statehouse; and we still lost it by half a point. Since then, it's gone hard left. The last time we won a statewide race was 2002; we hold no constitutional offices; and we're down to less than a third in the statehouse and a third in the state Senate. There's no way you can win Oregon."

Trump smirked. "I don't need to," he said. "I can win California."

"No, you really can't," Rove chuckled, wondering whether the candidate was being facetious. Judging from Trump's expression, he was not. "You're down seventeen points in the RCP average," Rove told him. "It's a giant suck of time and money. There's no way you can win California."

Trump was growing irritated. "Well, I'll win New York."

Rove sighed. "No, you won't. Bernie Sanders got more votes by himself than all the Republicans combined. Two and a half times the number of people voted in the Democrat primary than the Republican primary. You're losing to Hillary by twenty-six points in the RCP average, and it's a waste of time. If you spend a day trying to win votes in a place like California or New York or Oregon, it's a day you can't spend trying to win votes in Pennsylvania or Iowa."

Trump looked puzzled. "I can win Iowa?"

"Oh yeah," Rove cooed, building the candidate back up after tearing down his illusions. "You didn't win the caucuses, but those farmers in the western part of the state, they hate her guts. And there are a bunch of blue-collar workers in the eastern part of the state that are worried about their jobs. You can win Iowa. But not if you're spending your time in Oregon, California, and New York."

Trump turned to Wynn. "Why aren't people in my campaign talking to me about this?"

(Three days later, Trump gave a speech naming the "fifteen states" that he would campaign in. Among them: New York and California.[5])

As the conversation progressed, Trump grew less defensive. He seemed to recognize that Rove, however patronizingly, was trying to help him succeed. Trump's clutch of advisers talked little of long-term strategy or historical voting trends; mostly, they urged him to concentrate on animating the base with his rhetoric and policy positions. He had long dismissed the complaints from his adult children that Corey Lewandowski, the campaign manager, was doing him a disservice. But now, as he soaked up a briefing of unprecedented depth, Trump was beginning to wonder.

The meeting spilled into its third hour. Rove coached Trump on everything he could think of, from campaign finance to parochial swing state policy disputes. When the conversation turned to Pennsylvania, Wynn complained about Chinese steel, and Trump sounded off on the country, saying the United States should never have allowed it to join the World Trade Organization. "Actually, we should have," Rove corrected him, "because that binds them to an international set of trading norms, and if they violate them, we can take action in front of the WTO. It takes a little time to do it, but in 2015, the Obama administration filed like one hundred and fifteen actions against China and other actors, and if history is any guide, we'll win almost every one of those actions and recoup money for affected industries."

Trump arched an eyebrow. "Really? We can do that?"

Rove nodded. *This guy has been talking about trade for thirty years,* the Republican Svengali thought, *and he doesn't know the basic tools at the president's disposal.*

The Republican Party's new leader was curious about one more thing. His team had been preparing a list of vice-presidential selections, but he felt that everyone advising him on the decision was pushing an agenda. He wanted to know what Wynn and Rove thought.

"Kasich, no question," Wynn volunteered.

Trump frowned. "He doesn't say nice things about me. Who else?"

"Well," Rove said, "I think your battlegrounds are going to be between Pennsylvania and Iowa, and if you're going to break the Blue Wall, you need someone with midwestern sensibilities and someone who has evangelical appeal. There's one guy who fits that description: Mike Pence."

It was the strangest of smoke-filled rooms, a Central Park château populated by the renowned party strategist alternately called "Boy

Genius" and "Turd Blossom" by his former boss; the financier and casino tycoon who would soon become a high-profile casualty of the country's sexual harassment crackdown; and the rookie politician who had
heckled and hoodwinked his way to the Republican nomination for
president. It wasn't quite how Jack and Bobby had picked LBJ, or how
Reagan had settled on George Bush Sr., but a seed was planted that day.

Trump allowed a smile at the suggestion of Pence. "He says nice
things about me."

ON THE EIGHTH FLOOR OF THE MARRIOTT MARQUIS IN TIMES SQUARE,
Marjorie Dannenfelser stabbed anxiously at a plate of salad while offering a series of defensive answers.

It was June 21, and Dannenfelser, a social conservative titan and
president of the anti-abortion group Susan B. Anthony List, was one of
nearly a thousand Christian activists who had traveled to New York City
for an afternoon summit with the presumptive Republican presidential
nominee. She also was one of roughly fifty people to join him for a VIP
meeting beforehand. Many of these leaders, including Dannenfelser,
had vigorously opposed his candidacy throughout the campaign.

Yet, much like Ryan—who had finally dropped his objections earlier
that month, endorsing Trump in a piece for the *Janesville Gazette*—they
were beginning to feel as though they had no recourse.

"All along the way, he was our last choice," Dannenfelser said. "But
when you get to the end, to the point of having a binary choice, you must
choose."

This sentiment echoed around the Manhattan hotel's ornamented
hallways. Some prominent Christian leaders, including Liberty University president Jerry Falwell Jr., went out of their way to lavish Trump
with praise despite his sui generis secularity. (After introducing Trump
at the New York summit, Falwell Jr. posed for a photograph alongside
the candidate back at Trump Tower, with a *Playboy* magazine cover on
the wall behind them.) But for most of the faith leaders in attendance,
Trump represented the manifestation of their fears about societal decline. Here was a man who had paraded his mistresses through the tabloids; who had bantered with Howard Stern about the size of his own

daughter's breasts; who had previously taken extreme pro-abortion positions; who seemed to marinate in coarseness and cruelty; and who had nonetheless won the GOP nomination for president.

These concerns were not necessarily allayed during the VIP meeting. Speaking to the group of spiritual influencers, Trump said of Christianity, "I owe so much to it in so many ways." He then proceeded to explain that he wouldn't be standing before them without it, not because of how the faith shaped his life or informed his worldview, but "because the evangelical vote was mostly gotten by me." The attendees walked out of the room in a daze.

The general session went somewhat better, thanks to the lively introductions of Falwell Jr. and Franklin Graham, another descendant of American Christendom royalty. Graham remarked on how God had used deeply flawed men throughout history to shape the world for good, drawing parallels between Trump and David, the giant slayer and Israeli king who ordered the husband of his mistress killed in battle. The comparison left some in the room feeling queasy.

Trump spent much of his remarks acting as though he were before any other audience, giving a self-glorifying rundown of the latest polls and his recent media coverage. But at some point, either because of his own observation or due to a planned transition, Trump switched gears. He deployed carefully curated phrases, including "pro-life judges." The attendees, in decades of hearing from Republican political figures, had never heard someone so bold as to use that terminology; a typical conservative politician would use coded language to assure voters of such a priority. Trump also broke new ground when he raised, unsolicited, concerns about a fifty-year-old law implemented under President Lyndon Baines Johnson that could threaten the tax-exempt status of churches that spoke out on social issues. Prompted by members of his newly formed Evangelical Executive Advisory Board, Trump warned about the "Johnson Amendment," and promised to fight on behalf of Christians in a way that no political leader had before.

This was like David's harp to Saul's ears. Eight years of Barack Obama's presidency had left the white evangelical community feeling besieged, not just from the forces of big government, which approved same-sex

marriage and mandated contraceptive coverage, but from a godless, violent, overdrugged, hypersexualized culture that was chewing through
the fabric of their Judeo-Christian civilization. "Evangelicals had been
used over and over by Republicans. And there was something different
about his interaction with us," recalled Tony Perkins, the president of
the Family Research Council. "You could describe it as transactional.
He wanted our votes, and he made promises that most Christian candidates would never, ever make."

Ever since Indiana, prominent evangelicals had advised Trump that
he needed to do two things to win their voters. The first was to emphasize a commitment to conservative judges. In the wake of Scalia's death,
and McConnell's refusal to allow a hearing for Obama's nominee, Mer-
rick Garland, the looming Supreme Court vacancy was the ultimate
mobilizer for Christian conservatives. Trump did them one better: In
mid-May, after running a wildly unconventional idea past several allies
(including Leonard Leo, president of a GOP lawyer association called
the Federalist Society, and Don McGahn, the future White House counsel), he released a list of conservative judges he would pick from for Scalia's seat.

"I had no idea how important Supreme Court judges were to a voter,"
Trump admits. "When I got involved, deep into it, I realized that there
was tremendous distrust of me because they didn't know—was I a conservative? Was I a liberal? They didn't know anything about me."

He pauses, sensing how this might sound demeaning to his celebrity.
"They knew me very well. *The Apprentice* was one of the most successful
shows on television by far. They knew me; they got to know me very well,
they knew me long before *The Apprentice*. That's why I was chosen to do
The Apprentice, right?"

He continues, "But what they didn't know, is he going to like conservative judges? Or is he going to like liberal judges?"

The judicial roster won glowing reviews from the religious right. But
there was another box for Trump to check.

"We want to see," Perkins said after the New York summit, "who he
picks as his running mate."

WHILE THE PRESUMPTIVE REPUBLICAN NOMINEE WAS HARD AT WORK attempting to heal the lacerations suffered in his primary, Democrats were still swinging knives.

The contest to succeed President Obama atop the Democratic Party was underwhelming. It drew only six declared candidates, three of whom withdrew before the voting began. Of the remaining group, Martin O'Malley, the former Maryland governor, promptly exited the race after a distant third-place finish in Iowa.

That left just two contenders for the nomination: Hillary Clinton and Bernie Sanders.

Everyone in politics recognized that the two parties had been systematically weakened since the turn of the century. What no one could have predicted was that the two candidates who most energized the party bases in 2016, Trump and Sanders, did not actually *belong* to the parties.

Though Sanders caucused with the Democrats during his quarter century in Congress, first in the House and later in the Senate, he was an independent and self-described socialist, a left-wing version of the erstwhile GOP presidential candidate Ron Paul: a ruffled, doctrinaire, septuagenarian zealot. Much like Paul, whose brand of strident libertarianism struck a chord with portions of the post–George W. Bush Republican Party, Sanders initially seemed more energized by influencing the post-Obama Democratic Party's direction rather than winning its nomination.

And then, the familiar flaws of his opponent resurfaced.

More than any figure in American political life—more than Obama, who had helped birth the Tea Party, and more than her own husband, who had been impeached—Clinton had a knack for eliciting congenital hatred from the right. It dated back several decades to her time as First Lady, and the perception of her complicity in all her husband's scandals dating back to his days as the governor of Arkansas. Her approval ratings peaked after the president admitted to a sexual relationship with a young White House intern, Monica Lewinsky, but even then, many conservatives viewed her as dishonest and politically calculating.

Clinton went on to be popular and highly effective as the junior senator from New York. She was also well liked during her tenure as secretary of state, with a 66 percent approval rating that topped Obama's own standing. Even so, and yet again, Clinton found herself under fire from the right: her failed "reset" with Russia; her conflicts of interest related to the Clinton Foundation; her response to the terrorist ambush that killed four Americans in Benghazi; and, as the House GOP's Benghazi probe uncovered, her use of a private email server to conduct government business.

Underscoring all these vulnerabilities was the most basic of political defects: a failure to connect with people. It had been a defining moment of the 2008 primary when Obama, smirking during a debate, remarked, "You're likeable enough, Hillary," highlighting the charisma gap between the rival candidates.

Bernie Sanders was no Barack Obama, but like the forty-fourth president, he had tapped into something unique on the left. Sweating through his oversize suit, blades of white hair shooting in every direction, jabbing a finger in the air and talking of the *yuge* gap between the one-percenters and the rest of the country, Sanders was the Doc Brown of the Democratic Party, and the issue of economic inequality was his flux capacitor.

He became a cult hero to the progressive base. Clinton couldn't hope to match his raw enthusiasm, but she boasted the one thing Sanders lacked: support from within the party institution. Democratic nominating contests had come to rely heavily on so-called superdelegates, the elected officials and party heavyweights given automatic votes at the party's convention. Clinton's virtual monopoly on superdelegates angered Sanders supporters and fueled allegations of a fixed election, even though she won nearly four million more votes and would have prevailed on the strength of her regular delegate count versus his.

A defiant Sanders remained an active candidate all the way through the final primary contest on June 14, well after Clinton's victory was assured, and he did not endorse her until July 12.[6]

The divisions exposed by their unexpectedly competitive and prolonged race loomed large as the Democrats prepared for their convention in late July. She was the prohibitive favorite heading into the general election; Trump lacked the raw numbers to win a high-turnout election. What he did have, however, was a passion in his base that Clinton could only dream of.

The flame that Trump carried—populism, nationalism, nativism—was beginning to light up the entire Western Hemisphere. Over the next several years, far-right parties advocating strict immigration crackdowns and protectionist economic policies took Europe by storm, some sweeping into power and others becoming the primary opposition voice in national governments.

The surest sign of the revolutionary times: On June 12, two days before the conclusion of the Democratic presidential primary, residents of the United Kingdom stunned the global community by voting to leave the European Union. "Brexit," as the move was dubbed, represented to some a return to sovereignty; to others, it was a misguided rejection of the century's geopolitical realities.

Brexit was strongly opposed by the White House. Unsurprisingly, it had a staunch ally in Trump.

RIGHT AROUND THE TIME FALWELL JR. WAS POSING IN FRONT OF THAT cover of *Playboy*, the news reached Pence: Trump was seriously considering him for the vice presidency.

A month earlier, Pence's longtime pollster, Kellyanne Conway, had visited Trump Tower for lunch with Jared Kushner and Ivanka Trump. Having spent the previous year leading a pro-Cruz super PAC, Conway was now a free agent. Kushner was keen on bringing her aboard and asked Conway who she thought made the most sense as his father-in-law's running mate.

Conway replied that it wasn't about "who," but rather, "what," and laid out her criteria: someone with appeal in Middle America, someone trusted by conservatives, someone who added stability, not excitement—"because we've got all the excitement we need"—to the ticket. She was making the case for Pence, just not by name.

When informed soon after that meeting that Trump's campaign wanted to vet him, Pence scoffed. No two human beings could have less in common, the governor joked to friends. Pence was a lifelong free-trader; Trump wanted to rip up NAFTA. Pence supported a path to citizenship for many illegal immigrants; Trump had floated the idea of a "deportation force."[7] Pence was a devoutly religious mid-westerner who refused to attend alcohol-related functions without his wife or work alone in a room with female staffers; Trump was a thrice-married Manhattanite who worshipped at the shrine of his magazine covers.

And yet, as time passed, the governor had grown more intrigued. The wholesome, aw-shucks, milk-drinking routine mastered by Pence belied the beating heart of a shrewd and ferociously ambitious politi-cian, and he saw in Trump someone who had achieved a preternatural connection with the electorate, channeling voters' anxieties in a way he had never witnessed. The longer Pence watched, the more he gravitated toward this source of power.

There was also the matter of self-preservation. Pence's reelection was looking bleak: Public polling showed the race neck and neck, but private surveys conducted that spring showed the governor's numbers looking dreadful all across the state. The religious liberty debacle had cost Pence a shot at the White House, and now it might cost him a sec-ond term. If Trump might rescue him from his predicament in Indiana, was the governor in any position to refuse?

Dazed by this set of circumstances, Pence reached out to a number of friends for advice. One of them was David McIntosh, the former Indiana congressman who was now president of the Club for Growth, an orga-nization that had spent millions of dollars attacking Trump during the primary. "What if he offers me the position?" Pence asked.

"That's a no-brainer," McIntosh replied. "The most likely result is you don't win in the fall, but you're probably the next presidential nom-inee. Or, who knows—you might even be vice president."

"You don't think it'll be damaging to my career to be associated with Trump?" Pence pressed.

"No," McIntosh said. "You're still going to be Mike Pence."

The Indiana governor decided to make a request of Trump's cam-paign. Before proceeding any further—and certainly before answering, if the offer were extended—Pence wanted his family to spend time with Trump's family. He assumed that such an ask was unrealistic given the time constraints on a presidential campaign; if Trump could not accommodate him, Pence figured, he would know that it wasn't meant to be.

Almost immediately, however, Trump responded in the affirmative. His campaign invited Pence's family to spend the July Fourth weekend at his private golf club in New Jersey. On his way to the airport, Pence placed an anxious call to Conway—who, it so happened, had been formally hired by the campaign one day earlier. All the concerns he had about Trump were flooding over him. She wouldn't hear it. "You crossed the Rubicon. Now that you've gone this far, there's no going back," Conway told Pence. "I'm going to make sure you get it."

The access he was given to Trump that weekend proved surprising—and surprisingly reassuring. "Morning, noon, and night, we got to be around them," Pence recalls. "That first time we got together, I was really struck by what an inquisitive person he is. He literally leads by asking questions. The first time we were together, we had breakfast and played a round of golf. Then we had lunch and dinner together. He must have asked me a thousand questions."

About what?

"Everything," Pence says. "My background. Politics. People. Policy. I mean, we were talking through things. But he never stops. And I've learned from him, it's a leadership style in which he's constantly asking questions."

Trump was also fun to be around—unpredictable, comfortable in his own skin, and often, hilarious. Picking up the phone as he sat with Pence on Saturday, Trump dialed Steve Scalise, the House majority whip. "Steve, question for you," he said. "I'm thinking of making Mike Pence my vice-presidential pick. What do you think about him?"

Scalise gushed with positive feedback on Pence, his friend and fellow alumnus of the Republican Study Committee. "Well, that's good, real good, Steve," Trump said. "Because he's sitting right here!"

As the weekend wore on—and especially after a breakfast in which Trump charmed the Pences' twenty-three-year-old daughter, Charlotte, who had accompanied her parents on the visit—Pence found

himself smitten with Trump. The Indiana governor began to believe
that his friends in the governing class had gotten their nominee all
wrong. No longer would he be the pursued; Pence became openly desir-
ous of the position. (Boasting to reporters that Trump "beat me like a
drum" on the golf course was a good start.[8])

By the time he departed New Jersey with his wife and daughter,
Pence felt sure that he wanted the job. He was less certain that Trump
would offer it.

PENCE'S FRIENDS WERE FLOORED TO HEAR OF HIS HUNGER TO JOIN THE
Republican ticket. There were the obvious differences: Pence was a
known foreign policy hawk and democracy promoter, while Trump had
spent much of the campaign flattering foreign strongmen, most conspicuously Russia's Vladimir Putin. Yet stranger still, to the governor's
old friends and allies, was how Pence could bring himself to ignore the
man's behavior. Trump's history of ad hominem ridicule, of sexual innuendo, of routine deception, was well established. And he seemed intent
only on adding new chapters to this legacy.

In June, as Pence found himself coming around to the campaign's en-
treaties, Trump found himself embroiled in a fresh controversy. A federal judge named Gonzalo Curiel, an American by birth whose parents
were naturalized U.S. citizens from Mexico, was presiding over multiple
court cases related to Trump University. The plaintiffs alleged they had
been conned into paying tens of thousands of dollars for an education
that never materialized. After the judge repeatedly ruled against him
in the various proceedings, Trump criticized Curiel for having "an inherent conflict of interest" in the case.[9] The reason: Trump was campaigning on a pledge to build a wall along the Mexican border, he said
on CNN, and the judge was "of Mexican heritage, and he's very proud of
it."[10] Trump repeated the claim at his rallies: Curiel could not rule fairly
because of his Mexican roots.

Republicans rushed to denounce their nominee.

"It's time to quit attacking various people that you competed with,
or various minority groups in the country, and get on-message," Mitch
McConnell told reporters.[11]

South Carolina senator Tim Scott called Trump's remarks "racially toxic." Scott's home-state colleague, Lindsey Graham, one of Trump's former rivals for the GOP nomination, told NBC News, "It's pretty clear to me that he's playing the race card."[12] Nebraska senator Ben Sasse, an outspoken critic of the GOP's nominee, tweeted, "Public Service Announcement: Saying someone can't do a specific job because of his or her race is the literal definition of 'racism.'"

And then there was Ryan.

The Speaker had urged Trump, during their RNC détente, to stop attacking fellow Republicans. In the weeks thereafter, Trump had mocked Romney (for being a "choker" and walking "like a penguin"), Rick Perry (for initially opposing him and then reversing course), Jeb Bush (for not having the "energy" to endorse him), South Carolina governor Nikki Haley (for opposing him in the state's primary), and New Mexico governor Susana Martinez (for "not doing the job" well). The day after the Martinez putdown, Ryan blew up at Trump during a private phone call, explaining that Martinez was a friend—and the GOP's most prominent Latina elected official. Ryan suggested that it would behoove the Republican nominee to focus his fire on Clinton. Instead, two days later, Trump picked a new target: Judge Curiel.

Standing with community leaders outside a drug-rehabilitation house in one of Washington's poorest neighborhoods, Ryan winced as he looked out at the assembled press corps. Here he was, attempting to promote the GOP's solutions to fighting the endemic scourge of poverty, and all anyone wanted to ask about was Trump's attacks on a judge for his "Mexican heritage." Making matters worse, Ryan had finally given in and endorsed Trump just days earlier.

"Claiming a person can't do their job because of their race is sort of like the textbook definition of a racist comment," Ryan said. "I think that should be absolutely disavowed. It's absolutely unacceptable."

If Ryan assumed that such a forceful response—*the textbook definition of a racist comment*—would satisfy the reporters, he was mistaken. The next question came: Did Ryan worry that Trump's inflammatory rhetoric would "undercut" the House GOP's agenda? Yes, Ryan said; their exchange was proof that Trump was overshadowing their "Better Way" proposal, a blueprint for governing the country. The third question was also Trump-related; so were the fourth, fifth, sixth, and seventh. When a reporter finally asked about the minimum wage, Ryan let out a laugh. "Thank you so much."

Among the more tepid rebukes, Pence called Trump's commentary "inappropriate," then added, "But that being said, if I wanted to comment on everything that's said in the presidential campaigns, I would have run for president. I'm focused on the state of Indiana." (Incidentally, Judge Curiel had been born, raised, and educated in Indiana.)

Pence was wise to tread carefully. Any slight of Trump, real or perceived, could mean the difference between running mate and historical footnote. Two other VP finalists, former Speaker Newt Gingrich and Tennessee senator Bob Corker, had rebuked Trump for his Curiel comments. But a fourth candidate, Chris Christie, had distinguished himself from the field.

"People are always gonna express their opinions," the New Jersey governor said in response to the uproar.[13] "Those are Donald's opinions and he has the right to express them."

THE "SHORT LIST" OF POTENTIAL TICKET MATES GOT SHORTER IN A hurry, thanks to a revamped campaign operation manning the controls inside Trump Tower.

On June 20, at last hearing the pleas of his adult children, Trump fired his campaign manager. Corey Lewandowski had been a disruptive presence for good and for ill, encouraging Trump's primal political instincts but never refining them. Replacing him atop the campaign was Paul Manafort, the veteran scoundrel who'd sworn to friends that he was joining Trump's team solely to oversee the convention mechanics.

The following week, Trump tapped a new communications director, Jason Miller. It made for an interesting interview: Miller had spent the past sixteen months helming Cruz's messaging machine and was responsible for a flurry of brutally negative tweets directed at the GOP front-runner. Trump worried about the operative's allegiance. In a conference room on the twenty-fifth floor of Trump Tower, the presumptive nominee squinted at Miller with a mischievous sneer.

"You just came over from Cruz? I guess you want to join the winning team, right?" Trump said. "Ted is a little nasty. Sometimes he's nice."

Miller didn't speak.

"Let's see where your loyalties lie," Trump continued. "Tell me something negative about Ted. Give me some dirt."

"I can't do that," Miller replied.

"No?" Trump said. "C'mon. You have to give me something."

Miller still refused. After two more rounds of this, Trump abruptly turned angry. "Okay, I'm not fucking around anymore," he told Miller. "Give me something on Cruz or you're outta here."

The room went silent. The assembled cast—Manafort, Donald Trump Jr., Ivanka Trump, and her husband, Jared Kushner, who had extended the job offer to Miller—wore concerned looks. Miller sat speechless, expecting to see security coming for him at any moment.

Then Trump broke into a grin. "Right answer!" he cried, pounding the table. "Jared, did you coach him?" (If this smacked of a mafioso scene, it wasn't coincidental: Trump had learned at the knee of legendary New York City fixer Roy Cohn, who was famous not just as Senator Joseph McCarthy's general counsel but as consigliere to some of America's biggest mobsters.)

Finally, a few days later, Trump hired Conway, the veteran pollster who had been waging a stealth lobbying campaign on behalf of her longtime client, Pence. As it happened, Trump and Conway were already well acquainted; she had polled on his behalf in 2011, when he was flirting with a 2012 presidential run. They were a natural pairing: Conway had spent her career pushing the party establishment to ditch its concerns about "electability" and embrace outsider candidates who could reach new voters. In this sense, although he'd defeated her preferred candidate in Cruz, Trump's vanquishing of the GOP was the realization of her life's work.

"The Republican Party was always looking for the next Ronald Reagan, but it kept picking Bushes," Conway says.

Trump reveled in such assessments, feeling disrespected even after spanking a sprawling field of sixteen well-regarded Republican opponents. That summer, as he neared a decision on his running mate, he agreed to meet with a small group of GOP-friendly corporate kingpins. They represented a range of industries, from banking to energy, and were convened by Jeff Sessions for a private get-to-know-you at Trump's new hotel in Washington, DC.

The property, once home to the historic Old Post Office, was still under construction, and laborers in hard hats milled about as the conversation commenced. After the Republican heavyweights introduced themselves, and Trump broke the ice by grilling an automotive executive about the productivity of Mexican workers, he surveyed his audience with a question: How many of them had supported him during the primary?

Nobody raised their hand. The men looked around nervously. Trump, leering in a way that implied some combination of delight and disgust, went around one by one, demanding to know whom they had voted for and why. Most of the attendees said Jeb Bush, out of loyalty to the family; a handful said Marco Rubio, believing he was best equipped to beat Hillary Clinton.

A long silence hung in the air. "Well," Trump finally told them. "At least none of you supported Lyin' Ted Cruz."

AT THE URGING OF HIS NEW AND PROFESSIONALIZED CAMPAIGN STAFF, Trump began weeding out the field of prospective running mates.

He had floated the idea of "America's mayor," Rudy Giuliani, giving the Republican ticket a pair of tough-talking New Yorkers. But Rudy was, among other things, pro-choice, a nonstarter with the already wary evangelical community. He was out.

Trump saw similar benefit in selecting Christie. The New Jersey governor would reinforce his strengths—brassy, unflinching, in-your-face leadership—while adding valuable executive experience. But Trump didn't need reinforcement; he needed balance. Moreover, the "Bridgegate" scandal had blown up back home, plunging Christie's approval ratings to all-time lows.[14] ("Why not save Christie for attorney general?" Manafort asked Trump. "Because," Trump replied, "that guy would prosecute my own kids and not think twice about it.") Christie was out.

Seeking a wild-card option, Trump began whispering to allies that he was high on a retired Army lieutenant general, Michael Flynn. Trump had made a habit of mocking the efficacy of the U.S. armed forces, even going so far as to say, "I know more about ISIS than the generals do." Yet beneath the bluster, Trump, having attended a military school, was enamored of the institution. He believed the military embodied a toughness that was fast diminishing in American society. He loved the imagery of a soldier on the ticket with a businessman, a tandem of unbeholden outsiders taking Washington by storm.

Except that Flynn *wasn't* unbeholden. After feuding with Obama administration officials and being forced into early retirement in 2014, Flynn launched a consulting firm that soon won contracts with companies linked to the Russian government. Taking a cursory glance at the general's workload since joining the private sector, Trump's lawyers warned that Flynn's ties to the Kremlin would be deadly for a campaign already accused of being pro-Putin. Flynn was out.

By the Fourth of July, it was apparent that Trump had only three choices: Pence, Gingrich, and Corker. Then, a day later, Corker withdrew from consideration.

The Tennessee senator, a mannerly southerner and chairman of the Foreign Relations Committee, had been circumspect about Trump since their first meeting back in May. Arriving at the candidate's skyscraper, Corker, he later told friends, thought the entire spectacle odd: the characters milling around, the corded rope guarding Trump's magazine covers like priceless artifacts, the candidate's insistence on sitting behind his desk for their entire conversation, a gesture that Corker found uncouth.

Still, Corker, like many who encounter Trump, was strangely charmed. And like many Republicans coming to grips with his perch atop the party in the middle of 2016, he craved influence over the campaign. He began communicating regularly with Trump, offering advice on matters of international affairs and providing feedback on foreign policy–themed remarks given by the candidate. By June, Corker's team had submitted vetting paperwork to Trump's legal team. Then, on July 5, he was summoned to New York for an official interview and a joint campaign trip with Trump to Raleigh, North Carolina.

But even before the plane departed for Raleigh that afternoon, Corker knew he wasn't the right fit. Trump needed someone more political—and more loyal. When Corker broke the news, while the plane was taxiing on the runway of LaGuardia, Trump took it well; he, too, thought it a poor pairing. One problem remained: Trump's aides wanted Corker to introduce him in Raleigh. Corker told Trump that it was a bad idea. "I'm not auditioning. I'm not going out on that stage," the senator said. "It will look like I'm auditioning for a job that we both know I'm not going to do."

Swarming to insecurity like a fly to dung, Trump delighted in calling Corker onto the stage in North Carolina despite the senator's repeated demurrals. But this small victory moved Trump no closer to naming a running mate. With the GOP convention less than two weeks away, its presumptive nominee was down to two choices: Gingrich and Pence.

IT WAS 8:30 P.M. ON JULY 12 IN WESTFIELD, INDIANA, WHEN PENCE launched into his tryout. With six days until the start of the convention, and three days until Trump's self-imposed deadline to name a running mate, Pence sought to answer the whispered questions of whether he possessed the intestinal fortitude for what was shaping up to be a nasty, low-down, watch-through-the-slits-in-your-fingers campaign.

"To paraphrase the director of the FBI," Pence declared, "I think it would be extremely careless to elect Hillary Clinton."[15] The crowd ate it up. Their governor, basking in the noise, then introduced "the next president of the United States of America, Donald J. Trump."

Trump climbed onto the stage with a satisfied smile, mouthing "wow" and pointing to Pence. Concluding his speech nearly an hour later, Trump said, "I don't know whether he's gonna be your governor or your vice president—who the hell knows?"

Such uncertainty didn't sit well with Manafort. He was adamantly opposed to Gingrich as the nominee, believing his loud mouth and self-important streak would become a distraction to a campaign already swimming in such traits. But he had yet to convince Trump, who arrived in Indiana that Tuesday night with lingering doubts about Pence's toughness.

As the rally drew to a close, Manafort pulled a rabbit from his hat. After coordinating with the candidate's traveling personnel and offering a few modest bribes, he informed Trump that his plane was suffering "mechanical problems." They would have to stay the night in Indiana.

Meanwhile, Manafort schemed with Kushner, who booked a flight into Indianapolis along with Ivanka, Don Jr., and Eric Trump, as well as with Pence's top advisers. They were to make certain that the governor capitalized on the additional time. Pence promptly invited the extended Trump clan to breakfast the next morning at the governor's mansion.

With news crews camped outside on the lawn, the Pences and Trumps broke pastries and sipped coffee inside the Tudor-style edifice. Sensing his final opportunity, Pence, as the *New York Times* reported, "delivered an uncharacteristically impassioned monologue," describing to the Trump family "his personal distaste for Hillary Clinton and her husband, the former president, and spoke of feeling disgusted at what he called the corruption of the 1990s."[16]

A short while later, with the "mechanical problems" fixed, Trump was wheels up to New York. He was still unsure of whom to choose, but the candidate's children were not. With Manafort and Kushner egging them on, they made the hard sell. Pence was deferential. He would attract evangelicals. He was polished. And he looked the part—an invaluable asset in the eyes of their father.

Finally, after another twenty-four hours of unceasing cajolery, Trump was convinced—or, as convinced as he was ever going to be.

His campaign flew Pence to New Jersey on Thursday night and then ferried him to a Manhattan hotel. As Pence settled in, with reports surfacing that he would be announced the next day, Trump went on Fox News and said he hadn't made his "final, final decision." Pence chuckled and turned off the television; this seemed like some last-minute showbiz suspense. Except that it wasn't. Trump, in California for a fundraiser, was furious after learning that Pence's trip to New York had been leaked—apparently by Manafort, in an attempt to lock in the selection. Stranded on the West Coast, away from the action in Trump Tower, the candidate spent much of the night on the phone with his friends and family, agonizing over the circumstances and complaining that he felt "backed into a corner" by Manafort. He even took a call from Christie, who made an emotional closing argument for himself.

Trump bought himself some time by pushing back the formal announcement of his running mate until Saturday. This was out of respect for the victims in Nice, France, where an ISIS-inspired jihadist had rammed a truck into a crowd celebrating Bastille Day, killing eighty-six and injuring hundreds more. But with reports swirling of his uncertainty over the VP selection, Trump decided on a plot twist.

Flying back from California on Friday morning, he tweeted, "I am pleased to announce that I have chosen Governor Mike Pence as my Vice Presidential running mate. News conference tomorrow at 11:00 a.m."

It was the unlikeliest of pairings, and arguably the smartest political decision Trump ever made. "Pence was exactly what he needed, because he was the antithesis of Trump: a solid Christian conservative who the evangelicals loved," says John Boehner, who had handpicked Pence to join his own leadership team years earlier. "And Pence needed Trump. Here's a guy who's about to lose his reelection, then Trump picks him, puts him on the ticket, and gets him out of his troubles in Indiana. He's been a loyal soldier ever since Trump threw him that lifeline."

On Saturday the sixteenth of July, in front of a friendly audience in New York City, Trump spent nearly half an hour introducing his ticket mate, though much of the homily had nothing to do with Pence. Trump did make sure to mention how the Indiana governor hadn't endorsed him a few months earlier, noting that it was due to pressure from donors and GOP hacks scared of his candidacy. He then devoted a considerable stretch of time to recounting his primary conquests, detailing his methodical destruction of the Republican primary field.

Somewhere amid the soliloquy, Trump stopped himself. "One of the big reasons I chose Mike is party unity," he said. "I have to be honest."

CHAPTER FIFTEEN

JULY 2016

"Stand and speak and vote your conscience."

NINE DAYS BEFORE HE ANNOUNCED HIS CHOICE OF MIKE PENCE IN the name of "party unity," Donald Trump found himself locked in a tense verbal confrontation with a Republican senator. His name was Jeff Flake, and for the past fifteen years he had been Pence's closest friend in politics.

They were ideological soul mates. Both ran conservative think tanks in their states in the 1990s; both had been elected to Congress in 2000, at one point occupying neighboring offices; both were lonely leaders of intraparty rebellions during the big-spending tenure of George W. Bush; both had left the House of Representatives in 2012 to run successfully for statewide office; and above all, both strove to be regarded as gentleman conservatives, known for a personal decency that infused their relationships and reputations in the nation's capital.

And yet both men knew, in the summer of 2016, that their friendship might never be the same. Trump's ascent would leave a wreckage of relationships in its wake—friends, neighbors, families divided—but there was no more dramatic divergence than that of Flake and Pence.

It began on the afternoon of July 7. Trump was visiting Washington for a series of meetings aimed at coalescing the party's elected officials behind him: one with House Republicans, one with Senate Republicans, and a third, private conversation with Ted Cruz, who remained unsupportive of Trump two months after quitting the GOP race. The House meeting went swimmingly; the same rank-and-file renegades who had spilled John Boehner's blood over a lack of conservative bona fides emerged spouting praise of Trump. And the Cruz sit-down was a qualified success: Though he wasn't ready to endorse, he did accept an invitation to speak at the convention later that month.

It was Trump's meeting with Senate Republicans that went off the rails. After some introductions and polite banter, the niceties came to a sudden halt when Trump singled out one of the senators, Flake, for having criticized his candidacy.

"Yes, I'm the other senator from Arizona, the one that wasn't captured," Flake responded, referring to Trump's infamous attack on John McCain the previous summer. "I want to talk to you about statements like that."

"You know, I haven't been attacking you," Trump snapped back. "But maybe I should be. Maybe I will." He glowered at Flake, warning that his dissension would cost him his Senate seat.

"I'm not even up for reelection this cycle," Flake snorted, rolling his eyes.

Flake had already decided he would not be attending the Republican National Convention in Cleveland beginning July 18. In fact, as Trump accosted him that day, the Arizona senator took comfort in knowing he wouldn't be sitting in the convention hall as Trump completed his hostile takeover of the GOP. But his subsequent selection of Pence—which, Flake says, left him in a state of "shock"—forced the senator to reconsider. He wouldn't just be turning his back on Trump and the GOP by shunning the convention; he would be betraying his dear friend.

Ultimately, it wasn't enough to change Flake's mind. He stayed away from Cleveland, in protest of Trump, while his old pal was crowned heir apparent.

For Flake, this was a matter of "principle over party." He could understand why some Republicans might hold their noses and vote for Trump against Clinton as a lesser-of-two-evils choice. What he couldn't understand was the categorical cheerleading of someone whose candidacy was antithetical to much of what modern conservatism was supposed to stand for. Both stylistically and substantively, Flake believed, Trump was poisoning the Republican Party. The senator would not blindly pledge his allegiance for the sake of winning one election.

This earned Flake no shortage of abuse from the right, including from many longtime allies. How ironic, they snickered, that Flake would lecture about fidelity to principle, given his own professional metamorphosis. Once a cutthroat conservative in the House, Flake had become just another wallflower Republican in the Senate, refusing to join the likes of Cruz and Mike Lee in ramming at the establishment's barricades.

"We have had some enormous departures, some rather stark political divergences, with Jeff over the recent years," Trent Franks, a former Arizona congressman and confidant to both Pence and Flake, said. "We've been disappointed with some of the things Jeff's done."

Flake knew he would face special criticism for breaking rank in 2016. "This wasn't a situation where I woke up a month ago and thought, hey, I'm out of step with my party," he said. "I was uncomfortable with Trump before he got in the race. And then on day one, it was Mexican rapists. And before that, over the past years, it was the birtherism, which I thought was just the most vile, rotten thing you could do to President Obama. And then he just seemed to carry forward from there."

Franks, an original member of the House Freedom Caucus, had no such concerns. "As I've gotten to know the guy, I've seen a heart, and kind of a John Wayne valiance in him that is compelling to me," he said of Trump. "I'm convinced he came along at a time when the country needed someone to punch government in the face."

WHEN THE HOUSE FREEDOM CAUCUS FORMED IN 2015, TURBOCHARGing the anti-leadership engine once driven by the likes of Flake and Pence, its organizing principle was to speak on behalf of forgotten Americans. Jim Jordan, the group's founding chairman, believed Washington worked on behalf of big entities (banks, corporations) and parochial interests (the poor and unemployed) but not the "second-shift workers" and "second-grade teachers" like the ones in his 88 percent white district, where only 18 percent of residents earned college degrees.[1]

This is not to suggest Jordan was racist, or even using racially coded language; he was simply speaking to the realities of north-central Ohio. Many of his white, working-class constituents felt that they were falling behind and that the federal government didn't much care. This was a sentiment reflected in the membership of the Freedom Caucus: During the 114th Congress, spanning the years 2015 and 2016, the group had thirty-nine members. On the whole, their districts were 75 percent white (higher than the national average) and 27 percent college-educated (lower than the national average), according to data culled from *The Almanac of American Politics.*

Despite these demographic profiles, and their own stated mission to represent the forgotten voters of flyover country, the Freedom Caucus members had long trafficked in ideological orthodoxy. They believed this was what their constituents demanded: less spending, more trade, restructured entitlement programs, and above all, limited government.

And then Trump came along.

One Freedom Caucus member described the "oh shit" moment in the spring of 2016, when he and his comrades realized what was happening. Marauding across the country, Trump was delivering an anti-Washington message rooted not in any narrowly philosophical approach, but in the belief that politicians had failed voters. Back home, the conservatives saw their constituents responding in force, much as they had in 2010. But Trump was no Tea Party purist selling a small-government creed. He was selling outrage at the status quo.

Trump, they realized, had co-opted and broadened their message. He wasn't merely attacking the establishment; he was attacking *them.* After promising major changes (repealing Obamacare, rolling back Dodd-Frank, reining in executive actions) and failing to deliver, *they* were now part of the broken political class Trump was railing against.

Watching in horror as he won more than two-thirds of their districts, the Freedom Caucus members, most of whom had endorsed either Cruz or Rand Paul, wondered how their voters could reconcile supporting a Tea Partier for Congress and a totalitarian for president.

Thomas Massie, the Kentuckian who says he was excluded from the Freedom Caucus for being "too crazy conservative," said it best in an interview with the *Washington Examiner.*[2] "All this time, I thought they were voting for libertarian Republicans," says Massie, who backed Ron Paul in 2012 and his son four years later. "But after some soul searching, I realized when they voted for Rand and Ron and me in these primaries, they weren't voting for libertarian ideas—they were voting for the craziest son of a bitch in the race. And Donald Trump won best in class."

Reaching that same conclusion in May 2016, Freedom Caucus members debated what could be done about it. Only one of their members had endorsed Trump in the primary: Scott DesJarlais, the physician who a few years earlier was discovered to have carried on sexual relationships with multiple patients and pressured both a mistress and an ex-wife to have abortions. (He was reelected multiple times thereafter.) His endorsement of Trump drew sneers from his colleagues, many of whom believed that neither man reflected the values of their club. But now DesJarlais was looking prophetical. One by one, the archconservatives who had spent the past five months snickering at Trump in their closed-door gatherings took turns announcing their support for his candidacy.

Tribal bitterness often lingers after party primaries. But the reflections on Trump's conquest from leading right-wingers spoke to the extraordinary mistrust they felt for him—even as they endorsed him for the highest office in the land.

Nobody captured this mood better than Mick Mulvaney. "As a conservative, my confidence level in Trump doing the right thing is fairly low," the South Carolina congressman said. He laughed. "But, hey, my confidence level in Hillary Clinton doing the wrong thing is fairly high!"

Mulvaney, the mouthiest of the conservative rebels, couldn't help himself. He had supported Paul, the Kentucky senator. He would have settled for Cruz or Marco Rubio. He found the specter of Trump's nomination laughable, though not necessarily unsettling. "Don't worry, we're not going to let a President Donald Trump dismantle the Bill of Rights," Mulvaney said prior t the convention. "For five and a half years, every time we go to the floor and try and push back against an overreaching president, we get accused of being partisan at best and racist at worst. When we do it against a Republican president, maybe people will see that it was a principled objection in the first place."

"It might actually be fun," Mulvaney added, "being a strict constitutionalist congressman doing battle with a non-strict-constitutionalist Republican president."

House conservatives would spend much of that spring and summer blaming the GOP's establishment for enabling Trump's victory. In one sense, this was fair. When presented with the dichotomy of Trump versus Cruz, many of the party's graybeards, from John Boehner to Bob

Dole, had voiced their preference for the former, believing that he could be controlled whereas Cruz could not. ("Crazy, I could deal with," Boehner says. "But not pathological.") Yet this obscured a more fundamental question: Why hadn't the House hard-liners, the custodians of party purity, done more to thwart Trump's rise in the first place?

Jordan, the two-time collegiate wrestling champion, had turned a roster of ragtag back-bench congressmen into a scrappy, disciplined, productive unit. His followers had mastered the use of technique and leverage to defeat opponents of superior size; lacking in seniority and campaign cash, the Freedom Caucus often outmaneuvered the rest of the majority, pushing leadership relentlessly to the right and refusing any compromise that would chafe the grass roots.

The group also had symbolic momentum. Two of its newer members, Warren Davidson of Ohio and Dave Brat of Virginia, occupied the seats once held by Boehner and Eric Cantor, respectively. The two most prominent casualties of the Tea Party era had each been replaced by members pledging allegiance to the Freedom Caucus.

But the House conservatives did nothing to slow Trump's march to the nomination.

There had been no press conferences, no rallies on the Capitol lawn, no coordinated exercises with outside groups to signal opposition to the GOP front-runner. Half of the Freedom Caucus members had endorsed rival candidates, but the other half had endorsed no one at all. One of those who had remained neutral was Jordan. Watching Trump's rise, he spent the summer of 2016 pondering not the failures of the past five months, but the failures of the past five years.

"The one thing I do reflect on is what could we, as a Republican Congress, have done differently to avoid creating this environment that was conducive to someone like Donald Trump becoming the nominee?" Jordan said in late June.

It was less than a month before the party's convention, and Jordan and his fellow conservatives spoke of Trump's nomination as a foregone conclusion. This was misleading. A faction of Republican activists and officials, under the banner of #NeverTrump, was organizing furiously ahead of the proceedings in Cleveland to defeat the GOP's presumptive nominee. Their effort revolved around a change to the party's rules, allowing delegates to vote their conscience rather than for the candidate to whom they were bound by their state's results.

It was a long shot. But a number of respected conservatives, including Lee, the Utah senator, were involved in the plotting. They believed the reward of preventing Trump's nomination was worth the risk of a backlash from his supporters.

The Freedom Caucus did not. Of its thirty-nine members, none would publicly support the rule change ahead of the convention.

"What people hate most about Washington is backroom deals, and that would be the ultimate backroom deal," John Fleming, a Freedom Caucus board member, warned. "I think it would destroy the party."

Mark Meadows, a former Cruz supporter, said prior to the convention that he was sympathetic to the #NeverTrump effort. Ultimately, however, he could not abide such an affront to his constituents.

"If I question their judgment on who they have as a nominee, I have to question their judgment on the fact that they continue to put me back in," Meadows said. "That becomes very problematic when you think they're smart in reelecting you but perhaps not as informed on a presidential nominee. So, you've got to trust the will of the people, even though sometimes you disagree with it."

CRUZ HAD KEPT HIS HEAD DOWN EVER SINCE DEPARTING THE RACE. IN public settings, he projected stoicism, a certain peace about the result that kept questions at bay. Beneath the surface, however, he was boiling with resentment—toward his fellow senators for disowning him, toward Ben Carson for milking what should have been a one-day story, toward Marco Rubio for refusing to join his ticket, and toward Donald Trump for, well, *everything*.

Replaying the events of the previous year in his mind, Cruz grew only more upset with his adversary. Trump hadn't been content to beat him politically; he had tried to butcher him personally. Calling him ineligible for the presidency? Suggesting that his wife was ugly? Implicating his father in the JFK assassination?

In Cruz's mind, Trump had crossed lines that couldn't be uncrossed.

Nothing—certainly not some half-assed kumbaya session in DC—could change that. In preparing for his July 7 meeting with Trump, anticipating an invitation to speak in Cleveland, Cruz had gathered his kitchen cabinet of advisers and close friends. He believed there were three options: speak and endorse; speak and don't endorse; or don't speak at all.

Jeff Roe, Cruz's campaign manager, disputed the premise. He told Cruz that giving a convention speech without endorsing the nominee could be disastrous. For a man who still harbored burning ambitions for the presidency, there was too much risk. Roe believed Cruz should speak and, at the very least, assure the convention delegates that *he personally* would be voting for Trump. But Roe was in the minority. Most of the members of Cruz's inner circle, movement conservatives with decades of ideological skin in the game, were too acutely offended by Trump to entertain the possibility of an endorsement. They encouraged Cruz to accept the invitation to speak; once it arrived, they lobbied him to withhold his support for the nominee.

It wasn't a difficult decision for Cruz. While he usually hung on Roe's advice, and had come to appreciate his manager's pragmatic streak, he told his confidants that there was "no way in hell" he was prepared to subjugate himself to Trump in front of tens of millions of viewers. "History isn't kind to the man who holds Mussolini's jacket," Cruz told friends while crafting his speech.

The Republican nominee had insulted his wife, his father, his family. An endorsement would make Cruz look weak—and worse, it would make him look like the soulless, calculating swindler his detractors painted him as. He would not endorse Trump in Cleveland, and he was confident that the convention delegates would respect his decision.

He was wrong.

Cruz walked onto the stage Wednesday evening, July 20, to a thunderous ovation from the party faithful. It was the most anticipated speech of the convention, in prime time, and the packed house inside Quicken Loans Arena delivered a lengthy, raucous salute to the 2016 runner-up. The senator lifted a hand to the masses and nodded his head, basking in a moment that he believed should have been his and his alone.

"I congratulate Donald Trump on winning the nomination," Cruz said, earning booming applause.[3] The audience expected an endorsement, and understandably so: It was inside that very arena, the previous August, where all the Republican candidates (save for Trump) had agreed that they would support the eventual nominee.

Instead, it was the last time Cruz would mention Trump's name. The senator's address, which emphasized the theme of "freedom," was sharp, steady, and well received until its closing minutes. "We deserve leaders who stand for principle, unite us all behind shared values, cast aside anger for love. That is the standard we should expect from everybody," Cruz said.

As the arena began to buzz, Cruz delivered two fateful lines. First: "And to those listening, please, don't stay home in November." The audience erupted with cheers. Then, Cruz added: "Stand and speak and vote your conscience. Vote for candidates up and down the ticket who you trust to defend our freedom and to be faithful to the Constitution."

It was a stunning turn of phrase. "Vote your conscience" had been the anti-Trump rallying cry all summer, only for Reince Priebus and his allies inside the RNC to crush the rebellion in Cleveland just days earlier— with Mike Lee, Cruz's closest friend in the Senate, leading the last gasp of the mutiny. Cruz would later swear that he didn't appreciate the implications of his wording, but Trump's supporters inside the convention hall weren't about to give him the benefit of the doubt.

Tipped off in advance by Paul Manafort, who had seen a copy of Cruz's speech and knew he wouldn't be endorsing, the sprawling New York delegation, which sat front and center in the arena due to Trump's native son status, detonated with boos. The ruckus tore across the convention floor and climbed all the way up to the second and third decks.

Meanwhile, Trump himself had just entered the arena on a cue from his staff, hoping to mess with Cruz by gawping at him from an offstage wing. Necks craned to see him. The decibel level spiked all the higher. For Trump, a longtime fan of professional wrestling, this was a page out of Vince McMahon's playbook: the hero emerging just as the crowd turned against the villain.

Cruz had four short paragraphs left in his speech, words that paid homage to his mother and father and to a slain Dallas police officer. But

they were difficult to hear. It was anarchy on the convention floor: The
heckiers became shriller and nastier; in response, pockets of Cruz loyalists began shouting back in a futile attempt to drown them out. Cruz
continued on, voice shaky, as the noise swallowed him whole.

When he had uttered his final words—"God bless each and every one
of you, and may God bless the United States of America"—he was show-
ered with deafening, cascading boos that seemed to rain all the way
down from the rafters. The senator stepped away from the lectern yet
remained on the stage for several moments, waving and smiling awk-
wardly, trying not to appear paralyzed by the unmitigated nightmare
playing out before him. His wife, Heidi, had to be escorted off the convention floor by security officials concerned for her safety. The senator
and his team quickly bunkered down in a hotel suite, assessing the extensive damage and plotting his next move.

Back in February, standing inside a pole barn at the Iowa state fairgrounds, Cruz had previewed his acceptance speech: "This July, in
Cleveland, you will hear these words spoken from the podium of the
unified Republican convention," he said. "'Tonight, I want to say to every member of the Democratic Party who believes in limited government, in personal opportunity and the United States Constitution, and
a safe and secure America, come home.'"[4]

Nearly six months later, Cruz had the opportunity to heal divisions in
the party and help create a "unified Republican convention" on behalf of
his former rival. He declined. And it didn't go over well.

Several of Cruz's biggest financial backers turned on him, saying
the senator had broken the promise he had made to support the party's nominee. Among them were Robert Mercer and his daughter, Rebekah, who had pumped more than $10 million into a flotilla of super
PACs supporting Cruz.[5] In a show of their anger, the media-shy Mercers
upbraided Cruz in a statement to Maggie Haberman of the *New York
Times*.[6] The article quoted Kellyanne Conway, the pro-Cruz strategist
turned Trump adviser, who said of the Mercers, "They supported Ted
because they thought he was a man of his word who, like them, would
place love of country over personal feelings or political ambition."

The morning after his convention speech, Cruz was booed and jeered by members of the Texas delegation when he arrived at their breakfast. They called him a liar and a sore loser. "I am not in the habit of supporting people who have attacked my wife and attacked my father," Cruz told them. "And that pledge was not a blanket commitment that if you go slander and attack Heidi, then I'm not going to nonetheless come like a servile puppy dog and say, 'Thank you very much for maligning my wife and maligning my father.'"

It was a paradox: Never had Cruz been so authentic, yet never had he been so despised.

FOR ALL TRUMP'S FAMILIARITY WITH SHOW BUSINESS, HIS CONVENTION wasn't the smoothest production. There was plagiarism and pettifoggery; grudge matches and goonery; ugly exchanges and awkward embraces. Just hours before Trump took the stage to deliver his acceptance speech, a pro-Clinton super PAC obtained and leaked the transcript. It was a fitting capstone to a convention defined by the party's squabbling disunity, enhanced by the Trump campaign's disorganization and repeated political miscalculations.

Ohio governor John Kasich's decision to skip the convention prompted Manafort to open the festivities on Monday by accusing the home-state governor of "embarrassing" his constituents.[7] But Kasich wasn't alone in steering clear of Cleveland. Of the five living Republican presidential nominees, just one, Bob Dole, attended the convention. The notable absences of Mitt Romney, John McCain, and both Bush presidents set the tone for a week of intraparty bickering that came to a head with Cruz's refusal to endorse Trump.

For an hour and fifteen minutes on Thursday night, July 21, it was Trump who brought a modicum of normalcy to the proceedings. He delivered acceptance remarks that were smart and tightly scripted. Taking the stage wearing a luminous red tie, the nominee waved triumphantly as the delegates on the floor broke out into a chant: "Trump! Trump! Trump!"

Stepping into character as America's strongman, he cast President Obama as feckless and weak, blaming his administration for everything from the murders at the hands of illegal immigrants to the protests against law enforcement on city streets. "The crime and violence that today afflicts our nation will soon come to an end," he said. "Beginning on January 20, 2017, safety will be restored."[8]

Trump also assailed Obama—and the Democratic nominee, Hillary Clinton—for sowing turmoil around the world. From the Iran nuclear deal to the nonenforcement of the Syrian "red line" to the killings of four Americans in Libya, the United States had been neutered on the international stage, he said. When Trump made mention of Benghazi, the crowd began to chant, "Lock her up!" A few nights earlier, from the same stage, retired general Michael Flynn joined in the chant, declaring of Clinton, "If I did a tenth of what she did, I would be in jail today!" But Trump, showing restraint, raised an index finger to silence the crowd. "Let's defeat her in November," he said. The audience roared.

The Republican faithful got what they came for. Tony Ledbetter, a first-time delegate from Florida who had volunteered for Trump during the primary, said the GOP was united "except for a small minority of people" and that the party was better off without them. "Rubio, Bush, all these establishment insiders, I don't care if they're here," Ledbetter said on the convention floor after Trump's speech. "They can stay home— Romney and Kasich, too. This is not their Republican Party anymore."

Trump couldn't resist taking a parting dig at his detractors. As his family joined him onstage, with red, white, and blue balloons falling from the rafters and confetti dancing through the air, a Rolling Stones tune began blasting over the loudspeakers.

"You can't always get what you want . . ."

AS REPUBLICANS DEPARTED CLEVELAND, WATCHING FROM AFAR AS their Democratic counterparts gathered in Philadelphia, Trump could have found any number of weaknesses in the opposition to pick apart. He might have focused the country's attention on Bernie Sanders getting stonewalled by the Democratic establishment; or on Hillary Clinton being outshone by the speeches given by Barack and Michelle Obama; or on the liberal base's lukewarm reaction to her pick of Tim Kaine, the Virginia senator and committed Catholic with a pro-life past, as her running mate.

Instead, Trump found himself feuding with a pair of Gold Star parents, Khizr and Ghazala Khan, whose Army captain son, Humayu Khan, had lost his life to a suicide bomber in Iraq. They were so offended by Trump's rhetoric toward Muslims that they agreed to appear at the Democratic National Convention in late July. Paying tribute to his son, Khizr Khan waved his pocket-size copy of the Constitution and questioned whether the Republican nominee had ever read it. "Go look at the graves of brave patriots who died defending the United States of America," Khan said. "You will see all faiths, genders, and ethnicities. You have sacrificed nothing and no one."[9]

His speech quickly became a viral news sensation. Trump could not resist punching back. Appearing on ABC's *This Week*, he observed that Khan was "very emotional" in his speech. Instead of leaving it there, the Republican nominee began to speculate as to why Khan's wife, Ghazala, who stood silently next to her husband during his speech, had not said anything. Trump wondered aloud whether she was not allowed to speak, presumably because of subservient gender roles in the Muslim tradition.

Just as with his earlier attacks on Judge Curiel, Trump found himself engulfed by criticisms from within his own party—from the likes of McCain, Romney, Lindsey Graham, and of course, Speaker Ryan.

"As I have said on numerous occasions, a religious test for entering our country is not reflective of [our] fundamental values," Ryan said. "Many Muslim Americans have served valiantly in our military, and made the ultimate sacrifice. Captain Khan was one such brave example. His sacrifice—and that of Khizr and Ghazala Khan—should always be honored. Period."

Trump seemed to take particular umbrage with Ryan's rebuke. He threatened to withhold his support for the Speaker in his Wisconsin primary that August, and began saying positive things about Ryan's challenger, an anti-Semitic buffoon named Paul Nehlen. (Trump, on the advice that he would look foolish when the Speaker prevailed in the primary, later issued a halfhearted endorsement. Ryan won 84 percent of the vote against Nehlen.)

Fortunately for Republicans, they had not cornered the market on

intraparty warfare. Days ahead of the Democratic convention, the website WikiLeaks—which was later shown to be working in concert with a
Russian campaign to interfere in the U.S. elections—had dumped tens of
thousands of hacked emails from the Democratic National Committee.
The emails showed, among other things, a clear preference for Clinton
over Sanders among DNC staffers who were obligated to remain neutral.
DNC chair Debbie Wasserman Schultz resigned ahead of the convention
and was replaced by vice chair Donna Brazile, who later confessed that
the party committee had unethically conspired to aid Clinton in the
primary.[10]

In a continuation of the Campaign That Nobody Wanted to Win, the
Republican nominee kept finding ways to make his opponent a sympathetic figure, even as her own party's progressive wing was burning with
resentment toward her.

On August 9, Trump seemed to suggest that Clinton could be assassinated if she won the White House. "Hillary wants to abolish—essentially
abolish the Second Amendment," he said in North Carolina. "By the
way . . . if she gets to pick her judges, nothing you can do, folks. Although
the Second Amendment people, maybe there is. I don't know."[11]

Having sparked a national frenzy—another one—Trump ran straight
into the comforting arms of Sean Hannity. "Obviously you're saying
that there's a strong political movement within the Second Amendment,
and if people mobilize and vote, they can stop Hillary from having this
impact on the court," the Fox News host said.

"Well, I just heard about that," Trump replied, playing dumb, "and it
was amazing because nobody in that room thought anything other than
what you just said."

Except that some people did. Darrell Vickers, a local Republican
and Trump supporter who sat directly behind the candidate onstage,
had his shocked reaction captured on live television. "I was just absolutely taken aghast," Vickers later told CNN.[12] "Down here in the South,
we don't curse in front of women, we don't drink liquor in front of the
preacher, and we don't make jokes like that in public." (Vickers said he
would still be voting for Trump.)

A day after the "Second Amendment people" stunt, Trump blamed

Obama for creating a power vacuum by withdrawing troops from Iraq—but in less diplomatic terms. "He's the founder of ISIS. He's the founder of ISIS. He's the founder. He founded ISIS," Trump said of the president. "I would say the co-founder would be crooked Hillary Clinton."[13]

As Trump flailed, his numbers spiraled sharply downward. He had consistently trailed Clinton by healthy margins, both in national polling averages and in battleground state surveys. But as Labor Day approached, signaling the final sprint of a presidential campaign, things were looking bleaker than ever. As of the middle of August, the Real-ClearPolitics averages showed Clinton leading Trump by 9 points in Pennsylvania; by 7 points in Michigan; and by 9 points in Wisconsin. He was closer in North Carolina and Florida, and his campaign felt good about Ohio and Iowa. But the keys were Pennsylvania, Michigan, and Wisconsin. Without a sweep in those "Blue Wall" states in the Rust Belt, Trump's team feared, he wouldn't have a prayer.

Whatever nominal bounce Trump had received from the convention was long gone. The party's fissures were fresher by the day and showing no signs of repair. His campaign was treading water, understaffed and out-organized: Some media estimates reported that the Democratic nominee had nearly three times the number of field offices as her Republican opponent.[14]

There were two saving graces for Trump. The first was Priebus and his infrastructure. Since taking over the Republican National Committee in early 2011, the chairman had completely revamped its operations. The party had raised record amounts of money and spent heavily to strengthen the field programs of its affiliates in the key battleground states. In the realm of technology and voter targeting, where Obama's Democratic Party was once hopelessly ahead of its counterpart, Republicans had all but caught up. Priebus had, in the span of five years, turned the RNC from a punch line into a powerhouse.

And not a moment too soon: Trump had virtually no campaign organization to speak of. In many of the crucial nominating contests, while Cruz commanded a sprawling ground game and a data-driven turnout machine, Trump countered with small, ragtag teams of volunteers. This made his primary conquest all the more impressive, but it rendered him woefully unprepared to compete in the general election. Without a strong national party doing the blocking and tackling on behalf of his campaign, Trump's chances might have slipped from slim to none.

The second silver lining for the Republican nominee was his opponent. Trump was the most unpopular major-party nominee in modern American history, but Clinton wasn't far behind. Controversies had dogged her candidacy from day one: Benghazi, the Clinton Foundation, a private email server that had been wiped of potentially damning messages. Even after then-FBI director James Comey cleared Clinton in July, rebuking her use of the server but recommending no criminal charges, the allegations of her slipperiness remained, with large majorities of voters throughout the year telling pollsters that she was "untrustworthy."[15] By August, Clinton's popularity had reached an all-time low. The ABC News/*Washington Post* poll showed that 59 percent of registered voters viewed her unfavorably—compared to 60 percent for Trump.[16]

Perhaps even more detrimental was her campaign's strategic obliviousness. It was plainly apparent by late summer that Trump's only path to 270 Electoral votes ran through the states of Pennsylvania, Michigan, and Wisconsin. And though Clinton poured time and resources into Pennsylvania, she had a decidedly lighter footprint in Michigan, and was completely MIA in Wisconsin. Republican officials in the latter two states sensed that she was vulnerable but feared that Trump and his amateurish campaign were incapable of capitalizing.

In mid-August, the Republican nominee announced a dramatic shakeup of his operation. Manafort was relieved of his duties as campaign manager, replaced by Conway, the messaging maestro. Priebus, having all but relocated from Washington to New York, was taking on a broader role as unofficial chief strategist. In the most newsworthy move, Trump hired Breitbart News honcho Steve Bannon as the campaign's chief executive.

While no single person's influence on American politics has ever been more overstated—journalists would spend parts of 2017 penning stories suggesting that Bannon was creating a shadow party to take down the GOP establishment—his hire was of enormous symbolic value.

Trump had spent the three months since clinching the nomination attempting to conform himself to the party: firing Corey Lewandowski, bringing on veteran operatives, playing nice (for the most part) with GOP leaders, dialing back (when possible) his rhetorical superfluities. The addition of Bannon, whose website had championed Trump's "America First" policies and lashed out at his establishment critics, suggested that the Republican nominee was going to finish the campaign *his* way—win or lose.

The Clinton camp could barely contain its euphoria. Having long debated the timing of hitting Trump explicitly over his ties to the "alt-right," a marginal internet movement of nationalists and Neanderthals, she saw Bannon's hiring as the ideal opportunity. "The de facto merger between Breitbart and the Trump campaign represents a landmark achievement for this group, a fringe element that has effectively taken over the Republican Party," Clinton announced the following week during a speech in Nevada. She warned that Trump was campaigning in concert with "the rising tide of hardline, right-wing nationalism around the world."[17]

Trump, a firm believer in the "all publicity is good publicity" mantra, saw Clinton's speech as a net positive for his campaign. So did many of his friends and advisers. Even those skeptical of the Bannon move now felt that Clinton's attacks could help Trump bring home the base.

There was another layer of intrigue. Roger Ailes, the longtime Fox News chieftain who had recently been fired amid spiderwebbing allegations of sexual harassment, had begun advising Trump in an informal capacity. By bringing Bannon aboard the campaign, Trump was now guided by the leaders of the two most loyal media outlets on the right. It was all gravy for Ailes and Bannon. If Trump won, their kindred spirit would occupy the Oval Office. If he lost, the possibilities for a new, nationalist-branded, Trump-inspired media empire were boundless.

As Labor Day approached, a nation alienated from itself over issues of politics, culture, and identity found fresh ammunition for its intrasocietal cold war. It came from the unlikeliest of places: the sidelines of a football game.

Colin Kaepernick, the biological son of a black father and a white mother, was given up for adoption and raised by an affluent white family in California. A second-round pick in the 2011 NFL draft by the San Francisco 49ers, he spent his rookie season on the bench before gaining stardom a year later, replacing the team's starter halfway through the season and leading the 49ers all the way to the Super Bowl. (In his first career playoff game, Kaepernick ran for 181 yards, setting the NFL's single-game record for rushing yards by a quarterback.) He took the 49ers back to the conference championship game in 2013 and was rewarded with a princely $126 million contract. His next two years, however, were plagued by injuries and inconsistency, and by the start of the 2016 season, Kaepernick was the 49ers' designated backup.

Although he didn't see the field until the team's sixth game, Kaepernick was the talk of the NFL. On August 26, after staying seated on the bench during a rendition of the national anthem, the quarterback told a reporter with NFL.com, "I am not going to stand up to show pride in a flag for a country that oppresses black people and people of color. To me, this is bigger than football and it would be selfish on my part to look the other way. There are bodies in the street and people getting paid leave and getting away with murder."[18]

Two days later, Kaepernick expanded on his explanation. "I'm seeing things happen to people that don't have a voice, people that don't have a platform to talk and have their voices heard and effect change. So I'm in the position where I can do that and I'm going to do that for people that can't," he told the local media. "I have great respect for the men and women that have fought for this country. I have family, I have friends that have gone and fought for this country. And they fight for freedom, they fight for the people, they fight for liberty and justice—for everyone. That's not happening. People are dying in vain because this country isn't holding their end of the bargain up."[19]

Kaepernick also noted of the two presidential nominees, "You have Hillary who has called black teens or black kids 'super predators,' you have Donald Trump who's openly racist."

The anthem protest blitzed America's consciousness in a way that no sports-related story had since the turn of the century. Within a week of the NFL.com interview, just about every media outlet in the country was covering Colin Kaepernick—and everyone, it seemed, had an opinion.

Sensing an opportunity to further rile his base, Trump pounced. "Well, I have followed it and I think it's personally not a good thing. I think it's a terrible thing," the Republican nominee told a conservative talk radio show in Seattle. "Maybe he should find a country that works better for him."[20]

Kaepernick was an imperfect messenger. On August 31, amid the national uproar over his protest, a local reporter tweeted out a photograph of the quarterback at practice a few weeks earlier wearing socks that showed pigs dressed like police officers. Months later, after the election concluded, Kaepernick revealed that he did not vote, drawing harsh criticism from liberal commentators who questioned his seriousness as a social activist.

Yet he had launched a national dialogue virtually overnight. And though he wasn't backing down, Kaepernick did make an attempt to refine the contours of that dialogue. Before the team's final preseason game in San Diego, he met with Nate Boyer, a former Green Beret who had played briefly in the NFL. The two decided that kneeling, rather than sitting on the bench, was a more respectful gesture. At the game, while being pelted with boos, Kaepernick was joined on a knee by his teammate Eric Reid.

By September, Kaepernick was on the cover of *Time* magazine. Dozens of professional athletes across other sports had joined in the demonstration. Stories popped up across the country of black high-schoolers kneeling before their games as well.

Trump could not have asked for anything more. The controversy was perfectly suited to his campaign's narrative of a culture in rebellion against the country's traditional values, with anyone holding said values made to feel backward and bigoted for rebelling against the rebellion. Even sweeter for the Republican nominee: His opponent played right into it.

"You could put half of Trump's supporters into what I call the basket of deplorables. Right?" Clinton said on September 9 at an LGBT for Hillary gala in New York City, "The racist, sexist, homophobic, xenophobic, Islamaphobic—you name it."

Arguing that Trump had "given voice" to those elements, she continued, "Some of those folks—they are irredeemable. But thankfully, they are not America."

THE FIRST GENERAL ELECTION DEBATE, HELD SEPTEMBER 26 AT HOFstra University in New York, served as a ninety-minute microcosm of the Trump-Clinton contrast. With a record eighty-four million people tuning in, Clinton was steady if unspectacular, giving safe and crisp answers while keeping her cool throughout. She was clearly well prepared. Trump, on the other hand, was out of his depth. He was baited into damaging sound bites and fitful, long, rambling responses. He was blindingly unprepared.

Clinton was scored the consensus winner. Even the Republican's cable news advocate struggled to spin his performance.

"The good news for Donald Trump is that he discussed serious issues for ninety minutes," Howard Kurtz, the conservative media reporter, said on Fox News afterward. "But Hillary Clinton won the night on points. She was aggressive out of the gate, and in basketball terms, she controlled the ball. He started to talk louder, faster, trying to compete with her. And as time went on, it seemed to me that he got a little more disjointed."

To Trump, the verdicts of his debate defeat were reflective of nothing more than a biased jury of journalists. The Republican nominee had used the media as a foil throughout the campaign, tapping into decades of percolating distrust of (and bitterness toward) the press corps among conservatives. He called out and derided individual reporters by name. He blacklisted certain publications—the *Washington Post*, BuzzFeed, the *Des Moines Register*—refusing to grant them access to cover his campaign events. He accused the Fourth Estate of peddling "fake news" to deceive the masses (a perversion of the term used to describe attempts by foreign troublemakers to sow chaos in the electorate by propagating deceptive information online).

In an era defined by friction over "snowflakes" (overly sensitive people) acting "woke" (highly attuned to political correctness) in response
to "microaggressions" (perceived slights to marginalized persons or
communities), Trump's hostility toward the press, increasingly perceived as the arbiters of American dialogue, made him a hero to the
right.

He was not always wrong with his charges of bias or hysteria.
Twitter-happy reporters and click-drunk newsrooms and advertising-
mad cable news shows turned no small number of molehills into mountains. In early August, during a rally in Virginia, Trump teased a mother
about her crying baby, flippantly remarking, "You can get the baby out of
here."[21] After dozens of outlets reported that he'd ejected an infant from
his event, PolitiFact was forced to weigh in: "Donald Trump accurately
says media wrong that he kicked baby out of rally."

Even so, the GOP nominee deserved the historic number of negative
headlines dropped on his campaign. Trump told hundreds and likely
thousands of provable lies during the 2016 campaign, falsehoods both
big (his supposed opposition to the Iraq War) and small (his endorsement from Immigration and Customs Enforcement). He routinely said
things far outside the mainstream of political discourse, be they personal insults or pointless boasts or menacing threats.

Even when he stood to benefit from a news cycle, such as when the
New York Times and other prominent outlets reported on the details
of the FBI's probe into Clinton's emails (tough press coverage the right
never seemed bothered by), Trump had a knack for snatching defeat
from the jaws of victory. The same day that the State Department's inspector general released a report excoriating Clinton for her email habits, he stole headlines by lashing out at the Republican governor of New
Mexico.

Professional press-bashers on the right, such as the Media Research
Center, waged a campaign to delegitimize the coverage of Trump. The
group's signage—"DON'T BELIEVE THE LIBERAL MEDIA!"—was
ubiquitous around Cleveland the week of the convention, "LIBERAL
MEDIA" written in bloody red. And yet it was the MRC's president, the
longtime conservative activist Brent Bozell, who had been among the

most strident essayists in the infamous *National Review* issue, calling Trump "the greatest charlatan" he'd ever seen in politics.[22] Bozell, a Cruz supporter in the primary, also called Trump a "huckster" and a "shameless self-promoter" in one Fox News appearance, concluding, "God help this country if this man were president."

By the first week of October, the Republican nominee's lack of support from the establishment media, including its most conservative elements, came into sharp focus. Trump was the first presidential nominee in history to receive no major newspaper endorsements. The traditionally conservative editorial pages of the *Dallas Morning News*, the *Arizona Republic*, the *Houston Chronicle*, and the *Cincinnati Enquirer* backed Clinton; others, including the *Detroit News*, the *New Hampshire Union-Leader*, and the *Richmond Times-Dispatch* supported the Libertarian Party's nominee, Gary Johnson. *USA Today*, which had never endorsed in its history, threw its weight behind Clinton, calling Trump "a serial liar" who was "unfit for the presidency."[23]

The poll numbers were no more encouraging. As of early October, Trump still trailed Clinton by 9 points in Pennsylvania, according to the RCP average; by 7 points in Michigan; by 6 points in Wisconsin; and by 3 points in both Florida and North Carolina.

With the writing on the wall, and the post–Election Day repercussions to consider, some of Trump's frenemies in the GOP began circling the wagons. Cruz finally offered an endorsement in late September. And Ryan, who had gone out of his way never to be photographed with Trump, fearful that it would be used to tarnish his image, invited the nominee to join him at "Fall Fest," an annual rally in his Wisconsin district on October 8.

There would be a reckoning among Trump's supporters after he lost in November, and his Republican rivals were acting preemptively to avoid any blame.

CHAPTER SIXTEEN

OCTOBER 2016

"Mother is not going to like this."

ONE BY ONE, THEY HAD TRICKLED OUT OF THE CONFERENCE ROOM ON the twenty-fifth floor of Trump Tower. It was Friday, October 7, two days before the second presidential debate, and the Republican nominee's brain trust had spent the morning running a carefully simulated rehearsal session. Chris Christie, playing the role of Hillary Clinton, was seated adjacent to his opponent at a conference table; Reince Priebus, acting as the moderator, was positioned directly across from Trump. The rest of the observers—Hope Hicks, Steve Bannon, Kellyanne Conway, David Bossie, Jared Kushner, and the nominee's children, among a few others—listened critically, offering occasional feedback.

Hicks had left the room first. The others, more glued to their smartphones than usual, began taking turns excusing themselves. Priebus, Christie, and Trump pushed onward with the debate prep. Finally, looking up and realizing that it was only the three of them remaining, Priebus paused the proceedings. "Okay," he told Trump. "When the entire staff leaves the room, something's up."

Trump hadn't noticed, either. Now he glanced from side to side. To his right, through the glass-plated doors, he could see the members of his team huddled outside the conference room, arguing in hushed tones. "Yeah," Trump said, breaking from his practiced debate cadence and barking toward the glass. "What the hell's going on out there?"

A few agonizing moments passed before the door opened. In walked Hicks, carrying a stapled packet of papers. She handed them silently to Trump. A former Ralph Lauren model known for her sharp looks and confident mien, Hicks was now ashen-faced. Trump eyed the top sheet and began reading. "Uh huh," he said, flipping to the next page. "Mmm hmmm."

Priebus was growing impatient—and fearful. "What is it?" he said. "Tell me what's happening."

Trump ignored him. Turning to a new page, he scanned the print and then stopped suddenly, his expression and tone shifting at once. He looked up at Hicks. "This doesn't sound like me."

Priebus raised his voice in uncharacteristic fashion. "Someone tell me something, please!"

Trump looked at him, put the packet on the table, and slid it across. The party chairman began to read, the room now filling around him with the rest of the team. They had all seen it: an email exchange with *Washington Post* reporter David Farenthold, who claimed to have an old audio recording of Trump making exceedingly lewd remarks about women and boasting of his ability to get away with sexual assault. Farenthold had sent over the alleged quotes and was requesting comment from the campaign for a story that would run later that day.

"Wow, this isn't good," Priebus said, his eyes fixed on a single line. "This is really, really bad."

The group was paralyzed with silence. Finally, Kushner piped up. "You know, I don't think it's all that bad."

"Jared, what are you talking about?" Priebus said, burying his head in his hands. "This is as bad as it gets."

Trump, talking to no one in particular, repeated himself. "This doesn't sound like me."

Two of the nominee's advisers spoke up in support of that theory. Conway and Bossie vouched for Trump, saying they had never heard him use any such language to describe women. This wasn't his style.

Priebus was struck by an impossible bolt of optimism. He told Trump that maybe it was all a mistake; he recalled the time he was misquoted after a speech, when the chairman had used the phrase "hates us" and a reporter wrote that he had said "racist." Tape recordings were tricky things, Priebus said. Maybe this entire situation was a foul-up.

Just then, Bossie pulled out his iPad. Farenthold, the *Post* reporter, had sent the audio file. With the nominee's team clustered around him, Bossie pressed play. They listened. And then, Trump spoke up. "Well," he said, "that's me."

The room fell hushed. "It was a moment of humility and vulnerability," Conway recalls. "He legitimately did not remember saying that."

IT WAS JUST BEFORE 4:00 P.M. IN THE NATION'S CAPITAL WHEN THE *Washington Post* published an "October surprise" for the ages.

Farenthold's story told of an exclusively obtained audio recording of Trump, eleven years earlier and newly married, boasting of his sexual exploits to television host Billy Bush. The two were riding together on a bus, preparing to shoot a segment for the NBC show *Access Hollywood*, when Trump recalled how he'd once tried to sleep with Bush's cohost, Nancy O'Dell.

"I moved on her and I failed. I'll admit it. I did try and fuck her. She was married. And I moved on her very heavily," Trump said on the tape.[1] "In fact, I took her out furniture shopping. She wanted to get some furniture. I said, 'I'll show you where they have some nice furniture.' I took her out furniture—I moved on her like a bitch. But I couldn't get there. And she was married."

Then, when the two men on the tape spotted a young woman awaiting them outside the bus—actress Arianne Zucker—Trump told Bush, "I've got to use some Tic Tacs just in case I start kissing her. You know, I'm automatically attracted to beautiful—I just start kissing them. It's like a magnet. Just kiss. I don't even wait."

Trump added, "And when you're a star, they let you do it. You can do anything. Grab 'em by the pussy. You can do anything."

The fallout was apocalyptic.

Paul Ryan had been scheduled to make his first joint appearance with Trump the next morning at Fall Fest, the annual beer-and-bratwurst political rally in his district. Preparing to speak at a fund-raiser for a congressman in Cleveland, Ryan was pulled aside by his longtime aide, Kevin Seifert, who showed him the story. Ryan, the Boy Scout, burst into a fit of cursing just outside a roomful of wealthy donors.

He phoned Priebus immediately. "He cannot come here," Ryan said. "You need to tell him."

Priebus relayed this to Trump, who promptly shot the messenger. "Oh no," the Republican nominee replied, "I'm coming."

The party chairman called Ryan back with Trump's reaction. "You're gonna have to publicly disinvite him, Paul."

"Fine, then he's disinvited. He ain't coming," the Speaker said, raising his voice to Priebus for the first time in their decades-long relationship. "This isn't something I'm intimidated by."

A short while later, Ryan's office blasted out a press release saying he was "sickened" by Trump's remarks and announcing his banishment from the Wisconsin event.[2] Priebus understood but was nonetheless distraught. He had started Fall Fest years ago as the Wisconsin GOP chairman. Saturday's event was supposed to be a homecoming for him and a harmonious breakthrough for the party. All of them—Priebus, Trump, Ryan—were meant to take the stage together, at long last projecting a united front entering the final weeks of the campaign.

Up until that point, despite Trump's self-destructive antics, Priebus believed his party had a chance. Clinton was so deeply flawed, and the Democratic base had been made so complacent by the combination of her candidacy and eight years in power, that Priebus clung to the belief that Trump somehow, in some way, might just win the White House.

Everything changed, however, when he heard the *Access Hollywood* tape. And it wasn't just the party chairman's own gut reaction. Over the next thirty-six hours, Priebus fielded scores of phone calls from the most prominent people in Republican politics: congressmen and senators, governors, donors, activists, and his own RNC members. Every single person told him the same thing: Trump was doomed. The party needed to replace him with Mike Pence atop the ticket.

Reconnecting by phone later that night, Ryan demanded that the national party take action to excommunicate Trump. "This is fatal," he told Priebus. "How can you get him out of the race?"

Priebus had to explain—to Ryan and to everyone else—that there was no mechanism for removing Trump. But this answer proved inadequate. The voices on the other end of the line demanded that something be done. Many suggested that he, the RNC chairman, publicly renounce Trump and ask for him to step aside as the nominee for the good of the party. (Even some of the people endorsing such an ultimatum knew how silly it sounded. Trump cared nothing for the party; he had not belonged to it until signing his name to a piece of paper a year earlier.)

For his part, Trump had agreed after some cajoling to offer a non-apology apology, issuing a statement to the *Post* that read, "This was locker room banter, a private conversation that took place many years ago. Bill Clinton has said far worse to me on the golf course—not even close. I apologize if anyone was offended."

But his team quickly realized this would not suffice. By Friday evening, Trump's campaign appeared on the brink of collapse. There were rumors of an imminent mass exodus of Republican officials who would publicly withdraw their support for the party's nominee. The first departures came that very night.

Senator Mike Lee of Utah, whom Trump had recently named to an extended list of potential Supreme Court nominees, called on Trump to drop out. So did his colleague Senator Mark Kirk of Illinois, who asked the RNC to "engage rules for emergency replacement." Jason Chaffetz, the Utah congressman and chairman of the House Oversight Committee, told a local TV station, "I'm out. I can no longer in good conscience endorse this person for president. It is some of the most abhorrent and offensive comments that you can possibly imagine."[3]

It had begun as just another day in the Trump campaign: He was getting hammered after stoking his longest-simmering racial controversy, telling CNN that he still believed the "Central Park Five" were guilty of their alleged 1989 rape despite the DNA evidence that had overturned their convictions.

But no amount of fire-extinguishing done by Trump's team over the past year had prepared them to fight the inferno of October 7. The only silver lining was a choreographed counterpunch against his opponent's campaign: Within an hour of the *Post* story's publication, the Moscow-friendly group WikiLeaks released more than two thousand emails that had been hacked from the personal email account of John Podesta, Clinton's campaign chairman.

WikiLeaks announced that it had hacked some fifty thousand of Podesta's emails, and soon set about publishing them in staggered fashion.

The content verified much of what was already suspected. There were complaints about Clinton's terrible political instincts; fears about ethical conflicts within the family foundation; traces of ideological shape-shifting in those pricey paid speeches she'd been criticized for giving after leaving the State Department.

None of it was exceptionally damning. If anything, the mere fact that Clinton's campaign chairman had been hacked—*while* the country was fixated on the matter of her clandestine use of a private email server— seemed a bigger story than any of Podesta's individual emails. Still, the timing couldn't have been worse: Minutes after it looked like the Democrats had caught their biggest break of the year, they were forced into a defensive crouch.

TRUMP'S CAMPAIGN WAS IN SCRAMBLE MODE. HE AND CLINTON WERE scheduled to debate Sunday night in St. Louis. Convinced that a thorough, videotaped apology was their only chance to survive the weekend, his senior aides set about staging the production. Some tinkered with the text, debating how much emphasis to place on the Clintons' past scandals with women. Others prepared for the most important video shoot of the celebrity's career, choosing between four background screens: daytime Manhattan, nighttime Manhattan, campaign signage, or a flat, unassuming blue.

Trump seemed mystified by the blur of manic activity. "I've never taken anyone furniture shopping!" he laughed, throwing up his arms. His staff members traded disoriented looks.

Just after midnight, on Saturday the eighth of October, the campaign posted a ninety-second video clip to Trump's Facebook page. Against a dark superimposed horizon of illuminated skyscrapers, Trump looked directly into the camera. "I've never said I'm a perfect person, nor pretended to be someone that I'm not. I've said and done things I regret, and the words released today on this more than a decade-old video are one of them. Anyone who knows me know these words don't reflect who I am. I said it, I was wrong, and I apologize," he said.

Trump added, "I've said some foolish things, but there is a big difference between the words and actions of other people. Bill Clinton has actually abused women and Hillary has bullied, attacked, shamed, and intimidated his victims. We will discuss this more in the coming days. See you at the debate on Sunday."

Not for a moment would Trump consider quitting the race. He was unmoved by the rebukes of the Republican lawmakers who were piling on with excoriating statements; most of them, he scoffed, were the same people who had opposed his candidacy from its inception. Trump cackled as one of his aides read aloud the rolling list of disavowals from the likes of Ryan and Romney. He could not have cared less what they had to say.

There was one politician whose reaction Trump worried about: Pence.

It had been a shotgun marriage, one of convenience more than love. Yet Trump had grown unusually fond of Pence. There was a sincerity to his running mate that he thought rare and endearing. Certainly, Trump found Pence a bit alien: the way he was always praying; the way he referred to his wife, Karen, as "Mother"; and the way the couple was constantly holding hands. ("Look at them!" Trump would tease. "They're so in love!") But he appreciated the earnestness with which Pence seemed to believe, as so few in the party did, that Trump was a decent person. Trump had worked hard to earn that faith. On the night of the October 4 vice-presidential debate, he even left a voice mail for Pence letting him know that he would be saying a prayer for him.

Speaking in Ohio just after the *Access Hollywood* bombshell dropped, Pence had initially dismissed the news as just another media hatchet job. Yet soon after, he called Trump from the road, checking in as he did daily, sounding upset. He advised Trump to offer a sincere apology. That was the last anyone had heard from the VP nominee. Pence had gone back to Indiana and bunkered down, cutting himself off from the outside world, praying with his wife about what to do next and telling his advisers that he wasn't sure he could continue with the campaign.

To the extent Trump felt regret, it was over disappointing the Pences.

"Oh boy," he said Friday afternoon after hanging up with his running mate. "Mother is not going to like this."

THE APOLOGY VIDEO DID LITTLE TO STANCH THE FLOW OF DEFECTIONS. On Saturday morning, another tranche of Republicans—congressmen, senators, governors, former primary rivals—announced their renunciations of Trump.[4] The list also included GOP luminaries such as Bill Bennett, the former education secretary, and Condoleezza Rice, the former secretary of state, whose name was being tossed around inside the RNC as a potential substitute running mate if Pence took over the ticket.

By midday Saturday, October 8, more than two dozen Republican elected officials had abandoned Trump (counting only those presently in office). Many were calling for Pence to replace him as the GOP nominee. Among them were Senator John Thune, a member of the GOP leadership, and Ann Wagner, the Missouri congresswoman and a former co-chair of the national party committee.

Priebus continued to swat away the suggestion. As the former general counsel of the RNC, he knew better than anyone that no trigger existed for forcing out the party's nominee—especially not at this late stage. When he received a call Saturday morning from Wisconsin's national committeeman, Steve King, informing him that some RNC members were mulling an organized mutiny, the party chairman told King the same thing he was telling everyone else: "It's not going to work. We need to ride this out."

But Priebus worried, as did just about everyone else he spoke with, that another shoe was soon to drop. There had been rumors in recent weeks that a lethal opposition-research blast was imminent. Now that it had occured, Republicans felt certain there were more to follow; that somewhere there existed a veritable treasure trove of old tapes revealing Trump's greatest hits: misogyny, racism, and all sorts of other uncouth talk from the set of his NBC show. (Reporters raced unsuccessfully to reach Mark Burnett, producer of *The Apprentice*, sensing that he possessed the power to swing an election.)

While the party chairman saw no path to removing Trump, he wasn't ruling out the possibility of Trump stepping aside on his own accord.

Having gotten the sense from Pence's advisers that the Indiana governor would be willing to take over if Trump quit, Priebus talked into the wee hours Friday night with trusted allies—Ryan and McConnell, as well as top staffers and party lawyers—discussing the logistical hurdles to replacing a nominee one month before Election Day. It wouldn't be easy: Early voting had begun in some states, and ballots had been printed in most others.

The biggest obstacle, of course, was Trump. It would be tricky enough rejiggering the ticket to pair Pence with a new running mate; doing so without Trump's blessing would be impossible.

Shortly before 11:00 a.m. Saturday, the Republican nominee convened the campaign's high command in his residence on the sixty-fourth floor of Trump Tower. Everyone looked withered. Giuliani wore a Yankees cap low over his eyes. Priebus hadn't shaved. Christie, dressed in jeans and a Mets jacket, had already informed the group that he needed the rest of the weekend off and would not fly to Sunday's debate with Trump as planned.

"So," Trump began, looking to Priebus. "What are you hearing?"

The RNC chairman had spent the past day defending Trump's rightful claim to the party's nomination, dismissing calls to expel him and urging calm amid the commotion. But Priebus was not going to sugarcoat the situation. He had long been nauseated at watching all the nominee's sycophants telling him whatever would keep him happy and upbeat. Trump needed to hear the truth for a change.

"I'll tell you what I'm hearing," Priebus said. "Either you'll lose in the biggest landslide in history, or you can get out of the race and let somebody else run who can win."

Nobody said a word. Trump's many loyalists who had gathered—his children, Hicks, Bannon, Conway, Christie, Bossie, Giuliani—were shocked by the blunt assessment. Yet none was eager to push back on it. When Trump went around the room, asking what people thought his chances were, he heard a lot of throat-clearing. Even Bannon, who made it a habit of always saying "one hundred percent" whenever Trump asked the question, dodged it this time.

Trump tried humor. "So, what's the good news?" he said.

Nobody laughed.

The meeting lasted another thirty minutes, most of which was spent pushing Trump to sit for an interview that afternoon with David Muir of ABC News. His team said it would be best to discuss the comments fully, and repent for them, ahead of the debate. Trump agreed and the meeting broke up. But then he abruptly changed his mind. Complaining that he would look "weak" by subjecting himself to a journalist whose sole purpose would be extracting as many apologies as possible, he told Hicks the ABC interview was off.

The Republican Party was going to live or die with Trump; if his team couldn't persuade him to do a network television interview, they certainly weren't going to convince him to step aside as the nominee. Whatever fantasies of a Pence-Rice ticket danced through the heads of party elders were officially dashed on Saturday afternoon. "The media and establishment want me out of the race so badly," Trump tweeted. "I WILL NEVER DROP OUT OF THE RACE, WILL NEVER LET MY SUPPORTERS DOWN! #MAGA."

Pence himself was nowhere to be found. Ryan had asked his old friend to attend the Saturday rally in his district in lieu of Trump. Pence had accepted. Accommodations were made; a Secret Service checkpoint, waved off at the news of Trump's disinvitation, was re-erected outside the event in Elkhorn, Wisconsin. But then Pence didn't show up. There was no notice, no courtesy call from the VP nominee's staff. Ryan dialed his old friend's cell number and got voice mail. Pence was AWOL.

Instead of returning Trump's calls, or Ryan's calls, or flying to his friend's district, the Indiana governor spent Saturday at home. He mostly prayed with his wife, Karen. She was apoplectic, warning her husband that she would no longer appear in public if he carried on as Trump's running mate. He, in turn, hinted to his advisers that his time on the trail might be up. Feeling moved to communicate his inner anguish, Pence wrote Trump a letter describing what hearing that audio had done to him and his wife. When two of Trump's advisers learned of the letter, they worried they had seen the last of his running mate.

Meanwhile, Ryan was left to fly solo in Elkhorn—no Trump, no Pence, and no Priebus.

"There is a bit of an elephant in the room," the Speaker said, taking the stage in Wisconsin.[5] He referenced his statement from the previous day and how "troubling" the situation was. Then, announcing that he wasn't there to talk about said elephant, Ryan pivoted to his homily about "ideas" and "conservative principles" and his vision for being a "proposition party."

But it was hard to hear over the boos. Chanting the nominee's name, Trump's supporters in the audience heckled Ryan throughout his speech. "Shame on you!" they shouted.

THE WOMEN FLANKED TRUMP, TWO OF THEM ON EACH SIDE, SEATED BE-hind rectangular folding tables draped in olive fabric. The small conference room, on the campus of Washington University in St. Louis, was barren save for the tables, some black coffee mugs, bottles of water, and an American flag. Reporters rushed into the room. Cameras started rolling. Jaws hit the floor.

It was less than two hours until the start of the October 9 presidential debate, a spectacle that would draw tens of millions of eyeballs, and the GOP nominee was putting on a surprise pregame show. Without advance warning, Trump held an impromptu press conference alongside a group of women who had publicly accused Bill Clinton of sexual misconduct.

There had been speculation for months that he could invite one of the former president's accusers to a debate, perhaps having them sit in the front row to unnerve Clinton's wife. Trump's campaign always dismissed the rumors. Priebus, who joined Trump on the flight to St. Louis to help with last-minute debate prep, had heard nothing about the planned stunt; the only gossip from the plane ride was Trump railing against Giuliani's performance on the Sunday shows, yelling repeatedly through the cabin, "What the fuck is Rudy doing? Get this guy off the television!"

Inside the debate hall, when co-moderators Anderson Cooper and Martha Raddatz introduced them, Trump and Clinton entered from opposite wings of the auditorium looking steeled for a street fight. They approached one another, only to stop abruptly and stand several feet apart. There would be no handshake—a first, it was believed, in the annals of presidential debating.

After a schoolteacher in attendance asked the opening question, about whether the candidates felt they were modeling good behavior for the nation's children, Cooper sensed a natural segue to ask about Trump's remarks. The Republican nominee offered an answer rehearsed again and again on the plane ride from New York: "I'm not proud of it," he said, "but this is locker room talk." Pressed on what his comments meant, Trump replied, "I have great respect for women. Nobody has more respect for women than I do." There were audible groans from the audience.

When the moderators turned to Clinton, she, like Trump, commenced with a clearly practiced soliloquy. "With prior Republican nominees for president, I disagreed with them . . . but I never questioned their fitness to serve," Clinton said. "Donald Trump is different."

Trump attacked and counterattacked throughout, bringing up Bill Clinton's history of being "abusive to women" and aggressively prosecuting Clinton's use of a private email server while secretary of state, an issue he had failed to raise during the first debate. "If I win," Trump declared, "I am going to instruct my attorney general to get a special prosecutor to look into your situation. Because there have never been so many lies, so much deception."

"Everything he just said was absolutely false," Clinton responded when given the floor, adding, "It's just awfully good that someone with the temperament of Donald Trump is not in charge of the law in our country."

"Because you'd be in jail," Trump shot back. Some audience members gasped. Others cheered.

When the moderators asked Clinton to explain, given her statements about some Trump supporters being "deplorables" who are "irredeemable," how she could unite the country, she expressed some remorse. "My argument is not with his supporters," Clinton said of her opponent, "it's with him."

"She has tremendous hate in her heart," Trump replied.

It was, without question, the ugliest and most vitriolic presidential debate in the mass-communication era. And it was exactly what Trump needed. Facing pressure unlike any White House hopeful in memory, the Republican nominee didn't just get off the mat; he came up swinging. "What were the odds? Like fifty-fifty, will he show up?" Trump says. "That debate won me the election."

RYAN FELT VALIDATED BY THE *ACCESS HOLLYWOOD* TAPE, EVEN AS HIS worst fears were being realized.

It was nearing the one-year anniversary of his swearing-in as Speaker of the House, and he'd spent much of that time sounding the alarms about Trump. He worried about an opposition research attack that could cripple the nominee and do serious collateral damage to the party. In fact, it almost seemed inevitable. Trump had been in the public eye for decades and rarely missed an opportunity to raise eyebrows. He was a regular guest on shock jock Howard Stern's radio show, often to discuss the female anatomy. He had been "roasted" in the crudest of terms on Comedy Central. And there had long been talk that Trump, between owning the Miss Universe pageant and starring in NBC's *The Apprentice*, had left a documented trail of raunchy talk and devious behavior.

The *Access Hollywood* tape didn't just present a crisis for Trump's candidacy. It threatened to torpedo Republicans down the ballot in contests across the country. All throughout the weekend, McConnell lobbied Priebus to redirect the RNC's cash earmarked for the presidential race toward his Senate campaigns. His argument: If they didn't maintain their majority in the Senate, President Hillary Clinton would remake the federal courts for a generation.

Ryan wanted to take more dramatic action. On an emergency leadership conference call, the day of the St. Louis debate, Ryan floated the idea of withdrawing his endorsement of Trump. He would kill their majority, the Speaker said; cutting him off might be their best hope of saving the House. It was Kevin McCarthy, the majority leader and Trump's favored member of the GOP leadership, who talked Ryan down. Withdrawing their support, McCarthy argued, would backfire by depressing turnout in Trump-friendly districts and states.

Ryan found himself agreeing. He would not go so far as to renounce his endorsement of Trump. He would, however, tell members that he planned to do nothing to help the nominee over the final month of the campaign, focusing solely on protecting their House majority. And he would advise them to do what they felt was best to survive in their districts, whether that meant defending Trump or running away from him.

On Monday morning, October 10, Ryan convened a conference call with all 246 House Republicans. According to audio that was later leaked to Breitbart.com—a sign of how Ryan's far-right members reacted—the Speaker said of Trump, "His comments are not anywhere in keeping with our party's principles and values. There are basically two things that I want to make really clear, as for myself as your Speaker. I am not going to defend Donald Trump—not now, not in the future."[6]

Ryan added, "Look, you guys know I have real concerns with our nominee. I hope you appreciate that I'm doing what I think is best for you, the members, not what's best for me. . . . I talked to a bunch of you over the last seventy-two hours and here is basically my takeaway. To everyone on this call, this is going to be a turbulent month. Many of you on this call are facing tough reelections. Some of you are not. But with respect to Donald Trump, I would encourage you to do what you think is best and do what you feel you need to do."

As the Speaker finished, stepping back to let his members weigh in, he felt uneasy. Ryan had wanted to unendorse Trump; McCarthy had convinced him not to. Now Ryan worried that he hadn't gone far enough, that his members would be upset about his merely saying he would no longer defend Trump.

Listening in, the Speaker was stunned to realize that the opposite was true: He had gone *too far*. Some members were furious that Ryan had dared to publicly condemn Trump. They felt he was abandoning the party by abandoning its nominee. In their eyes, he was waving a white flag of surrender.

They weren't alone in this view. Just before noontime, the AP blasted out a bulletin: "House Speaker Paul Ryan is all but conceding Hillary Clinton will be the next president." Soon after, Trump tweeted, "Paul Ryan should spend more time on balancing the budget, jobs and illegal immigration and not waste his time on fighting [the] Republican nominee."

Ryan's office rushed to clean up the perception of his comments, but it was too late. The grass roots were ablaze with indignation. Congressional phone lines exploded with irate GOP constituents calling for Ryan's head. Some members privately began questioning the sustainability of his position atop the party; later in the month, when leaders scheduled Ryan's internal speakership election, some pro-Trump lawmakers lobbied for the vote to be postponed, which would give them more time to assess whether Ryan should remain Speaker.

The Freedom Caucus sensed an opportunity. In a secret meeting later that month at Meadows's downtown DC apartment, the group's board members devised a plan to deny Ryan the 218 votes needed to retain his speakership. The strategy called for Jim Jordan to serve as the right's sacrificial lamb, running against Ryan not to win, but to collect enough votes to force a second ballot. The idea was that Ryan, who talked often (and annoyingly, to some members) about how he'd never wanted the job to begin with, would step aside to avoid the spectacle. Conservatives had been searching for a Ryan alternative from outside their narrow ranks, someone who, unlike Jordan, could appeal to the rest of the conference. They decided that Mike Pompeo, the dry-witted defense hawk from Kansas, would be their top choice.

As Republicans schemed against their Speaker, the underlying assumption was that Trump would lose and the conservative base would be out for blood, resulting in an overthrow of Ryan. Either that, or Trump would win a shocking upset and kick the Speaker—"Our very weak and ineffective leader," the nominee tweeted after their *Access Hollywood* altercation—to the curb. Either way, Ryan would be finished.

TRUMP'S DIAGNOSIS AFTER *ACCESS HOLLYWOOD* WAS TERMINAL—UNTIL IT wasn't.

Public polls showed Trump collapsing in the two weeks following the *Washington Post* report, and those numbers squared with the internal data being collected by the Trump campaign. Yet after two weeks, his numbers began climbing back to where they had been previously, eventually plateauing and leveling off. He still trailed Clinton in the key states and was still hopelessly unpopular with the broader electorate, but it was remarkable nonetheless. The man nicknamed "Teflon Don," who had weathered firestorms no other politician could have survived, had done it again. His candidacy was like a stress ball: No matter how hard the squeeze, it always returned to form.

The immortality of Trump, as demonstrated by his survival of Grab-'Em-by-the-Pussy-Gate, owed principally to three explanations.

The first was reflexive distrust of media. The overwhelming majority of conservative voters, even those not enamored of Trump in 2016, had come to see the press as a partisan combatant. Whether it was the paper-thin *New York Times* report[7] insinuating John McCain's affair with a lobbyist in 2008 or the countless petty pile-ons that dogged Romney in 2012, years of negative coverage had alienated Republican voters from the mainstream media. As a result, many on the right tuned out the traditional gatekeepers, preferring to get their information from Fox News or the conservative wing of the internet, places where critical coverage of Trump was hard to find. Among those conservatives who *did* still drink from the mainstream media's well, there was a desensitization to outrage: After being told every other day that Trump's latest infraction was calamitous, they became numb to the instances that really were.

Second, it was impossible to overstate the depth of disdain for Clinton on the right, even in an age of hysterical, hypertribal politics. That disdain had been cultivated for the past quarter century; there was no softening her image or persuading detractors to give her a fresh look. Trump may have been a shameless deviant, but in the eyes of conservatives, he was running against the first family of perversion. He may have been unethical, but so was she—hence his "Drain the Swamp!" motto, which became the closing chant at his October rallies. There was no sharp contrast for Democrats to draw. Trump was the most unpopular nominee in recent memory, but he was running against the *second-most* unpopular nominee in recent memory.

"We have perhaps two of the most flawed human beings running for president in the history of the country," Mick Mulvaney said in South

Carolina shortly before Election Day, in comments reported by *The State* newspaper.[8] "Yes, I am supporting Donald Trump, but I'm doing so despite the fact that I think he's a terrible human being."

The third and most significant reason for Trump's survival: the un-flinching support of the Christian right. Where many evangelical leaders had once expressed an open contempt for the primary candidate, they became his staunchest, most faithful allies during the general election campaign—including in the aftermath of *Access Hollywood*. There were notable exceptions. On the evening of the tape's release, Russell Moore, the head of the Southern Baptist Convention's political arm, tweeted in response to his high-profile peers, "What a disgrace. What a scandal to the gospel of Jesus Christ and to the integrity of our witness. . . . The po-litical Religious Right Establishment wonders why the evangelical next generation rejects their way. Today illustrates why." The next day, after Trump defended his transgression as "just words," Moore tweeted: "No contrition. 'Just words.' How any Christian leader is still standing be-hind this is just genuinely beyond my comprehension."

But Moore was an outlier. In case after case, over the final five weeks of the election, prominent Christian leaders rallied around the Repub-lican nominee. "The crude comments made by Donald J. Trump more than eleven years ago cannot be defended," Franklin Graham, son of the famed evangelist Billy Graham, wrote on his Facebook page. "But the godless progressive agenda of Barack Obama and Hillary Clinton like-wise cannot be defended." Added Jerry Falwell Jr., the other spiritual dynasty scion, "We're never going to have a perfect candidate until Je-sus Christ reigns forever on the throne."

Their principal rationale in standing by Trump: the Supreme Court.

Judicial appointments traditionally have been a more effective ral-lying cry for the right than for the left; every four years, GOP officials and activists have endeavored to mobilize the base by describing a Su-preme Court on the precipice of a liberal occupation. But 2016 was dif-ferent. The death of conservative legal giant Antonin Scalia, and the subsequent decision by Mitch McConnell to block hearings on President Obama's nominee, had placed the issue of Supreme Court appointments front and center unlike during any election in modern history. With an automatic appointment waiting to be filled, Justice Anthony Kennedy hinting at his pending departure, and a pair of other justices past the age of mandatory corporate retirement, conservatives believed the ends of a sympathetic high court justified the means of supporting Trump.

As Hugh Hewitt, the radio host and constitutional law professor who'd butted heads with Trump, had written in the *Washington Examiner* that summer, "It's the Supreme Court, stupid!"[9]

To the credit of the political newcomer, Trump possessed an innate understanding of this constituency's control over his destiny. If white Christian turned out to vote en masse, he had a chance to upset Clinton; if they didn't, he would be roadkill. This explains the speech at Liberty University, the summit in New York City, the release of two lists of Supreme Court candidates, the formation of a faith-based advisory board, and the selection of Pence as his running mate.

There was one final thing they needed: to hear Trump speak their language. The nominee's Christian-ese was stiff and rehearsed, often laughably so. For as horrified as they were of a Clinton-controlled Supreme Court ruling on everything from abortion to guns to religious liberties, conservatives still harbored justified skepticism of Trump's conversion. If they were going to turn a blind eye to his odious behavior in the name of Supreme Court appointments, they at least wanted assurance—real, heartfelt, unscripted assurance—that he would deliver.

They got it in Sin City, of all places.

During the third and final presidential debate in Las Vegas, on October 19, Trump hammered the significance of the high court. After praising the *Heller* decision, which protected the individual's right to keep and bear arms, Trump pledged to appoint justices who would overturn the landmark ruling in *Roe v. Wade* that had legalized abortion.

Then, he went even further. After Clinton defended her Senate vote protecting the practice of partial-birth abortion, Trump pounced. "If you go with what Hillary is saying, in the ninth month, you can take the baby and rip the baby out of the womb of the mother just prior to the birth of the baby," he said.[10] "Now, you can say that that's okay. And Hillary can say that that's okay. But it's not okay with me."

It was a seminal moment in his candidacy. Ralph Reed, the longtime

Christian conservative honcho and president of the Faith and Freedom Coalition, was effusive afterward. "Trump just sealed the deal with evangelicals," he predicted.

It was the consummating feat Trump needed, especially as it distracted from his otherwise lackluster debate performance, which included continued allegations of a "rigged election" and a sinister veiled threat not to accept the results of November 8.

He was still the underdog. But two weeks after *Access Hollywood* threatened to kill his candidacy, Trump had life.

CHAPTER SEVENTEEN

OCTOBER 2016

"Thank God."

WE SMACKED INTO THE RUNWAY AND FELT THE WHEELS CLAWING for traction on the rain-slicked tarmac. The Boeing 737 finally lurched to an ungraceful standstill, at which point we laughed and exchanged jokes. Mike Pence was grinning a minute later as he approached from the front of the plane. "Everybody okay?" he asked me and six other reporters.

Yes, we replied, no big deal. Except that it was: The plane had slid off the runway altogether and sliced through a collapsible concrete track designed to stop us from spilling into the East River. Rescue vehicles were now screaming across LaGuardia's tarmac, sirens blaring in the brisk October night; first responders would soon climb the back stairs and shout for us to evacuate immediately. "I didn't realize it," Pence told us of the accident, "until I saw mud on the front windows."

Alas, it was impossible to survey the wreckage from inside the plane.

Such was the story of his final four weeks as Donald Trump's running mate. The release of the *Access Hollywood* tape was a traumatic event for the VP nominee. He was initially inconsolable, retreating to Indiana, signaling to some friends that he might not stay with the campaign. Convinced by advisers that his only real option was to run through the tape—no pun intended—Pence dutifully resumed his role as Trump's wing man.

Soon, however, he went into a different sort of shell. Having emerged from hiding after forty-eight hours and spoken candidly with Trump, the VP nominee began to feel certain that his running mate—a man he'd prayed with, golfed with, become friends with—was genuinely contrite, was truly a different person than the one on the decade-old recording, yet was being victimized by a bloodthirsty liberal media.

This conclusion afforded Pence the luxury of becoming willfully oblivious to perception. He ignored Trump's critics and retreated deeper into the safe confines of the campaign's echo chamber, blocking out the antagonism and gloom. After returning to the trail, he rarely interacted with the embedded reporters who traveled with him. In sporadic interviews, he responded to questions highlighting Trump's behaviors and inaccuracies with a foreign gaze. Pence had insulated himself—from the possibility that Trump may have committed sexual assault; from the harshest critiques of his decision to join the GOP ticket; and from the reality that its defeat was likely.

At a rally in Fort Dodge, Iowa, he began by saluting his "great, great friend," the brazen race-baiter Congressman Steve King, who "does you proud every single day." He closed as he always did, by alluding to Scripture: "I truly believe what's been true for thousands of years is still true today," Pence said. "As the ancient words say, if His people who are called by His name will humble themselves and pray . . . He will hear from heaven and He will heal this land."

Problem was, none of Pence's traveling posse, a tight-knit group of loyalists, thought the GOP ticket had a prayer on November 8. It was nakedly apparent during my five days with them, on a swing through seven states in late October, that the VP nominee's team had shifted its focus from winning the election to protecting the image and preserving the future ambitions of Pence.

This was not especially surprising given that some of his top aides had been vehemently opposed to Trump in the first place. Marc Short, Pence's longtime consigliere, was the Koch brothers' lieutenant who quit after failing to convince them to finance an eight-figure assault on Trump; Nick Ayers, another trusted adviser, had warned Pence and his other clients throughout the primary season that Trump could bring down the entire party.

As we idled on the runway in Fort Dodge, awaiting clearance for takeoff during a lengthy delay, Pence's team ordered us off the plane, announcing that the VP nominee would quarterback an impromptu football game below. Pence led us away from the tarmac, positioning himself in front of a breathtaking backdrop of golden cornfields. As he cocked his arm to throw—sleeves rolled up, top of his shirt unbuttoned, tie loosened—you could smell his team salivating. This was a made-for-Iowa campaign commercial. Pence would be back, likely as the GOP front-runner, in a few short years.

The one man on the plane with other plans, the one who believed Trump was going to be the next president of the United States, was Pence himself.

The Indiana governor knew all too well the story of Dan Coats, his friend and fellow Hoosier. Back in 1992, Coats, a congressman, had joined then-Vice President Dan Quayle for a fly-around spanning the forty-eight hours before Election Day. When Coats climbed aboard the plane, Quayle told him, "It's done. We're going to lose. Bill Clinton is going to win. The next few days are going to be tough, and I just wanted someone here with me. I'm really glad you're here."

Pence was speaking no such fatalism to his traveling companion, Congressman Jeb Hensarling. (Jeff Flake was . . . unavailable. The Arizona senator refused to attend any events for the GOP ticket. Once, when Pence visited a church in Mesa, a Phoenix suburb, the senator texted to remind him that he would be campaigning less than a mile from Flake's home. "Can you help me trim some hedges?" Flake asked. Pence replied, "As long as we can carve 'Trump-Pence' in the hedge." Flake texted him back: "Small hedge. Only have room for 'Pence.'")

The VP nominee made a compelling case. Trump was going to win, Pence argued, not just because Clinton was a rotten candidate who would struggle to reassemble the Obama coalition, but because Trump represented an end to the party's civil war. It was an odd sentiment; Trump was the most polarizing Republican at least since Barry Goldwater, and probably ever. But Pence wasn't so much lauding his running mate's ability to unite warring factions. Having watched "a Republican party that had lost its way" during the Bush administration, and witnessed the years of internecine conflict thereafter, Pence believed that Trump was mobilizing a base of voters that had been abandoned—"The forgotten people" Republicans were long unresponsive to.

Trump's strength, Pence continued, was derived from his very rejection of party orthodoxy. Even in the instances where this made the

Indiana governor uncomfortable, such as with immigration and trade, he had begun to see the political genius behind it. "I've supported virtually every free-trade agreement that's ever come across my desk," Pence said. "But I just found his arguments very persuasive."

As he came around to understanding and eventually defending Trump's viewpoints, Pence also found himself convinced that the man himself was nothing like the outward caricature. He described how Trump had asked his wife, Karen, to lead a group prayer on several occasions, and insisted that Trump is a follower of Christ. "I respect the sincerity of his faith," he said.

This is when the BS detector starts to beep. Nobody who has spent time with Trump has ever walked away believing him to be a Christian. And, that aside, the notion of Trump being different when he's away from the bright lights—laid back, gentlemanly, "even humble," Pence joked—is mostly fantasy. If there ever had been a real distinction between his private self and his public persona, friends say, it receded from their view in 2016. While those who know Trump laud his humor and hospitality, they also say he is who he's always been: someone who values professional utility over personal relationships in the people he deals with, someone who shows regret for nothing he says or does, and someone who prizes loyalty above every other characteristic.

It's certainly possible that Trump felt remorse for his words on that old recording. But he had no choice other than to *tell Pence* that he was remorseful. Without the VP nominee standing loyally by his side, the campaign would have been finished.

"He took a little time. It's okay. I understand. Many people did," Trump says, acknowledging the letter Pence wrote him. "You know, a couple of days off, it didn't make an impact on me. Because I had people who took a whole *lifetime* off."

Pence's knee-jerk devotion to Trump upon his returning to the campaign trail was something to behold. Even some of his aides seemed uncomfortable with the degree to which Pence was going out of his way to profess his allegiance. It became problematic at one point in our conversation at 30,000 feet.

When I asked whether he would support Ryan remaining as Speaker, a simple question, Pence hesitated. It was unexpected. They had been friends for years; Ryan had introduced him at the convention that summer, and when Trump initially declined to endorse Ryan in his primary, Pence made a rare public break with his running mate, telling Fox News, "I believe we need Paul Ryan in leadership in the Congress." But in the time since, Ryan had denounced Trump after *Access Hollywood*, and Pence was visibly torn choosing between the two.

He declined three times to state his support for Ryan, which sparked an easily avoided tempest when our interview published a week later. "My respect for Paul Ryan is boundless," Pence said, repeating the phrase twice. "I'm not a member of the House Republican Conference anymore. I wouldn't presume upon what the members of the conference choose."

Just over an hour later, our plane walloped the runway at LaGuardia. The head Secret Service agent leapt from his seat, handed over his firearm, and crouched next to Pence, who quickly assured him that everything was fine. The media frenzy was every bit as exaggerated, with news crews (even TMZ) trailing Pence, his team, and the reporters to our Manhattan hotel. The only development of consequence was that Pence's plane would be garaged; in exchange, the next morning, we boarded a substitute aircraft that did not have Wi-Fi capabilities.

As we dipped below the clouds, descending toward the runway in Bensalem, Pennsylvania, on the afternoon of October 28, there was a different sort of commotion toward the front of the plane. Advisers to Pence were whispering to one another in shocked, kid-on-Christmas-morning excitement. We had dropped low enough for the cell towers to activate internet signals, and the news was at once coursing through all our smartphones: FBI Director James Comey had sent a letter to Congress reopening the investigation into Clinton.

"The big breaking news today, you may not have heard about standing in line, folks, is that we just learned that Hillary Clinton may have been a whole lot more than 'extremely careless' when it came to handling classified information," Pence declared at his rally in Pennsylvania.

His aides stood at the back of the crowd exchanging looks of comic bewilderment. Suddenly—and in some cases, for the very first time—they, too, believed Trump could win the presidency.

THE SPEAKER OF THE HOUSE SAT ALONE IN THE DEN OF HIS TWO-STORY Georgian-style home in Janesville, Wisconsin, savoring a few fleeting moments of quiet.

Election Day allowed him to dwell on the nightmare that had been the past year. First, he had accepted a job he never wanted, though he convinced himself it could be used for good. Then, he lost a struggle for the soul of his party to a demagogue with no experience in policy or governing. And now, Ryan had been told, Democrats would control the presidency for another four years. He had just concluded a series of phone calls with Priebus and other party elders. The exit polls released at 5:00 p.m. Eastern left no doubt: Trump was toast.

Huddled around their laptops in Trump Tower, the nominee's team felt blindsided. The data, collected for a consortium of major media outlets (the Associated Press, ABC News, CBS News, NBC News, CNN, and Fox News) suggested a blowout loss. It was impossible to dismiss these findings; the exit polling, based on surveys of more than twenty-four thousand voters nationwide, was generally thought to be reliable. Jared Kushner announced that he would call his father-in-law with the news.

"Everybody thought at five o'clock that I had lost the election, because the exit polling came out. And they're screaming, 'Did you vote for Trump? Or did you vote for Crooked Hillary Clinton?'" Trump says, offering his theory of the case. "But a tremendous number came out and said, 'It's none of your business.' Any of the 'It's none of your businesses' voted for Trump."

The RNC had modeled numerous Election Day scenarios, all of them resulting in a Clinton victory. Splicing their data with the exit poll figures, party officials predicted to Ryan that Trump would win 220 electoral votes; the House GOP majority would be cut in half; and Senate Republicans would lose control of the upper chamber. It would be a massacre—exactly what Ryan had feared with Trump atop the ticket.

He fumed in the backseat of his security detail's SUV as it ferried him across town to the Holiday Inn, where his campaign was hosting a party for supporters. It had all been so preventable. The Speaker had spent his first months on the job crafting a sweeping policy agenda for the GOP, one that projected inclusion and optimism from a party not often associated with either. Ryan hoped it would be an inspiration for the party's presidential field; instead, Trump sabotaged it by running a campaign based on fear and insecurity and exclusion. Boosted by unprecedented free media coverage and backed by millions of anti-establishment voters, Trump had successfully exploited the worst impulses of the electorate en route to winning the Republican nomination—and remaking the party in his own image.

Seething inside a first-floor conference room at the Holiday Inn, Ryan plotted his revenge.

Clinton's victory carried a silver lining: Ryan would be liberated, once and for all, to forsake Trump and purge the Republican Party of his insidious influence. The Speaker would waste no time. With members of the national media assembled in Janesville, he would give a speech blasting Trump and turning the page on a dark chapter in GOP history. He would be free of Trump and so, too, would be his party. Ryan would be its leader for another four years, and a top priority would be erasing the remnants of Trumpism.

Then the returns came in.

THE CLOSING DAYS OF THE CAMPAIGN HAD NOT BEEN KIND TO CLINTON. Between Comey's heavily criticized decision to make public the FBI's reopening of an investigation involving her emails and the torrent of WikiLeaks' hacked correspondences from Clinton's top advisers, Trump's misdeeds had faded from the front page.

This seemed to many a mere Band-Aid, something that might help at the margins, keeping some Republican House and Senate candidates from being washed out of office in a wave. In the forty-eight hours prior to Election Day, two of Priebus's top lieutenants, Katie Walsh and Sean Spicer, launched a furious preemptive spin campaign, putting the impending loss squarely on Trump and absolving the RNC of responsibility for a wipeout of the party.

And yet, the polls had been tightening for weeks—so much so that Priebus, who was famously stingy when it came to spending RNC money on television ads, bought airtime during Game 7 of the World Series in early November. (This earned him an earful from McConnell, who was still lobbying for party funds to be diverted away from Trump and toward competitive Senate races.)

Though Clinton still staked a comfortable lead in most of the key battleground states, there were signs that Trump was closing fast in several of them. The Republican nominee was working on an inside straight: If he held all the states won by Mitt Romney in 2012, he would need 64 additional Electoral votes to win the presidency. This was not inconceivable; internal polls showed North Carolina, the toughest state to hold, was trending toward the GOP. And it just so happened that the four states Trump had spent the most time targeting—Pennsylvania (20), Ohio (18), Michigan (16), and Wisconsin (10)—offered exactly 64 between them. Ohio was already in the bag; so, too, was Iowa, an Obama state whose 6 Electoral votes would provide insurance in the event that Utah slipped away due to a third-party conservative's effort there.

The 2016 election was coming down to Pennsylvania, Michigan, and Wisconsin. All of them were overwhelmingly white (Michigan's 2012 electorate was the most diverse of the three, at 77 percent white[1]). All of them were predominantly blue collar (a majority of voters in each of the three states lacked college degrees in 2012). None of them had been carried by the Republican Party in a presidential election since the 1980s.

Clinton was supremely confident, so much so that she lavished attention on Arizona in the hope of running up the score while ignoring Wisconsin in the belief that it was not truly competitive. Victory seemed certain: Even if one or two of the Rust Belt states slipped away, her campaign had invested tens of millions of dollars into North Carolina and Florida. If she took care of both, as expected, the election was over.

It was no surprise when, at around 10:40 p.m. Eastern, the networks called Ohio for Trump.

But when they moved Florida into the Republican nominee's column some fifteen minutes later, Democrats began to panic. It seemed premature, even to Trump's advisers. Was this going to be the inverse of 2000, when they called Florida early for Al Gore, only to take it back?

Sequestered away in an unfinished space on the fourteenth floor of Trump Tower, everyone was suddenly on their feet. The campaign had set up a makeshift war room where they wouldn't be bothered; the showcase was a hulking projector screen being updated from an RNC data feed. When the networks called Florida, Priebus ordered the staff to keep the state front and center, worried that Trump's lead would evaporate. Instead, it grew wider. Trump himself entered the war room, but nobody noticed: North Carolina had just been called for him, too.

The dominoes were falling in surreal fashion. Never, even under the sunniest of circumstances, had Trump's campaign considered a sweep of both North Carolina and Florida. They hoped for a split of the two, which would keep alive their hope for an inside straight in the Rust Belt. Now, with both states in the Republican column, it was *Clinton* who needed a sweep of Pennsylvania, Michigan, and Wisconsin.

Priebus pulled Trump aside. "You might win," the party chairman whispered.

Trump nodded. He suggested they move upstairs to the residence. The Republican nominee had not written a victory speech, and from the sound of things, he might just need one.

RYAN SAT IN HIS TEAM'S WAR ROOM AT THE HOLIDAY INN, ONE EYE ON Fox News and the other on a laptop spitting out sequences of numbers and projections.

His own race had been called early, and attendees waited patiently in the ballroom for his victory speech. But the Speaker was paralyzed, watching in silent disbelief as Trump surged past Clinton in Florida and North Carolina. The RNC's numbers, his advisers told him, as well as the toplines of the national exit polls, were badly flawed. The GOP's Senate majority was safe. Only a handful of House Republicans were losing. And if the current trends held, Trump was going to win the biggest upset in presidential history. The Republican Party was going to control the entire federal government.

Ryan called Priebus. Was this for real? The RNC chairman told him to prepare for a long night; the results in Pennsylvania, Michigan, and Wisconsin were so tight that anyone forecasting the outcome was guessing.

Shortly before 10:00 p.m. Eastern, Ryan finally took the stage and spoke for three minutes. He wore the look of a man who had escaped a burning building. "I've just been sitting there watching the polls," he told his hometown audience, shaking his head. "By some accounts, this could be a really good night for America. This could be a good night for us. Fingers crossed."

The Speaker returned to his bunker, still in a state of astonishment over what was unfolding. When the AP called Pennsylvania for Trump, just after 1:30 a.m. Eastern, Ryan phoned Pence. "I think you're going to win this thing," he said.

By that time, Trump and his team were finished revising his planned remarks. To the relief (and pleasant surprise) of everyone who had traveled upstairs to the residence, Trump was adamant about giving a gracious speech. "No bragging. Let's calm the waters," he announced. "That's what I want."

With the speech wrapped up, and Pennsylvania in the bag, Trump and his entourage set off for his Election Night party at the Midtown Hilton.

Pence, having long projected an unfaltering belief that Trump was destined to be a pivotal character in the American story, felt a certain absolution. Hours earlier, when the RNC officials and Trump aides had shared the exit-poll data, Pence ordered his team to ignore the noise. Then he sent them a photo, via text message, of the famous newspaper headline from the 1948 election: "Dewey Defeats Truman."

IT WASN'T OVER QUITE YET. BUT WITH TRUMP NOW AT 264 ELECTORAL votes, any one of the outstanding competitive races—Michigan, Wisconsin, or Arizona—would put him over the top.

He won all three.

When the final numbers were tabulated, Donald Trump had defeated Hillary Clinton in one of the strangest results in presidential history.[2]

Trump won the Electoral College with 306 votes to Clinton's 232 (officially 304 to 227, after seven pledged electors went rogue).

The margin of the GOP victory was found in three states—Pennsylvania, Michigan, and Wisconsin—which Trump won by a *total* of 77,744 votes, less than the capacity of some Big Ten football stadiums.

Meanwhile, Clinton won the popular vote by nearly three million.

All across the country, from the Rust Belt to the Great Plains to the spine of the Mississippi River, the Republican nominee flipped rural and exurban counties from blue to red on the potency of his appeal to middle- and working-class whites. (Clinton won just 37 percent of all white voters, per the exit polls, including 31 percent of white men; Trump was dominant among noncollege-educated whites, winning 66 percent of them to Clinton's 29 percent.)

But this recoloring of the map was not indicative of any enormous surge in voting among white men without college degrees. As the Brookings Institute reported, turnout for these voters "was markedly lower than it was in 2004, when George W. Bush beat John Kerry. It was also four points below that of white women without college degrees, and more than 20 points lower than white men or women with a college degree." It wasn't that Trump turned out historic new numbers of blue-collar whites; he simply won a far higher share of them than past Republicans had.

This was largely predictable. These voters had been trending toward the GOP for a generation, and Trump's candidacy was a known accelerant. The expectation was that Clinton would counter by mobilizing the groups central to *her* party's coalition: minorities, young people, college-educated women.

She did not. Nationwide, and particularly in the Midwest, Clinton badly underperformed among these constituencies relative to Obama's 2008 and 2012 campaigns. In the three decisive states, Pennsylvania, Michigan, and Wisconsin, Clinton won roughly 600,000 fewer votes than Obama had four years earlier, with particularly deep drop-offs in the urban precincts.

The irony wasn't lost on Priebus. Having spent the last four years laboring to build a party that wasn't solely dependent on working-class whites, he watched Trump win the presidency by prioritizing that very demographic in the narrowest possible way. "The dog caught the car," Priebus says. "Donald Trump had a good instinct. He knew he had the ability to excite people that haven't been excited in a long time. But what he didn't know, and what his campaign didn't know, is whether the numbers of those people would be enough to actually win."

The margin of victory erased any doubts about the Supreme Court's significance in shaping the outcome of the election.

Exit polls revealed that Supreme Court appointments were "the most important factor" for 21 percent of the electorate; Trump won 56 percent of those voters to Clinton's 41 percent.[3] Moreover, 26 percent of the people who voted for Trump called Supreme Court nominees "the most important factor" in their decision; only 18 percent of Clinton voters said the same. A total of 6,655,560 votes were cast for Trump in Pennsylvania, Michigan, and Wisconsin. Extrapolating from the exit poll numbers, that means 1,730,446 of them were primarily motivated by the Supreme Court—in states he carried by a combined 77,744 votes.

Any number of variables could tip the scales in such a tight election. But it's not difficult to deduce that without the Republican takeover of the Senate in 2014, allowing McConnell to block Obama's nominee, and thus dangling a vacant Supreme Court seat in front of reluctant conservatives, there would not have been a Republican takeover of the White House in 2016.

"I agree," McConnell says, grinning.

Jason Miller, the campaign's communications director, found Trump holed up backstage at the Midtown Hilton.

It was now 2:30 in the morning and a steady stream of friends, family members, and advisers had spent the past hour telling Trump that he was going to win, that the math had become impossible for Clinton. But the Republican nominee ignored them. No network had called the race, and he wasn't about to trust the delirious prognostications of his allies.

"Mr. Trump," Miller said. "The AP just called the race. You're going to be the president of the United States."

Trump turned to Miller. He looked neither happy nor sad, just surprised, wearing the expression of a student who earned the highest grade in the class despite not having studied for the test.

"*Really?*" he said.

Just then, a few feet away, Kellyanne Conway's phone rang. It was Huma Abedin, the longtime aide to Clinton. The Democratic nominee was calling to offer her concession. With his court of friends, family members, and advisers hugging one another and shouting in euphoria, Trump held the phone to his ear and stared ahead stoically. "I'm honored by your call," he told Clinton. "I'm very honored by your call."

Emerging onto the stage twenty minutes later, the president-elect sounded like a changed man. "Hillary has worked very long and very hard over a long period of time, and we owe her a major debt of gratitude for her service to our country. I mean that very sincerely," Trump said.[4] "Now it's time for America to bind the wounds of division. . . . I pledge to every citizen of our land that I will be president for all Americans. And this is so important to me. For those who have chosen not to support me in the past, of which there were a few people, I'm reaching out to you for your guidance and your help so that we can work together and unify our great country."

Later in his speech, Trump sang the praises of Priebus, calling him a "superstar" and inviting him to give remarks at the podium—the only person besides Pence to speak. The man who exactly one month earlier had warned the Republican nominee to either quit or suffer a historic loss was now standing at the lectern, in front of a frenzied crowd, saying, "The next president of the United States, Donald Trump!"

They shook hands. "God bless," Priebus announced. "Thank God."

INSIDE THE REPUBLICAN PARTY, REACTIONS TO TRUMP'S VICTORY RAN the gamut: delight and dread, mild surprise and utter shock, excitement at the idea of governing with control of all three branches and panic at the prospects of the president behaving in office as he had on the campaign trail.

For the party's Trump skeptics, there was plenty of dark humor. When a friend texted South Carolina's governor, Nikki Haley, expressing dismay at the night's outcome, she replied, "Cheer up. We just won the governor's races in Vermont, Indiana, and North Dakota."

Watching the returns down in Florida, Marco Rubio couldn't help but think that America was getting the president she deserved. "If our culture was as outraged by this stuff as some in the press seem to be, he wouldn't have been elected. It wouldn't have worked," Rubio says. "If people put this all on Donald Trump, they're making a big mistake. All you have to do is spend five minutes on Twitter and see some of the things that prominent people write about each other to realize this is the era we've entered into." (Having won reelection to the Senate, Rubio phoned Chris Christie to thank his old nemesis for making him a much-improved debater.)

As the granular details of the election's result came into focus, Republicans commenced a spirited debate that proved impossible to resolve.

Had Trump, by virtue of running up the score among working-class whites and flipping three "Blue Wall" states, shown that he was the *only* Republican capable of reaching 270 Electoral votes? Or had Clinton, thanks to her underperforming vis-à-vis Obama in urban areas and her failure to mobilize the Democratic base in Middle America, demonstrated that *any* Republican could have won the White House in 2016?

"It's hard to imagine that anybody else we nominated would have had the same kind of connection with working-class voters who, as *Hillbilly Elegy* pretty well laid out, felt that life had dealt them a bad hand," says McConnell, referencing the 2016 memoir by J. D. Vance about socioeconomic decline in Appalachia. "President Trump obviously was able to appeal to working-class people in Pennsylvania, Wisconsin, and Michigan, and he caught that lightning in a bottle. I'm not sure anybody else we nominated could have done that."

"They say anybody could beat her, yet we barely did, and we did with a candidate who uniquely spoke to people in northern Wisconsin and western Pennsylvania and mid-Michigan like none of the other sixteen candidates could have," Priebus says. "So, while people can wring their hands all day long about the nomination of Donald Trump, it turned out he was about the only person who could have won that race—even against a very weak Hillary Clinton."

The problem with such analyses is that they rely heavily on Trump's appeal to the white working class while ignoring other demographic groups with whom a less polarizing Republican nominee might have fared far better. While a Ted Cruz or a Marco Rubio or a John Kasich might not have done as well with Trump's core demographic, would they not have compensated by dramatically outperforming him among minorities and suburbanites and college-educated women, thus winning the same states (and possibly more), just with a different electoral coalition?

"He was running against somebody who was detested. We've never had an election in which one out of every five voters thought neither candidate was qualified by temperament or experience to be president. We've never had an election in which one out of every five voters who vote for a candidate doesn't like them," says Karl Rove. "It all came down to change. If you thought the country was headed in the right direction, you voted for her. But if you thought we were on the wrong track, you voted for him. And that was all tribal."

John Boehner, who says his former golfing buddy "never, ever expected to win" the White House, is more absolute. "The only Republican who Hillary Clinton possibly could have beaten was Donald Trump, and the only Democrat that Trump possibly could have beaten was Clinton," Boehner says. "Joe Biden would have run circles around him. Marco Rubio would have run circles around her." ("Three hundred and thirty million Americans," Boehner says of Trump and Clinton, sighing, "and we got those two.")

Boehner's successor in Congress, the Freedom Caucus member Warren Davidson, says he doubts that another Republican nominee could have won Ohio by an 8-point margin. But he believes that the raw numbers belie the disquiet many voters had to surmount before backing Trump—and the sense of compulsion they felt because of his opponent.

Davidson recalls talking with a young woman at his church who was eligible to vote for the first time. She was raised conservative and could never cast her ballot for Clinton. Yet she felt guilty about the idea of supporting Trump. Davidson told her that he personally viewed the election as "a binary choice," and urged her to pray about the decision.

Seeing her soon after the election, Davidson asked what verdict she had reached. "I prayed about it a lot. I got in the booth and prayed some more," she told him. "I voted for Trump. And then I prayed again to ask God's forgiveness."

RYAN HAD PHONED TRUMP AFTER WISCONSIN OF ALL STATES DELIVered the final verdict, the congratulatory call a blur of exhilaration and bafflement and trepidation.

He faced a legacy-shaping decision that night: Stay true to himself and step down as Speaker, or muzzle himself and serve alongside the new president. It was not a difficult choice. This was Ryan's chance to actually *achieve* the things he had spent decades fantasizing about. All those long commutes, all those nights missing family dinners and his kids' games and school events, would be worth it. Even if that meant getting in bed with the likes of Trump and Steve Bannon. Even if that meant accommodating behavior from a Republican president that he would never tolerate from a Democrat.

Then and there, Ryan knew what needed to be done. Having spent his entire adult life chasing the impossible goals of rewriting the tax code and reforming entitlement programs, here was his opening. He could now serve as Speaker of the House in a unified Republican government and pursue his legislative destiny—if only he were willing to go silent on Trump, beginning that night in Janesville. There would be no speech. There would be no more public blistering of Trump, period.

His friends called it "Paul's deal with the devil." And Ryan, like most Republicans, did not think twice about making it.

CHAPTER EIGHTEEN

NOVEMBER 2016

"You don't have to worry about my street credibility."

THE FOUR OF THEM STOOD ON THE SPEAKER'S BALCONY, GAZING OUT over the National Mall, pointing to some of the landmarks and making awkward small talk. In just over two months, Paul Ryan announced to the group, Donald Trump would stand in that very spot and be inaugurated as the forty-fifth president of the United States. He and his wife, Melania, took it all in. Mike Pence, the vice president-elect, wore the smile of a lottery winner.

Ryan hadn't slept one wink on Election Night. Instead, he lay in bed coming to grips with the arrangement he was about to enter into. "I felt a major onset of responsibility to help the institutions survive," Ryan recalls. "So, from the next day on, my mantra was 'Only one person can be Speaker of the House. I'm not a pundit, I'm not a think-tanker. Our job from now on is to build up the country's antibodies, . . . to have the guardrails up, to drive the car down the middle of the road, and don't let the car go off into the ditch.'"

Prior to the November 10 meeting, the Speaker shared with several friends that he planned to start by clearing the air, explaining to Trump why he had denounced him after *Access Hollywood*. They cut Ryan off: That was a terrible idea. He stood nothing to gain by reminding Trump, a known scorekeeper, of their past quarrels. Focus on the future, Ryan's friends warned him. Pretend the past didn't happen. Emphasize all the good things *you* can do for *him*. Kiss the ring, if necessary. To stand a chance of prospering in the new, post–November 8 Republican Party, one had to play the game by Trump's rules.

So, Ryan did precisely that, showering praise on the president-elect and acting as though they'd been allies from the get-go. Trump was gracious, willing to move past their beef (on the advice of Pence and Reince Priebus). But unlike Ryan, he couldn't pretend that nothing had ever happened.

"Paul's just a Boy Scout, that's all," Trump said to his wife unsolicited as they stood on the balcony, by way of explaining the past tensions between them. "He's like, a religious guy."

Ryan shrugged. "Well, I'm a devout Catholic."

"Oh, you're like Mike!" said Melania Trump.

Pence and Ryan exchanged looks. "Well, yeah, he's Protestant," the Speaker said. "But, you know, *yeah*."

When Trump visited with Mitch McConnell later that afternoon, the conversation was more direct. "Did you think I was going to win?" the president-elect asked.

"No," McConnell replied. "Frankly, I didn't."

Trump had a good laugh. Then the Senate majority leader got down to business. He and Ryan had already coordinated strategies to impress upon Trump that he would have a ready-made government on day one of his administration. The Speaker was handling the policy, putting together a comprehensive sequencing chart of the major legislative goals they would pursue over his first year in office. McConnell would be in the personnel business, running a tight ship in the Senate to confirm the new president's appointees in an expedited fashion.

"The first thing on my list," McConnell told Trump, "is judges."

MICK MULVANEY'S FRIENDS IN THE HOUSE FREEDOM CAUCUS COULDN'T believe what they were hearing. It was the Monday night following Election Day, and lawmakers were trickling back into Washington to resume their congressional duties. The next day, House Republicans would hold closed-door elections to choose their leadership for the upcoming 115th Congress, and no real drama was expected.

Ryan had angered many of the members by abandoning Trump's candidacy a month before the election. Some hoped the president-elect, after taking the stage just after 3:00 a.m. to give his victory speech, would suggesst retribution against the holier-than-thou Speaker of the House.

When Trump did no such thing, the Freedom Caucus members watched for a smoke signal, expecting tacit permission to launch their revolt against Ryan.

But the Speaker was a step ahead of his adversaries. Even before the race was called, Ryan had moved swiftly to solidify his standing in Trump's orbit.

Mulvaney was eager to do the same. The South Carolina congressman wore his ambition as subtly as a Mike Tyson tattoo. A lawyer with degrees from Georgetown, Harvard, and the University of North Carolina, Mulvaney, upon coming to Congress in 2011, made few doubt that he was the smartest man in Washington—and that he was destined for more than the House of Representatives. First, he had wanted the Senate seat vacated by Jim DeMint. When it went to Tim Scott, Mulvaney shifted his focus to running for governor at the end of Nikki Haley's second term. Now, with Trump's upset victory, Mulvaney's plans had changed again. Having distinguished himself as one of the party's fiercest fiscal hawks, winning admiration for his intellectual consistency even from those GOP elders who detested his ego, he set his sights on a dream job: director of the Office of Management and Budget.

He had not exactly been a Trump booster; between calling the nominee "a terrible human being" and suggesting House Republicans might be required to teach him about the Constitution, Mulvaney made a strong case to be excluded from the new administration. But the congressman was a close observer of Trump. Watching him, reading *The Art of the Deal*, studying his relationships, Mulvaney developed a theory of how to ingratiate himself. He would do what Ryan had done: Sell the president-elect on the value he brought to the team.

The only difficulty was, Mulvaney didn't know how to approach Trump. So, he went to Ryan. Their conversation was transactional. Mulvaney detailed the plotting by Freedom Caucus members against the Speaker. Ryan asked for Mulvaney to nominate him for reelection in the House GOP's upcoming meeting. In exchange, Ryan would talk to Pence, who had taken over the transition team, about bringing Mulvaney to Trump Tower.

As the Freedom Caucus board gathered for its preliminary briefing, held prior to the weekly meeting with the full membership, Jim Jordan, the group's chairman, broke some awkward news: Mulvaney, a board member, would formally nominate Ryan for Speaker the next day. Some colleagues thought Jordan was joking; he assured them he was not. Word quickly spread to the entire group, and when Mulvaney, who was running late, finally entered the room, he was greeted with a chorus of angry expletives. When they demanded to know why he'd agreed to nominate Ryan, the cagey Mulvaney replied, "Because he asked me to."

His comrades threw up their hands. "What else would you do if he asked you to?" Justin Amash, Mulvaney's friend and a fellow board member, bellowed at him.

To the disgust of some House conservatives, Ryan was reelected in a near-unanimous vote of the conference one day later.[1] (Thomas Massie, the Kentucky scamp, was the lone dissenter.) The melodrama was about more than just Ryan and his past squabbles with the president-elect; it spoke to something fundamental about how the insurgent forces in American politics had been emboldened by Trump's ascent and were eager to capitalize on a moment of upheaval.

Beginning in the dawn hours of November 9, many Republicans came to believe they were entering a metamorphic period in the party's history, one in which their loyalties and ideologies and dogmas could be scrambled and realigned. Conservatives in particular tended to believe this was a good thing, and rejoiced in the reality that Trump, while not philosophically flush with them in a few areas, nonetheless represented the culmination of their years-long jihad against the establishment.

Indeed, eight days after Trump's victory, the Conservative Action Project, an umbrella group comprising the right's most prominent activist leaders, held a celebratory gathering at the Ritz-Carlton in Tysons Corner, Virginia. Some of the attendees had been vehemently opposed to Trump throughout 2016. They were surprised to hear the Heritage Foundation's president, Jim DeMint, talk about how the president-elect had finally unified the party; and they were downright stunned at the glowing remarks about Trump from Ed Meese, the former attorney general under Ronald Reagan and an icon in the conservative movement.

There was a similar giddiness pulsing through the veins of

Republicans on Capitol Hill. Lawmakers who had been openly hostile to Trump's candidacy were suddenly aglow at the prospects for the next four years. Even Ted Cruz was genuinely excited. He would forever nurse a grudge over the insults levied against his family, but the Texas senator wasn't going to let his rivalry with the president-elect get in the way of steering the government sharply rightward. Thrilled by the GOP takeover of Washington, and facing his first reelection to the Senate in 2018, Cruz met with Trump in December and volunteered to be the president-elect's battering ram in the new Congress, abandoning his identity as an intraparty instigator and adopting the role of party-line enforcer.

Most of the Republicans in Congress, including all the Tea Party products, had known nothing but the suppression of serving with a Democrat in the White House. Now awoken to the realities of an incoming Republican president and a unified Republican government, their reservations about Trump melted like snowcones in the Sahara.

AS FOR THE CONSERVATIVES WHO HAD HELD THEIR NOSES IN VOTING for Trump, well, they could be excused for feeling a sense of relief at his victory. As far and fast as the GOP had lurched to the right over the past several years, there were signs of an equal and opposite reaction on the left. Much of the angst over Trump's victory was understandable, particularly within communities that felt threatened by the president-elect's policies. Yet the broader cultural trajectory of progressivism was sufficiently startling to assure even the most reluctant Trump supporters that they had made the right call.

The month after the election, Lena Dunham, a leading feminist voice of the new left and creator, writer, and star of HBO's popular show *Girls*, recalled on her podcast how she had visited a Planned Parenthood clinic in Texas and felt guilty that she could not relate to the women she was speaking with there. The reason: She had never had an abortion. "Now, I can say that I still haven't had an abortion," Dunham said on the show, "but I wish I had."[2]

It was Bill Clinton who called for abortion to be "safe, legal, and rare." In 1996, the Democratic Party adopted a platform[3] that sought to make abortions "less necessary" and "more rare," concluding, "we respect the individual conscience of each American." Twenty years later, Dunham, who was given a speaking slot at Hillary Clinton's convention, was expressing regret at never having had an abortion.

At the turn of the century, the ranks of antiabortion Democrats in Congress numbered nearly fifty. By the time Trump won the presidency, they were seven and dwindling.

This reflected a hollowing out of the middle on myriad issues for which Republicans were not solely culpable. Obama had won the presidency by declaring marriage to be between one man and one woman. He had spent his first term deporting record numbers of illegal immigrants. He had refused to give the single-payer health care advocates a seat at the table when drafting Obamacare. All those positions were considered antiquated by the base of the new, post-Obama Democratic Party, and now that Hillary Clinton's centrism was out of the way, it would drift even harder and hastier to the left.

America's two parties were moving farther away from the middle in part because Americans of different party affiliations were moving further away from one another.

David Wasserman of the *Cook Political Report* newsletter uses an ingenious method to track the twin trends of ideological and geographical clustering in America. Using corporate brands as a proxy for the cultural tilt and socioeconomic profile of a given part of the country, Wasserman has concluded that the most likely brand to be found in a Republican county is Cracker Barrel while the most likely brand to be found in a Democratic county is Whole Foods.[4]

It makes perfect sense: Cracker Barrel restaurants are most often found in rural and exurban areas with less population density, less diversity, lower incomes, and lower education rates. These are the areas, on the whole, hit hardest by the transformation from a manufacturing economy to a tech-based economy; far more people are moving out than moving in.

Whole Foods grocery stores, meanwhile, tend to concentrate in upscale urban and suburban settings with diverse populations and high numbers of college graduates. These are the areas, on the whole, that have thrived in the postindustrial age, drawing mass migrations of new residents seeking jobs in high-skilled fields.

In 1992, the first year Wasserman tracked the results, Bill Clinton won 61 percent of counties nationwide that had a Whole Foods and 40 percent that had a Cracker Barrel. The 21-point "culture gap," as Wasserman calls it, grew wider in every successive presidential election.

By 2000, the culture gap was 32 points: George W. Bush won 75 percent of Cracker Barrel counties and 43 percent of Whole Foods counties.

By 2008, the culture gap was 45 points: Barack Obama won 80 percent of Whole Foods counties and 35 percent of Cracker Barrel counties.

In 2016, Donald Trump won 76 percent of Cracker Barrel counties and just 22 percent of Whole Foods counties. The culture gap was 54 points.

FOR CONGRESSIONAL REPUBLICANS, THE ECSTASY OF THE MOMENT WAS inversely proportional to the expectations of the previous several months. They had spent so much time bracing for the aftermath of a Trump defeat that the sudden trappings of a Trump victory were exhilarating: staffing the administration, passing big bills, and of course, stocking the federal courts.

Conservatives had all the more cause to rejoice when Chris Christie was axed as the head of the transition team, apparent payback from the new crown prince, Jared Kushner, whose father had been prosecuted and sent to prison by Christie on tax-evasion charges years earlier. Christie was replaced by Pence. A longtime affiliate of Washington's professional right, the vice president-elect was, in effect, starting the transition process from scratch and given broad latitude to fill critical positions in the cabinet and throughout the new government (with Trump's perfunctory approval, of course). Pence did not disappoint conservatives. He tapped his old friend, Congressman Tom Price, to run Health and Human Services. He picked his fellow charter school champion Betsy DeVos, the GOP megadonor with no experience in the public schools, to lead the Education Department. And he saw to it that Mulvaney was given the keys to run OMB.

With much of the attention focused on the headliner appointments (secretary of state, attorney general), Pence was cunningly effective in leaving his imprint on the administration. Time and again, when loyalists came to him expecting a job in the VP's immediate orbit, Pence surprised by asking them to fill a role elsewhere, one from which they could report back to him. To be an influential vice president, he would need eyes and ears across the government.

Previous opposition to Trump was not disqualifying. Marc Short, Pence's closest adviser and the former Koch operative who had been bent on stopping the GOP front-runner, was named the White House's director of legislative affairs. Certain allies who had Trump's ear, including former campaign manager Corey Lewandowski, were incensed that Short was given such a prominent position. Occasionally, Trump could be stirred by these concerns; for example, he nixed the hiring of former Bush 43 official Elliott Abrams as deputy secretary of state due to Abrams's past critiques.[5] But this was the exception. If Trump were to rule out every Republican who had combated him, the administration would cease to function for want of staffers.

Naturally, he reserved the right to have *some* fun with his former foes.

Stringing along the media, Trump delighted in tormenting Mitt Romney by dangling the job of secretary of state. Having taken a call from Pence while vacationing with his family in Hawaii, Romney raced back stateside to interview for the job. Trump was never going to give it to him. This was no "team of rivals" exercise; it was the continuation of a reality show, and in this episode, the host craved the spectacle of his most prominent detractor groveling at his throne. In a perfect distillation of this dynamic, a photo was taken of the two men during a dinner in which Romney was ostensibly interviewing for the position. Romney resembled someone caught on *Candid Camera*, his pursed lips and furrowed brow screaming mortification. Trump, seated next to him, wore a waggish grin and a thought bubble that read, *Who's the phony now?*

There was another quality Trump craved in his appointees: They had to look the part. When it came to choosing a director of the Central Intelligence Agency, nobody auditioned quite like Mike Pompeo. The Kansas congressman, first in his class at West Point, had in his brief time in Washington made a strong impression on the full spectrum of his fellow Republicans. Built like an offensive lineman, with a barrel chest and thinning silver hair swept across his forehead, Pompeo was straight out of central casting. He came with a forceful recommendation from Pence, and the president-elect hired him on the spot after a meeting in New York. Trump had apparently forgotten all about the Kansas caucuses: the biting remarks from Pompeo, the stare-down from the wings of the stage. When Cruz's campaign manager, Jeff Roe, called Kushner to have a laugh about it, Kushner put the call on speaker so Trump could hear. "No! That was him? We've got to take it back!" he cried. "This is what I get for letting Pence pick everyone!" (Trump did not take it back; Pompeo served as CIA director and later as secretary of state.)

Some of the president-elect's appointments were products of patronage. Back in January 2016, South Carolina's lieutenant governor, Henry McMaster, became the first statewide official in any of the three early-nominating states to endorse Trump. McMaster went all in, traveling with the campaign and becoming close to the future president, never wavering in his support. A few days after the election, Trump called McMaster and said, "Henry, what do you want? Name it."

McMaster told him he wanted to be governor.

"That's it?" Trump replied. "Well, that should be easy. You're already the lieutenant governor!"

McMaster explained that it wasn't that simple. Elections were uncertain things. The only way to ensure his promotion would be for Nikki Haley to go away. Within days, seemingly out of left field, Trump announced Haley as his pick for ambassador to the United Nations. McMaster was sworn in on January 24.

The only thing that seemed to bother Trump during the transition was the occasional rejection of his job offers. The president-elect felt as though he were making knights of commoners, extending to them a prestige unattainable in other walks of life. In reality, many Republicans who interviewed for administration jobs knew they would be taking pay cuts to work tough, thankless jobs that carried the indelible stigma of serving under President Donald Trump. Most interviewees nonetheless found the fragrance of power too strong to resist. Of those who did not, Ken Blackwell's rejection of Trump became the stuff of legend.

Formerly the mayor of Cincinnati and the Ohio secretary of state, Blackwell had spent decades as a shot caller in the conservative movement, serving on the boards of the Family Research Council and the National Rifle Association. When Pence took over for Christie, Blackwell jumped in as the head of the domestic transition team. As Trump hunted for a secretary of Housing and Urban Development, Blackwell was a natural fit. He had worked under a previous HUD secretary, Jack Kemp. He was experienced. He was knowledgeable. And he was, well, *black.* (Diversity was a stated goal in filling many positions, but none more so than at HUD.)

The problem was, Blackwell didn't want the job. He was knocking on seventy's door and didn't need the headache of working in government. When Trump learned of his disinterest, he demanded that Blackwell be summoned to New York. Sitting across from him days later, Trump asked Blackwell to accept the job. Blackwell declined. "So, you're afraid of the challenge?" Trump asked.

Blackwell said that he wasn't afraid. He simply wasn't interested in the position.

"Maybe you don't have the street credibility we need," Trump said.

Blackwell arched an eyebrow. "You don't have to worry about my *street credibility.*"

"Oh yeah?" Trump replied. He picked up his phone and began dialing. The others in the room, including Reince Priebus and Steve Bannon, exchanged looks. "Hey, I'm trying to kick the tires on a guy from Ohio," Trump said into the receiver. "I'm wondering if you know him. His name's Ken Blackwell."

Everyone heard the voice singing on the other end: "Kennn-aaaaay!" It was Don King, the legendary (and black) boxing promoter.

Blackwell shook his head. "Like I said," he told Trump, "you don't have to worry about my *street credibility.*"

All things told, the transition process was orderly compared to the anarchy of Trump's campaign. The RNC, flush with Priebus's longtime staffers, was a natural farm system for mid-level hires. (One of them, twenty-five-year-old Madeleine Westerhout, broke down crying on Election Night, inconsolable over Trump's victory. To the amusement of her RNC peers, she was later chosen as the president's executive assistant, and now sits just outside the Oval Office.)

Pence's ties to the conservative movement, and to so many members of Congress and Republican leaders around the country, were instrumental in filling out the administration. So, too, was a project by the Heritage Foundation years in the making that sought to provide an incoming Republican president with an exhaustive file of ready-made appointees to federal jobs from secretary of defense to White House speechwriter. Heritage, once the mighty engine of the right, had seen its influence wane in recent years. Rumors had circulated about the board's displeasure with DeMint, who had antagonized many of the think tank's allies and mismanaged the foundation from the top. The ambitious staffing project bought Heritage some goodwill, but it seemed unlikely to save DeMint's job.

Any other Republican president might have sent the base into open revolt by tapping a pair of veteran Goldman Sachs executives, Steven Mnuchin and Gary Cohn, for treasury secretary and National Economic Council director, respectively. Yet the rapture of the postelection period, on top of Trump's promises of hiring "the best people" to help the government run more like a business, bought him plenty of leeway. This was equally true for his eventual secretary of state choice: Exxon-Mobil's Rex Tillerson, who enjoyed a warm relationship with Russian officials that would traditionally have sent the GOP's hawks into a tizzy.

There was plenty of slack being cut in part because the new president, aided by Pence, was filling out his roster in ways that were largely energizing to conservatives. Jeff Sessions, the Alabama senator and immigration hard-liner, was picked for attorney general. Rick Perry, the former Texas governor (who'd called Trump "a cancer on conservatism"), was tapped to lead the Energy Department. And Ben Carson, the storied heart surgeon whose political ascent began with a viral rebuke of Obama, ultimately accepted the position at HUD.

The most reassuring hire, for many Trump fans and skeptics alike, was Jim Mattis. The retired four-star Marine general, lauded for his intellect and beloved by his subordinates, was appointed secretary of defense. Nicknamed "Mad Dog" for his array of plucky quotes ("Be polite, be professional, but have a plan to kill everybody you meet"⁶), Mattis was better known within the military as a warrior monk. He was married only to the Marine Corps, a general known for taking watch shifts alongside young grunts and requiring moving vans to relocate his vast collection of books.

One hire did give party officials heartburn: Michael Flynn, the retired general who had joined a chant of "Lock her up!" while addressing the GOP convention, would be Trump's national security adviser. Flynn was qualified on paper, but his temperament and judgment were suspect; in December 2015, he had attended a dinner in Moscow honoring the television network Russia Today (RT), a state-run propaganda outlet. Flynn's seatmate at the gala dinner? None other than Vladimir Putin.

The most symbolic selection for Trump was his White House chief of staff.

The candidate's general election victory had been, to paraphrase the young private from *Platoon*, a child born of two fathers. On the one side, the energy and grassroots support behind Trump's candidacy owed largely to the base, as embodied by Bannon, the combative former head of Breitbart. On the other side, the infrastructure and organizational support were lent primarily by the party's establishment, whose avatar was Priebus, the mild-mannered RNC chairman.

The jockeying began no sooner than the race was called. Everyone on the right saw Trump as malleable to their ideas, if only they controlled the flow of information. That job belonged to the chief of staff; the competition to fill it became a proxy war for the soul of Trump's presidency.

After five days of suspense, the president-elect decided to split the baby. He named Priebus his chief of staff and Bannon his chief strategist and senior counselor. (Bannon received top billing in the press release, sending gasps through the tea leaf readers in Washington.) Trump had not yet been sworn in, but already he had created warring power centers in his White House.

THOUGH NOT AN ELECTED OFFICIAL, THE WHITE HOUSE CHIEF OF STAFF has long been considered the second-most powerful figure in Washington.

Traditionally, the chief is given supremacy to organize, authorize, hire, fire, and speak on behalf of the president. The position is that of manager, decision shaper, and ultimate gatekeeper, filtering the flow of people and information reaching the Oval Office so that a time-constrained president is met only with the most pressing matters.

Priebus knew that would not be his job description.

Having spent the past four months traveling with Trump and observing his management style, Priebus realized that the president-elect would never empower someone to run such a structured enterprise. Anyone who read his books or watched his television show knew that Trump thrived on turmoil and dissent, competing viewpoints and warring personalities. He hated to be overbooked; he wanted to go into the office with a wide-open schedule each day and see what happened.

No staff member, regardless of title, was going to change that.

To the extent it was possible to curb Trump's instincts toward chaos, the chief of staff position required a strong hand, someone who could go nose to nose with the president and talk him down if necessary. But Priebus was never going to be that person. Meek and mild-mannered, he had thrived as party chairman precisely because of the job's accommodating nature. He spent most of his days doing maintenance: donors, RNC members, elected officials, activist groups. Priebus's job as chairman had been, above all, to raise money, keep the peace, and win elections. By those metrics, he had been a historic success.

Recognizing all this, the chairman's friends warned him not to take the chief of staff's job. Ride into the sunset, they urged him. Give some paid speeches. Write a book. Go make a million bucks a at some law firm or lobbying office. Steer clear of the shitshow.

The warnings were always the same. And so was Priebus's response: "We need a sane voice in the Oval Office," he told friends. "There has to be a reasonable person in the room with him."

Sane and reasonable, Priebus was. But he lacked the authority, the swagger, the piss-and-vinegar personality needed to rule Trump's White House. And he knew it.

Shortly after Thanksgiving, Priebus sat down for a private dinner with former Bush 43 chief of staff Josh Bolten. They were at Bolten's downtown office, in a conference room overlooking Lafayette Square and the White House. Carefully arranged around the table were four-by-six cards with the titles of the key assistants to the president as well as some of the deputy assistants whom Bolten considered important—a system nearly identical to the one used by the Obama White House.

As they munched on takeout food, Bolten explained all the positions to Priebus and advised him on which were the most critical for him to fill personally—jobs where he needed experienced people, not just Trump loyalists, who could fit into a manageable structure. "Either you create the org chart and you fill in these boxes, or someone else will," Bolten warned. "And you'll have a very hard time running the White House."

Bolten also described the "Andy Card Principle," named for his predecessor as Bush's chief: "There's a difference between *wanting* to be in a meeting and *needing* to be in a meeting." It would be his role, Bolten told Priebus, to direct traffic and dictate an efficient schedule.

Priebus listened politely. But he seemed distant, even disinterested. It wasn't that he didn't appreciate the advice. But he knew that much of what Bolten was prescribing was implausible. Priebus had been allowed to hire a deputy, his RNC chief of staff, Katie Walsh, as a security blanket who could reaffirm him and look out for his interests. But most of the other positions Bolten was describing would be filled by Trump or by members of his inner circle.

"He had already, I think, relegated himself to an executive assistant role rather than the chief of staff, the person that actually organized and ran the White House," Bolten recalls of Priebus. "He did not treat himself as the chief of staff, and it was probably because his boss was unwilling to treat him as chief of staff."

The one person excited for Priebus was his old friend from Wisconsin, the Speaker of the House. They went back decades and had served as mutual sounding boards and grief counselors throughout the 2016 campaign. With changes to their party gusting all around them, Ryan and Priebus clung to each other, a buddy system that did not escape the watchful eye of Trump.

After ensuring his own survival on Election Night, Ryan now saw as his new concern the perching of angels and devils on the new president's shoulders. He was horrified at the prospect of Bannon running the White House. As a self-proclaimed figurehead of the "alt-right," an internet movement of knuckle-dragging misfits who rejected the classical liberal philosophies that underpinned modern conservatism, Bannon had used Breitbart to stoke the embers of xenophobia that smoldered beneath the tinder of nationalism.

Not only that, but Bannon had led a ruthless onslaught against the GOP itself, with Ryan occupying an honored place in Breitbart's crosshairs. On editorial calls with the outlet's reporters during the 2016 campaign, Bannon would refer to Ryan as "the enemy," according to reporting by journalist Jonathan Swan, and plot for his ouster as Speaker.[7] Swan quoted one former Breitbart staffer who said Bannon "thinks Paul Ryan is part of a conspiracy with George Soros and Paul Singer, in which elitists want to bring one world government."

Even though they had pretended to make up and play nice after the election, Ryan could not stomach the idea of Bannon as chief of staff. The selection of Priebus, then, gave the Speaker great comfort. He would have an ally inside the Oval Office who could help him to influence the president's thinking.

None of this was lost on the House Freedom Caucus. They had long resented Ryan for his undermining of the GOP nominee. Now they feared the Speaker, whom Trump likened to "a fine wine" after their postelection rapprochement, would be steering the president's agenda while they, who had stood publicly behind Trump through his tribulations, would be treated as second-class legislative citizens.[8]

This was foreshadowed by a December incident in which Jordan informed Ryan of his intention to proceed with an effort to impeach the IRS commissioner. Ryan's office objected, and when Jordan ignored them, the Freedom Caucus chairman got a sudden call from Priebus (whose phone number Jordan didn't recognize), asking him to please hold back. Jordan pushed ahead, all the more motivated after Ryan's apparent decision to enlist Priebus to stop him. (Jordan's resolution was rejected on the House floor and referred back to committee.)

The thought of being sidelined by a Ryan-Priebus axis was especially irksome to Mark Meadows. The North Carolina congressman had, in private, been as skeptical of Trump as anyone. The month before the convention, Meadows told friends in the Freedom Caucus that he was considering not going to Cleveland, despite being a delegate, because he feared living with the legacy of nominating the erratic Trump. As the campaign progressed, Meadows was instrumental in stifling criticisms of the GOP nominee that brewed within the Freedom Caucus. He told his comrades that when Trump lost—not *if*, but *when*—the base would be out for blood. Did they want to be blamed for Trump's loss? Or did they want Ryan to own it?

All the while, Meadows nestled closer to the center of power. He introduced himself to Trump and his team, and by fall was campaigning with him regularly during the GOP nominee's trips to his battleground state. It was during these visits that Meadows became acquainted with Bannon. The two men could not have been more different; Bannon was hyper and disheveled, Meadows equable and polished. But Bannon respected what Meadows and Jordan had built with the Freedom Caucus. More important, the two men had a common enemy: Ryan.

As Ryan celebrated the placement of his close friend as White House chief of staff, Meadows toasted his ally's selection as the president's senior counselor and chief strategist.

The alliances had formed, spanning both ends of Pennsylvania Avenue: It would be Ryan and Priebus, the establishment insiders, versus Meadows and Bannon, the populist outsiders.

WHEN MICK MULVANEY WAS NAMED DIRECTOR OF OMB, THE POWERful agency that supervises and coordinates the government's financial planning, Freedom Caucus members—and Ryan, notably—issued statements lauding Mulvaney's selection as a sign of Trump's commitment to fiscal responsibility.

That was one way of looking at it. Another way: Trump had sidelined one of the House's most outspoken conservatives, someone who repeatedly stood up to Republican leadership, thereby weakening potential intraparty resistance to his administration's initiatives.

Republicans had spent the past eight years complaining of executive overreach and abuses of power by the Democratic administration.

They referred to Obama as "an imperial president," a continuation of the Bush-era expansion of executive authority that showed little regard for the primacy of the legislative branch. They pledged, after Trump's election, to reassert themselves as an aggressive check and balance on the new administration in hopes of a return to limited government. "We saw Republicans stray away from the core principles during the Bush 43 presidency," Texas congressman Bill Flores, the outgoing chairman of the Republican Study Committee, warned during a December forum at the American Enterprise Institute.

But as Trump prepared to take office, the question wasn't whether he would stray from the party's core principles. It was whether he would redefine them altogether.

This presented something of an early existential challenge to the Freedom Caucus. They worried about standing up to Trump, but they also wondered whether his election was an implicit rebuke to their own hard-line philosophical stances. Conservatives had learned a hard lesson over the previous year: Anger at Washington was not a mandate for ideological purity. This was apparent in Trump's rise, but also in the elimination of one of their own.

Since the dawn of the Tea Party, no primary challenger had defeated a Republican incumbent by running to their *left*. That changed in 2016: Tim Huelskamp, a leading instigator of the 2010 class, lost his seat to obstetrician Roger Marshall, who campaigned on the message that Huelskamp was representing a rigid ideology rather than the people of Kansas. This had been preventable: In the agriculturally dependent "Big First" district, Huelskamp had made himself vulnerable by voting against the Farm Bill in 2013—*after* he'd already been kicked off the Agriculture Committee for other protest votes.[9] Marshall, who promised to make the government more responsive to the interests of the district, beat Huelskamp by 13 points, a giant margin against an incumbent with no ethical or legal baggage.

The episode put a scare into conservatives. They saw establishment Republicans emboldened after claiming their first Freedom Caucus scalp and wondered who would be targeted next. Sensing opportunity, Meadows convinced Jordan to step aside as chairman of the Freedom

Caucus. Its members had little cash in their campaign accounts and were therefore susceptible to primary challenges from better-financed, establishment-backed candidates; Jordan was persuaded to throw himself into growing the House Freedom Fund, his leadership PAC, with the aim of defending those members.

That left Meadows at the controls of the Freedom Caucus. It was the culmination of a meteoric rise. Feted as the man who felled John Boehner, Meadows became a cult celebrity on the right, keynoting dinners and receiving awards. Four years after arriving in DC as an obscure businessman turned realtor from rural North Carolina, he was the incoming president's conservative point man on Capitol Hill and the chairman of Congress's most influential faction.

Not everyone in the Freedom Caucus thought this was a positive development. Raúl Labrador and Justin Amash, two founding board members, raised repeated concerns about Meadows's coziness with the president-elect and questioned how aggressively the chairman would position the group to Trump's right. They were also wary of Meadows's proximity to Bannon; some of the members believed both men to be more interested in celebrity than conservatism. Three weeks after the election, there was a shouting match between Meadows and some of his members during a Freedom Caucus meeting. The reason: Breitbart had published a story with the headline "Exclusive—Rep. Mark Meadows: House Conservatives Ready on Day One to Help Donald Trump."

The issue wasn't merely about whether Meadows had the stomach for a principled fight with the new administration. It was about the tactical orientation of the Freedom Caucus, a group that had been founded on the notion of placing ideological consistency ahead of partisan unity. Meadows was taking over the group at a time of transition. Mulvaney was gone; so, too, were board members Scott Garrett, who had lost his New Jersey seat in November, and John Fleming, who lost his bid for Louisiana Senate. Meanwhile, an incoming board member, Dave Brat, the Eric Cantor slayer, was nicknamed "Bratbart," for his love of the far-right website and his determination to stay in its good graces.

Labrador found it all a bit unnerving. But he, too, had reason for caution. The congressman was preparing to run for governor of Idaho in 2018, and he couldn't afford a nasty tiff with Trump.

Against this backdrop, the reactions to Trump's first domestic policy splash were telling.

In December, the incoming administration made a show of offering Carrier, the heating and air-conditioning giant, $7 million in tax breaks and incentives to keep roughly a thousand jobs in Pence's home state of Indiana. Ten months earlier, just days after Trump won the New Hampshire primary, a viral video taken by a Carrier employee in Indiana showed a corporate executive announcing to hundreds of employees that their jobs were being shipped to Mexico. Trump had seized on the video and now saw an obvious opening to deliver on a symbolic promise to protect American workers.

The Carrier deal was a clear example of the "crony capitalism" conservatives had railed against, and part of a propaganda campaign in which Trump attempted to demonstrate before taking office that his election was already benefiting the domestic workforce. Yet the response from Republican leaders, including Ryan, who for years had warned that the government should not pick winners and losers, was to celebrate the deal. Most conservative leaders kept quiet, too. One notable exception was Sarah Palin, who, scoring points for intellectual seriousness, criticized Trump and Pence. Within the Freedom Caucus, the only vocal critic was Amash. "More corporate welfare and cronyism," the Michigan congressman tweeted. "Equal protection is denied when one company receives favors at the expense of everyone else in Indiana."

David McIntosh, the Club for Growth president and former Indiana congressman who had been Pence's friend for two decades, said the Carrier deal set "a terrible precedent." Having listened in disbelief as Pence defended the deal, saying the free market had failed to protect Hoosier workers from their jobs being shipped overseas, McIntosh began to question whether Pence would be true north in the administration. "What I saw him do during the campaign was kind of reinterpret 'Make America Great Again' into a list of conservative initiatives," McIntosh recalled. "The Carrier thing was disappointing because he didn't do that, and it kind of seemed like they were giving up on the free market and talking about tariffs instead."

McIntosh hoped that Carrier would be a "one-off thing," but there was evidence suggesting otherwise. Ten days after the election, Bannon put the party on notice in an interview with the *Hollywood Reporter*. "We're going to build an entirely new political movement," he boasted. "The conservatives are going to go crazy. I'm the guy pushing a trillion-dollar infrastructure plan. With negative interest rates throughout the world, it's the greatest opportunity to rebuild everything. Ship yards, iron works, get them all jacked up. We're just going to throw it up against the wall and see if it sticks. It will be as exciting as the 1930s, greater than the Reagan revolution—conservatives, plus populists, in an economic nationalist movement."[10]

Bannon was correct that traditional conservatives wouldn't support the agenda he described. But in the era of Trump, the very definition of conservatism was up for grabs. *Populism* had become the new buzzword on the right; a few days after the election, Jordan made repeated references to "populist-conservative policy," advocating the suddenly chic notion of a marriage between Trump's Everyman appeal and the Tea Party's ideological exactitude.

Yet it was never clear that such a merger was even possible. "Populism as an ideology is not ideological," Arthur Brooks, president of the American Enterprise Institute, said before Trump took office. "Populism basically says, 'There's a parade coming down the street and I'd better get out there because I'm their leader.'"

Trump's threat to penalize companies that shipped jobs overseas might have excited a blue-collar worker in rural, red America, but the idea was fundamentally incompatible with the precepts preached by the elected Republican who represented that worker's district. The politician in question might agonize over the violation of conservative orthodoxy, but when regular people are forced to choose between their livelihoods and a set of abstract principles, it's a no-brainer. To that point: If Pence, who was once arguably the most ideological Republican in Congress, could be persuaded by Trump to stop supporting multinational trade deals while offering tax breaks to Carrier, it wasn't hard to

imagine Republican lawmakers writ large adapting to a new and different mandate from their constituents.

To combat this, Ryan had a plan: He would pack the GOP government's schedule so full that Trump wouldn't have time to deviate from
party orthodoxy.

In mid-December, the Speaker arrived at Trump Tower carrying
a Gantt chart with a meticulously detailed agenda for the year ahead.
With the help of McConnell, Ryan had laid out on paper the policy initiatives, the key players, and the deadlines that would guide the GOP's
lawmaking process in 2017. He spent nearly three hours walking Trump
and his senior staff through the chart, and to his surprise, the president-
elect was engaged throughout.

Bannon, no fan of Ryan's, spoke up to warn the president-elect of
what he was committing to. "You realize that if you sign onto this, this is
what we'll be doing for the next year," he said.

"I got it, I got it," Trump replied. He looked at Ryan and shrugged.
"Okay. Let's do it."

FOR MUCH OF THE YEAR PRIOR TO TRUMP'S ELECTION, JOURNALISTS,
donors, lobbyists, and political professions had heard rumblings of the
candidate's shady association with Russia. The thrust of the speculation centered on his business dealings—namely, the attempt to build
a Trump Tower in Moscow—and on the notion that he was hiding his
tax returns because they would show a pattern of bribes and kickbacks
involving foreign nationals. The theory of a Trump-Kremlin nexus was
further fueled by his litany of head-snappingly suspicious comments,
such as when he declared at a July 2016 press conference, "I will tell
you this, Russia: If you're listening, I hope you're able to find [Clinton's]
30,000 emails that are missing."

Trump aides knew that reporters and political rivals were investigating these questions. What they didn't know was that a former British
MI6 agent, Christopher Steele, was secretly compiling a dossier of intelligence reports on Russia's relationship with Trump.

A respected veteran of undercover operations in Moscow, Steele had
been contracted twice by the American political research firm Fusion

GPS. The first time, in October 2015, his work was underwritten by the *Washington Free Beacon*, a conservative media outlet financed by Republican megadonor Paul Singer, a patron of Rubio's campaign. The second time, in April 2016, Steele's services were purchased through Fusion GPS by a lawyer working on behalf of Clinton's campaign and the Democratic National Committee. In both cases, Steele's objective was the same: getting to the bottom of the Trump-Russia connection.

What Steele's sources told him was so startling that he contacted American law enforcement to pass along the intelligence: Trump was in the pocket of Moscow. The Republican nominee's team, Steele's sources said, was actively coordinating with the Russian government, which had compromising information to wield against Trump. According to sworn testimony by Fusion GPS employees and interviews given by Steele's associates, he believed his findings constituted a national security threat, hence his decision to share them with old counterparts in U.S. intelligence.[11]

As Steele's warning slowly worked its way through the American law enforcement apparatus, then-CIA director John Brennan was busy launching his own investigation into the Republican nominee's ties to Russia. He suspected that the Kremlin was not just interfering in the U.S. election but was actively boosting Trump, possibly with assistance from the Republican nominee's campaign.

Despite mounting speculation around Washington about the existence of these inquiries, nothing was made public prior to Election Day. Democrats would later groan that Obama had bottled up the news of Brennan's probe, fearing the optics of a politically motivated leak that would fuel Trump's theorizing about a "rigged election."

On January 10, ten days before Trump was to take office, CNN reported that both he and Obama had been briefed on classified documents that "included allegations that Russian operatives claim to have compromising personal and financial information" on Trump.[12]

CNN did not publish the allegations, but BuzzFeed did.

Among the other findings in his dossier, Steele reported that Russia had been "cultivating, supporting, and assisting" Trump for at least five years; that his team had accepted "a regular flow of intelligence from the Kremlin" on his political opponents; that several of Trump's lieutenants had acted as intermediaries; and that the Russian government possessed compromising information, or *kompromat*, on Trump himself.

At the heart of the *kompromat* were allegations of "perverted sexual acts" that had been recorded by the Russian government. One particularly salacious claim was that back in 2013, while staying in the presidential suite of the Ritz-Carlton in Moscow, Trump had paid Russian prostitutes to urinate on a bed that the Obamas had slept in.

Trump, for his part, seemed more bemused than angry by the details of the Steele Dossier. "Does anyone really believe that story?" the president-elect said at a January 11 press conference. "I'm also very much of a germaphobe, by the way."[13]

Trump's team was less sanguine. On the evening BuzzFeed published the dossier, Priebus and Bannon cornered Michael Cohen inside the president-elect's personal office on the twenty-sixth floor of Trump Tower. The dossier reported that Cohen had in August 2016 met with "Kremlin officials" in Prague on behalf of Trump to discuss coordinated efforts against Clinton. It was mortally dangerous intelligence, if true—and Priebus and Bannon thought it might very well be.

Priebus, a trained lawyer, sat Cohen down and began deposing him. It was a vivid scene, with members of the transition team frozen outside the office watching the confrontation unfold. Priebus interrogated Cohen on his specific whereabouts for the entire month of August 2016, and demanded to know every country he'd ever visited in Europe. Cohen grew increasingly heated during the exchange, swearing that he had never been to Prague in his entire life.

"Prove it," Priebus said. "Go get your passport and show us."

Cohen, a tenant of Trump Tower, obliged them. Returning a short time later with his passport, he handed it to Priebus. There was no stamp from the Czech Republic.

Priebus, Bannon, and other top incoming White House officials were satisfied that Cohen was telling the truth. But they remained deeply wary of him. Everyone who had spent time around Trump had heard him complain about the recklessness of his personal lawyer. "Michael's supposed to be the 'fixer,'" Trump liked to say. "But he causes more problems than he fixes."

CHAPTER NINETEEN

JANUARY 2017

"Did you see my tweets?"

THE WEATHER WAS ALL TOO APPROPRIATE. WHEREAS EIGHT YEARS earlier vivid sunshine had illuminated Barack Obama's inaugural address, storm clouds moved in over Washington as Donald Trump took the oath of office to become America's forty-fifth president. Not a minute into his speech, the skies dimmed and rain began to fall. His would be fairly described as the angriest, the gloomiest, the most ominous inaugural address ever delivered.

"Today, we are not merely transferring power from one administration to another or from one party to another, but we are transferring power from Washington, D.C., and giving it back to you, the people," Trump declared.[1]

"The forgotten men and women of our country will be forgotten no longer. Everyone is listening to you now. You came by the tens of millions to become part of a historic movement, the likes of which the world has never seen before."

Trump continued, "Americans want great schools for their children, safe neighborhoods for their families, and good jobs for themselves. These are just and reasonable demands of righteous people and a righteous public. But for too many of our citizens, a different reality exists. Mothers and children trapped in poverty in our inner cities; rusted-out factories scattered like tombstones across the landscape of our nation; an education system flush with cash but which leaves our young and beautiful students deprived of all knowledge; and the crime and the gangs and the drugs that have stolen too many lives and robbed our country of so much unrealized potential.

"This American carnage," the president said, "stops right here and stops right now."

TRUMP COULD BE EXCUSED FOR NOT DIVING INTO POLICY SPECIFICS IN an inaugural address. But the sweeping condemnations and blanket pronouncements were startling given the lack of nuance. While he no doubt connected with many Americans on an emotional level, the intellectual corruption of his remarks was breathtaking.

As of January 2017, violent crime rates had dropped precipitously from their modern high in 1991.[2] More people had jobs in the United States than ever before. Inflation-adjusted wages were higher than at any point in the country's history. The United States remained the wealthiest nation in the world by gross domestic product. And while there certainly were some "rusted-out factories" blighting the landscape of middle America, the manufacturing sector had come roaring back in the years since the Great Recession. As of 2017, U.S. manufacturing exports were at an all-time high, thanks in no small part to the Bush-Obama bailout of Detroit's automakers, which had more than doubled their exports between 2009 and 2014.[3]

Other key sections of the president's speech were similarly lacking in context.

When he decried "the very sad depletion of our military," Trump failed to mention the role of the Republican-authorized sequestration cuts, preferred by conservatives to the alternative of a major budget compromise with the White House that could have raised tax revenues by closing loopholes for the wealthiest earners only.

When he said, "Protection will lead to great prosperity and strength," warning against "the ravages of other countries making our products, stealing our companies, and destroying our jobs," Trump denied not just his personal history of developing products overseas, but also the net benefits of international commerce. Global prosperity had contributed tremendously to American wealth, and while trade deals had hurt a certain segment of the population, they were hardly the chief driver of domestic job loss. In December 2016, the *Financial Times* reported that of the estimated 5.6 million manufacturing jobs lost between 2000 and 2010, "85 percent of these jobs losses are actually attributable to technological change—largely automation—rather than international trade."[4]

And when he said, "We have defended other nations' borders while refusing to defend our own," Trump ignored the fact that Obama deported more illegal immigrants than any president in U.S. history and "more than the sum of all the presidents of the 20th century," according to ABC News.[5] Also missing: the history of how conservatives rejected the 2013 Senate bill, which offered an unprecedented influx of border agents, without offering any alternative in the House. Neither party had been innocent when it came to playing politics with immigration.

Trump was selling plenty of evocative sound bites but few fact-based assessments—and even fewer practical solutions.

The speech was, however, coherent in presenting a worldview that had remained consistent from the moment Trump first began flirting with a White House bid three decades earlier. "From this day forward, a new vision will govern our land. From this day forward, it's going to be only America first," the president said. "Every decision on trade, on taxes, on immigration, on foreign affairs will be made to benefit American workers and American families."

The phrase "America First," the rallying cry of noninterventionists resisting entry into World War II, had been off-limits in the generations since due to its anti-Semitic intimations. The speech was crafted by Steve Bannon as well as Trump's incoming policy adviser, Stephen Miller, who had been a longtime immigration staffer to Jeff Sessions. Deftly, Miller inserted a phrase to rebut interpretations of xenophobia: "When you open your heart to patriotism, there is no room for prejudice." Yet given the rhetoric of Trump's campaign, his associations with the likes of Alex Jones and the alt-right, and his incessant pitting of Americans versus non-Americans, it rang somewhat hollow.

Sitting on the dais behind the newly inaugurated president, George W. Bush couldn't help but hear the "isms" he had warned of eight years earlier: isolationism, protectionism, nativism.

When the speech concluded, Bush made his way off the stage. "That was some weird shit," he said aloud, according to journalist Yashar Ali.[6] (Bush's spokesman did not dispute the report.)

It was a sentiment shared by many on the dais—not just the Democrats whom Trump had spent the past year bashing (Obama, Bill Clinton, and Hillary Clinton, whose demeanor during her assailant's inauguration was the stuff of hostage videos), but also the Republicans who had been encouraged by Trump's post-election performance. They had heard him talk of unity in the wee hours of November 9. They had watched him assemble a generally respected cabinet. They were cautiously optimistic, on the eve of the inauguration, that the incoming president would feel the weight of his office, abandon his trademark bombast, and adopt a more thoughtful, deliberative approach.

And then came "American carnage."

Trump would not be relinquishing his penchant for provocation—or his appetite for conflict. It wasn't outwardly apparent at first. He floated through his first hours on the job: After finishing the inaugural address, speaking to a VIP luncheon inside the Capitol (feeling so magnanimous that he singled out Hillary Clinton for a standing ovation), and completing the parade down Pennsylvania Avenue, the new president had been paralyzed by wonder upon entering the Oval Office for the first time. "Wow," he said to Reince Priebus, turning in circles and glancing from carpet to ceiling. "Can you believe it?"

Everything was perfect—until he learned of the crowd-size comparisons.

Days earlier, the incoming president had predicted "an unbelievable, perhaps record-setting turnout." But while Obama's 2009 inauguration had been record-setting; Trump's had not. Obama's crowd had swelled to some 1.8 million people; using the most generous estimate, Trump's was one-third that size.

The new president could not suffer this indignity. On the occasion of his coronation, the man who had once felt compelled to vouch for the size of his penis during a televised debate would not stand for unfavorable comparisons to his reviled predecessor.

The next day, in what the White House called his first official act in office, the president visited CIA headquarters in Virginia. It was meant as an olive branch: Trump had frequently derided the intelligence community, including ten days earlier, when he compared American spies to Nazis for their role in disseminating the Steele Dossier. The president was met with applause upon his arrival, and he was careful to emphasize his support for the CIA and its officials. But his appearance went off the rails thereafter. Standing in front of the agency's sacred memorial to its fallen officers, Trump boasted of his election win, bashed the media for its coverage of him, and claimed that his crowd a day earlier had surpassed one million people.[7]

Meanwhile, Trump asked his new press secretary, Sean Spicer, to go even further.

Spicer was a curious choice to be the administration's mouthpiece. As much as any official in the party, he had objected to and actively opposed the new president's ascent. Even after Trump won the primary and Priebus worked to rally the GOP apparatus behind him, Spicer remained cool to the prospect of associating with the presumptive nominee. He did not trust Trump or any of the characters around him. More than once during the campaign, Spicer warned people heading to Trump Tower for meetings to watch what they said; he believed the inside of the building was wiretapped. (Whether he thought the recordings were made by the candidate himself or by the government investigating a possible crime was unclear.)

Spicer's tepidness was not a state secret. During the transition, some of Trump's allies took to calling Spicer a "November Ninth Republican" or a member of the "November Ninth Club," in reference to those longtime skeptics who were reborn as loyalists the day after the election. Trump knew this. Also, as a stickler for appearances, he wasn't big on the idea of putting a short, pale, provincially dressed party hack in front of the world's cameras as his emissary. But the pickings were slim. None of the television veterans Trump envisioned in the role wanted to work for him. Kellyanne Conway thought the job beneath her erstwhile status as campaign manager. And Sarah Huckabee Sanders didn't have enough experience in front of the cameras.

Trump reluctantly agreed to install Priebus's longtime spokesman. The president, however, told friends that he would be watching carefully to gauge the depth of Spicer's allegiance. When the crowd-size dispute grabbed headlines, Trump saw a perfect opportunity to test his new flack. He wanted Spicer to issue a definitive, on-camera statement from the White House press podium declaring the 2017 inauguration to be the biggest in U.S. history.

This struck many in the West Wing as an unequivocally awful idea. The administration was less than twenty-four hours old. It was a pointless and losing fight to pick, Priebus told Trump. Shouldn't they be concentrating their energies elsewhere?

Trump was adamant, giving Spicer the chance to prove himself. Confronting the White House press corps for the first time, on the evening of Saturday, January 21, Spicer proclaimed, "This was the largest audience to ever witness an inauguration, *period*, both in person and around the globe."[8]

Priebus was across the street. With loads of his extended family flying in from Greece to witness the inauguration, he and his wife seized the occasion to have her baptized in the Greek Orthodox Church. Having already been late to the ceremony, Priebus tried to shut out all distractions at the dinner reception afterward. It wasn't until some of the other attendees, including Jared Kushner and Ivanka Trump, eyes transfixed on their smartphones, alerted him that Priebus caught wind of what was happening at the White House.

Wearing an ill-fitting pinstriped suit and sonorous bags under his eyes, Spicer barked his nearly six-minute statement, spawning a devastating *Saturday Night Live* parody featuring actress Melissa McCarthy. One day into his presidency, Trump had chosen to squander the White House's capital on a decidedly unimportant and easily disproven argument. It set a troubling tone: Trump had lied and misrepresented facts at an astonishing clip on the campaign trail, and his administration, it appeared, would treat the truth with similar disregard.

That same day, as the president girded for a clash over crowd sizes, the "Women's March" attracted more than half a million protesters to Washington in a show of opposition to Trump. Hundreds of thousands of women were also demonstrating in cities around the country (and around the world), an unprecedented show of antagonism toward the one-day-old administration.[9]

Then, on day three, Kellyanne Conway went on NBC's *Meet the Press*. The winning campaign manager had wanted the chief of staff's job but had settled on the title of "counselor to the president." Instead of counseling Trump, it was her duty to clean up a needless mess of his making.

The host, Chuck Todd, asked why Trump had asked Spicer to "utter a falsehood" in his first statement from the White House press podium.

"You're saying it's a falsehood," Conway responded. "Sean Spicer, our press secretary, gave . . ." She hesitated. "Alternative facts."

Finally, on the fourth day of his presidency, Trump used his first meeting with congressional leaders to complain that he would have won the popular vote had it not been for some three to five million ballots being cast illegally. The baseless claim drew a fresh round of harsh media coverage; election officials around the country, both Republican and Democratic, said there had been no indications of meaningful voter fraud, much less on a massive scale.

By any metric, this was a baneful start for the new administration.

IT WAS LATE ON FRIDAY AFTERNOON, AND A WEARY WASHINGTON WAS looking forward to the weekend. The first seven days of Trump's presidency had been no calmer than his seventeen months as a candidate. With an approval rating of 45 percent in January 2017, Trump was the most unpopular new president in modern American history, according to Gallup.[10] It would not rise based on the week's developments: the Women's March, the politicized appearance at the CIA, the lies about crowd size, the "alternative facts." Everyone, including and especially the members of his administration, needed to catch their breath.

No such luck. At 4:39 p.m., during a visit to the Pentagon, Trump signed an executive order that vowed to keep "radical Islamic terrorists out of the United States of America." Effective immediately, anyone with an immigrant or nonimmigrant visa coming from seven majority-Muslim countries (Iraq, Iran, Libya, Somalia, Sudan, Syria, and Yemen) was prohibited from entering the United States for 90 days.[11] The order also banned all refugees worldwide from entering for 120 days and placed an indefinite ban on refugees from Syria, where millions of people were reported to have requested asylum into the United States to escape the civil war that had already claimed more than four hundred thousand lives.

Trump's executive order provoked a furious backlash. Lawsuits were filed in numerous jurisdictions. Protests erupted at international airports all around the country. Democratic lawmakers, and a vocal minority of Republicans, excoriated the administration. Even those Republicans who supported the policy were alarmed by the process behind it, which had sown mass confusion and plunged the nation's customs operations into chaos.

Conceived by Miller, the president's far-right policy adviser, Trump's executive order was impulsive and half-baked. There had been no vetting of the language by John Kelly, the retired four-star Marine general who was Trump's secretary of homeland security, or Jim Mattis, the secretary of defense, or Rex Tillerson, the secretary of state. Not only had these cabinet heads not reviewed the executive order, but they had known practically nothing about it before the president's signing. There had been no coordination from the White House communications shop, no soliciting of input, no answering of questions, no rehearsal of talking points. The secretaries and their staffs, as well as key congressional players, including leadership officials and chairmen of relevant committees, were left grasping for an understanding of the policy and an explanation of why it had been so hastily implemented.

Meanwhile, the nation's airports were seized by turmoil. Customs agents had received conflicting directives on how to enforce the directive. Airplanes were landing, carrying visitors from the countries on the list, as the order was being distributed around the government. The confusion resulted in the detention of travelers arriving at U.S. airports in a number of major cities.

By Sunday, Republican critics of the administration were out in force. Senators John McCain and Lindsey Graham issued a statement saying the policy "may do more to help terrorist recruitment than improve our security."[12]

Congressman Will Hurd of Texas, a former undercover CIA officer, called the policy "the ultimate display of mistrust," saying it would "erode our allies' willingness to fight with us" against terrorism overseas.[13]

One person was conspicuously silent that weekend: Paul Ryan.

A botched policy like the so-called Muslim ban would dominate the legacy of any other administration. But in the age of Trump, bonfires of controversy burned hot and fast, their oxygen stolen by the inevitable next inferno. Two weeks after the executive order fiasco, Trump announced the forced resignation of Michael Flynn, his national security adviser. The cause? Flynn had lied to Vice President Pence and other administration officials about his conversations with the Russian ambassador during the transition. As if that weren't enough scandal for one week, Trump asked James Comey the next day to shut down the investigation into Flynn's web of misdeeds. "I hope you can let this go," the president told the FBI director of his ongoing investigation.

Trump had campaigned as a managerial whiz who would surround himself with "the best people" and run the federal government like a high-functioning Fortune 500 company. Instead, he was proving to be a clumsy chief executive with a toxic weakness for staffing.

WHILE THE FRENZIED ACTIVITY AND BREAKNECK PACE OF THE NEWS cycle unnerved much of official Washington, the conservative base had cause for optimism. In his first thirty days, Trump had, among other things, withdrawn the United States from the Trans-Pacific Partnership trade agreement, signed an executive order requiring that two existing regulations be eliminated for every new regulation adopted, and canceled a meeting with Mexico when its president reiterated that his country would not pay for Trump's promised border wall.

Over the ensuing months, as concerns mounted on the right about the prospects for reforming the tax code, building the wall, and repealing Obamacare, Trump went above and beyond in delivering for one special constituency: evangelicals.

The president reinstated and toughened the Mexico City policy, which eliminates U.S. funding for international nongovernmental organizations that perform or promote abortions. He rescinded Obama's protections for transgender students to use preferred bathrooms in public schools. He signed legislation that routs federal money away from Planned Parenthood. And he cut off funding to the UN Population Fund, which critics had long accused of supporting coercive abortions in China and other countries. He accomplished these items, and others, with the help of pro-life Christians whom Pence had stockpiled throughout the administration.

Trump also benefited from the vigorous assistance of Ted Cruz. The Texas senator had reinvented himself at the dawn of the Republican government as a team player, one freshly intent on torturing the opposing party rather than his own. In a Senate GOP luncheon that January, McConnell stood before the room beaming with pride, praising "the new Ted Cruz."

"Look, Donald Trump was not my first choice to be president, but he's who the American people elected," Cruz says. "I faced a choice. I could choose to have my feelings hurt. He said some very tough things about me and my family. It would have been easy and natural for me to take my ball and go home. But I also think that wouldn't have been doing the job I've been elected to. I'm not going to defend the indefensible, but I'm going to fight for principles and values that matter."

The crown jewel of Trump's presidency, in the eyes of conservatives, was Neil Gorsuch. On January 31, Trump nominated the archconservative federal appellate judge to replace the late Antonin Scalia on the U.S. Supreme Court, thrilling the full spectrum of the Republican Party and validating the decision made by so many conservatives the previous November to hold their noses and punch the GOP ticket.

"It was a leap of faith. Trump was untested," Marjorie Dannenfelser, the antiabortion leader, said after the Gorsuch pick. "It became very hard to stand [by him]. But all that disruption, all that anxiety, all that tension—it was worth it. Because he has turned out to be a man of his word."

Trump had kept a promise of monumental importance to his base. Now it was time for the GOP-controlled Congress to keep one of its own.

PRESIDENT TRUMP HAD HEARD ENOUGH ABOUT POLICY AND PROCESS. It was a Thursday afternoon, March 23, and members of the House Freedom Caucus were peppering the president with wonkish concerns about the American Health Care Act: language that would leave Obamacare's "essential health benefits" in place; the community rating provision that limited what insurers could charge patients; and whether Speaker Paul Ryan's supposed master plan was even feasible. Trump suddenly cut them off.

"Forget about the little shit," the president said. "Let's focus on the big picture here."

The group of roughly thirty lawmakers, huddled around an immense conference table in the Cabinet Room of the White House, exchanged disapproving looks. For the past seventeen days, House Republicans had labored to unite around a health care bill that satisfied the complex and often conflicting demands of members representing different congressional districts and both poles of the party's ideological spectrum. The president did not particularly care what the bill looked like. He just wanted a victory. As they talked, Trump emphasized the political ramifications of a defeat; specifically, he said, it would derail his first-term agenda and imperil his prospects for reelection in 2020.

The lawmakers nodded and said they understood. They knew that Trump was not a policy maven but were disturbed by his dismissiveness nonetheless. For many of the members, the "little shit" meant the details that could make or break their support for the bill—and have far-reaching implications for their constituents and the country.

"We're talking about one-fifth of our economy," Mark Sanford, the South Carolina congressman, scoffed after the meeting.

Of the president's hecklers in the GOP, none had become as truculent as Sanford. Once an ascendant superstar and the party's most compelling contender for its 2012 nomination, the South Carolina governor's career was set ablaze in 2009 by an extramarital romance that was discovered while he claimed to be hiking the Appalachian Trail. Sanford would later suggest, somewhat astoundingly, that he hoped to get caught in the affair because of his reluctance to seek the presidency. "I've oftentimes wondered," he said, "was there some weird subconscious element that just wanted to derail the train and get off the train?"

Sanford's career in politics seemed finished. And then, a butterfly flapped its wings; Jim DeMint resigned from the Senate, Tim Scott was appointed to succeed him, and a special election was held to replace Scott in South Carolina's First District, formerly represented by none other than Sanford. After winning back his old seat, Sanford haunted Trump throughout the campaign, calling for the release of his tax returns and questioning his knowledge of the Constitution. Three weeks into the new president's term, Sanford could no longer hold back.

During an interview in his office, he described how Trump "represents the antithesis, or the undoing, of everything I thought I knew about politics, preparation, and life." Sanford added, "All of a sudden a guy comes along where facts don't matter? Look, we're in the business of crafting and refining our arguments that are hopefully based on the truth. Truth matters. Not hyperbole, not wild suggestion, but actual truth."

Sanford knew these comments might cost him his job. "I'm a dead man walking," he said, smiling. "If you've already been dead, you don't fear it as much."

Sure enough, the following month, after the Freedom Caucus meeting with Trump, Mick Mulvaney pulled Sanford aside. "The president wants me to let you know," he told his friend, "that he's going to take you out next year."

While many of the Freedom Caucus members shared Sanford's concerns, few were so bold as to air them publicly. Besides, in their fight over health care, Trump wasn't the problem. For all their frustration with the mixed messages and strategic ineptness coming out of the White House, conservatives didn't blame the president for their predicament. They blamed Ryan.

The Speaker had approached the health care effort with all the finesse of a forklift operator. Believing that House Republicans were uniformly supportive of the policy sketches in his "Better Way" agenda, which Ryan had promoted as the blueprint for a Republican government, he rushed headlong into drafting the American Health Care Act without the consultation of his conference—or any advice from the think tanks, lobby shops, activist groups, and media outlets that would render judgments of the legislation sooner or later. It seemed a no-brainer to proactively meet with these interests, answer their questions, accept their criticisms, and preempt any attacks on the legislation itself. Republicans had spent seven years promising to repeal and replace Obamacare; a few weeks of selling the product wouldn't hurt one bit.

Ryan didn't feel such preventative measures were necessary. And he was in a hurry, fearing that Trump was a ticking tweet-bomb, always one tantrum away from ruining the party's best-laid plans. After days of drafting the bill in secretive locations at the Capitol—Senator Rand Paul exposed the absurdity by bringing reporters along as he hunted door to door for a copy[14]—the text was leaked, and then unceremoniously released, without any clearly coordinated media strategy between Ryan's office and the White House. Conservatives around Washington, including some of the Speaker's friends, were stunned. "The bill has had the worst rollout of any major piece of legislation in memory," Rich Lowry, editor of *National Review* and a longtime Ryan ally, wrote in his *Politico* magazine column on March 15.

Leading health care experts on the right, such as Yuval Levin and Avik Roy, trashed the bill. Conservative outside groups and their media allies immediately branded it as "Obamacare Lite." Only then did Ryan move to mitigate the damage, convening a group of conservative journalists in his office and doing interviews with the likes of Sean Hannity and Laura Ingraham. But it was too little, too late.

At the heart of the opposition to Ryan's effort was the fact that he was not pursuing a *full* repeal of the Affordable Care Act. This ignored the realities at hand. Republicans had, while Obama was still in office, voted to eliminate the law in its entirety. But that was a statement vote on something that stood no chance of being signed by Obama. Now that they controlled the government, the circumstances were more fraught. For starters, Republicans didn't have a filibuster-proof majority in the Senate; they could only repeal the parts of the bill that touched on taxation, which required 50 votes through the reconciliation process. There were also the politics of the matter: voting to strip health coverage from millions of people, with no ready replacement, had been a whole lot easier to do when a presidential veto loomed as the backstop. Now there were real consequences to consider; it was no longer an empty ideological exercise.

As Ryan pushed to close ranks around his embattled legislation, he got little assistance from Trump. The president had never been keen to wade into the quagmire of health care, despite his promises on the campaign trail to get rid of Obama's signature law. Some of his advisers encouraged him to start with a bipartisan infrastructure push; others thought he should secure money and begin construction on the border wall as quickly as possible.

But Ryan was insistent. Republicans had spent the better part of a decade promising to repeal and replace Obamacare, he told the president. They had no choice but to do this, and the closer they got to the midterm elections, the harder it would be for members to take such a difficult vote. "We get this done early," Ryan warned Trump, "or we don't get this done at all."

THE PRESIDENT KEPT THE SPEAKER'S HEALTH CARE BILL AT ARM'S length for more than a week after its unveiling on March 6, offering a smattering of favorable remarks but never fully embracing it. Ryan's rivals in the Freedom Caucus, sensing daylight between the president and the Speaker, moved quickly to exploit it.

In the middle of March, during a budget meeting at the White House, Mark Meadows and Jim Jordan repeatedly diverted the discussion to health care, much to the annoyance of Budget Committee chairwoman Diane Black. When the meeting broke, Meadows and Jordan swiftly sought an audience with the president to discuss Ryan's bill. Trump granted them the meeting. The conservative ringleaders complained to the president that Ryan was presenting members with a take-it-or-leave-it proposition that was doing the entire party a disservice. Trump replied that he was open to negotiation and new ideas, and Meadows and Jordan left the White House believing they had pulled the president into their corner.

When word got back to Ryan that Trump had undercut him—saying he wasn't married to the current product after Ryan had spent the past two weeks telling members he was—the Speaker boiled over. He had gone out of his way to maintain a solid working partnership with the president. He had looked the other way and had bitten his tongue time and again over the first two months of the administration, hoping to preserve his influence over policymaking. Ryan knew that chewing out Trump would be counterproductive. The way to persuade the president, he had concluded, was to frame things in a way that sounded beneficial *for Trump*—not necessarily for the country and certainly not for the party.

"The Freedom Caucus isn't your ally," the Speaker told the president, taking deep breaths. "I'm the one trying to help you get a win here. These guys will find a reason to vote against anything we produce."

That weekend, a few days after their impromptu meeting at the White House, Meadows flew down to Florida to spend time with Trump at Mar-a-Lago, the president's Florida resort. The Freedom Caucus chairman lobbied aggressively for changes to Ryan's package, capping a week of wrangling about making alterations to the House bill. Trump eyed Meadows warily, remembering what Ryan had told him. By the end of the weekend, however, the president was on board, pledging to push for the conservative modifications.

But Ryan had learned another lesson in dealing with the president: Always be the last voice in his ear.

With Trump set to speak Tuesday morning at the House GOP conference meeting, Ryan spent Monday night working the president, reminding him of the fragile dynamics within the party, urging him to deliver the message that there would be no negotiating the details of the bill. When he rose to address the lawmakers, Trump had a simple message: There would be no further changes to the health care package. He expected Republicans to rally around Ryan's version.

Meadows was dumbstruck. For months, he had boasted about his relationship with Trump; more than once, he had arranged for the president to call him during one of the weekly Freedom Caucus meetings, making a show of answering and thereby wowing a collection of members who had never enjoyed real proximity to power. Upon returning from Mar-a-Lago, Meadows had triumphantly informed them that Trump was on their side. Now they were all staring at him.

Suddenly, so, too, was the president. Implying that there would be consequences for disloyalty to the party, Trump called out the Freedom Caucus chairman by name. "Stand up, Mark," he announced, half-smiling and half-leering at the congressman, who rose weak-kneed from his chair. "Mark, I'm gonna come after you if you don't support us on this." Then Trump turned to the rest of the room. "I think Mark Meadows will get on board."

It was a crucial misreading of the North Carolina congressman's situation. Months into his chairmanship, some of his colleagues in the Freedom Caucus still feared Meadows was too cozy with Trump and would hesitate to defy the White House. The health care fight was shaping up as a test of Meadows's independence from Trump; the moment the president called him out, the Freedom Caucus chairman was boxed in. If he gave even an inch now, he would confirm the whispers of the skeptical members in his group.

Meadows, thoroughly chastened by Trump's routine in the conference meeting, rushed to leave the room once it adjourned. But he was stopped by Patrick McHenry, his colleague in the North Carolina delegation and the leadership's chief deputy whip. "He's gonna come after you, Mark!" McHenry said, practically squealing with glee.

Meadows's face, already flush, was now glowing red. "You're not helping, Patrick!" he growled. He turned and took several steps away, leaving McHenry and a small crowd of gawkers gaping. They had never seen Meadows lose his customary cool.

Meadows spun back around. The creases in his brow had vanished; the amber in his cheeks was gone. Placing his hand on McHenry's shoulder, he said, "But I still love you." The onlookers, including several of the Freedom Caucus members, traded looks of incomprehension.

Back-channeling with the administration in the hope of changing the president's mind, Meadows and Jordan landed what they thought was an invitation to the White House the next day, Wednesday, March 22. Instead, they found themselves hauled into the less-than-inspiring Executive Office Building for a pep rally with Pence, Priebus, Bannon, and other administration staffers—but not the president himself. The Freedom Caucus members realized there would be no more negotiating. Pence tried to pump them up, saying the fight was theirs to win and that they needed to help Trump and Ryan score a victory for the new administration. The plea landed on deaf ears.

"You need to take one for the team, guys," Bannon said, growling like a sergeant instructing a roomful of privates. "You have no choice but to vote for this bill."

Joe Barton, a conservative elder statesman from Texas, couldn't handle being lectured to by the likes of Bannon. "The last time someone ordered me to something, I was eighteen years old, and it was my daddy," Barton told the chief strategist. "I didn't listen to him, either."

The room filled with uncomfortable silence. Bannon backed down and the meeting went on. (Barton eventually announced his support for the legislation; all told, Trump was responsible for moving upward of 10 votes over the course of the month.) After several hours, the members returned to the Capitol feeling frustrated. Several complained to Meadows that the meeting had been a waste of time and wondered if he had lost the president's ear for good.

That night, however, the White House sent word to the Freedom Caucus that one thing they had been pushing—reforms to the "essential health benefits" provision under Title I of the Affordable Care Act—could be negotiated. Excitement spread throughout the group. But there was also confusion: Some members believed that such a concession would be enough to win their vote, while others felt it was only a step in the right direction. As they sought to clarify their internal disagreements, there was another meeting scheduled for the next morning, Thursday, March 23—this one at the White House and with the president himself.

Renewed with hope, Freedom Caucus members were once again promptly disappointed. The next day's meeting was yet another "take one for the team" seminar. The atmosphere was friendly enough; the president had the group laughing with irrelevant riffs and stories of negotiations past. But it became clear, as soon as he made the "little shit" comment, that no serious changes were going to be made.

The problem was coming into focus. Trump possessed the requisite tools of a salesman; he had converted a handful of holdouts with late-night phone calls, using a blend of profane jokes, veiled threats, and appeals to loyalty. But the president was handicapped by his inherent disinterest in the specifics of the bill. He didn't have a sufficient grasp of the policy, or of the legislative dynamics in Congress, to know what could or couldn't pass.

Ryan, conversely, knew every nook and cranny of the legislative text. Having served as the chairman of two relevant committees—Budget, and Ways and Means—the Speaker was deeply versed in the details of his proposal. Unfortunately, he had no marketing skills to complement his command of the subject matter. He had alienated many of his members with his assertion of a "binary choice," and not just the conservatives. As they drew closer to a scheduled vote in the House on Friday, a growing number of moderate Republicans signaled their opposition to the bill, expressing frustration that Ryan and his leadership team were cramming it down the conference's throat.

As the reality of the bill's likely defeat set in on Thursday afternoon, Trump's team began to assign responsibility to Ryan, most notably feeding quotes to a *New York Times* story that questioned the Speaker's approach.[15] Ryan's team was prepared for this. They had already begun pushing the blame toward Trump; subtly at first, calling him "the closer," then more overtly, emphasizing that it was the president's job to deliver the Freedom Caucus.

On Thursday night, Mulvaney, the OMB director who had been deputized as a bridge between the administration and his former Freedom Caucus bandmates, stood before the House Republican Conference and issued an ultimatum: Trump was ready to move on from health care after Friday's vote. It was a timeless negotiation tactic, and one that didn't work very well. Republicans walked out of the meeting chuckling about Mulvaney, whom they'd known as a whiny backbencher, now lording it over them with such a threat.

The next morning, March 24, Trump made a final attempt to bully the conservatives into submission. "The irony is that the Freedom Caucus, which is very pro-life and against Planned Parenthood, allows P.P. to continue if they stop this plan!" the president tweeted. It didn't work; if anything, it may have backfired, just like his singling out of Meadows three days earlier. The conservatives certainly feared Trump, but if they were to suddenly switch their positions after a tweet on the morning of the vote, the president would own them for good.

It wasn't just the conservatives who sank Ryan's effort. By the time the Speaker arrived at the White House for an emergency meeting with Trump that afternoon, more than two dozen moderate and centrist members were also opposed. Lawmakers care about policy and process, and between the two, there was no clear upside in backing Ryan's bill. It left too many people without coverage and failed to drive down premiums; it had been hastily rewritten to accommodate changes and felt rushed for no good reason. Nearly seven years to the day after Boehner gave his "Hell no!" speech protesting the forced passage of Obamacare, a bill that was discussed, debated, and dissected for over a year, House Republicans were attempting to pass a replacement that they had introduced eighteen days earlier.

While Ryan met with Trump, the Freedom Caucus members filed into a private room at the Capitol Hill Club. They wanted to plot their next move in secret; to avoid leaks, no aides or White House officials were told of their location. Not long after they had gathered, however, the door flung open and in marched Pence accompanied by Priebus. Neither man was smiling. The vice president pleaded with his fellow Tea Partiers to reconsider their opposition.

"I was the Freedom Caucus before the Freedom Caucus existed," Pence told them, his voice rising, letting loose an uncharacteristic flash of anger. "Don't try to tell me this bill isn't conservative enough."

Pence then abruptly stormed out. Several of the members, grown men, broke into tears, fearful less of disappointing the vice president than of winding up on the business end of a Trump tweet.

Inside the Oval Office, Ryan explained that his team lacked the votes to pass the bill and wanted to pull it from the floor to avoid an embarrassing defeat. But the president wanted the vote to proceed, telling the Speaker that the GOP dissenters should be publicly shamed for their disloyalty to the party. Ryan talked him down, arguing that it was early in the Congress, that they would need those members' votes down the line. Trump conceded the point, though it didn't stop him from doing some shaming of his own. Feeling personally betrayed by Meadows, Jordan, and Labrador, the president called them out by name in a tweet the following week, and also posted a separate message encouraging the defeat of Freedom Caucus members in 2018. All across Washington, card-carrying members of the GOP establishment were elated.

Returning to Capitol Hill from his meeting with Trump, the Speaker canceled the vote and informed reporters in a somber press conference that Obamacare remained "the law of the land." He sighed, adding that the House GOP was still learning how to be a "governing body."[16]

It was a revelation. Despite controlling the White House and both chambers of Congress, the Republican Party was no more cohesive than it had been while out of power.

Watching the party implode from a new and unique vantage point, his home on the back nine of Wetherington Golf and Country Club in suburban Cincinnati, John Boehner felt one part liberated and one part guilty.

He certainly didn't miss the day-to-day shenanigans of Capitol Hill, and he was somewhat amused by how Trump had deepened the party's paralysis. "*Dysfunction* is a relative term," the former Speaker said that spring. "Right now, it looks like I was a genius."

But Boehner was worried for Ryan. The new Speaker had never wanted the job to begin with, and now he found himself buffeted by the same forces of factionalism within the conference, all while dealing with a deeply incompetent White House. Boehner didn't like the way things were headed, not for the institutions of government and certainly not for the GOP. Asked what he thought historians were going to make of his legacy, and that which he had bequeathed to Ryan, Boehner replied, "They'll be talking about the end of the two-party system."

The policy hopes of the unified Republican government rested on Ryan's shoulders. He was the man with the charts, having wowed everyone at Trump Tower in December with a detailed presentation of target dates and vote estimates for executing the party's legislative agenda.

Thus far, however, things had not exactly gone according to plan— and Ryan bore the blame.

Shortly after the House GOP's health care bill failed, Boehner received a text message from his close friend George W. Bush. They were always "two peas in the same pod," as Boehner says, a pair of even-keeled gents who didn't take themselves too seriously. When Bush, while still in office, refused to join the exclusive Burning Tree Club in Washington, due to the optics of golfing someplace where women weren't allowed, Boehner told the president, "You're a pussy." Years later, when Bush left the White House and became a member, promising the Speaker that he was going to whup his ass on the course, Boehner responded, "You're still a pussy."

"Hey. Are you still talking to Ryan?" Bush texted Boehner. "Are you giving him advice?"

"Yeah," Boehner typed back. "If he calls, I give him advice."

"He needs to call you more," Bush replied.

THINGS WERE GOING NO SMOOTHER ELSEWHERE IN THE GOVERNMENT. While health care was hogging the domestic policymaking spotlight, Washington was increasingly fixated on a drama of international intrigue: Russia's meddling in the 2016 U.S. presidential election.

On March 2, Trump's attorney general, Jeff Sessions, recused himself from any investigation into Russia's interference in the election, citing conflicts of interest given his once-undisclosed contacts with the Russian ambassador in 2016. Trump was incensed. He had expected Sessions, as the nation's top law enforcement official, to double as his personal protector. Allegations of collusion with the Russian government during the campaign, and the corollary talk that his presidency was illegitimate and potentially compromised, were gnawing at the president.

Two days later, still stewing over Sessions's recusal and raging about a "deep state" of government bureaucrats angling to take him down, Trump rose early at Mar-a-Lago. It was Saturday morning and there was no staff around. Clicking on his television and finding the previous night's edition of *Special Report* on Fox News, the president was stunned to hear a discussion between Bret Baier and Speaker Ryan about a "report" that accused the Obama administration of wiretapping Trump Tower the previous summer. Baier seemed uncertain of the report's specifics, and Ryan appeared visibly baffled by the questioning.

Trump raced to his phone. "Terrible! Just found out that Obama had my 'wires tapped' in Trump Tower just before the victory. Nothing found. This is McCarthyism!" the president tweeted. He followed up:

"How low has President Obama gone to [tap] my phones during the very sacred election process. This is Nixon/Watergate. Bad (or sick) guy!"

Five minutes later, at 6:40 in the morning, the president dialed Priebus. The chief of staff, hoping for an uneventful weekend with Trump out of town, was jolted out of his sleep. "Did you see my tweets?" came an excited voice on the other end.

Priebus leapt from his bed and opened Twitter on his iPhone, quickly finding Trump's pair of statements.

"Who told you this?" he asked the president.

"It's all over the place," Trump replied. "Listen to this!"

The president, a longtime fan of the TiVo recording device, rewound and played for Priebus the *Special Report* clip, a muddled exchange that offered nothing but confusion for most viewers.

"See! Did you hear that?" Trump asked Priebus.

It wasn't unusual for the president to begin his day with predawn tweets inspired by whatever he had seen or heard on Fox News, making *Fox and Friends* the most influential bit of programming in the world. Priebus could live with that. It was unusual, however, for the president to publicly accuse his predecessor of spying on him—without a shred of evidence to support the allegation.

The chief of staff felt sick. He hung up and called Ryan in Wisconsin. He was an hour behind, in the central time zone, and still asleep. "Paul, what the hell is going on?" Priebus asked. "What the hell is he talking about?"

Ryan, too, jumped out of bed and located the president's tweet. When Priebus explained that Trump's charge against Obama was based on the Baier clip, Ryan burst into maniacal, almost punch-drunk laughter. "I didn't even know what Bret was talking about," the Speaker exhaled. "I just BS'd my way through the question!"

It was a needed moment of levity for Ryan, but Priebus couldn't find the humor.

APRIL BROUGHT A BRIEF INTERLUDE OF TRANQUILITY. BUT THE MONTH of May saw fireworks the likes of which Americans hadn't witnessed since Watergate.

At the beginning of the month, Trump told his top aides that he'd made up his mind: He wanted to fire James Comey. They warned him that this was a very bad idea; that the FBI was investigating Russia's alleged meddling in the 2016 election, a probe that would be looking closely at him, his family, and his campaign. Firing Comey would make the president look suspicious. But Trump didn't care. In fact, it was Comey's very handling of the Russia case that irked him: Three times, the president claimed, Comey had assured him privately that he was not *personally* being investigated (which the FBI director later confirmed in congressional testimony), and yet he refused to say so publicly.

Desperate to stop Trump from acting impetuously, Priebus and White House counsel Don McGahn persuaded him to wait until at least getting an opinion from Rod Rosenstein, the deputy attorney general who was overseeing the Russia inquiry. They felt certain that Rosenstein, an even-keeled career prosecutor, would help them talk the president down. Instead, when he arrived in the Oval Office, Rosenstein blindsided them by agreeing with Trump: Comey deserved to be fired, he said, based on his handling of the Clinton email investigation in 2016.

Trump sacked Comey on May 9, publicly citing Rosenstein's reasoning for doing so. Senior White House officials, including Pence himself, insisted to reporters that Trump had acted on the recommendation of Sessions and Rosenstein. They swore up and down that the president's decision had nothing to do with the Russia probe. Trump, however, would quickly undermine those claims—and sabotage his own stated rationale for dismissing the FBI director.

In the Oval Office a day later, Trump hosted two top Russian officials, Foreign Minister Sergey Lavrov and Ambassador Sergey Kislyak. The president called Comey "a real nut job," according to the *New York Times*,[17] and told them of the FBI probe, "I faced great pressure because of Russia. That's taken off." (Trump also disclosed highly classified information about an operation targeting the Islamic State, according to the *Washington Post*.[18] The only photos of the meeting were shared by a Russian state photographer; no American media were permitted.)

The next day, May 11, the president continued to stray from his original story. Sitting down with Lester Holt of NBC News, Trump said of the Comey firing, "In fact, when I decided to just do it, I said to myself, I said, 'You know, this Russia thing with Trump and Russia is a made-up story, it's an excuse by the Democrats for having lost an election that they should have won.'"

Trump then committed a presidency-defining mistake the next day. Fittingly, it started with a tweet.

"James Comey better hope that there are no 'tapes' of our conversations before he starts leaking to the press!" the president wrote on the morning of May 12.

Trump would later admit that he possessed no such tapes. But that wasn't the point anymore. Prompted by the tweet, Comey, who had written contemporaneous memos after his meetings with the president, shared the memos with a law professor friend, authorizing him to leak them to the press. Comey's goal was to trigger the appointment of a special counsel to continue the investigation into Russian meddling. The ploy worked. One day after the *New York Times* published a story detailing Comey's claims about the president's request for lenient treatment of Flynn, raising questions about obstruction of justice, Rosenstein named a special counsel.[19]

Robert Mueller, the highly respected former FBI director who had served under Presidents Obama and George W. Bush, wouldn't merely be picking up where previous investigators had left off. He would be expanding the probe into places it might never have ventured before.

Trump's impulsive dismissal of the FBI director, his self-contradictory statements, and his taunting tweets had conjured a nightmare that would haunt the first term of his presidency.

CHAPTER TWENTY

JUNE 2017

"Rainy Sunday afternoons are the devil's play shop."

MEAN.

That's how the Republican president of the United States described the Republican House majority's hard-fought legislation to finally, at long last, deliver on the seven-year promise of repealing and replacing the Affordable Care Act: "Mean."

Facing the wrath of the base following the March 24 debacle, with Fox News and conservative talk radio leading the way in lampooning a congressional party that was ruining Donald Trump's prospects for a successful presidency, Republicans got their act together. And it started with Mike Pence.

From the moment Trump picked him, through the first few months of 2017, the vice president had been all but invisible in the parade of palace intrigue stories detailing the rivalries, alliances, backstabbing, self-promoting, and stock watching inside Trump's reality TV–inspired White House. That was no accident: Pence had gathered his team, first after the VP announcement and then once more on the eve of the inauguration, and warned them that the spotlight belonged to Trump. Leaking, speaking out of turn, or doing anything to upstage the president, he said, would not be tolerated.

All the while, Pence got busy piloting the administration in ways few vice presidents ever had. Unlike the other West Wingers who nurtured narratives of their own indispensability—Steve Bannon, Jared Kushner, Kellyanne Conway, among others—it was the vice president who pulled the levers during the early months of 2017. Pence figured prominently in Trump's selection of Gorsuch for the Supreme Court. He convinced the president to take specific actions on abortion and religious liberty. He stocked the cabinet agencies with longtime allies and kindred spirits. And he said nothing about any of it, deflecting all credit to the commander in chief. Whereas Bannon had put a target on his own back, giving countless interviews and even appearing on the cover of *Time* in early 2017 ("The Great Manipulator"), Pence understood how to survive and thrive in the Trump White House: Get the job done and avoid all acclaim in the process. The boss's ego allowed for nothing else.

Trump quickly came to trust his second in command above all others, prizing Pence's unwavering fidelity and discretion. And yet the vice president's camp continued to operate in a continual state of apprehension, having been handed enormous latitude by a president known for his insecurities and his acute sensitivity to being overshadowed. Ken Blackwell, who ran the domestic policy wing of Pence's transition team, put it this way in early 2017: "Mike Pence has a very full and complex portfolio in his briefcase. And he has to carry it like there's a bottle of nitroglycerin inside."

Pence had spent the final days of March coaxing the president in private conversations. The vice president explained to Trump, ever so gingerly, that while he didn't want to second-guess his decision to move on from health care, it would hurt him politically. Republicans on the Hill, Pence said, would be eager to negotiate after the backlash from their constituents. He asked for permission to spearhead a new repeal-and-replace effort. And he assured the president, in so doing, of the myriad benefits it would have for him.

Pence had learned the same lesson as Ryan: Trump responds to what's good for Trump.

With the president's blessing, Pence met with warring factions of the House GOP—the Freedom Caucus and the moderate "Tuesday Group"—to pitch them on his proposal. It would allow states to opt out of certain Obamacare requirements. The debate over what insurance plans would be required to offer had been a sticking point in past negotiations, and Pence's idea was a waiver to give states flexibility.

Both sides expressed interest. Pence's office drafted language, and he got busy selling it to both tribes—first, Mark Meadows and his Freedom Caucus, and then, in a separate meeting, to New Jersey congressman

Tom MacArthur, the leader of the Tuesday Group. After a joint gathering to iron out details, Pence had one request: He asked them to stop referring to the idea as the "Pence amendment." He didn't want or need any recognition.

Sure enough, the compromise on state-based waivers became known as the "MacArthur amendment," and it led to the House of Representatives passing the American Health Care Act on May 4. Trump was ecstatic. Calling Ryan to congratulate him, the president told the Speaker, "Paul, you're not a Boy Scout anymore. Not in my book."

Ryan was taken aback, finally realizing that it had never been a term of endearment to begin with. "It's like a dupe, a stupid person," Ryan says, rolling his eyes. "Boy Scouts are stupid because they don't cut corners, they're not lethal, they're not *killers*."

The mood was festive in the Rose Garden a few hours later as House Republicans assembled behind the president for a celebratory press conference. Rarely had Trump seemed to enjoy his new job. But that afternoon, with the smell of his first significant presidential victory wafting through the springtime air, was an exception.

"How am I doing? Am I doing okay?" he said, laughing. "I'm president. Heh! Hey, I'm president!"

Pence kicked off the victory lap, which lasted nearly forty minutes and included speeches from no fewer than ten people, with a simple message. "Welcome to the beginning of the end of Obamacare," the vice president declared.

It was a tad premature.

ALMOST IMMEDIATELY, REPUBLICAN SENATORS SIGNALED THEIR OPPOsition to the House legislation. Mitch McConnell announced that he and his colleagues would devise their own health care bill, with the aim of merging it with the House version in a conference committee down the road. Trump was mystified by this. The intricacies of the legislative process and the complexities of the interchamber relationship—any aspect of governance that did not fit neatly into a tabloid headline, really—did not interest him. All he knew was that House Republicans had finally passed a bill repealing Obamacare, and now his advisers were telling him it wasn't good enough for Senate Republicans. No wonder everyone hates Congress, the president groaned.

McConnell hadn't invested much energy in the anti-Obamacare effort up until that point. For one thing, privately, he saw the benefits of the Affordable Care Action back home in Kentucky. The uninsured rate there had plummeted over the past six years thanks to the law's Medicaid expansion, a provision that had become enormously popular in the deep-red state.[1] The success stories in Kentucky were so plentiful that the new governor, Republican Matt Bevin, decided to leave the state's Obamacare-driven Medicaid program alone after promising its demise as a candidate.

More to the point, McConnell had believed it was highly unlikely that House Republicans would pass a bill. There was no point driving into a legislative cul-de-sac, he told colleagues, when there were dozens of federal judicial vacancies in front of them waiting to be filled. When the House version passed, and Trump relayed his displeasure to McConnell at not having the Senate prepared to immediately follow suit, the majority leader scrambled to get to work.

The Senate's bill-drafting process made the House's look honest by comparison. In the nearly seven weeks between House passage on May 4 and the Senate GOP leadership's release of its bill on June 22, hardly a soul on Capitol Hill outside of McConnell and his staff knew what was being written. This was infuriating to many of the rank-and-file senators, particularly conservatives such as Ted Cruz and Mike Lee, who felt the party was being hypocritical given its vilification of the process behind Obamacare. ("This massive piece of legislation that seeks to restructure one-sixth of our economy is being written behind closed doors, without input from anyone, in an effort to jam it past not just the Senate but the American people," McConnell had told reporters in December 2009.)

One person who didn't seem to mind was Trump. The president made it clear to McConnell and his team that his only concern was the end result. The process was unimportant, and frankly, so, too, was the policy. Having heard from liberal friends who decried the House legislation as hostile to low-income Americans, Trump had only one request for the senators as they fashioned their version. "We need to be more generous, add more money to help the people. We need to have heart," Trump told senators over lunch at the White House, CNN reported. "Their bill is just mean."

But the Senate bill wasn't exactly warm and fuzzy. While less draconian than the House version in some respects—offering more in subsidies to the working poor, for instance—it also made steep cuts to Medicaid while ending the Obamacare tax hikes used to pay for it, amounting to a windfall for the wealthy while millions of working-class Americans stood to lose coverage.

There was something for everyone to dislike in the legislation. Promptly, a bloc of conservatives voiced their opposition, followed by a chorus of centrist Republicans. The bill was going nowhere fast. Having originally scheduled a vote before the July Fourth recess, without any committee hearings or extended floor debate, McConnell was forced to postpone any such action into July due to schisms in his party.

Things didn't get any easier after the recess. Republicans spent the first three weeks of July bickering over changes to the bill, only to discover that any alteration that gained two votes was costing them three. The balancing act appeared impossible. At one point, McConnell threw up his hands and declared the "replace" part of their mission dead, announcing that Senate Republicans would vote on a repeal-only bill. (For a man touted as the canniest governmental operator since Pericles, it was something of a reputation-deflating summer.)

McConnell's decision only fanned the internal angst. His colleagues trashed the notion of giving up—or of stripping coverage from millions of Americans without offering a substitute. The GOP leadership went back to the drawing board, soliciting final suggestions for a modified bill. Finally, on July 25, McConnell directed the Senate to begin debate on legislation to repeal and replace the Affordable Care Act.

Enter the Hollywood scriptwriters.

Earlier in the month, John McCain, who'd been missing from the Senate due to what his office described as eye surgery, announced that he had cancer. It was glioblastoma, the same lethal brain tumor that had killed his friend Ted Kennedy. Washington was shaken by the news. There was no spinning what it meant. McCain, the man who seemed immortal after surviving five and half brutal years as a prisoner of war in Vietnam, was going to die soon.

Returning to the Capitol for the first time since his diagnosis, McCain stormed the Senate floor with the last major speech of his distinguished career.

"Our responsibilities are important, vitally important, to the continued success of our Republic. And our arcane rules and customs are deliberately intended to require broad cooperation to function well at all," McCain said. "The most revered members of this institution accepted the necessity of compromise in order to make incremental progress on solving America's problems and to defend her from her adversaries."

He went on: "That principled mind-set, and the service of our predecessors who possessed it, come to mind when I hear the Senate referred to as 'The world's greatest deliberative body.' I'm not sure we can claim that distinction with a straight face today."

The crux of his complaint, with McConnell and the health care push, was that the GOP had abandoned "regular order," the process of writing bills in committees, debating them, allowing amendments to be offered, and then subjecting them to scrutiny on the floor of the Senate. McCain had spent three decades in the upper chamber advocating this practice of transparency. He was not going to forsake it in the twilight of his life. The Arizona senator announced that he would vote against the bill, and urged McConnell to start from scratch, working with Senate Democrats on a bipartisan solution.

McConnell did not heed this advice. After failing to advance the repeal-and-replace bill on July 25, and failing to pass a repeal-only bill on July 26, the Senate majority leader brought forth a new piece of legislation on July 27. (To be clear, the procedural hypocrisy here was gobsmacking: McConnell and other Republicans had spent the previous eight years accusing Democrats of shoving ill-considered legislation down the public's throat, only to spend their first year in charge of Washington crafting laws in the dead of night and voting on them without hearings, markups, debates, polling, or input from constituents.)

Nicknamed "skinny repeal," the newest legislation from McConnell represented a scaled-down version of the Senate's earlier efforts. The contents were unimportant at this point: McConnell told Senate Republicans they simply needed to pass a health care bill, any health care bill, so that they could enter into conference committee negotiations with the House. At that point, they would worry about the final details.

Buying into this strategy, 49 of the 52 Republicans were in favor. Two were opposed: Alaska's Lisa Murkowski and Maine's Susan Collins. The deciding vote belonged to McCain. If he backed the effort, Pence, the constitutional tie-breaker, would approve passage of the bill. A conference committee would be created. Ryan and McConnell would craft a final product and have Trump throw the full weight of his presidency behind it. Obamacare repeal would be on the one-yard line, and Republicans, facing the fury of their voters, wouldn't dare refuse to push it over the plane.

The senator from Arizona was nowhere to be found. He had disappeared from the chamber. As senators murmured of his whereabouts, McCain was on the phone with Ryan. He respected the young Speaker and wanted his commitment—his word—that the conference committee bill would be negotiated in the open and debated in the daylight, unlike the current bill before the Senate, which had been scraped together in a matter of hours and tossed onto the floor. Ryan promised him that it would be so. McCain, in turn, told Ryan he planned to vote yes.

It was midnight. Ryan called McConnell to confirm McCain's support. "He's good," Ryan said. Thinking it was a done deal, the Speaker went to sleep.

A short while later, McCain walked onto the Senate floor. He chatted with Pence. He walked over to the Democrats' side and held court. Then, after the final vote was called, McCain disappeared again. With all of Washington holding its collective breath, the Arizona senator reemerged, walked toward the clerk's desk, and raised his hand flat in the air. He held it steady for several moments, then gave an emphatic thumbs-down.

Gasps filled the room as the aging senator turned and ambled to his desk. The Republican drive to repeal Obamacare was dead. McCain had preserved the legacy-defining achievement of the very man who had defeated him for the presidency.

THERE WAS PLENTY OF BLAME TO GO AROUND FOR REPUBLICANS' FAILure to deliver on the promise that had animated the party since the spring of 2010. Ryan had botched the rollout of the original bill. McConnell had been unprepared to act when the House passed its later version. Conservatives and moderates alike had failed to pressure their leadership to bring the process out of the shadows, fueling charges of hypocrisy from the left. And of course, McCain, a favorite scourge of the conservative base, had dealt the fatal blow to the repeal-and-replace crusade.

The one politician seemingly spared of all culpability in the eyes of Republican voters was Trump, whose approval rating with the base didn't budge. Polling showed GOP voters blaming Congress, far more than the president, for the failure to repeal Obamacare.[2] Reported vignettes from across the land revealed a grace period for a president who was predated by dysfunction in his party. "I really don't think people are trying to help Trump," Melinda James, a supporter from Broadview Heights, Ohio, told PBS after the failed House effort in March. "We need to unify. We need to give him a chance."[3]

Trump had tilled the field for this. Having spent the past two years railing against do-nothing politicians who never follow through on their promises, the president walked into a win-win. If Republicans delivered, he would be hailed as the redeemer of the party; if Republicans fell short, he would be excused as a sympathetic figure, another victim of an undrainable swamp.

Leaning into the anti-politician sentiment that continued to buoy him, the president made a show of acting unilaterally to keep his commitments. In his first six months in office, he signed a flurry of executive orders touching on everything from energy regulations to religious liberty, abortion to deportation policy. He began filling the federal courts with conservative jurists, and his White House ran a professional operation overseeing the confirmation of Neil Gorsuch to the Supreme Court. Perhaps most visibly, he withdrew the United States from the Paris climate accord, a global-warming pact with other nations that Trump said threatened American economic sovereignty. "I was elected to represent the citizens of Pittsburgh, not Paris," he decreed.

From thirty thousand feet, and certainly from inside the Beltway, these developments were obscured by the president's unpopularity. An ABC News/*Washington Post* poll in mid-July showed Trump's approval rating at 36 percent, "the lowest six-month approval rating of any president in polls dating back 70 years."⁴ And yet that same poll showed Trump's approval rating at 90 percent among conservative Republicans.

In many ways, the new president's flaws and failures—and the harsh judgments thereof—endeared him to the GOP base. Conservatives, and especially churchgoing Christians, could identify with someone dismissed by the political elite, disrespected by the mainstream media, delegitimized by the American left. Feeling ostracized in a culture that no longer reflected their core values or tolerated their most polarizing principles, the religious right came to feel a kinship with Trump that defied all reasonable expectations.

In early June, at the Faith and Freedom Coalition's annual gathering in Washington, the president offered an extraordinary sentiment when pledging his continued support to Christian conservatives. "We're under siege. You understand that," Trump said. "But we will come out bigger and better and stronger than ever."

It was a stroke of polysemantic genius from the president and his speechwriters. As heads nodded in agreement across the hotel ballroom, media outlets seized—as the White House knew they would—on the phrase, "We're under siege." After all, at that very moment, just six miles from where Trump was speaking, former FBI director James Comey was testifying under oath in front of the Senate Intelligence Committee about his unseemly interactions with the commander in chief. These were the tensest hours of Trump's young presidency, and here he was, acknowledging a defensive posture. But he was also expressing solidarity with an audience that could relate to feeling victimized.

Of all the early surprises offered by the Trump presidency, none proved more enduring than his alliance with Christian conservatives. Trump thrived on transactional relationships, and in white evangelicals—81 percent of whom voted for him in 2016—he discovered an ideal trading partner. He would give them the policies and the access to authority that they longed for. In return, they would stand behind him unwaveringly. "Those fucking evangelicals," Trump mused in a meeting with GOP lawmakers, smiling and shaking his head at the depth of their devotion.

That he was not *one of them* was beside the point. "I've been at the White House for meetings more in the first four months of the Trump administration than I was during the entire Bush presidency," Tony Perkins, the Family Research Council president, said that spring.

Ralph Reed, the Faith and Freedom Coalition's chairman, put it thusly: "Jimmy Carter sat in the pew with us. But he never fought for us. Donald Trump fights. And he fights for us."

Naturally, this marriage invited skepticism, if not outright scorn. It was Jesus who posed the question, "For what does it profit a man if he gains the whole world and forfeits his soul?" And yet many of his followers, students of a faith that stresses the eternal, not the ephemeral, pledged their uncritical allegiance to an earthly leader in exchange for political gratification. To vote for Trump as the lesser of two evils while holding him to a high standard in office was more than defensible; but those who stood indiscriminately supportive of him despite behaviors that would have been intolerable from a Democrat opened themselves to every charge of shameless opportunism.

"In my experience over the last thirty or so years of political life, there's hardly any group in American politics that is as easily won over or seduced by power as Christians," says Pete Wehner, a Trump critic and one of the prominent evangelicals who served in the Bush 43 White House.

The most frequent rebuttal from faith leaders supportive of Trump amounted to a fascinating concession: Their idyllic visions of virtuous leadership in government had been a mirage. They had railed against Bill Clinton's philandering, but came to realize afterward that America was past the point of prioritizing morality in its leaders. The country was changing too much, too quickly, for their old expectations to be realistic. In 1976, Jerry Falwell had crucified Jimmy Carter for giving an interview to *Playboy*; forty years later, Jerry Falwell Jr. posed with Trump in front of a framed *Playboy* cover featuring a nearly nude woman. This was to be the new normal. "It's not our job to choose the best Sunday school teacher, like Jimmy Carter was," Falwell Jr. told CNN in 2016.[5]

In fairness, it wasn't just the religious right revising its standards. This was a time of transition for conservatism writ large. The Tea Party insurgency had redrawn the battle lines inside the GOP. The Freedom Caucus had neutered the House Republican leadership. John Boehner and Eric Cantor had been pushed out of power. Mitch McConnell found his chamber increasingly ungovernable. In May, Jim DeMint was fired by the Heritage Foundation, concluding a ruinous experiment that had further sullied the think tank's reputation. And Reince Priebus, the White House chief of staff who remained Trump's strongest link to the party establishment, was on his way out.

That summer, having breakfast with a blissfully retired Boehner, I asked him whether the Republican Party could survive Trumpism. "There is no Rep—" He stopped himself.

There is no Republican Party?

He shrugged. "There *is*. But what does it even mean? Donald Trump's not a Republican. He's not a Democrat. He's a populist."

What Trump was demonstrating, however, is that political labels were less relevant than ever. And to the extent that they still mattered, their very definitions were changing.

Any doubts of this were erased by the Conservative Political Action Conference of 2017. One year earlier, there had been threats of a mass walkout if the GOP front-runner came to speak, leading Tump to cancel his appearance. Now, CPAC had turned into "TPAC," as Kellyanne Conway told the audience to wild cheers. She was right. To attend the event was to witness an ideology conforming to an individual rather than the other way around. Trump brought down the house with a speech that made no mention of "liberty" or "constitution," choosing instead to champion "our movement" as one that would embrace protectionist, cronyist, big-spending policies in the name of shielding Americans from the menace of a global economy.

Meanwhile, traditional mainstays such as Paul Ryan and Marco Rubio and Rand Paul were nowhere to be found. In their place was Steve

Bannon hawking "economic nationalism" and his old media company, Breitbart (once banned from CPAC) sponsoring the event with its logo slapped across the stage. The event's organizers had even invited Milo Yiannopoulos, the alt-right carnival barker with no serious claim to conservatism, to speak. (He was disinvited only after video surfaced of him making approving remarks about pedophilia.)

In certain respects, the conservative support for Trump was understandable. He had delivered on important promises. At the same time, he had done things—facilitating the Carrier deal in Indiana that smacked of government favoritism; bullying private corporations and individual citizens; asserting a moral equivalence between the U.S. government and Vladimir Putin's—that would traditionally have put any politician in the conservative movement's guillotine.

Trump did not suffer for these apostasies. And the best explanation wasn't that voters were ignoring his rejections of Republican orthodoxy; it was that they accepted his rebranding of that orthodoxy. In a poll taken at the event, a full 86 percent of CPAC attendees approved of Trump's job performance. Moreover, 80 percent agreed with the notion that he was "realigning" conservatism. "In many ways, Donald Trump is the conservative movement right now," Jim McLaughlin, the Republican pollster who conducted the survey, announced to CPAC attendees. "And the conservative movement is Donald Trump."

THE BASE'S DEVOTION TO TRUMP WAS LIKE BALM TO A PRESIDENCY covered in third-degree burns.

For all the upheaval of the first few months, nothing had prepared the White House for the hell unleashed by Trump's firing of Comey and the subsequent appointment of Robert Mueller as the special counsel.

As the summer wore on, it became evident that Mueller was looking into not just Russia's attempts to influence the presidential campaign, but also potential coordination between Trump team's and the Kremlin. An assortment of criminal activities carried out by the president's former associates, including Paul Manafort, Michael Flynn, and an obscure policy adviser named George Papadopoulos, was bringing the investigation inside Trump's campaign and inside his White House.

Even more threatening to the president, Mueller's investigation was also circling his family.

On July 8, the *New York Times* reported that in June of 2016, Donald Trump Jr. had met with a Kremlin-linked lawyer at Trump Tower.[6] Also present were Manafort and Jared Kushner. The story rocked the West Wing. The president, flying on Air Force One, dictated a statement saying the meeting had been about a Russian adoption program and nothing more. (Trump's lawyers initially denied his involvement in issuing the statement, only to admit later his role in writing it.)

The next day, the *Times* followed up with a far bigger blockbuster: Trump Jr. had arranged the Tower meeting in response to promises of receiving dirt on Clinton from the Russian lawyer.[7]

On June 3, 2016—just as the general election campaign was commencing—Trump Jr. had received an email from one of his dad's former Russian business associates. This person had been contacted by a Kremlin official who was offering information that "would incriminate Hillary and her dealings with Russia and would be very useful to your father," the email read, adding, "This is obviously very high level and sensitive information but is part of Russia and its government's support for Mr. Trump."

Minutes later, Trump Jr. replied, "If it's what you say I love it especially later in the summer."

The White House had repeatedly and vehemently denied that there had been any contact in 2016 between Trump's inner circle and the Russian government. And the president himself had claimed on numerous occasions that there was no evidence whatsoever to hint at collusion—or even an *intent* to collude. Suddenly, thanks to his son and namesake, there was reason to believe otherwise. Blood was in the water.

All the while, the White House was confronted with the normal volatility of running the country: judicial appointments and confirmations, votes on health care, legal challenges to the travel ban, a missile strike on Syria, and the continued military offensive against the Islamic State.

Republicans were also dealing with the continued trauma of a large-scale assassination attempt. On the morning of June 14, a gunman named James Hodgkinson opened fire at a baseball diamond in northern

Virginia where House Republicans were practicing ahead of the annual congressional baseball game. Steve Scalise, the House majority whip, was playing second base. He thought a tractor had backfired. Instead, it was Hodgkinson, a liberal activist and Bernie Sanders campaign volunteer who kept a list of the Republicans he wanted to mow down, shooting at him with 7.62-caliber SKS semiautomatic rifle. (As intriguing as the thought experiment "Imagine if Obama did what Trump just did" was the question of "How would we react if a Ted Cruz devotee tried to murder a dozen House Democrats?")

Scalise was hit. The bullet traveled through his hip, shattering the femur and wrecking the pelvis, with bullet fragments lodged in his muscle tissue and organs. As he crawled from the infield dirt into shallow right field, Scalise was bleeding to death. Four others were wounded, but Scalise's situation was the direst. Once the shooter was neutralized—by Scalise's security detail, who almost certainly prevented a massacre that morning—the other congressmen ran to their colleague from Louisiana. The first one to reach him was Brad Wenstrup, a little-known third-term Ohio lawmaker who had served as a combat medic in Afghanistan. Wenstrup tied a perfect tourniquet. It would later be credited with saving Scalise's life. "When I got to the hospital, they said I was within a minute of death," Scalise said.

Trump's handling of the situation was strangely reassuring. He issued a standard statement saying, "We are deeply saddened by this tragedy." He visited the hospital where Scalise and Capitol Police officer Crystal Griner were being treated. He and the First Lady spent a prolonged period of time with Scalise's wife, Jennifer, whose husband was unconscious. He gave Scalise's children a personal tour of the West Wing at the congressional picnic a week later.

There were no provocative tweets, no divisive rhetorical salvos aimed at Sanders or anyone else. (The Vermont senator said he was "sickened" upon hearing of Hodgkinson's devotion, and said on the Senate floor, "Real change can only come about through nonviolent action.") The only unusual part of Trump's response was his fixation, in discussions with doctors at the hospital and later with Scalise himself, on the size of the bullet. There was also the question he posed to friends and aides in the days following the shooting. "Should we do gun control?" the president asked. "Steve can lead the way. He's got street credibility now."

ADMINISTRATION OFFICIALS WERE MYSTIFIED AT TIMES SUCH AS THIS. The incident had shown that Trump had the capacity for calm, the feeling for normalcy, when the circumstances so demanded. Still, they wondered why the president refused to carry himself like this regularly; why his handling of the Scalise shooting was the exception and not the rule.

For much of 2017, Trump's top aides, as well as the rest of Republican Washington, found themselves twitching every time their smartphones twinkled to life. The president had used Twitter during the campaign to rewrite the rules of mass communication in politics, circumventing the media gatekeepers and reaching tens of millions of people instantly. But Trump's social media addiction was proving debilitating to his presidency. Like clockwork, his impulsive rants distracted from the White House's efforts to advance its agenda. Most administrations do whatever possible to avoid controversy; Trump specialized in creating it.

Though White House officials joked constantly about stealing his smartphone or changing the password—really, they weren't joking at all—nothing could be done to stop the president from tweeting. The pleas from his staff did no good. He told friends that without Twitter, he never would have won the presidency to begin with; now he told aides that without Twitter, his presidency would be derailed by a hostile press corps. Ironically, both his friends *and* his aides believed the greatest threat to his presidency was Twitter itself.

While enduring nightmares of self-implicating tweets and dreading the morning ritual of checking to see what the president had already posted, Trump's aides looked forward to one period of relative peace: Sunday afternoons. It was then that the president went golfing, leaving his smartphone behind. (Trump, who savaged his predecessor for golfing on the job, would play roughly twice the number of rounds as Obama in his first two years in office.)

The president's staffers lived in fear of one thing: bad weather. Some spent Saturday nights praying for clear skies the next day, knowing a tweet-free afternoon would give them a window of uninterrupted tranquility, time to spend with their families and decompress from a job known to be demanding under the most normal of circumstances.

"Rainy Sunday afternoons," Priebus told Ryan, "are the devil's play shop."

Of all the president's men, Priebus was the most fatigued. He had been a doormat from day one. Trump emasculated him, calling him "Reincey" and bad-mouthing him behind his back. Ryan, his longtime friend, who had expected a close alliance with the chief of staff, learned quickly that Priebus had no power in the White House, and he went around him to engage Trump directly.

Priebus did his best to exert authority. He would declare key meetings off-limits to certain staffers and stand guard at the Oval Office doorway, wary of unannounced visitors. But it was of no use: People poured in past him, and meetings spilled over with uninvited guests. Omarosa Manigault, a former contestant on *The Apprentice* whom Trump had brought into the administration to serve in part as a liaison to the black community, did little work that anyone ever saw but proved adroit at finding her way into ultra-important conversations in the West Wing. It later paid off: She had been secretly recording them to gather material for a tell-all book.

There was no structure, nothing resembling an organized commercial enterprise, much less a disciplined presidency. The West Wing was the Wild West—strangers roaming free, Trump springing new surprises on his staff at every turn, meetings running hours behind schedule, Secret Service agents scrambling to keep wanderers out of the Oval Office. Priebus aimed to stick near the president whenever possible, fearful that his influence was diminished by every moment he was not by Trump's side. Yet he also had a White House to run, and spent his days sprinting between meetings, unable to trust many of the people whom he should have been delegating to. Friends compared the chief of staff to a battlefield medic hustling between patients but never able to stop the bleeding.

It wasn't working for anyone. Despite his own penchant for such commotion, the president knew he needed an infusion of order. As for Priebus? He just needed a nap and a cold Miller Lite.

Spicer could tell that the chief's days were numbered. After a rocky and bizarre tenure as White House press secretary, the former RNC flack had stepped down on July 21. His exit owed in part to the hiring of a hot-blooded political neophyte named Anthony Scaramucci as White House communications director, an addition that Priebus vigorously opposed. It was his last battle as chief of staff—and unsurprisingly, it was a losing one.

A week later, on July 28, Priebus was replaced by John Kelly, the retired four star Marine general who had previously run the Department of Homeland Security. In firing his chief of staff and the party chairman who'd helped him get elected, Trump was severing his last major tie to the Republican establishment. Looking around at the president's inner circle, one saw Scaramucci, who had previously donated to Obama and Hillary Clinton; Bannon, who had used Breitbart to try to burn the GOP to the ground; National Economic Council director Gary Cohn, a life-long Democrat; the director of strategic communications, Hope Hicks, who had zero history with GOP politics; and Ivanka Trump and Jared Kushner, a pair of self-professed Manhattan progressives. Of Trump's closest advisers, only Pence had any association with the Republican Party.

The staff makeover was only beginning.

One day before the chief of staff's departure, the *New Yorker* reported on a phone call between Scaramucci and Ryan Lizza, one of its reporters, in which the White House communications director went on an expletive-laden tirade.[8] During the conversation, Scaramucci called Priebus "a fucking paranoid schizophrenic" and said of the president's chief strategist, "I'm not Steve Bannon, I'm not trying to suck my own cock. I'm not trying to build my own brand off the fucking strength of the President. I'm here to serve the country."

The White House communications director was promptly fired—after six days on the job.

Capping an extraordinary stretch, a few weeks after Scaramucci's dismissal, Trump fired Bannon. The president had long since grown tired of his taste for celebrity: the endless string of out-of-school interviews, the *Saturday Night Live* sketch depicting him as the Grim Reaper, and most recently, the book *Devil's Bargain*, by journalist Joshua Green, painting Bannon as the president's puppeteer.

When Kelly took over as the new White House chief of staff, his first priority was to jettison Bannon, whom he viewed as a destabilizing force inside the government and a self-promoting clown to boot. Trump acted swiftly on his new chief's recommendation.

For a moment—for his first few weeks on the job, really—it appeared that Kelly was equipped to finally bring order to the White House. It wouldn't last.

CHAPTER TWENTY-ONE

AUGUST 2017

"God made me black on purpose."

THE MURDER OF NINE BLACK PARISHIONERS INSIDE THEIR CHARLESton church in June 2015—by a white gunman, Dylann Roof, who told police he wanted to start a race war—marked a genuine inflection point in the American argument over race, culture, and politics.

South Carolina's subsequent removal of the Confederate flag from its statehouse grounds, spearheaded by Governor Nikki Haley, triggered a sweeping campaign to purge the nation of symbols that spoke to the dark echoes of its past. These efforts were concentrated in the South. All across Old Dixie, tributes to Confederate soldiers and their causes remained ubiquitous 150 years after the surrender at Appomattox. Coinciding with the ascent of Donald Trump, the fault lines were drawn: Some Americans argued that radicals were attempting to erase the nation's rich and complex heritage; other Americans argued that radicals were attempting to preserve the nation's historical architecture of oppression.

The conflict came to a head in Charlottesville, Virginia.

A neatly manicured city, home to Thomas Jefferson's Monticello plantation and the pristine campus of the University of Virginia, Charlottesville was an unlikely backdrop for race rioting. Yet the proposed removal of Robert E. Lee's statue from Lee Park—the city council had voted to rename it Emancipation Park—was becoming a fight of the utmost symbolic importance. In the heart of the Confederacy, in a settlement built by a slave-owning (and slave-impregnating) president of the United States, the South's most famous general still sat high atop the city on his bronze steed, daring the forces of progress to topple him.

Charlottesville had braced for a planned demonstration the weekend of August 12, nicknamed "Unite the Right," meant as a rallying point

for the scattered cells of white supremacists and neo-Nazis nationwide. Counterprotesters from around the country mobilized quickly,
descending on Charlottesville. With the city's police and government
officials in preparation mode, a group of several hundred right-wing
demonstrators staged a surprise march through the UVA campus,
hoisting torches and shouting racist slogans. Video of the event, captured with chilling precision by *Vice News*, showed them chanting, "You
will not replace us! Jews will not replace us!" It was an ode to the theory
that a Jewish-dictated demographic makeover of the United States was
meant to dilute the power of the white race.

At the end of a chaotic Saturday full of skirmishes for which the local
law enforcement was visibly underprepared, tragedy struck. A twenty-
year-old Nazi sympathizer from Ohio, James Alex Fields, rammed his
Dodge Challenger into a crowd of counterprotesters downtown. The
terrorist attack injured twenty-eight people and killed Heather Heyer,
a thrity-two-year-old waitress and paralegal.

The demonstrators drawn to Charlottesville represented disparate
cogs of an ideological machine—one emboldened by the election of the
forty-fifth president. They were neo-Nazis, neo-Confederates, right-
wing militia members, and Klansmen. The former grand wizard of the
KKK, David Duke, was among other white supremacist luminaires in
attendance. "We are determined to take our country back," Duke said at
the rally. "We are gonna fulfill the promises of Donald Trump."

The collection of counterprotesters was more diverse. There were
interfaith leaders, locking arms and singing hymns. There were college
kids and faculty members. There were local residents. There were members of Black Lives Matter, the nascent organization that used aggressive nonviolent tactics (such as street demonstrations and "die-ins") to
combat police brutality. Problematically, there were also disciples of
Antifa ("antifascist"), a conglomeration of groups, some with a history
of inciting violence.

Two hours after Fields plowed his car into a throng of people, Trump
delivered prepared remarks from his golf club in New Jersey. "We condemn in the strongest possible terms this egregious display of hatred,

bigotry, and violence"—he looked up from his script—"*on many sides*. On many sides."

The president also urged the country to "come together as one," saying, "The hate and division must stop." But these remarks were peripheral to his opening statement, which had drawn a moral equivalence between the white supremacists marching in Charlottesville and the counterprotesters opposed to their ideology. Horrified by Trump's language, Republicans raced to condemn him, strafing the White House with an unprecedented barrage of criticism.

"We should call evil by its name," Utah senator Orrin Hatch tweeted. "My brother didn't give his life fighting Hitler for Nazi ideas to go unchallenged here at home."

The White House tried to put out the fire, issuing a statement on Sunday that read, "The President said very strongly in his statement yesterday that he condemns all forms of violence, bigotry, and hatred. Of course that includes white supremacists, KKK Neo-Nazi and all extremist groups."

But this would not suffice. After collecting decades' worth of racial baggage and running a campaign during which he insulted Mexicans, called for banning Muslims from the country, and played footsie with extremists, Trump would need to clean up this mess himself.

He tried. "Racism is evil," the president said on Monday, August 14, during an impromptu appearance before the White House press corps. "And those who cause violence in its name are criminals and thugs, including the KKK, neo-Nazis, white supremacists, and other hate groups that are repugnant to everything we hold dear as Americans."

But less than twenty-four hours later, on Tuesday, the fifteenth, he reverted to form. Facing a scrum of reporters inside Trump Tower, the president defended his original remarks from Sunday, repeating his assertion of "blame on both sides." He also lashed out at the media, which he claimed had unfairly attacked the rally's participants. "Not all of those people were neo-Nazis, believe me. Not all of those people were white supremacists by any stretch," he said. Saying there had been "very fine people on both sides," Trump argued that the campaign to remove Confederate monuments was an attempt to "change history."

The outpouring of anger was nearly unparalleled, likely bested only by the reactions to the *Access Hollywood* recording. Dozens of elected Republican officials past and present condemned Trump's rhetoric and called on him to emphasize, once and for all, that there were no "very fine people" marching with torches in Charlottesville.

One voice pierced the din of familiar outrage that had come to shadow Trump. "What we want to see from our president is clarity and moral authority," Tim Scott, the Senate's lone black Republican, told Vice News.[1] "And that moral authority is compromised when Tuesday happened."

HAVING BEEN DEPLOYED AS A "PROP" BY HOUSE REPUBLICANS AFTER his election in 2010, and having heard the grumblings about being an affirmative-action hire following his appointment to the Senate in 2012, Scott was determined to avoid being defined by his blackness. In the absence of words, the new senator tried to lead with actions. He assembled one of the most diverse offices on Capitol Hill, led by his black, single-mother chief of staff. He also poured time and resources into mentoring programs in distressed communities back home.

But his silence on cultural events became deafening. "Scott has said little on the racial controversies and civil rights issues of the last four years, from the killings of Trayvon Martin and Jordan Davis to the death of Michael Brown and the explosion of anger and rage in Ferguson," Jamelle Bouie wrote in *Slate* in November 2014.[2] Soon enough, circumstances made the senator's reticence unsustainable.

In April 2015, Walter Scott (no relation) was fatally shot in the back by a policeman in North Charleston. Shaken by the hometown incident, the senator called for the officer's prosecution. (In 2017, he introduced the Walter Scott Notification Act, which would create a database of police shootings.)

Ten weeks after Walter Scott's murder, Roof opened fire at Emanuel AME Church. Scott delivered an emotional address on the Senate floor. ("God cares for His people," he said, quoting the son of one of the victims he had spoken with. "God still lives.")

The next summer, in July 2016, amid another rash of police shootings, Scott returned to the Senate floor. In a spellbinding speech, the senator described how he had been victimized by racial profiling—pulled over in his car seven times *in one year* as an elected official, and twice forced to show identification to Capitol Police despite wearing his members-only Senate pin.

In a town filled with empty rhetoric, Scott's remarks got everyone's attention. Democrats applauded the GOP senator's courage. Older, white Republicans found themselves guilt-ridden by their colleague's story. There was no turning back for Scott. He had never wanted to be identified as the Black Republican. But there was too much at stake and too few voices in his party capable of speaking to the moment.

"We have a significant percentage of people in this country who feel they're treated differently because of their background or their color. And we need to talk about it," says Rubio, Scott's close friend. "No one else could have done that. No one else could have given that speech."

Every Republican lawmaker was bound to walk a tightrope when Trump took office, forced to weigh their moral and philosophical objections to him against fears of losing influence in Washington and support back home. But nobody had it worse than Scott. Any rebuke of the president would expose him to lunacy from the right; he became a frequent target of menacing calls and messages, with a Georgia man arrested for threatening to kill him. ("He said he wanted to be the next Dylann Roof," Scott says.) But biting his tongue when it came to Trump's behavior opened Scott to unyielding vitriol from the left, with accusations of being an "Uncle Tom" or worse. When one Twitter user called him a "house nigga" in early 2017, after his vote to confirm Jeff Sessions as attorney general, Scott tweeted back a one-word reply: "Senate."

In the age of Trump, the senator from South Carolina came to terms with an uncomfortable truth: The fixation on his color was a feature, not a bug. No matter his achievements or aspirations, Scott was sentenced to exist in America's collective political subconscious as a black man first and everything else second.

The senator did, however, come to see a silver lining. A deeply religious individual who twice nearly pursued preaching as his vocation, he rejected the notion that Trump had been chosen by the Almighty. But Scott did believe that *he himself* had been chosen, placed in a unique position at a unique time in history, to help "the American family" navigate some "painful, ugly, embarrassing" conversations about race and other combustible subjects that had simmered for generations.

"God made me black on purpose. For a specific reason," Scott says. "I am not pretending that this characteristic, this Earth suit that I'm in"—he pinches the skin of his arm—"isn't being evaluated. It requires a response, or a reaction, to the situations at my level of government. I am fully aware of that. I just don't want to play a game with it."

Cerebral and deliberate, Scott was known in Congress to speak very little until he was sure there was something worth saying. His assertion that Trump had forfeited the "moral authority" of his office after Charlottesville, then, got everyone's attention, including Trump's.

Reached on his cellphone by the new press secretary, Sarah Huckabee Sanders, the senator agreed to sit down with the president and explain his displeasure. What ensued inside the Oval Office a few days later was a lengthy, Scott-led seminar on America's history of institutional racism and systemic discrimination. He talked of the socioeconomic hurdles facing young black men in his native streets of North Charleston. He described the hopelessness, the lack of opportunity, that had long suffocated the potential of minority youths in America. He told the story of his grandfather, Artis Ware, who left a segregated school in the third grade to pick cotton for fifty cents a day. Scott remembered his role model scouring the newspaper each morning, impressing upon his grandsons the importance of reading; it wasn't until years later that Scott realized his grandfather was illiterate.

The White House, for its part, released a photo of Trump listening intently to a senator identified as "Tom Scott."

Explaining that the resurgent racial tensions in America owed in part to anxieties over a dramatic cultural and demographic transition, Scott urged Trump not to prey on them.

"I know what fear looks like. I think fear typically comes with anger and hostility. You're afraid that you're losing something, that you won't have something that you used to have," Scott said later, looking back on

Charlottesville. "I think people who march with torches—who want to resurrect a thankfully dead part of who we were—these are people who are afraid. Afraid of the changes happening in the country. Afraid of the other man who doesn't look like them."

Trump took all this in, rarely interrupting. "What can I do to be helpful?" he finally asked.

The senator was prepared with an answer. If Trump was getting a nice photo-op out of their meeting, Scott was going to get something, too.

AUGUST WAS AN UNPLEASANT TIME FOR REPUBLICAN LAWMAKERS. Charlottesville aside, the summer recess was filled with heated inquiries from irritated constituents wanting to know why, after seven years of promising, they still had not repealed and replaced the Affordable Care Act. There was no dodging the question, no spinning the answer. The GOP had once again failed to deliver on a core pledge to its voters, and legislators were shouldering far more blame than Trump. Whereas the president had installed a new Supreme Court justice and manufactured a bevy of unilateral wins from the executive branch, congressional Republicans had no major legislative accomplishment to show for their first eight months of unified government.

It was around this time that a new push, for a rewrite of the tax code, became the all-consuming obsession of the Republican Party.

There was no guarantee of success. In fact, lawmakers widely believed that making changes to America's tax policy would be infinitely harder than altering its health care system. Whereas the Obamacare fight had implications for only certain people, groups, and industries, the tax code touched every aspect of life—and there were the lobbyists to prove it. Every deduction, every loophole, every footnote of the existing law was safeguarded by special interests.

But congressional Republicans had no choice. It wasn't just that they needed to get results; it was that the party appeared to be in disrepair, with Trump's performance increasingly a cause for alarm.

In the final four months of the year, the president pardoned the infamous Arizona sheriff Joe Arpaio, who had been convicted of criminal contempt; called North Korean dictator Kim Jong-un "Rocket Man" in a speech to the United Nations; attacked kneeling NFL players, advising owners to "Get that son of a bitch off the field right now!" during a speech in Alabama; ridiculed a Puerto Rico mayor who had criticized his administration's ham-fisted response to Hurricane Maria, which killed some three thousand Americans; accepted the resignation of his first cabinet official, health and human services secretary Tom Price, who *Politico* revealed had traveled on private and government planes at a cost of more than $1 million to taxpayers; falsely accused Obama of not contacting the families of fallen troops; and feuded with a Democratic congresswoman over his condolence call to a soldier's widow.

Trump also decided to publicly defend Roy Moore, the Alabama Senate nominee, against allegations of child molestation. Many Republican leaders were calling for Moore to quit the race; instead, buoyed by the president's backing, he stayed in, losing to Democrat Doug Jones in bloody-red Alabama and cutting the GOP's Senate majority to 51 seats.

BENEATH THE POLITICAL MAELSTROM, THE WHEELS OF CULTURAL INsurrection kept on turning.

In early October, the *New York Times* and the *New Yorker* published stories that detailed nearly thirty years of claims of rape and sexual misconduct against Hollywood filmmaking mogul Harvey Weinstein.[3] Responding to these reports, which had long been an open secret in show business, women across the world joined a social media movement by tweeting a simple phrase to demonstrate the scope of the epidemic: "Me too."

The floodgates thrown open, #MeToo triggered a cascade of accusations against some of America's most prominent men. Among them were television host Charlie Rose; actor Kevin Spacey; comedian Louis C.K.; political journalist Mark Halperin; music mogul Russell Simmons; and *Today* show host Matt Lauer. Some of these men, and many dozens more, saw their careers irreparably ruined by their misdeeds.

The world of politics was not immune to this reckoning. For some lawmakers, Capitol Hill had long functioned as a frat house, teeming with attractive young women to keep them company while living away from their wives four nights a week. By the spring of 2018, seven male lawmakers had been accused of misconduct; five of them resigned, including Al Franken, the popular Minnesota senator, and John Conyers, the Detroit Democrat and Congress's longest-serving member.

While speculation stalked the careers of many additional legislators on Capitol Hill, it was impossible to ignore the man on the other end of Pennsylvania Avenue. More than a dozen women had come forward with accusations of sexual assault against Trump during the presidential campaign, charges that were freshly relevant in light of the #MeToo movement.

Trump escaped the #MeToo tsunami, aided partially by the scale of accusations sweeping across industries and trashing the reputations of innumerable other men. But he could not rid of himself of another existential threat: Robert Mueller.

As the year drew toward closure, indictments and guilty pleas were piling up in the special counsel's investigation, with four of the president's associates caught up in the Russia probe. Two of the catches were particularly big fish. Michael Flynn, the president's original national security adviser, pleaded guilty of making false statements to the FBI. And Trump's former campaign chairman Paul Manafort was indicted by a federal grand jury on charges of conspiracy against the United States. (Manafort had told his old friend, Scott Reed, that everything he would be doing for Trump would be legal. That may have been true; but by joining the campaign, Manafort unwittingly exposed his past criminality to the scrutiny of the special counsel's office.)

The sense of downward spiral, politically and otherwise, further emboldened some of the president's critics within the party. Secretary of State Rex Tillerson called Trump "a moron" after a meeting at the Pentagon with top national security officials, according to NBC News, a report that Tillerson did not deny. Senator Bob Corker, the Foreign Relations Committee chairman whom Trump had considered for his vice-presidential pick, likened the White House to an "adult day care center" and told the *New York Times* that the president's actions could set the nation "on the path to World War III."

Around that same time, in mid-October, a most unexpected voice chimed in with censures of his own: George W. Bush.

Having remained dutifully silent on political matters throughout all eight years of Obama's presidency, Bush was struggling to bite his tongue in the opening months of Trump's first term. He had never been terribly bothered by the attacks on himself and his family; there was no personal grudge keeping him up at night. What Bush could not stomach—what he found increasingly intolerable, he told friends—was the president of the United States using his office to demonize immigrants, abuse his political opponents, and divide the nation for partisan gain.

After nine months of stewing, Bush broke his silence. Noting how "bigotry seems emboldened" and how "our politics seems more vulnerable to conspiracy theories and outright fabrication," the former Republican president delivered a speech in New York City that landed like an asteroid on Washington.

"Bullying and prejudice in our public life sets a national tone, provides permission for cruelty and bigotry, and compromises the moral education of children. The only way to pass along civic values is to first live up to them," Bush said. "Our identity as a nation, unlike other nations, is not determined by geography or ethnicity, by soil or blood. This means that people from every race, religion, ethnicity can be full and equally American. It means that bigotry and white supremacy, in any form, is blasphemy against the American creed."

His audience erupted with applause. Bush never said Trump's name. He didn't need to. Everyone who heard the remarks understood their purpose and their gravity. It was no longer just a bunch of liberal Democrats and biased journalists accusing the Republican president of gross misconduct; it was the previous Republican president.

Less than a week later, Jeff Flake, Trump's original foil in the Senate, announced that he would not seek reelection in 2018. Trump's prophecy had come true: Flake's dissent had tanked his poll numbers back home. Still, like Bush, he wasn't going quietly into the night.

"Reckless, outrageous, and undignified behavior has become excused as telling it like it is when it is actually just reckless, outrageous, and undignified," Flake said when announcing his retirement on the Senate floor. "And when such behavior emanates from the top of our government, it is something else. It is dangerous to a democracy."

All this was beginning to take its toll. The president's numbers trended sharply downward in late 2017, with a number of polls showing his approval rating dropping below the 40 percent mark. One such survey, from ABC News and the *Washington Post* in November, pegged Trump at an anemic of 37 percent.[4] Tellingly, however, 91 percent of his voters said they approved of his job performance.

The president was continuing to throw bones to his base: ending the DACA program that protected undocumented minors from deportation, issuing a new travel ban as previous iterations made their way through the courts, and, in December, moving the U.S. embassy in Israel to Jerusalem, fulfilling a campaign promise of spiritual significance to his evangelical Christian supporters, many of whom believe the "eternal capital" of God's chosen nation will be the site of the Messiah's return.

But there was still no landmark legislative victory to show for Trump's presidency and for the Republican Party's total control of government. The clock was ticking. Only a few months remained until the calendar turned to 2018, an election year, during which Congress would be hard-pressed to pass anything big.

To the shock of just about everyone on Capitol Hill, tax reform wound up being a breeze—relatively speaking.

With enormous pressure to produce, and without the emotional baggage of the health care fight—it's easier to cut someone's taxes than take away their health insurance—Republicans defied all expectations in passing a significant overhaul of the tax code through both the House and the Senate and putting it on the president's desk before Christmas.

None of this was to say the bill made for good policy. If anything, even as they rushed it through Congress—once again taking a number of expedited votes, once again violating the promises of "regular order" made by the party's leaders—Republicans never seemed enamored of the bill itself. It would deliver the disproportionate bulk of its benefits to corporations and the wealthy, undermining Trump's pledge of targeted relief for the middle class. It would offer less assistance to working families than many in the party hoped; Rubio had to hold the legislation hostage just to receive the slightest bump in the child tax credit.

Most distressing to conservatives, the bill would blow an enormous hole in the deficit, according to numerous nonpartisan projections. Republicans responded by pushing the disproven theory that economic growth would make up for the lost revenue by slashing rates for businesses and top earners, but the math was never going to add up. An analysis from the Joint Committee on Taxation showed that the GOP tax bill would add $1 trillion to deficits, *after* accounting for estimated growth.[5]

This presented something of an intellectual quandary for the party. A milestone legislative win was desperately needed, and it was now within their grasp. But the bill was nothing what many of them had envisioned when Ryan described harpooning his white whale of tax reform. Even as the Speaker muscled it through the House, he recognized that the bill did more cutting than reforming. There would be the same number of tax brackets and many of the same loopholes; there would be no filling out one's tax returns on a postcard, as the party had once advertised.

Still, Ryan felt an urgency unlike any in his career. For the past quarter century, since his days as a think tank staffer working for Jack Kemp, his dream had been to rewrite the tax code. However imperfect, this legislation represented their best chance in three decades to do so.

The Speaker knew that his legacy was on the line, and it wasn't just about tax reform. For the past year, he had justified his silence in the face of Trump's behavior as means to an end. Determining that public confrontations would only result in deeper intraparty fractures that would stall policymaking efforts, Ryan restrained himself. The executive branch had already gone off the rails, he told colleagues; they would gain nothing by turning the legislative branch into a comparable circus.

Yet this was a profoundly naïve perspective. In the House under Ryan's stewardship, the only thing missing were the bearded ladies. The already-fraught relationships between the parties—leaders, key committee personnel, rank-and-file members—became irreperable. Norms of process and procedure, already neglected, deteriorated to an unrecognizable degree. The House Intelligence Committee, once a paragon of congressional maturity, had devolved into a schoolyard taunting contest. The chairman, Devin Nunes of California, had turned one of Capitol Hill's most esteemed panels into a partisan food fight. (Not that no one saw this coming; Nunes had once referred to a colleague of Arab descent, Michigan's antiwar congressman Justin Amash, as "al-Qaeda's best friend in Congress.") By the time Ryan stepped in to address the dysfunction of Nunes's committee, the institutional damage had been done.

The Speaker simply could not afford to let tax reform fail. As strange as the conditions were, they were ripe for success. And as cynical as he knew it was, Ryan had come to view Trump's manic activity as advantageous to a rushed legislative process that would have received far more scrutiny had the president of the United States not been tweeting his vendettas before sunrise. If they couldn't muscle a bill through under these circumstances, Ryan figured, it was never going to happen.

Ryan seized control of the process in the House, angering members of the Ways and Means Committee, who complained that they didn't see legislative text until just days before voting on it. The complaints echoed those from the original Obamacare push: The Speaker was running the House like an autocrat, ignoring the input of members and breaking the very vows he had made when ascending to the position. "He's more controlling than Boehner . . . and I voted against John Boehner and worked with Mark Meadows to vacate the chair," Walter Jones, a frumpy older congressman from North Carolina, said during the tax reform campaign. "I'm very dissatisfied. I've been here twenty-two years, and this is the most closed shop I've ever seen."

Ryan heard the criticism. He also heard the rumors that the Freedom Caucus, feeling increasingly disenfranchised, was discussing another overthrow of another Speaker. But he paid these distractions no mind. Because for Ryan, in the fall of 2017, there were only two things worth thinking about: tax reform and retirement.

He was sick of Congress, tired of spending five days a week away from his family, and most of all, fed up with babysitting Trump. His solid working relationship with the president had done nothing to convince him of the man's suitability for the office he held; rather, Ryan had become a professional counter to ten, convincing himself each morning anew of the counterproductive nature of a pissing match with the commander in chief. Sitting in the Speaker's office, knowing the institutional chaos that would flow from a game of chicken between the two most powerful officeholders in the federal government, Ryan had convinced himself that it was better in the long term, even if emasculating and hypocritical in the short term, to keep his composure with Trump.

The two men had developed a surprisingly strong working relationship, thanks largely to this acquiescence from Ryan in the pursuit of policy victories. But the Speaker shuddered at the thought of sticking around for 2019 and 2020, when Trump would be actively running for a second term. The idea of enduring another Trump election cycle was nauseating to Ryan. Retirement was the only escape hatch. What better way to go, he thought to himself, than with a signature tax reform law?

ACROSS THE CAPITOL, FOUR REPUBLICANS WERE RESPONSIBLE FOR drafting and redrafting the Senate version of the tax bill. But it was Tim Scott, the South Carolinian, who emerged as the key player.

Having worked on the issue since his arrival in Congress in 2011, and bringing a wealth of real-world experience from his career as a successful insurance salesman, Scott labored for months to align the conflicting visions of his colleagues. He sat at the intersection of the ideological tribes, someone trusted by both conservatives and centrists alike. Time and again, throughout the fall of 2017, Scott was tasked by McConnell with putting out the latest fire that threatened to engulf the entire effort. He became the indispensable actor among Senate Republicans.

As he brokered settlements between his colleagues, Scott had but one priority himself: "Opportunity Zones."

A few years earlier, Scott had gotten to know Sean Parker, cofounder of the Napster music streaming service and the original president of Facebook. Parker described to the senator a pair of problems that he saw as interrelated: a staggering amount of capital sitting on the sidelines of the economy, and an uneven recovery in which investments rarely found their way to the communities most in need. The government, Parker argued, should offer incentives for venture capitalists and entrepreneurs to invest in Opportunity Zones, blighted areas (determined by poverty level and median household income) desperate for economic renewal.

Scott was sold. Working with his friend, Democratic senator Cory

Booker, and with a bipartisan group of lawmakers in both chambers, Scott led the charge to incorporate the policy into the GOP's tax bill.

But he needed an ally. Influential as he was in the tax reform negotiations, Scott did not have the juice to strongarm the Senate into adopting a policy that few of his colleagues were familiar with. Securing the Opportunity Zones would require a lot of weight to be thrown around. He had just the person in mind.

"Well," the senator had replied to the president during their post-Charlottesville conversation. "You can support the Investing in Opportunity Act."

Trump wasn't familiar with the policy, but he gave Scott his word that he would support it—and he did. The president endorsed the legislation the very next day and remained loyal until it became law as part of the GOP's tax reform package.

Scott was exuberant. So was Ryan. The final, compromise legislation passed through both chambers of Congress in mid-December. No Democrat in either the House or the Senate voted for it, a sign of the polarized climate but also of the GOP's lack of outreach across the aisle. (Tax cuts aren't exactly a tough sell in purple states, but there was virtually no pressure placed on vulnerable Democrats to support the Republican bill.) The partisan nature of the end product didn't much bother Scott or Ryan. Both had notched crowning, legacy-making victories.

When Republicans gathered on the South Lawn of the White House on December 20 to celebrate their triumph, Ryan stepped to the lectern and uttered one of the defining observances of his speakership. "Something this big, something this generational, something this profound," he declared, "could not have been done without exquisite presidential leadership."

Throughout the ceremony, Scott stood right next to Ryan, flanking the president. The extent of his influence was on full display. But that's not what everyone saw. Just minutes before Trump invited Scott to speak at the lectern, Andy Ostroy, a *HuffPost* blogger, tweeted, "What a shocker . . . there's ONE black person there and sure enough they have him standing right next to the mic like a manipulated prop. Way to go @ SenatorTimScott."

When the event ended, and Scott opened Twitter and spotted the comment, he felt defeated. "Uh probably because I helped write the bill for the past year, have multiple provisions included, got multiple Senators on board over the last week and have worked on tax reform my entire time in Congress," he responded. "But if you'd rather just see my skin color, pls feel free."

The senator had more to say—so much more. He was weary of being targeted, weary of being the Republican spokesman on race, weary of trying to thread an impossible needle in the era of Trump.

Still, he held back. For all the flowery talk of how he represented a breakthrough—the first African American ever to serve in both chambers of Congress—Scott had begun to realize he was no such thing. He would always be a "prop" for his party. But the next wave of black Republicans wouldn't have to be—not if he did his job and did it well, conducting himself with dignity and turning the other cheek to his abusers. Generations of Scotts had suffered and sacrificed to send a cotton picker's grandson to the halls of Congress. Now it was his turn.

"I've been frustrated. And angry. *Man,*" Scott says, his voice trembling. "It's too easy to be angry. And too natural. And also, too unproductive for me. But I get it. I get it. I'm not at a point where my grandfather was. He could say nothing. He had to eat his anger. Or the next generation, who harnessed their anger and led marches. I'm on the inside track. I have a very different responsibility. It cannot be about me."

CHAPTER TWENTY-TWO

JANUARY 2018

"I said, 'That's Eva Perón. That's Evita.'"

TEMPERATURES HAD PLUNGED INTO THE SINGLE DIGITS AND CAMP David was coated with a fresh snowfall when they huddled inside a secure meeting room on Saturday morning. Donald Trump had arrived at the presidential retreat in Maryland one day earlier, along with members of the congressional Republican leadership, to chart the party's legislative priorities for the coming year. Now the leaders had joined Trump for a partial meeting of his cabinet, including Secretary of State Rex Tillerson, CIA director Mike Pompeo, and Defense Secretary Jim Mattis. Notably absent was Attorney General Jeff Sessions, who remained chained up inside the president's doghouse for the trespass of recusing himself from the Russia probe.

There were other disloyalties weighing on Trump that weekend. Friday had seen the official release of Michael Wolff's book *Fire and Fury: Inside the Trump White House*, which used salacious stories to paint the president as a narcissistic nitwit whose administration was drowning in its own incompetence. Despite Wolff's carelessness with basic facts and his prior reputation as someone of lax journalistic standards, the book nonetheless contained enough accuracies to throw the White House into a panic. Most bothersome to Trump were quotes attributed to Steve Bannon, his former chief strategist, who told Wolff that Trump Jr.'s meeting with the Russian lawyer in June 2016 was "treasonous" and "unpatriotic," predicting, "They're going to crack Don Junior like an egg on national TV." Bannon, having returned to the helm at Breitbart after leaving the White House, did not dispute these quotes.

Trump was in a tizzy. On Thursday night, with excerpts of the book trickling out, the president tweeted of Wolff's work, "Full of lies, misrepresentations and sources that don't exist. Look at this guy's past and watch what happens to him and Sloppy Steve!"

On Friday morning, Trump tweeted again about "Sloppy Steve Bannon," this time in reference to how the megadonor Robert Mercer and his daughter, Rebekah, had publicly severed their relationship with Bannon as a result of his remarks in *Fire and Fury.*

Then, late on Friday night, Trump tweeted again: "Michael Wolff is a total loser who made up stories in order to sell this really boring and untruthful book. He used Sloppy Steve Bannon, who cried when he got fired and begged for his job. Now Sloppy Steve has been dumped like a dog by almost everyone. Too bad!"

The GOP leaders worried about Trump's state of mind. To a man, they had spent the past year trying their best to ignore the president's Twitter feed, wanting plausible deniability when reporters inevitably asked for comment on Trump's latest social media salvo. But many of his statements were impossible to miss. Just a few days earlier, the president had tweeted, "North Korean Leader Kim Jong Un just stated that the 'Nuclear Button is on his desk at all times.' Will someone from his depleted and food starved regime please inform him that I too have a Nuclear Button, but it is a much bigger & more powerful one than his, and my Button works!"

Republicans in Congress had been feeling a sudden rush of optimism as they turned the calendar to 2018. After sputtering for much of Trump's first year in office, energizing the base of the Democratic Party in a way that no Democratic politician seemed capable of, Republicans had secured a momentous victory in the twilight of 2017. When the president signed the new tax law into effect a few days before Christmas, Republicans went home for the holiday recess feeling equal parts elation and relief. No longer would their constituents skewer them for failing to do their part to Make America Great Again; no longer could the Democrats paint them as dysfunctional and unproductive. (Dysfunctional? Sure. Unproductive? Not anymore.)

Thanks to an initial burst of glowing headlines (businesses making new hires, corporations giving out bonuses) tax reform looked like a political winner. House Republicans hoped it could be a majority-saver. A new president's party often takes a pounding in his first midterm election, an average loss of 32 House seats, and Democrats needed a net gain of just 23 to win back the chamber. Given the visceral surge of energy and the considerable financial support flowing to a large supply of high-caliber Democratic challengers, Republicans were fighting an uphill battle to keep the majority.

The tax bill, Paul Ryan and Mitch McConnell felt, represented their chance at salvation. Kevin McCarthy showed Trump polling during their Friday meeting at Camp David to demonstrate the party's months-long declining popularity, particularly among college-educated suburbanites, and argued that these same voters would be most rewarding of tax reform. "We have something to run on now," the Speaker told Trump. They all took turns urging the president to keep the country's attention trained on the benefits of their shiny new tax law.

It was hopeless. For Trump, the political concept of "message discipline" meant nothing more than listening to the very advice he'd ignored all the way en route to the White House. Besides, he didn't much care about the tax law. It was a *policy* accomplishment, surely, but the president was more consumed with the *personal*: allegations of collusion with the Russians in 2016, mounting chatter in the media about his fitness for office, and now, damning remarks from a supposedly staunch ally about his family's potential criminality.

On Saturday morning, as they prepared for their meeting with Trump, some of the Republican leaders and agency heads were alerted by their aides to a barrage of sunrise tweets from the president. At 7:19 a.m., he began: "Now that Russian collusion, after one year of intense study, has proven to be a total hoax on the American public, the Democrats and their lapdogs, the Fake News Mainstream Media, are taking out the old Ronald Reagan playbook and screaming mental stability and intelligence . . . Actually, throughout my life, my two greatest assets have been mental stability and being, like, really smart. Crooked Hillary Clinton also played these cards very hard and, as everyone knows, went down in flames. I went from VERY successful businessman, to top T.V. Star . . . to President of the United States (on my first try). I think that would qualify as not smart, but genius . . . and a very stable genius at that!"

The tweets were the elephant in the room when everyone gathered a short while later around a colossal conference table. Trump seemed his typical self, no more or less animated than usual. He was relatively engaged during remarks from several of the cabinet secretaries and seemed exceptionally interested by a classified briefing from Mattis on clashes with ISIS fighters. As the Pentagon chief spoke, the president scribbled wildly on a sheet of paper in front of him, all the while nodding and looking up to make eye contact. The others in the room took this as an encouraging sign: The Pentagon had released a report just weeks earlier claiming that ISIS had lost 98 percent of its territory.

When Mattis finished, the president lifted the piece of paper while gesturing, just high enough for several people to see it. He had drawn a flight of bullet points on the page, all of them underneath an all-caps header that was clearly visible: "SLOPPY STEVE."

THE WHITE HOUSE HAD GROWN ADEPT AT WEATHERING RHETORIC-based storms, with Trump showing a breathtaking capacity for turning the page on statements that would have been definitional for any other president. Whether it was his boasting about the size of his nuclear button, or calling himself "a very stable genius," or questioning an immigration policy that would allow people from "shithole countries" like Haiti into the United States, Trump spent the first month of 2018 defying conventions in the same way he had throughout his first year in office—and paying no real political price for it.

It was a different story when it came to the outward, existential threats to his presidency. Try as he might, Trump could not counter or distract from the growing perception of unscrupulous, and possibly unlawful, activity emanating from his inner circle. For much of 2017, this had been narrowly focused on the twin questions of collusion with Russia and obstruction of justice in his dealings with (and eventual firing of) FBI Director James Comey. But 2018 brought a different and potentially more damaging set of revelations.

On January 12, the *Wall Street Journal* reported that Michael Cohen, the president's longtime attorney and fixer, had paid $130,000 to a pornographic film star during the 2016 campaign to prevent her from sharing details of their past romance.[1] Stephanie Clifford, who went by the professional name of Stormy Daniels, alleged that she and Trump had had sexual intercourse soon after meeting at a celebrity golf tournament in Lake Tahoe back in 2006—when the future president was newly married to Melania Trump, who had just birthed the couple's first child.

The story was ground-shaking. The moral implications of Trump cheating on his new wife with a porn star aside, the reported outlay of hush money would represent a flagrant violation of federal campaign finance laws. The *Journal* was reporting that Cohen had paid off Daniels in October 2016, the same month Trump's candidacy went on life support due to the *Access Hollywood* tape. If true, the Republican nominee's campaign had bought the silence of someone whose disclosures could have altered the outcome of the presidential election.

The White House denied the report. So did Cohen, who provided a signed statement from Clifford that read, "Rumors that I have received hush money from Donald Trump are completely false." Yet, the next month, in a statement to the *New York Times*, Cohen admitted that he *had* paid the porn star $130,000—and insisted that it came out of his own pocket.[2] "Neither the Trump Organization nor the Trump campaign was a party to the transaction with Ms. Clifford," Cohen said, "and neither reimbursed me for the payment, either directly or indirectly."

This was a lie—one of many that would come back to torment both Trump and his associates.

Then, in March, both Stormy Daniels and Karen McDougal, the former *Playboy* model whose own story had been bought and buried by Trump's friends at the *National Enquirer*, sued to invalidate the nondisclosure agreements they had signed. (Cohen had played a role in negotiating both.) Topping it all off, at month's end, Daniels appeared on *60 Minutes*. She gave a highly credible account of the unprotected sexual intercourse she had with Trump in 2006 and told of how, five years later, after sharing her story with the gossip magazine *In Touch*, a man approached her with a physical threat: "Leave Trump alone."

Trump didn't take the bait—at least not right away. "So much Fake News. Never been more voluminous or more inaccurate," the president tweeted the morning after *60 Minutes* aired. "But through it all, our country is doing great!"

TRUMP WAS NOT DOING GREAT.

The First Lady was seething at the humiliations suffered by the Stormy Daniels discoveries, he confided to friends. Far more bothersome, the president's children were being pulled into Mueller's probe, and his legal team seemed more concerned with each passing day about the scope of the special counsel's investigation. For all aggravations Russia-related, Trump held one person responsible: Jeff Sessions.

Once a darling to the president's team—Sessions was the first senator to endorse his 2016 campaign—the attorney general had become Trump's enemy number one since recusing himself from the Russia inquiry. (According to the *New York Times*, Trump asked White House counsel Don McGahn to lobby Sessions against the recusal and "erupted in anger" when Sessions would not comply, "saying he needed his attorney general to protect him."[3]) When Mueller was installed as the special counsel, a direct result of Trump's firing of Comey, the president made Sessions his whipping boy: mocking his southern drawl, tweeting insults at him, and telling reporters that he never should have chosen the former Alabama senator to lead the Justice Department. On at least two occasions Trump requested his attorney general's resignation only to be convinced by White House aides that Sessions's dismissal would only compound his myriad legal problems.

Sessions wasn't the only cabinet member who had taken up residence on Trump's bad side.

Rex Tillerson had come to annoy the president in ways big and small. Trump found the secretary of state to be dreary and slothful; not far, he laughed with friends, from the "low energy" caricature he'd slapped on Jeb Bush. Because of Tillerson's deliberate speaking style, Trump joked about the secretary of state being slow, and was therefore bemused at reports of Tillerson calling him not just "a moron" but a "fucking moron."

"I think it's fake news, but if he did that, I guess we'll have to compare IQ tests," Trump told *Forbes* magazine in response to the alleged quote. "And I can tell you who is going to win." (Tillerson never denied the reports.)

Increasingly isolated from the White House and removed from decision-making processes—to the extent that such processes existed—Tillerson found himself hopelessly out of sync with Trump. In the fall of 2017, the secretary of state told reporters that the United States was looking to negotiate with North Korea; one day later, the president tweeted, "I told Rex Tillerson, our wonderful Secretary of State, that he is wasting his time trying to negotiate with Little Rocket Man . . . Save your energy Rex, we'll do what has to be done!"

Six months after that episode, Tillerson condemned Russia for its poisoning of an ex-Russian spy and his daughter with a military-grade nerve agent, calling the attack in England "a really egregious act" that "clearly" had been ordered by the Kremlin. The next morning, Trump fired Tillerson via Twitter.

More than three dozen senior administration officials had either resigned or been fired in the president's first fourteen months on the job. Tillerson was the fifth to exit in a span of five weeks.

The other recent departures included H. R. McMaster, the Army lieutenant general who was axed as Trump's second national security adviser after repeated clashes with the president; Gary Cohn, the director of the National Economic Council, who quit after a dispute with Trump over his tariffs on steel and aluminum; Rob Porter, the staff secretary, who resigned amid public allegations of abuse from both ex-wives; and Hope Hicks, Porter's girlfriend and a longtime aide to the president, who quit one day after an eight-hour testimony before the House Intelligence Committee in which she admitted to telling white lies on Trump's behalf but pleaded ignorance of any Russian connections.

On the morning of Monday, April 9, the president's circle shrank even smaller. FBI agents raided the offices of Michael Cohen, acting on a referral from none other than Mueller himself.

Trump had witnessed enough legal battles to recognize that this would end in one of two ways: Cohen would either take a legal bullet for him or turn state's witness and leverage damaging information to reduce his own sentence. Enraged by the lawful looting of his attorney's privileged materials, the president decided to weigh in on an active federal investigation.

"So, I just heard that they broke into the office of one of my personal attorneys," Trump told reporters soon after reports of the raid surfaced. "It's a disgraceful situation," the president continued. "It's a total witch hunt."

After suggesting that he might fire Mueller, an atomic recourse that his lawyers, staffers, and allies uniformly warned against, Trump added of the Cohen situation, "It's an attack on our country, in a true sense. It's an attack on what we all stand for."

During an interview on Fox News a few weeks later, the president's new personal lawyer, former New York City mayor Rudy Giuliani, told Sean Hannity that Trump had reimbursed Cohen the $130,000 used to pay off the porn star. It was a dizzying admission, contradicting Trump's past statements of knowing nothing about the hush money transaction and jolting Hannity, who appeared bewildered at having unwittingly made news on his program.

"We finally got our side of the story," Giuliani told the *Wall Street Journal* after Hannity's show aired. White Houses aides were apoplectic. Trump wasn't angry; he was just mystified. *Why the hell was Rudy always on television?*

THE PRESIDENT'S PREOCCUPATION WITH QUESTIONS OF LOYALTY, attempting to distinguish those people truly supportive of him from those cozying up for the sake of expedience, increasingly informed his dealings with Capitol Hill. As Republican lawmakers sidled up in search of favor or influence or professional gain, Trump eyed them warily, wondering who among them would remain steadfast if things went south.

Improbably, one person about whom the president had no such doubts was Paul Ryan.

Once upon a time, there had been few tougher Trump censors in the GOP than Ryan, who felt duty-bound to combat the Republican front-runner's dark rhetoric and the party's nativist drift. Yet there had hardly been anyone softer on Trump since Election Day 2016 than the Speaker, who, intent on delivering the policy promises made to voters, calculated that doing so meant ignoring the ad hominem savaging of private citizens, the payoffs to porn stars, the assaults on private businesses, the undermining of institutions, and the innumerable other acts for which Barack Obama would have been impaled by the right.

Theirs was a fragile marriage, no doubt, one born out of mutual practicality: Ryan needed a president to make his legislative dreams a reality, and Trump needed a Speaker to deliver wins as quickly as possible. In time, however, the alliance proved stronger than anyone in either camp could have anticipated. Ryan carefully avoided criticizing the president while offering frequent, elementary tutoring sessions on policy and process behind closed doors, grumbling about the task only to a handful of close friends; Trump reciprocated the Speaker's restraint and spared him the sort of public shaming doled out to other top Republicans, including McConnell.

"I told myself, I gotta have a relationship with this guy to help him get his mind right," Ryan recalls. "Because, I'm telling you, he didn't know *anything* about government. So I thought, I can't be his scold, like I was. . . . I wanted to scold him all the time. What I learned as I went on, to scratch that itch, I had to do it in private. So, I did it in private—all the time. And he actually ended up kind of appreciating it. We had more arguments with each other than pleasant conversations, over the last two years. And it never leaked."

Encircled by loose-lipped self-promoters almost every waking moment, the president came to appreciate Ryan's discretion. Knowing that their private discussions would remain private freed the two men to speak candidly in a way that Trump found refreshing. He also recognized that unlike many of the other Republicans kissing his ring, the Speaker had nowhere to climb; he was the second-most powerful man in Washington and hadn't wanted that job to begin with. Even when the Speaker didn't share his priorities, the president found himself more trusting of Ryan's motives than those of most of the ambition-drunk politicos in DC. In spite of himself, Trump had come to like, and rely heavily on, a person whom he had once accused of trying to sabotage his campaign.

All this made it painful when Ryan called Trump, in the early hours of April 11, to inform the president that he would retire at year's end.

Five months earlier, when *Politico* reported that Ryan was telling confidants of his decision to leave Congress, the president had reacted angrily, calling the Speaker to solicit assurances that no such departure would be made, that he would serve all four years of Trump's first term. At the time, Ryan told the president what he wanted to hear. But privately, his mind was made up. He was ready to go home. When he explained the decision to Trump five months later, the president said he understood.

Ryan's departure left a power vacuum in the congressional wing of the party. Kevin McCarthy, the heir apparent as majority leader, had failed once before to earn a promotion to Speaker; Steve Scalise, whom Trump had nicknamed "the Legend from Louisiana," was now a household name with designs on the top job himself. Of course, Ryan's retirement only lent to the perception that the midterm elections were shaping up to be ugly for Republicans—so ugly that there might not be a GOP Speaker in 2019.

Whether it was McCarthy or Scalise who wound up replacing Ryan atop the House GOP, they would follow in the Speaker's footsteps, subverting their own political identities to appease Trump. It was his party now, without question or caveat.

THE PRE-TRUMP GOP HAD BEEN SPLINTERED ALONG ANY NUMBER OF asymmetrical boundaries: libertarians and neocons, evangelicals and cultural moderates, big-spending pragmatists and small-government purists. All these dichotomies existed within the broader construct of conservative versus moderate, or even outsider versus establishment, spurring incessant talk of a "Republican civil war" from 2008 to 2016.

But the civil war was over now—or, at least, the battle lines had shifted dramatically. Trump's conquest effectively ended the squabbling that had defined the GOP in the post-Bush era, replacing disputes over policies and principles with a simpler question that spoke to the dueling identities in the party. "Are you with Trump or not?" said Corry Bliss, the executive director of the Congressional Leadership Fund, the super PAC charged with protecting the House GOP's majority in 2018. "It's not about ideology anymore. It's only about Trump. Are you with him or are you against him? That's the only thing that matters to voters in the Republican base."

This dynamic played right into the president's obsession with loyalty; even most of the Republicans in the latter camp had no choice but to block any daylight between themselves and the president, fearful as they were of alienating his cultlike following among their own constituents. If there was space for an anti-Trump Republican to flourish in federal races, nobody running in the party's 2018 primaries had found it. Seeing this, and watching the political gymnastics of onetime critics now claiming true allegiance to Trumpism, the president relished his role of kingmaker.

In the Florida governor's race, Adam Putnam, a former congressman and the state's agriculture commissioner, was leading the Republican primary by 7 points in his internal polling. That was until Trump endorsed his opponent, Congressman Ron DeSantis, a graduate of Yale and Harvard Law School who had worked as a JAG Corps prosecutor before deploying to Iraq during the troop surge as a legal attaché for a team of Navy SEALs. (Trump, a sucker for a good résumé, backed DeSantis after a brief courtship.) The next poll conducted by Putnam's campaign showed him down 11 points. "An eighteen-point swing in the space of a few weeks," says Terry Nelson, a veteran GOP consultant working for Putnam. "I've never seen anything like it. *Ever.*"

Not everyone was lucky enough to land Trump's support. Diane Black, the Tennessee congresswoman, was running in a crowded GOP primary to become the state's governor. During a meeting with several House Republicans in the Cabinet Room early in 2018, she pulled the president aside. "You really need to endorse me," she told him, stabbing a finger at his chest. Trump found her rude and presumptuous. "She got in my personal space," he told aides afterward. "Big mistake." The White House Office of Political Affairs threw a bone, having Mike Pence endorse her. But Black kept at it, badgering the White House political director, Bill Stepien, for a presidential vote of confidence. Stepien asked an intern to aggregate a full record of everything Black had ever said about Trump, good and bad. The list was printed out and carried over to the Oval Office. Trump scanned the document, picking out the negative remarks, then pulled out a Sharpie. "Diane," he wrote. "This is NOT good!" He furiously underlined the word "NOT," then asked Stepien to hand-deliver the document to Black.

It was a similar story in the Idaho governor's race. Raúl Labrador, the congressman and Freedom Caucus cofounder, touted his alliance with Trump during the Republican primary, but the president's official endorsement had yet to surface. One of Labrador's opponents, Lieutenant Governor Brad Little, employed consultants who heard that the congressman's friends (namely, Mick Mulvaney) were putting the squeeze on Trump to endorse Labrador. Certain that such a development would tip the race against them, Little's team cut a highlight reel that showed Labrador criticizing the president and sent it to the Political Affairs shop, hoping it would reach Stepien. Instead, the video made it all the way to the president, who upon seeing it resolved once and for all not to intervene on the congressman's behalf. After Little won the GOP primary, he received a phone call from Trump. "I can't believe all these people wanted me to endorse Labrador. Why would I do that?" the president said. "He said a lot of nasty things about me. He's a really nasty guy."

In one case, Trump endorsed as a means of punishment. Having heard that Minnesota congressman Erik Paulsen was distancing himself from the White House in the hope of holding his seat in the Twin Cities' suburbs, the president stewed and asked that the political shop send a tweet of support for Paulsen—thereby sabotaging the moderate Republican's efforts. When his aides demurred, Trump sent the tweet himself, issuing a "Strong Endorsement!" of the congressman in a late-night post that left Paulsen fuming and his Democratic opponent giddy.

No single result gave Trump as much satisfaction as that of the Republican primary in South Carolina's First District. Having made promises of retribution against the GOP incumbent, Mark Sanford, the president itched to announce his support for Katie Arrington, who was running as a Trump-inspired populist and highlighting Sanford's critiques of the president. White House aides were vehemently opposed to the idea: Sanford was going to win, they told him, and when he did, Trump would look weak and ineffectual.

The president stood down, but he kept tabs on the race. Sanford had taken Arrington for granted. The congressman's numbers in the district had dipped, and the contest tightened as he hoarded campaign cash instead of unloading on his challenger. On the day of the primary, flying back from Singapore aboard Air Force One, Trump decided to roll the dice. "Mark Sanford has been very unhelpful to me in my campaign to MAGA. He is MIA and nothing but trouble. He is better off in Argentina," the president tweeted, referencing Sanford's transnational affair. "I fully endorse Katie Arrington for Congress in SC, a state I love. She is tough on crime and will continue our fight to lower taxes. VOTE Katie!"

Sanford's friends in the Freedom Caucus were livid. "He's one of the most principled, consistent, and conservative members of Congress I've ever known," Congressman Justin Amash tweeted in response to Trump. "And unlike you, Mark has shown humility in his role and a desire to be a better man than he was the day before."

When the ballots were tallied that night, Arrington finished with 50.5 percent of the vote, just clear of the 50 percent needed to avoid a runoff. She had edged Sanford by the narrowest of margins, and Trump's last-minute tweet, which her team immediately turned into a robocall and blasted around the district, was very likely responsible.

Just as in the case of Arizona senator Jeff Flake, whose denunciations of the president sent him to an early retirement, Trump had foretold the demise of one of his harshest intraparty critics. Sanford and Flake, two longtime conservative stalwarts with deep philosophical moorings and voting records well to the right of most Republicans in Congress, were exiled from the GOP for the high crime of dissenting from its new leader.

THE REPUBLICAN PARTY WAS NOT ALONE IN ITS REVOLUTIONARY CONvulsions.

Joe Crowley woke up on June 27 harboring aspirations of becoming the next Speaker of the House. As the fourth-ranking House Democrat, the congressman from Queens had spent years building alliances across his caucus and collecting favors to cash in. Whereas the number two and number three Democrats, Steny Hoyer and Jim Clyburn, were septuagenarians who offered no generational change, Crowley was a sprightly fifty-six years old. He was also a skilled straddler of the party's ideological divides, trusted and well liked by both moderates and liberals. With younger and newer members demanding a leadership change atop the party, Crowley was a virtual lock to succeed Pelosi one day as the top House Democrat—and a decent bet to become Speaker of the House with a Democratic takeover in November.

By night's end, however, Crowley was a household name for a very different reason. In an upset that shook Washington and foreshadowed the trajectory of its minority party, Crowley lost his primary in New York's Fourteenth District to a member of the Democratic Socialists of America. Her name was Alexandria Ocasio-Cortez. She was twenty-eight years old, a bartender, and a first-time candidate who had volunteered for Bernie Sanders's presidential campaign.

Leaning into the calls on the left for a dramatic makeover of the Democratic Party, Ocasio-Cortez used her youth, Latina heritage, and insurgent message to gain a cult following among progressives in the final months of her challenge to Crowley. Even then, and with a swelling number of liberal organizations and activist leaders supporting her, few took the rookie's candidacy seriously. A biographical video that she published on social media drew nearly a million views online, yet the *New York Times* ignored the contest in its own backyard.

One person who did take Ocasio-Cortez seriously: Donald Trump.

Watching television in the White House earlier that summer with some of his political advisers, the president says he caught a glimpse of the Democratic insurgent on a cable news program. "I see a woman, a young woman, ranting and raving like a lunatic on a street corner, and I said, 'That's interesting, go back.' It's the wonder of TiVo, right? One of the great inventions of all time. And I say, 'Go back, I want to see that again. Who was that?'"

Referring to his political advisers, Trump adds, "They say, 'It doesn't matter.' You know, I'm watching with some pretty good professionals. Semi-good. None of them are too great."

Watching the young woman—and learning that she was running against "a slob named Joe Crowley, who I've known for a long time, because I'm from Queens"—Trump became enamored. After soaking in her performance, Trump was starstruck. "I called her Eva Perón," he recalls. "I said, 'That's Eva Perón. That's Evita.'" (He places a comically exotic emphasis on the nickname: *Ah-vih-tah*.)

Trump says he told his team to call Crowley "and tell him he's got himself a problem; he better get off his fat ass and start campaigning."

The president says they laughed him off, promising him that Ocasio-Cortez had no chance. Later, when she won, he took the opportunity to remind everyone that they had similarly underestimated him.

"I'm very good at this stuff, believe it or not, even though I've only done it for a few years," Trump says. "And I'm good at talent. I spotted talent. She's got a certain talent."

Crowley's loss came four years to the month after Eric Cantor dropped his primary stunner to Tea Party activist Dave Brat. The symmetry of their expirations did not escape either of them. A few weeks after his defeat, Crowley paid Cantor a visit at his New York office to talk about life after Congress. (The former majority leader landed on his feet, making seven figures as an executive at a global investment firm.) The two men sat, once future Speakers of the House, exchanging their most unique condolences and comparing notes on the strange new realities of politics.

IF THE FIRST YEAR OF DONALD TRUMP'S TERM WITNESSED A PRESIDENT adapting to the philosophies of his party, the second year saw a party bending to the will, and the whims, of its president.

In early March, Trump had issued a sweeping set of tariffs on imported steel (25 percent) and aluminum (10 percent). Over the objections of a vocal minority of Republicans, the president said he was delivering on his promise to rejuvenate the American economy by overturning decades of free-market orthodoxy that had governed administrations of both parties. "Our factories were left to rot and to rust all over the place, thriving communities turned into ghost towns," Trump announced at the White House. "That betrayal is now over."

Though he initially exempted some of America's closest allies— Canada, Mexico, and the European Union—Trump soon extended the tariffs to affect those nations as well. All of them issued reprisal tariffs, effectively neutralizing whatever net economic gain the president had hoped for. Meanwhile, the administration slapped a tariff of 25 percent on more than eight hundred categories of Chinese exports. This sparked a separate and more damaging trade war that escalated throughout 2018. China retaliated by hammering U.S. agriculture exports, forcing Trump eventually to issue federal assistance to suffering American farmers.

This was Republicanism circa 2018: government bailouts to alleviate the burden of state-sanctioned market intervention.

Despite this obvious affront to the doctrine of conservatism, few Republicans on Capitol Hill were itching for a fight with Trump. Some convinced themselves, or at least said publicly, that the president was playing the long game and needed lots of latitude to negotiate. Others grumbled in private about the calamities that could ensue but dared not cross Trump publicly.

The display of Pharisaism was staggering. It wasn't simply that Trump was desecrating the GOP's free-market principles; he was brazenly flexing his executive authority to do so. After eight years of mocking Obama's "imperial presidency" and decrying his subjugation of the legislative branch, Republicans in Congress refused even to hold an up-or-down vote on their president's unilateral remaking of American trade policy.

There were some exceptions. On the House side, Warren Davidson, the Ohio conservative, grew so agitated during a meeting with Trump's two chief trade advisers, Peter Navarro and Larry Kudlow, that he flipped over a chair and stormed out of the meeting, cussing over his shoulder. While a passionate advocate of restructuring the nation's trade agreements, Davidson, a former manufacturing executive, told anyone who would listen that Trump's tactics were counterproductive and doing disproportionate harm to his own base.

"These are like the Trumpiest Trumpians, and they're telling me, 'We're getting killed here,'" Davidson says of his constituents. "I've got one county that's all [agriculture], 80 percent of them voted for Trump. . . . The administration is curious about how the aid for farmers is going over. So, I'm talking to this one guy back home, and he tells me,

'You know, I'm glad they're handing out Band-Aids, but I'd rather they just didn't shoot me.'"

In the Senate, Tennessee's Bob Corker distinguished himself as the loudest detractor of Trump's approach to trade. "We should vote on tariffs. But they're afraid. They don't want to poke the bear. I get it. But this is where we should call a floor vote and back the White House into a corner." When Corker and other senators pressed McConnell on this during a luncheon in the summer, urging him to at least call Trump and plead the party's case on the detrimental nature of his tariffs, McConnell scoffed. "You can call him," the Republican leader replied. "You want his phone number?"

When it came to intellectual consistency, an even greater departure for Trump-era Republicans was on spending and fiscal restraint.

Conservatives had renounced George W. Bush for his big-government policies. They had bloodied Obama for his bankrupting of America and tortured John Boehner for failing to stop it. Yet, in the spring of 2018, with the national debt having recently passed $21 trillion, Trump and his unified Republican government approved an omnibus bill that shattered Congress's budget caps and represented one of the largest spending increases in American history.

There were no widespread demonstrations, no marches on the Capitol, no Tea Party rallies. Less than a decade removed from the street protests that lit the party's populist fuse, scores of Republicans from that hard-charging 2010 class voted for more spending, more debt, and more government—without fear of consequence.

To their credit, some of the only lawmakers who lobbied Trump against the bill were House conservatives. In fact, the opposition campaign waged by the likes of Jim Jordan and Mark Meadows was so effective that Trump came to fear that the bill represented a betrayal of his base, what with its massive spending increases, barely any of which was for border security and none of which was going to the construction of the wall he had been promising. They urged him to veto the legislation.

On Thursday, March 22, the House of Representatives voted to approve the 2,232-page package less than twenty-four hours after GOP leaders unveiled it. The reason it passed? Republicans loaded the bill with record amounts of military spending to buy off defense hawks, and Democrats piled on generous increases to domestic discretionary programs to placate progressives. Freedom Caucus members, hoping to turn the president against the bill, complained to him that it did not provide funding to build his border wall and that it failed to defund Planned Parenthood, risking blowback from social conservatives. Buying these arguments, the president decided he would veto the spending bill—even though it would mean a government shutdown within forty-eight hours.

Catching wind of this, John Kelly, the chief of staff, called Ryan and urged him to get to the White House, pronto. The Speaker arrived to find Trump in the East Wing residence. Fully briefed on the Freedom Caucus's efforts, Ryan and Kelly arranged for the White House to block incoming phone calls from Jordan and Meadows. Then, the Speaker sat down with the president and launched into an urgent defense of the spending bill: It provided the major funding increase for military personnel and operations that he had long sought, and in order to secure that victory, Republicans had to give Democrats concessions in order for the bill to pass the Senate.

Trump wasn't satisfied. Namely, he wanted to know, where was his money for the border wall? Ryan told the president that they were getting a down payment, roughly $1.5 billion, and that they would get more later. A veto of the legislation, he warned, would only set them back.

Trump still wasn't convinced. He told Ryan he would probably veto the legislation. The Speaker exploded in anger, and a yelling match ensued. When each man had uncorked on the other—cathartic, surely, for them both—Trump told Ryan he would sign the bill on one condition: that Ryan give him room to build the suspense on Friday morning before announcing his blessing later in the day.

Sure enough, the next morning, Trump tweeted, "I am considering a VETO" and complained that his "BORDER WALL" was not being fully funded. By afternoon he was signing the spending bill at the White House, even as he called it "crazy," insisted that "nobody read it," and promised, "I will never sign another bill like this again."

Passage of the giant spending package signaled a defeat for the forces of small-government austerity that had been ascendant in the years predating Trump. It also completed an evolution within Ryan's own career. Once the party's most celebrated fiscal conservative, the Speaker had found religion on defense spending after his experience on the national ticket in 2012. He returned to Congress determined to help rebuild the military, even if it meant further ballooning the debt and the deficit. Ryan could find ways to rationalize his 2003 vote for the Medicare prescription drug benefit, or for the TARP bailout in 2008, but with his championing of the 2018 omnibus package the Speaker had willfully forfeited his reputation as a fiscal hawk.

This, on top of watching him turn a blind eye to Trump's ignominies, was too much for Ryan's friends and allies around Washington to bear. The Speaker's career was unfolding like a play in three acts. The first, from his election to Congress in 1998 until his vice-presidential run in 2012, starred the pushy, unpledged ideologue. The second, from that 2012 campaign until Trump's victory in November 2016, featured a more seasoned, mature legislator who sought compromise where necessary and felt obligated to enhance the party's image.

The third act, from Trump's election until Ryan's retirement from Congress, would not offer the happy ending he had once envisioned. His legacy would be defined by the fulfilment of a Faustian bargain in which he sold his soul to Trump in exchange for policy wins. The tragedy was, in the eyes of Ryan's friends, that those wins, from tax reform to the omnibus bill, weren't remotely worth the damage to his reputation. "He made a calculation that to get through the policies he cares about meant that he had to muzzle himself at certain times—many times—when it came to things that Trump said and did," says Pete Wehner, Ryan's longtime friend and former colleague at Jack Kemp's think tank. "I think it was an anguished time for him."

Of course, Pence had cut this very same deal with the devil—and was all the more insufferable in his observance thereof. The vice president, once among the most intellectually sovereign voices in all of Washington, had so pitifully subjugated himself to Trump that some of his longtime friends were left to wonder (only half-jokingly) whether the president had blackmail on him.

Pence's talent for bootlicking—he was nicknamed "the Bobblehead" by Republicans on Capitol Hill for his solemn nodding routine whenever Trump spoke—were at their most obscene during meetings at the White House. After Trump would open the floor to Pence, aides would suppress grins as the vice president offered his opening tribute to the president, exhausting his storehouse of superlatives and leaving the other attendees to wonder whether they, too, were expected to kneel.

The vice presidency is a supporting role. Being a team player is part of the job description. And Pence, a fervently religious man, draws from his faith, and from the military tradition in his family, a belief in "submission" and "servant leadership."[4] Yet there is a difference between submission and spinelessness; between deference and dereliction; between servitude and slavery. Nobody expected Pence to make a show of publicly rebelling against the president. What they did expect was a token of intellectual and ideological consistency rather than unabashed allegiance to all things Trump. Yet this was too much to ask.

In May 2018, the vice president visited Arizona for an event promoting the GOP's new tax law. In the audience he spotted Joe Arpaio, the recently pardoned convict and former sheriff of Maricopa County, who was now mounting a MAGA-inspired run for U.S. Senate. "A great friend of this president, a tireless champion of strong borders and the rule of law," Pence declared. "Sheriff Joe Arpaio, I'm honored to have you here."

To be clear: For the seat being vacated by his former best friend Jeff Flake, whose criticisms of the president made him unelectable after a career of conservativism by any objective metric, Pence was implicitly endorsing a man who had boasted of detaining Mexicans accused of no crime; run brutal prison camps that were allegedly responsible for men's deaths and women's miscarriages; and arrested journalists in the middle of the night for writing negative words about him. This, while calling him "a tireless champion" of "the rule of law."

It was perfectly in keeping with Pence's character in the Trump Show, and it was becoming too much for the vice president's onetime admirers to bear.

Of the growing critiques of Pence, the most blistering belonged to George Will, the preeminent conservative pundit. Writing in the *Wash-*

ington Post, Will recalled how the vice president "flew to Indiana so they could walk out of an Indianapolis Colts football game, thereby demonstrating that football players kneeling during the national anthem are intolerable to someone of Pence's refined sense of right and wrong." He asked, "what was the practicality in Pence's disregard of the facts about Arpaio? His pandering had no purpose beyond serving Pence's vocation, which is to ingratiate himself with his audience of the moment." And he said Pence's conduct "clarifies this year's elections: Vote Republican to ratify groveling as governing."

Will concluded, "Trump is what he is, a floundering, inarticulate jumble of gnawing insecurities and not-at-all compensating vanities, which is pathetic. Pence is what he has chosen to be, which is horrifying."

THE EVOLUTIONS OF PENCE AND RYAN DID NOT OCCUR IN A VACUUM. AS each man mutated to fit the age of Trump, so, too, did conservatism.

Jim DeMint's ouster from the Heritage Foundation had been the first domino to fall, triggering a sequence of reformation and realignment within the conservative movement.

The board of directors at Heritage, irate over the sullying of their once-venerable institution's brand, appointed Kay Coles James as the new president. If they wanted a sharp stylistic break from DeMint, then James, an alumna of the Bush 43 administration, was the perfect choice. Where DeMint was reactionary and doctrinaire, James was deliberate and studied. Moreover, whereas DeMint said Obama "took race back to the sixties" and blamed the Democrats for not putting "racism behind us," James, a black woman, said after taking over Heritage, "I don't think the Republican Party has ever had an honest conversation about race. And before we move forward, we need to have that conversation."

The problem with her appointment, to many on the right, was that it felt like an overcorrection: that Heritage was so consumed with rehabilitating its image and restoring its scholarly reputation that it would relinquish its mission to hold GOP elected officials accountable for their votes.

James was sending conflicting signals. One day she would assure hard-core conservative allies of her mandate to be nonpolitical, calling out the Trump administration for its forsaking of principle. The next day she would be currying favor with the administration, hoping to preserve influence like everyone else in DC. When James met with the new White House director of legislative affairs, Shahira Knight, the Heritage president kicked off the conversation by assuring Knight that Heritage was "working hard to win the president a second term."

Nobody was peddling the narrative of Heritage's unreliability harder than DeMint himself. Having been banished from the think tank, its former president assembled his core team of right-wing agitators to launch a new organization, the Conservative Partnership Institute. They went around town whispering to Heritage donors that the group had lost its nerve; that it was going soft to make nice with the establishment; that James was not a fighter for the movement; that she's a nice lady known for midday naps more than nighttime raids. This campaign was effective: Even before its official launch, DeMint's new venture was stealing major financiers away from Heritage, which led to an internal panic about James's capacity for going toe to toe with her predecessor.

The changes inside Heritage were encapsulated by the changes inside its lobbying arm, Heritage Action. For the previous eight years, the president of Heritage Action, Mike Needham, had made himself the most hated man on Capitol Hill. Constantly picking fights that Republicans couldn't win, and then fund-raising off their defeats, Needham became persona non grata even to some of the most conservative lawmakers in Congress. They viewed him, and his organization, as a parasite leeching off the anger toward the political class that Heritage was actively fueling. Not long after DeMint's departure as the think tank's president, Needham was relieved of his duties at Heritage Action. He was replaced by its COO, Tim Chapman, a well-liked veteran of the conservative movement who had a reputation for collegiality and coalition building.

But one of Chapman's first initiatives, spending money to help protect vulnerable moderate Republicans in their 2018 elections (on the theory that conservatives could make gains only if the GOP continued to hold its majorities), drew wailing and gnashing of teeth from the right. Veteran activists denounced Heritage in meetings. Big movement donors called, threatening to cancel their checks. Mark Meadows, the Freedom

Caucus chairman, told several Heritage officials in a meeting that summer that he wouldn't stick his neck out for them "given this new reputation of yours, and given that I've got my own reputation to worry about."

Needham, meanwhile, was finding religion. Official Washington was stunned when he joined Marco Rubio as the senator's chief of staff. No entity had brutalized him during the 2013 Gang of Eight fight like Heritage Action, often with attacks that were deeply personal. Now Rubio was hiring the gunner who had manned the heavy weaponry against him.

This looked, on the surface, to be a marriage of mutual necessity: Rubio needed to rebuild his street cred with conservatives, while Needham needed to repair his relationships with the Capitol Hill establishment. But there was something deeper at work. In the final months of DeMint's tenure, Needham had begun questioning the direction of Heritage and whether their absolutist approach of years past had backfired—and whether it was responsible for Trumpism. Joining one of the weekly conservative meetings for the first time since joining Rubio's staff, Needham's longtime comrades asked what the biggest surprise was in his new role. He replied that he had learned the importance, politically and economically, of sugar subsidies in Florida. He was laughed out of the room. Email in-boxes around Washington exploded with tales of how Needham had sold out. One of his longtime friends, worried about what he'd heard, called and teased Rubio's new chief of staff about going soft. "You know," Needham told him. "I'm not sure that the guy you think I am, the guy Washington thinks I am, really exists anymore."

This upheaval within the conservative movement occurred during a midterm election season that was unlike any in memory. Democratic challengers were out-raising Republican incumbents in record numbers, most of them smartly steering clear of Trump-related hysteria and focusing their campaigns on kitchen table issues: health care, jobs, economic inequality. Republicans, on the other side, were running unrecognizable campaigns, having largely ditched the ultraconservative messaging techniques of elections past and branding themselves as Trump loyalists playing to the issues that animated his base.

It was no accidental shift. The biggest donors in the Republican

Party made it known in 2018 that they would not write checks if their money would be wasted on ads hawking an academic sort of conservatism: lower spending, rising debt and deficit, uncontrolled entitlement programs, etc. Instead, they wanted an emphasis on cultural issues: immigration, the national anthem, whatever worked. "These weren't the activists. These were Wall Street types, the people who have spent years pleading with Republicans to avoid social issues and focus on the economy," Chapman, the Heritage Action president, says. "And now, with Trump, they want us to do the exact opposite."

There was a method to the madness. Over at the Club for Growth, where tens of millions of dollars had been spent over the past decade promoting a purist fiscal conservatism, their market research showed there was no appetite in the electorate for lectures on economics. "In this cycle, if you aren't talking about immigration and Trump, you aren't going to pick up that conservative base vote," says David McIntosh, the Club's president. "I've had donors and board members say, 'Why don't we just keep running the ads that worked before, about spending?' And I tell them because when we study it and poll it, it doesn't work."

Chapman concurs. "All the polling we get back shows the fiscal issues are a complete wasteland," he says. "And the donors know it."

"The Tea Party is gone. It doesn't exist anymore. There just aren't that many Republicans now who are that concerned about spending, about debt, about big government," says Justin Amash, the Michigan congressman elected in the 2010 wave. "Many people today think Trump is fiscally conservative because they see his tweets, they listen to him talk about trade deficits, and people fall for it. A lot of people are thanking Trump for getting our debt and deficit under control because they have bad information."

Ahem. Bad information?

"I think President Trump is one of a kind—you can't replicate what he's doing," Amash says. "It requires you to not feel shame. Most people feel shame when they do or say something wrong, especially when it's so public. The president feels comfortable saying two things that are completely contradictory in one sentence; or going to a rally and saying one thing and then holding a press conference and saying another. Most people aren't comfortable doing that. But because he is, it gives him this superpower that other people don't have."

Indeed, as the president's improvised trade war punished a disproportionate number of his own supporters across Middle America that summer, he gave a speech in Kansas City aimed at convincing those voters that they were not, in fact, being hurt by his policies. "It's all working out," Trump said. "Just remember, what you're seeing and what you're reading is not what's happening."

His words of reassurance could have been ripped straight from the pages of George Orwell's *1984*: "The party told you to reject the evidence of your eyes and ears. It was their final, most essential command."

THE FANATICAL DEVOTION TO TRUMP, OFTEN FOR FEAR OF REPRISALS from his cult following on the right, opened the GOP to attacks that were sometimes misleadingly simplistic.

It became highly *en vogue*, particularly in the cesspool of social media, for liberals to mock Republican criticisms of Trump as empty rebukes that weren't backed up by concrete actions to check the executive branch. ("Stop talking and DO something!" . . . "You have a vote!" . . . "Press releases are not oversight!" etc.)

The most fashionable of these arguments went something like this: Republican X, who spoke out in opposition to Trump over Y, was not sincere because he voted with the president on Z. Popular targets for such attacks were those GOP lawmakers who showed the gumption of habitually offering rebukes of the administration: Ben Sasse, Bob Corker, and of course, Jeff Flake. "When it comes right down to the nitty gritty—to casting votes—Flake usually toes the presidential line," columnist EJ Montini wrote in the *Arizona Republic*.[5] "According to a statistical analysis by the website FiveThirtyEight, Flake casts votes for the Trump position 83.3 percent of the time. The outrage over the president is, for the most part, all talk."

Such critiques were, for the most part, reductive and disingenuous.

When Trump entered office, he effectively contracted out all the policymaking decisions, the things that would require *votes*, to two people: Ryan and McConnell. They put forth a legislative agenda that,

while arguably flawed on the policy front and hypocritical on the process front, was broadly consistent with a contemporary Republican
platform: repealing Obamacare, cutting taxes, rebuilding the military,
slashing regulations, reforming the Veterans Affairs department, and
above all, confirming conservative judges to the federal courts.

To support these items was to vote not for Trump's position, but for
the party's orthodoxy. Expecting lawmakers to vote against their own
policy interests to make a statement of disapproval about Trump was
asking them to cut off their nose to spite their face.

"What we have is a president who was willing to sign what *we* wanted
done," Corker says. "Now, the tax bill to me could have been better, I had
trouble with it, and you know, it's a bet on America, and we took that bet.
But these things are what Republicans are: *We* believe in feeding the an-
imal spirits of business. *We* believe in conservative judges. *We* believe
in tax reform. And what we had was a president who was willing to sign
those things into law. That was our agenda—it wasn't his agenda."

But what of the president's agenda? This is where the notion of
craven acquiescence gains legitimacy. Whether it was his multiple attempts to implement a travel ban that he admitted on multiple occasions was targeted toward Muslims, or his signing of a morbidly obese
spending bill, or his launching multiple trade wars that hurt the American worker, Trump's abandonment of conservatism ("classical liberalism," as it was once celebrated) was met with little resistance from
the right. And many of those who *did* voice opposition were careful to
couch it in support for the president himself, fearful of provoking the
tweeter in chief.

Perhaps the most egregious example of Republican silence in the face
of Trumpism came in the late spring of 2018, when the administration
decided, on the advice of policy adviser Stephen Miller and his former
boss, Jeff Sessions, to enforce a "zero tolerance" policy at the southern
border. Meant to deter families from crossing into the United States illegally, the program resulted in nearly two thousand migrant children
being separated from their parents in one six-week stretch alone. The
images of crying toddlers and abandoned youths being detained in chain

link fence detention centers as their parents awaited sentencing were ghastly; worse was the bureaucratic ineptitude that caused months-long delays before some kids were reunited with their parents.

As had become customary, certain elements of the media played into the president's hand; at one point, a photograph of children sleeping in cages went viral online, annotated by journalists with sharp words for the White House, only for it to become clear that the photo had been taken when Obama was president. Such carelessness allowed Trump to falsely equate his enforcement with that of previous administrations and blame the opposition party for his manufactured crisis. "I hate the children being taken away," the president said from the White House. "The Democrats have to change their law—that's their law."

It was *not* their law. Previous presidents had used discretion to avoid splitting up families while adjudicating their cases; whereas Obama's administration had detained kids who came on their own, Trump's administration was actively separating children from their parents.

Many Republicans, including some of the fiercest immigration hawks in Congress, were nauseated by the scenes unfolding on the southern border. But most of them dared not criticize Trump. He had weaponized the issue of immigration too effectively in the past; with the midterm elections fast approaching and the conservative base showing signs of complacency, the last thing vulnerable Republicans wanted was to be called "soft" or "weak" by the president. Only when the pressure on him grew crushing—from party leaders, faith-based groups, and his own political advisers—did Trump relent, signing an executive order to end the zero-tolerance experiment.

The most lasting critiques of the president, and of his enablers, will extend far beyond policy. From the moment Trump took office, Republicans on Capitol Hill and throughout the administration would offer a common refrain: "Focus on what he *does*, not on what he *says*." For all Trump's bizarre behavior and inflammatory rhetoric, they explained, he was delivering on many policies for which the party had long hungered.

But this argument conveniently obscured a self-evident reality about the role of the presidency. Trump, as the American chief executive, is both the head of government *and* the head of state. His behavior and his rhetoric, therefore, were every bit as relevant as his policies. In certain instances, what the president *said* was actually more meaningful than what he *did*.

Take, for example, his relationship with Russia.

CHAPTER TWENTY-THREE

JULY 2018

---

*"If that means going on Fox News and lying
through their teeth about Trump, so be it."*

DONALD TRUMP AND VLADIMIR PUTIN STOOD SIDE BY SIDE IN HEL-
sinki, Finland, facing the world in a spectacle of unprecedented intrigue and unrivaled indignity.

Since his election, the American president had privately and publicly
expressed doubts about Russia's interference in the 2016 campaign. He
grumbled that the entire story was an attempt to delegitimize his presidency, perhaps not appreciating the serendipity of suffering in such a
manner after building his political brand on the foundation of delegiti-
mizing his predecessor. Nine months prior to the Helsinki summit, after meeting with Putin on the sidelines of an economic forum in Asia,
Trump told reporters, "He said he absolutely did not meddle in our
election."

This parroting of Putin's denial came after a unanimous assessment
from the U.S. intelligence community, in a report compiled by the FBI,
CIA, NSA, and Dan Coats, Trump's own hand-picked director of national intelligence, that concluded with "high confidence" that Russia
meddled in the campaign with the purpose of electing Trump. The Senate Intelligence Committee, chaired by a Republican, had reached the
same conclusion. (The House Intelligence Committee, consumed by
partisan grandstanding, could not reach consensus on whether the sky
was blue.) And just three days before the Helsinki summit, the Justice
Department indicted twelve Russian nationals as part of Special Counsel Robert Mueller's investigation. They stood accused of working at the
Kremlin's direction to hack Democratic emails and computer networks.

Yet when Trump spoke from the stage in Helsinki, he refused to identify anything for which the Russian government should be punished. He chose instead to focus on Mueller's "ridiculous" investigation, which he called a "witch hunt" that was preventing better relations between the two nations. When an American reporter, Jon Lemire, pressed him on whom he believed, Putin or his own intelligence officials, the U.S. president gave a response that will live in infamy.

"They said they think it's Russia. I have President Putin. He just said it's not Russia," Trump said. "I will say this: I don't see any reason why it would be."

With that utterance, Trump, taking the word of the KGB thug turned Russian strongman over that of his own intelligence community, had emasculated America on the international stage. He had also lent credence to the theory that Moscow had *kompromat* on the U.S. president. After all, the president had just spent the previous week disparaging NATO allies—condemning Germany, belittling the British prime minister while visiting her country, and referring to the European Union as a "foe." Why the accommodating treatment of Putin and his brutal, democracy-crushing, dissenter-slaying government?

The episode was jarring for many Republicans.

Some of the president's most steadfast defenders slammed the performance. Newt Gingrich called it "the most serious mistake of his presidency," and Fox News's Brit Hume said it was "a lame response, to say the least." Paul Ryan and Mitch McConnell offered unequivocal support for the intel community's findings without criticizing Trump by name. Dozens of other congressional Republicans offered harsher-than-usual rebukes. John McCain called it "one of the most disgraceful performances by an American president in memory," adding, "No prior president has ever abased himself more abjectly before a tyrant." And Will Hurd, the Texas congressman who had spent nearly a decade overseas working undercover for the CIA, went a step further, tweeting, "I've seen Russian intelligence manipulate many people over my professional career and I never would have thought that the US President would become one of the ones getting played by old KGB hands."

Sitting in his office a few days later, Corker, the Senate Foreign Relations Committee chairman, said it was necessary to place Trump's

Helsinki performance in the sweep of his upending of conservative Republican orthodoxy.

"Let me just go through the four things I believe," Corker said. "I believe that America is a force for good in the world, that the post–World War II institutions have been mostly very beneficial to the United States and our citizens; this president does not believe that. I believe that free trade has been an outstanding thing for the American people and for our country and for our GDP; this president is a protectionist. I believe the fiscal issues matter. He's not even close to being a fiscal conservative. And lastly, I think the domestic institutions that are fundamental to our democracy are important. We are conservatives, we are traditionalists, we are people that hold those things up, even though every institution needs oversight and can be improved. We believe that these institutions have helped make America great. Not him. He's willing to significantly undermine them if it benefits him politically."

There was one class of Republicans that approved of Trump's buddy-buddy routine with Putin: the Freedom Caucus.

In a Heritage-sponsored forum with some of the group's members, one day after the Helsinki summit, the conservatives spent an hour taking turns slamming Obama for his weak approach to Putin; Hillary Clinton for her failed "reset" of relations with Russia; reporters for daring to question Trump's belief in the U.S. intelligence community; and operatives of the "deep state" for attempting to undermine the president. ("The choice target was former CIA director John Brennan, who tweeted that Trump's showing in Helsinki "exceeds the threshold of 'high crimes & misdemeanors'" and was "nothing short of treasonous.")

"In order for something to be treasonous, it has to undermine who we are as a nation," Mark Meadows said of Brennan's charge. "I've never seen a press conference have that effect."

"Foreign policy-wise," Jim Jordan said, "the trip to China last fall was good, the Korean summit was positive, the [North Korean] hostages have come home, there's sanctions on Russia, the embassy is in Jerusalem, and we're out of the Iran deal. So, overall, people are pretty darn pleased."

"What was I disappointed in? I thought it was really odd that a reporter in Helsinki, Finland, after a conclusion of a brief summit, would ask President Trump the question that triggered this whole odd reaction that the summit was a failure because President Trump did not castigate and attack Vladimir Putin," said Andy Biggs of Arizona, blaming the "idiocy" of the media's questions.

"If I were the president," added Andy Harris of Maryland, "I wouldn't hold those press conferences anymore until the press decided to get serious about dealing with the world issues, as this president is. . . . It would have gotten [Trump] nowhere to get in Putin's face with election-meddling. There is no evidence of any collusion, but this is the main story of the liberal press."

On they went, up and down the dais, uttering not a critical syllable of the president less than twenty-four hours after he publicly sided with Putin over the U.S. intelligence community. It was surreal. Having covered many of these lawmakers for years—long before the Freedom Caucus existed—I knew for certain that had Obama said the same thing Trump had, they would have been preparing articles of impeachment.

Finally, the spinning and evading became too much. I raised my hand and asked, giving them a final opportunity, if any of them had any problem whatsoever with what Trump had said.

Davidson was the only one to speak up. "I think anybody that watched the press conference, including the president himself, would say that was not his finest hour," the Ohio Republican said, measuring his words. "But we support the fact that the president was there on the stage having the press conference and having the dialogue. . . . We should judge more about the deeds and less about the words."

It was the day the Freedom Caucus forfeited its credibility.

For much of the previous decade, House conservatives had been the most interesting members of Congress to cover. In an age of mindless tribalism, they were the independent thinkers, rejecting the party's hierarchy and challenging a system that rewarded blind loyalty and reflexive partisanship. But since Trump came along, they had become the most reflexively partisan Republicans on Capitol Hill, routinely brushing off actions from the executive branch that under Obama would have prompted talk of constitutional crises.

Walking out of the event in a daze, I ran into Matt Fuller, the *Huff-ington Post* reporter who had chronicled the rise of the Freedom Caucus closer than anyone. "That was the low point in my career covering Congress," Fuller said. I nodded in agreement. Moments later, a staffer for one of the Freedom Caucus members approached, shaking his head. "What a joke," he grumbled.

Notably, some of the core Freedom Caucus members had not been in attendance. Raúl Labrador, one of the group's cofounders, was absent, as was Mark Sanford, one of its leading voices. As it turned out, several members had deliberately skipped the event, not wanting to subject themselves to the humiliation incurred by their colleagues.

Justin Amash was one such person. Things had come full circle: Once considered annoying and eccentric for his principle-driven votes against the GOP leadership, the Michigan libertarian was one of the only lawmakers in the Republican Party remaining true to those principles—and one of the few willing to diagnose what was happening to conservatism in the era of Trump.

"THEY BELIEVE IN A COSMIC BATTLE BETWEEN THE RIGHT AND THE left, good and evil, and they think any criticism of Trump is helping the other side," Amash said. "So, they're willing to do whatever they need to. If that means going on Fox News and lying through their teeth about Trump, so be it."

The Michigan congressman was sitting in his office, the lights dimmed, C-SPAN flickering on a muted television. A few days had passed since the Helsinki summit, and Amash was still grappling with the subsequent defense mounted by his colleagues. The Freedom Caucus had shown signs of internal strain in Trump's first two years at the wheel, but this felt like a breaking point.

Amash had watched for the past eighteen months as his fellow House conservatives used their seats on key committees (Judiciary, Oversight) to wage a partisan war on Trump's behalf, neglecting the nonpartisan duties of checking and balancing assigned to the legislative branch and assuming a protective posture on behalf of the executive. Amash understood the anxieties about Democratic overreach and unchecked bureaucrats in the "deep state" going rogue out of political opposition to the president. He also shared the belief, held by many of Trump's defenders on Capitol Hill, that surveillance powers had been abused in proximity to the government's handling of the Trump-Russia case. But it was painful to watch his friends, his Freedom Caucus comrades, sacrifice their integrity in the service of shielding the White House from scrutiny it plainly deserved.

"Have you watched these committee hearings? They're all theater. Then they go on Fox News and continue their performance. And then they go home and say privately, 'Trump's such an idiot,' but the Fox News hit is all that matters," Amash said. "We've all fallen into tribes, and when they praise the president, they get instant gratification from their tribe."

He continues: "I think they're hurting themselves and they're hurting the country when they do this stuff. It's fine to say good things about Trump when you agree with him. I think Gorsuch could prove to be one of the best Supreme Court justices we've ever had. I agree with Trump on a number of regulatory issues. I agree with him when he's cut taxes—just not when he raised taxes by imposing tariffs. . . . But a lot of them have just fallen in line. And it's upsetting. It affects personal relationships. They are so obsessed with defending Trump, and the Russia stuff—I mean, they complain about the left being obsessed with Russia, but they're even worse. And it gets in the way of discussions on anything else. It makes it hard to relate. I can't understand it."

In a way, it was easy to understand. Politicians act out of self-preservation. For congressional Republicans—most of whom face no general election threat in their districts and all of whom fear Trump's fervent following in the party's base—the surest way to keep their power and enhance their influence was to stand by the Dear Leader.

Far easier than remaining intellectually consistent, applying critical thinking to the president's words and deeds regardless of party affiliation, was to enlist as one of his surrogates. The trappings of Trump's propaganda ministry were substantial: regular Fox News appearances, rides on Air Force One, invitations to the White House, phone calls with the leader of the free world. Many a GOP lawmaker fell prey to these perks. But none more odiously than Matt Gaetz.

Elected to Congress in 2016, Gaetz quickly distinguished himself as the Trumpiest lawmaker on Capitol Hill. He tried to hit a populist, anti-politician note out of the gate, announcing repeatedly at a press conference in 2017, "I don't speak Washington." (Gaetz's father, the former president of the Florida Senate, was instrumental in procuring the congressional seat for his son.) The rookie Republican quickly realized that his path to prominence wound through the good graces of Trump, and he set about becoming the president's most pugilistic supporter in Congress: railing against the "deep state" on Fox News, calling for Mueller's firing, even likening the special counsel's investigation to a "coup d'état." Before long, Gaetz was riding on Air Force One to Florida with the president and giving introductory remarks at an event.

In January 2018, Gaetz brought as his guest to the State of the Union address an alt-right troll and Holocaust denier, Charles Johnson, who, among his other claims to fame, helped raise crowdsourced money for the neo-Nazi website the Daily Stormer. This resulted in a brief hiatus from Fox News, but Gaetz was back before long, more frequently and more artificially bronzed than ever before, alternating between calling Mueller's probe a "witch hunt" and questioning the lack of investigations into the corruption of the Obama administration.

By the summer of 2018, Gaetz was on the president's speed-dialing list, talking with him regularly by phone and receiving constant feedback after his Fox News hits. This was not enough. The Florida congressman grew upset during one meeting with staff from the White House's Office for Legislative Affairs, dressing them down for not recognizing his "special relationship" with Trump. Gaetz argued that he should be getting more one-on-one time with the president. Not long after, he was aboard Air Force One for Trump's latest trip to Florida.

Gaetz had discovered a new path to power and influence for a freshman member of Congress. It was good for him but terrible for the institution of Congress—and for the Republican Party. "Matt Gaetz is not a legislator," Ryan says, shaking his head. "He's an entertainer."

Not everyone was so flamboyant as the Florida lawmaker. But then again, they didn't need to be. To remain relevant in Trump's GOP was to stick within his orbit. And to do so required little more than unyielding allegiance to the president. This meant never daring to oppose his policies, much less criticize him personally, all while defending him as a matter of instinct.

In lieu of any serious, substantive checking of the administration by its coequal branch of government on Capitol Hill, a class of professional Trump critics emerged on the right. Some, such as attorney David French at *National Review* and longtime talk radio host Charlie Sykes at the *Weekly Standard,* were thoughtful and measured. But most of the professionals were virtue-signaling reactionaries whose hysteria was surpassed only by their social media followings. Whether done by Ana Navarro on CNN (a "Republican strategist" who had strategized on behalf of no campaign that anyone could recall) or Jennifer Rubin at the *Washington Post* (a once-interesting blogger whose censures had become predictable to the point of self-parody), slamming the president's every syllable became a cottage industry with generous remuneration for those involved.

Indeed, throughout the fratricidal post-Bush era, few things got more clicks or better ratings than Republican-on-Republican violence. This trend exploded in the age of Trump. Newspapers competed to run columns by conservative detractors of the administration; cable news programs hustled to book guests whose broadsides against the president from the right would validate their own from the left. The scent of such intraparty treachery was so alluring that late in the summer of 2018, the *New York Times* ran an anonymous op-ed, which claimed to be authored by a "senior official in the Trump administration," that detailed how the president's own aides were "trying to do what's right even when Donald Trump won't," and said there had been secret discussions of involving the Twenty-Fifth Amendment to remove him from office.

The unrelenting torrent of condemnation—from the media, from celebrities, from the left, even from members of his own party—made Trump value those all the more who were dependable and subservient, those he could count on to advance his interests and defend him at all costs. Nobody had learned this better than Mark Meadows.

After the beating he took from Trump during the first, failed health care push in early 2017, the Freedom Caucus chairman groveled his way back onto Trump's good side. He stayed there by acting as the president's spy on Capitol Hill, reporting back the latest gossip and spinning everything he and his friends were attempting to do as benefiting the White House (as opposed to betraying the MAGA agenda, as Ryan and his leadership team were doing). The permanent perch Meadows earned atop Trump's shoulder was annoying even to the congressman's allies in the West Wing. Staff would regularly see Meadows walking the hallways uninvited and unannounced; White House phone logs from one month in the summer showed Meadows calling Trump at least twice as frequently as any other lawmaker.

This represented the apex of Meadows's ascent—from obscure freshman, to Defund Obamacare leader, to Boehner slayer, to Freedom Caucus chairman, to Trump whisperer—in just five years. The advantages were abundant. The North Carolina congressman, an avowed enemy of "the swamp," bought himself a lovely condo inside the Beltway and began living full-time in the DC suburbs. Rare was the exclusive party not attended by the congressman and his wife, sudden starlets of the capital's cocktail circuit. Meadows had made it.

Interestingly, despite all his earned goodwill, Meadows would not spend it standing up for one of his own members.

In late June, after Mark Sanford's loss in his South Carolina primary, Trump looked out over a meeting of the House Republican Conference and asked if Sanford was present. When members replied that he wasn't, Trump began taunting the congressman, calling him "a nasty guy" and saying sarcastically, "I wanted to congratulate him on running a great race!" Groans filled the room. Sanford had become a popular figure, especially among conservatives, for his policy knowledge and his plainspoken approach. Nobody appreciated Trump's routine.

The next day, however, Trump tweeted: "Had a great meeting with the House GOP last night at the Capitol. They applauded and laughed loudly when I mentioned my experience with Mark Sanford. I have never been a fan of his!"

Of course, nobody had laughed or applauded. The president was lying about an event to which there were more than two hundred witnesses.

Several of Sanford's colleagues in the Freedom Caucus came to his defense. Amash rebuked Trump in a tweet, calling out his "dazzling display of pettiness and insecurity." Labrador said it was "just wrong" what Trump had done to Sanford. But there was no such condemnation from Meadows. Despite Trump's continued insults of his colleague—including another shot at him while the Freedom Caucus was meeting one night—the group's chairman would offer no rebuke of the president, saying only that Trump was acting on "bad political advice."

Amash could no longer stomach the group's collective cowardice. Soon, he stopped attending the Freedom Caucus meetings and distanced himself from the organization he had cofounded.

"These guys have all convinced themselves that to be successful and keep their jobs, they need to stand by Trump," Amash said. "But Trump won't stand with them as soon as he doesn't need them. He's not loyal. They're very loyal to Trump, but the second he thinks it's to his advantage to throw someone under the bus, he'll be happy to do it."

Amash added, "It could be Mark Sanford today and Mark Meadows tomorrow."

THE SUMMER OF 2018 WASN'T EXACTLY A DAY AT THE BEACH FOR PRESident Trump. The family-separation crisis and the Helsinki disaster already promised to be legacy-defining blunders, and a surge of energy on the left was building what political pundits called a "Blue Wave" that appeared increasingly likely to wipe out the House GOP's majority in the fall elections.

There was also continued turmoil in his administration. In July, the embattled chairman of the Environmental Protection Agency, Scott Pruitt, resigned. This marked the seventh departure of a cabinet official in eighteen months; for watchdogs in Washington, it was the longest overdue. Pruitt had insulted taxpayers in ways that would make Tom Price blush, spending tens of thousands of dollars on a twenty-four-hour security detail; renting a DC town house from an industry lobbyist's wife for pennies on the dollar; taking private and first-class flights without approval; and building a soundproof phone booth in his office that cost $43,000, among other abuses.[1] It was fair to consider the swamp not yet fully drained.

Meanwhile, Trump was growing more preoccupied with the Mueller probe with each passing day, grousing to anyone who would listen about the alleged "deep state" and flying into profanity-laced rages about the orchestrated sabotage of his presidency. In the months of June, July, and August alone, Trump sent hundreds of tweets and retweets regarding the special counsel's inquiry, more than three dozen of them mentioning Mueller by name.

In one tweet, the president called the former FBI director, a decorated Marine Corps veteran who led missions in Vietnam before and after being shot in the leg, "Disgraced and discredited." He compared him to Joseph McCarthy. He described him as "totally conflicted" because of the registered Democrats working under him on the investigation. (Mueller, a Republican, had served presidents of both parties.)

For all the talk of a "witch hunt," Mueller proved incredibly skilled at finding hats and brooms. By the middle of July, according to a *Washington Post* tally,[2] the special counsel's team had collected "187 criminal charges in active indictments or to which individuals have pleaded guilty," while "another twenty-three counts against President Trump's former deputy campaign manager Rick Gates were vacated when he agreed to cooperate with Mueller." Additionally, thirty-two people and three businesses had been named in indictments or plea agreements, and Mueller had extracted "six guilty pleas from five defendants." Among the charges: "52 counts of conspiracy of some kind . . . 113 criminal counts of aggravated identity theft or identity fraud . . . Four guilty pleas for making false statements."

The biggest threat to Trump, it was becoming clear, was Michael Cohen. At first, the president's lawyer seemed unlikely to flip. Trump described him as a "good man" in the aftermath of the raid on his office. The two men talked by phone soon after. And Cohen said he would "rather jump out of a building than turn on Donald Trump." Yet, as the summer wore on and Trump playfully evaded questions about a pardon, the building jump was looking more and more appealing.

In mid-June, Cohen fired his existing legal team and brought on a new lawyer known for his deal-cutting prowess. A week later, Cohen resigned as the deputy finance chairman of the Republican National Committee, taking the opportunity to criticize Trump's family-separation policy at the southern border. Any remaining doubts about his allegiance were erased in early July, when he told ABC News that his first loyalty was to the country—not the president.[3]

On Tuesday, August 21, Cohen stood in a Manhattan courtroom and pleaded guilty to eight federal crimes: five counts of tax evasion, one count of making false statements to a financial institution, and two counts of campaign finance violations. On the latter two charges, Cohen testified that Trump—"Individual 1, who at that point had become the President of the United States," in court parlance—had directed him to make payments to Stormy Daniels and Karen McDougal during the 2016 campaign to prevent them from disclosing past sexual relationships.

The president's lawyer was implicating him in a major federal crime, one that had nothing to do with the Russia investigation he obsessed over. But the day was just getting started.

Minutes after Cohen fired his legal projectile, Manafort was found guilty on eight counts of tax fraud and bank fraud. The Virginia jury was unable to reach verdicts on ten other counts, resulting in mistrials, but it hardly mattered: Manafort was facing up to 240 years in prison, the severest conviction of a sitting president's former aide since Watergate. Having gone big-game hunting, Mueller was beginning to mount some serious antlers on the walls of Washington.

Rounding out a day unlike any other in recent political memory, Duncan Hunter, the California GOP congressman, was indicted on sixty counts of using campaign funds for personal purposes. Hunter had long been renowned as one of Capitol Hill's shadiest characters; stories of his hard partying and sexual exploits with staffers was the stuff of legend. He was also the second member of Congress to endorse Trump for president. As it so happened, the first, New York congressman Chris Collins, had been arrested by the FBI two weeks earlier and charged with insider trading.

(Soon after, Trump rebuked Sessions and the Justice Department for bringing charges against the Republicans ahead of the November elections. "Two easy wins now in doubt because there is not enough time. Good job Jeff," he tweeted. The law-and-order party's leader was asking the attorney general to play goalie for his political allies.)

The dazzling convergence of criminality surrounding Trump didn't seem widely bothersome to Republicans on Capitol Hill. Perfunctory statements of being "troubled" by the developments notwithstanding, few members of the president's party offered anything in the way of outward alarm at the events of August 21. Some, including John Cornyn of Texas, the second-ranking Senate Republican, even took the opportunity to point out that neither Cohen's pleas nor Manafort's convictions did anything to prove "collusion" with Russia.

Predictably, the president's base was even less cowed. Arriving in West Virginia that fateful Tuesday for an evening rally with the faithful, the president found himself surrounded by what could only be described as Fifth Avenue Republicans—the type who, as the president had once said, would stick by him even if he shot someone. The day's historic events went unappreciated by many in the crowd who, upon Trump's mention of Hillary Clinton, chanted, without an ounce of irony, "Lock her up! Lock her up! Lock her up!"

Their devotion was not without explanation. Despite all the struggles and setbacks of recent months, the president had delivered on more promises. He had withdrawn from the Iran deal. He had officially relocated the U.S. embassy to Jerusalem. He had brought North Korea to the negotiating table.

And most important, he had nominated another conservative to the Supreme Court. The retirement of Anthony Kennedy, the court's longtime swing vote on so many major decisions, had handed the new president a second appointment in as many years. Consulting once more with his conservative allies in Congress and his advisers at the Federalist Society, Trump had nominated an experienced judge with strong legal credentials and unquestioned conservative bona fides: Brett Kavanaugh.

CHAPTER TWENTY-FOUR

SEPTEMBER 2018

"We are better than this. America is better than this."

IT WAS A SEND-OFF BEFITTING A TITAN OF THE REPUBLIC: THE FLAG-draped coffin, the bagpipes, the angelic chorus, the stained-glass windows, and the gothic pillared arches encasing a sanctuary of some three thousand luminaries bidding a final farewell.

John McCain, the senator and statesman and prisoner of war who had spent five and half years in the Hanoi Hilton after refusing early release, had succumbed to cancer. He was eighty-one.

The Saturday-morning service, on September 1 at the National Cathedral, paid a grand homage to McCain. But it also felt like a memorial for Washington itself, a capital city that under President Trump no longer seemed capable, as the famed "maverick" was, of balancing fights with friendships, of divorcing disagreement from disrespect, of recognizing the basic difference between opponents and enemies.

With organ notes echoing throughout the cavernous complex before the ceremony, they mingled and shook hands and scanned the room for More Important People as they might at any black-tie affair. Former presidents and vice presidents elicited camera clicks. Senators compared notes with ambassadors. Military officials and government wise men and media personalities craned their necks. Jared and Ivanka held court with perfect strangers. The commotion outside—police escorts, a procession of black Cadillacs, hundreds of congressmen and senators being bused in, all with onlookers lining the surrounding sidewalks—made it a quintessentially DC occasion, a marriage of exclusivity and self-importance. The only thing missing from this meeting of official Washington was the chief executive of official Washington.

The president's absence testified to his rivalry with McCain; they had blistered one another relentlessly, in public and in private, ever since Trump infamously mocked the senator for having been captured while flying a combat mission in Vietnam. More fundamentally, though, Trump's absence reflected his tormented relationship with a town that purports to revere the virtues he was accused of lacking: courage, prudence, service, conviction, wisdom, humility, forgiveness, honor, and above all, a patriotism that transcends tribalism.

Trump could not be held solely responsible for the fractured nature of modern American politics. McCain's idyllic Washington, one defined by ferocious battles waged with mutual goodwill, had long been on life support. For much of Bill Clinton's presidency, and accelerating through the administrations of George W. Bush and Barack Obama, the electorate and its representatives were hardened by a combination of class warfare, zero-sum legislating, and cultural polarization that invited Trump's ascent. Having pulled the plug—and smothered the better angels of our nature with a pillow for good measure—the president found himself at once disinvited from a singular Washington gathering and yet dominating its consciousness.

The elephant in the room was the president not in the room.

Though his name was never mentioned, the eulogists invoked Trump with all the subtlety of a sledgehammer. It was only eleven minutes into the service when Meghan McCain launched the opening salvo with an emotional tribute to her father. "We gather here to mourn the passing of American greatness—the real thing, not cheap rhetoric from men who will never come near the sacrifice he gave so willingly, nor the opportunistic appropriation of those who lived lives of comfort and privilege while he suffered and served."

Ten minutes later, choking back tears, she added, "The America of John McCain has no need to be made great again, because America was always great."

The applause was at first tepid, and then thunderous; she was the only one of the five speakers to be so interrupted. Each of the subsequent eulogists lauded McCain in a manner that, even if unintentionally, contributed to what became a ceaseless rebuke of his party's current leader.

This was all very much by design. McCain had planned his memorial down to the last detail, making clear that Trump was not to be invited.

When the president learned of this from Kushner, his son-in-law, who had been tipped off by McCain's son-in-law, the conservative writer Ben Domenech, Trump projected nonchalance. Yet he privately seethed at the affront and remained so bothered by it that he refused to lower the White House flag to half-staff when the senator died. Only after spirited lobbying from the likes of Mike Pence and John Kelly did the president relent, ordering the flag lowered at what appeared to be the lone place in Washington where it wasn't already.

McCain, meanwhile, had arranged for his former rivals Bush and Obama to deliver remarks. There was no shortage of symbolism in the two-time presidential candidate's desire to be eulogized by the two men who had denied him the White House. When McCain spoke by phone with Obama a few months before his passing, and the forty-fourth president said that he would be honored to speak at his funeral, McCain described it to friends as one of the happiest moments of his life. It would be his parting message to America, he said, that patriotism has nothing to do with political affiliation.

Towering over the crowd from the cathedral's raised pulpit, Obama recalled the famous moment from 2008 in which McCain scolded one of his supporters for suggesting that the Democratic nominee wasn't an American. "I was grateful, but I wasn't surprised," Obama said. "He saw himself as defending America's character, not just mine."

Left loudly unsaid: Trump lying for years about Obama's birthplace. Leaving nothing to interpretation, Obama added, "So much of our politics, our public life, our public discourse, can seem small and mean and petty, trafficking in bombast and insult and phony controversies and manufactured outrage. It's a politics that pretends to be brave and tough but in fact is born of fear. John called on us to be bigger than that. He called on us to be better than that."

The other man who bested McCain for the presidency, George W. Bush, could afford to be less direct. It was just a few months earlier that he (along with brother Jeb) had insisted that Trump not attend the funeral of their mother, the former First Lady Barbara Bush. Still, he, too, got his point across. "John was, above all, a man with a code," Bush said, one who "lived by a set of public virtues," "detested the abuse of power," and "could not abide bigots and swaggering despots." Alluding to one of his own conflicts with McCain, over the use of torture as an interrogation technique, Bush noted, "At various points throughout his long career, John confronted policies and practices that he believed were unworthy of his country. To the face of those in authority, John McCain would insist: We are better than this. America is better than this."

Not that any of these critiques bothered Trump, who spent the morning of McCain's funeral tweeting about deep-state sedition and Canadian trade exploitation before heading to his Northern Virginia golf club. There were, after all, disparate realities to consider: one inside the holy halls of the National Cathedral where powerful people mourned the death of decency, and another in the surrounding city where many of those same powerful people drove nails ever deeper into its coffin on a daily basis. Indeed, the contrast between the McCain Washington remembered in death (valiant, virtuous) and the McCain Washington loathed in life (warmongering, irascible) was something to behold.

And there was a greater juxtaposition still: this one between the virtue-signaling, convention-worshipping insiders of the capital and the mad-as-hell, burn-it-down voters in the provinces.

McCain's funeral showed that Washington wasn't Trump's town. But it was still his country.

IN THE MIDDLE OF SEPTEMBER, WHITE HOUSE POLITICAL DIRECTOR Bill Stepien sat down in the presidential residence across from Trump and delivered a wake-up call.

Polling showed that Democratic voters were highly motivated ahead of the midterm elections, Stepien explained, while Republican voters were not—and Trump was feeding the complacency of his base by downplaying the threat in November. "Mr. President," Stepien told him, "please stop saying 'Red Wave.'"

Trump was perplexed. Having fully bought into the narrative of Republican invincibility, supported by boisterous crowds, a string of special election victories, and of course, his own experience defying the political prognosticators, the president thoroughly enjoyed turning the Democrats' "Blue Wave" mantra on its head. He struggled to imagine any scenario in which the nation delivered a rebuke to his government. Sensing this, and playing to his ego, Stepien and senior administration officials encouraged Trump to mobilize Republicans by making the election all about him. "Tell them that *you're* on the ballot," Stepien urged the president.

There was another pressing imperative, something White House aides were pounding into Trump's head as he prepared to travel the country campaigning on behalf of Senate candidates. "You cannot— absolutely cannot—attack Christine Blasey Ford," Kellyanne Conway warned him.

Just around that time, what had once seemed a pro forma confirmation process for Brett Kavanaugh to replace Anthony Kennedy on the Supreme Court—tipping its balance rightward for years, perhaps decades, to come—was being derailed by bombshell accusations that Kavanaugh had attempted to rape Ford when the two were teenagers. The president's allies knew restraint would not come easy: He had been accused during the 2016 campaign of sexual misconduct by at least fifteen women, and when Ford's accusations surfaced, his first response in private was to liken Kavanaugh's plight to his own.

To the shock of just about everyone at 1600 Pennsylvania Avenue, Trump stayed on message. At his first public appearance following the twin talks from Stepien and Conway, he addressed the Kavanaugh situation with the most delicate touch imaginable. "Brett Kavanaugh, and I'm not saying anything about anybody else, Brett Kavanaugh is one of the finest human beings you will ever have the privilege of knowing or meeting," the president said at a rally in Las Vegas. "So, we'll let it play out, and I think everything is going to be just fine. This is a high-quality person," he added. (Speaking to Sean Hannity before the rally, Trump did question why Ford's story hadn't been reported to the FBI "thirty-six years ago.")

Similarly, the president showed discipline in deleting "Red Wave" from his midterm lexicon, agreeing to make the election a referendum on himself. "Get out in 2018," he told a Missouri crowd a few days later, "because you are voting for *me* in 2018."

This strategy would prove helpful to protecting and expanding the

Senate majority. But it did nothing to slow the Democrats' stampede toward control of the House. After two years of roller-coaster news cycles driven by a president who thrived on tumult and governed with a showman's attention to shiny objects, Democrats were poised to regain the House majority by following a simple set of rules: Tailor the message to fit the district, talk about policy, and above all, don't take Trump's bait.

Whereas Trump sought to paint the opposition party as deviant radicals bent on the republic's destruction, many of the most effective Democratic challengers were running as centrists, emphasizing their affection for guns and objection to the growing debt. And whereas Trump sought to make the election about himself, Democratic candidates were methodical in focusing the electorate's energy on the alleged failures of his party: Republican tax reform that had exploded the deficit and disproportionately benefited the wealthy; Republican efforts to take away health care access from millions of people; and Republican politicians whose acquiescence to Trump had deepened the country's partisan divide and further diminished its faith in government.

With the GOP expecting a full-frontal progressive assault on the president, leading Democrats—from Minority Leader Nancy Pelosi, to Democratic Congressional Campaign Committee chairman Ben Ray Luján, to the party's biggest donors and elder statesmen—advised House candidates to run hyperlocal, nonhysterical campaigns that avoided Trump as much as possible while emphasizing independence from the national party. In dozens of cases, this meant pledging not to support Pelosi as Speaker.

It was working. All around the country, in supposedly safely red districts where Republicans had gone unchallenged for years, Democratic recruits had put the incumbents back on their heels.

The money and enthusiasm on the left had also scared dozens of other GOP incumbents into retirement, weakening the party's defenses. Of the forty-four districts vacated by Republicans who retired, resigned, or sought higher office, Democrats aggressively targeted half of them.

The most notable Republican to call it quits was Ryan, who nonetheless insisted on serving through the year's end to help protect the majority. The decision to stay as a lame-duck Speaker irked some in the party and uncorked a gusher of internal gossip. Kevin McCarthy felt exposed by the decision, believing that his best chance to succeed Ryan in the next Congress was to have a running start; sensing the same thing, allies of Steve Scalise whispered about McCarthy's vulnerabilities and suggested a stealth campaign to leapfrog him.

The tension among all three leadership officials, and their staffs, filled the water cooler talks on Capitol Hill that summer, especially as McCarthy and Scalise each jockeyed to find ground on the other man's right flank. (McCarthy aired radio ads in numerous congressional districts promoting his legislation to build a border wall, vexing local Republicans in tough races who were being outspent and didn't get so much as a shout-out from the majority leader while he was talking to their constituents.)

Ultimately, the concerns about Ryan sticking around were unfounded—he raised a record $200 million for the party in his time as Speaker—though his departure fed the narrative of Republicans surrendering in 2018. "There are a few folks that I tried to [convince] to stay in that didn't stay in," Steve Stivers, the Ohio congressman and chairman of the National Republican Congressional Committee, told *Politico* in late August. This qualified as the understatement of the cycle.

Stivers was no one's idea of a political powerhouse. In fact, he had become an expert at removing himself from GOP Christmas card lists. The previous fall, after Steve Bannon had left the White House and begun boasting of creating a shadow party to take down the establishment, the NRCC chairman traveled to Bannon's Capitol Hill town house—which doubles as Breitbart.com headquarters—and threw himself at Bannon's mercy, pleading with him not to target already vulnerable House moderates.

It was a bizarre maneuver, strategically and otherwise. The fact was, for all Bannon's talk, he had zero apparatus for actually recruiting and funding challengers to Republican incumbents. There was no money, no organization—just Bannon in all his rumpled, self-aggrandizing glory. Yet here was the NRCC chairman kissing the ring, and extracting promises from Bannon that his cabal would target only McConnell and his Senate members, not House incumbents.

McConnell nearly had a coronary when he heard of the meeting. Calling Ryan, who had no previous knowledge of Stivers's plans (and was himself irritated), the majority leader told the Speaker that Stivers was about to be persona non grata to the whole of the Republican Party.

Stivers got the message, but his performance as the campaign committee's leader was widely viewed as ineffectual bordering on incompetent. A record number of House Republicans had retired, and though much of that was due to expiring committee chairmanships and general Trump fatigue, Stivers was seen as part of the problem, rather than part of the solution.

It was time for the GOP to triage, cutting off doomed incumbents and steering its resources to those who still had a chance. Stivers wasn't helping things. It wasn't just that he was forced to make tough decisions; it was that some of his decisions were plainly idiotic. For instance, Barbara Comstock, a popular Virginia Republican representing the DC suburbs, had been trailing by double digits in every poll of the race throughout the entire summer. Yet the NRCC was continuing to pump money into her race, eventually spending $5 million on a seat that nobody believed could be held. (Comstock wound up losing by 12 points.)

Meanwhile, Kevin Yoder, a Republican representing the Kansas City suburbs, was running the best campaign of his career—and fighting hopelessly uphill in a district that Clinton had won in 2016. Brimming with frustration one Sunday in September, Yoder placed a phone call to Stivers. Word had just gotten out that the NRCC would be cutting $1.2 million in TV spending from his district, essentially conceding defeat. Yoder had learned of the development from press reports, not from the committee.

"When people ask me what I think of you, I can't decide whether to tell them you're a fucking idiot or a fucking liar," Yoder growled at Stivers. "But now I think you're both."

"TWO WORDS ARE GOING TO DEFINE THE NIGHT OF THE 2018 ELECTION in the next three weeks. One is 'Kavanaugh' and the other is 'caravan,'" Newt Gingrich told Sean Hannity. "I think the American people are going to reject both the way they treated Kavanaugh and the way they are dealing with the border, and I think those will end up being the reasons the Republicans keep the House and dramatically increase the number of senators they have."

It was the evening of October 17, twenty days before the midterm election, and as usual, the president of the United States was tuned in to Fox News. He loved what he was hearing.

Trump had all but pulled a hamstring taking victory laps since Kavanaugh was confirmed to the Supreme Court at the beginning of October. Moreover, just one day earlier, the president had tweeted about the "Caravan of people heading to the U.S." from Honduras. Now the conservative propaganda monster was following his lead on both.

The caravan issue was easily exploitable. Trump had threatened to cut off foreign aid to countries that did not thwart the advance of the estimated several thousand people moving through Guatemala toward the U.S. border. Caravans in Central America were nothing new; large groups of migrants have long banded together, traveling both north and south to escape violence and poverty. Earlier that year, in fact, a large caravan had been broken up by the Mexican government at its northern border. But this was different: With signs of complacency in the GOP base and immigration still its animating concern, Trump saw the latest mass migrant group as a prime political foil.

He had been thrilled when, the day before, Gingrich and Laura Ingraham spent part of her Fox News show discussing the issue. ("The largest caravan in a decade approaches our southern border," Ingraham warned of the people on foot roughly one thousand miles away.) Now the president was giddy at hearing Gingrich—the only man in politics, he felt, whose marketing talents rivaled his own—define the election in such crisp terms.

"I love it!" Trump told Gingrich by phone that night. "Caravans and Kavanaugh! That's my closing message!"

It seemed improbable that the president would stick to any single "closing message."

The previous week, on October 11, Trump had hosted Kanye West in the Oval Office for a meeting on criminal justice reform that turned into a surreal impromptu press conference. With a throng of reporters crowded around the Resolute desk (hewn from the timbers of a British Royal Navy barque and gifted to Rutherford B. Hayes by the famously austere Queen Victoria), the hip-hop artist who had once said that George W. Bush "doesn't care about black people" riffed for more than fifteen minutes on everything from the "welfare mentality" of African Americans to the jolt of masculine energy he felt when wearing the Make America Great Again hat, calling himself "a crazy motherfucker," to the delight of Trump.

A few days later—hours before Trump sounded the alarm about the caravan—the president tweeted the news that a judge had thrown out a lawsuit against him by Stormy Daniels. In so doing, Trump described his former sexual partner as "Horseface."

DESPITE THE DAILY CHURN OF SELF-IMPOSED DISTRACTIONS, TRUMP endeavored to echo Gingrich as often as possible. In the final weeks before the midterms, he regularly touted his appointment of not one but two Supreme Court justices, taking every opportunity to remind Republicans of the abuse Kavanaugh had been subjected to in his confirmation hearings, which had devolved into one of the nastiest partisan food fights Capitol Hill had ever seen.

Still, he was far more adamant about the caravan. Calling it "an invasion of our country" by "gang members," "very bad thugs," and "unknown Middle Easterners," Trump hammered the issue on a daily basis, even deploying five thousand troops to the southern border in what Pentagon officials later acknowledged to be a naked political stunt.

There were boasts of a booming economy and talk of tax reform's benefits in the kitchen-sink strategy used by the White House down the stretch. Trump also touted the recently renegotiated trilateral trade deal with Canada and Mexico that carried benefits for U.S. dairy farmers and automakers. But the thrust of his "closing message" was the same as it had been two years earlier in his pursuit of the presidency: fear.

Trump aimed to brand the election as a stark choice between two parties. Democrats were weak; Republicans were strong. Democrats were beholden to global interests; Republicans were prioritizing America's well-being. Democrats were motivated by malice and spite and an obsession with toppling the president; Republicans were motivated by patriotism and security and a desire to protect Americans from the wolves at the gate.

Interestingly, while most of the prized Democratic recruits around the country ran disciplined campaigns steering clear of these stereotypes, certain elements of the progressive base—and some of the party's most prominent figures—walked right into Trump's trap.[1]

In late June, Homeland Security secretary Kirstjen Nielsen was loudly confronted by protesters while dining inside a Mexican restaurant in Washington; members of the local Democratic Socialists of America chapter chanted "Shame!" in response to the family-separation policy at the southern border. The next week, Sarah Sanders was asked to leave a Virginia restaurant because of her work as White House press secretary. In response to these incidents, California congresswoman Maxine Waters told a crowd of her constituents, "If you see anybody from that cabinet in a restaurant, in a department store, at a gasoline station, you get out and you create a crowd and you push back on them."

Eric Holder, the attorney general under Obama, told a Georgia crowd while campaigning in the fall that he disagreed with the former First Lady's insistence on elevating the national discourse. "Michelle [Obama] always says, you know, 'When they go low, we go high,'" Holder said. "No. When they go low, we kick them. That's what this new Democratic Party is about."

The same week as Holder's remark, Hillary Clinton put a cherry on top of the civility debate. "You cannot be civil with a political party that wants to destroy what you stand for, what you care about," she said in an interview with CNN.[2] "That's why I believe, if we are fortunate enough to win back the House and/or the Senate, that's when civility can start again. But until then, the only thing that the Republicans seem to recognize and respect is strength."

All this played directly into Trump's argument that Democrats were the party of protests, of lawlessness, of hatred and hostility—all while he continued to embody those very things.

In October, Trump tweeted a campaign ad that was blatant in its deception and brazen in its racist innuendo: Rolling footage of brown-faced crowds funneling through fences, the ad highlighted a Mexican man, Luis Bracamontes, who had killed two police officers. "Democrats let him into our country," the caption read. "Democrats let him stay." But Bracamontes had been released "for reasons unknown" by none other than Sheriff Joe Arpaio, the right-wing guardian of law and order, before being deported under the Clinton administration and reentering the United States during Bush's presidency. This nuance was absent from the ad, which asked viewers, "Who else would Democrats let in?"

It was difficult to gauge the aggregate effect of the increasingly vitriolic national climate. In the battle for the Senate, with Republicans playing offense in a batch of predominantly conservative and more rural states, Trump's rhetorical firefight with the Democrats was a net benefit; in the contest for control of the House, with Republicans defending dozens of moderate, suburban-based congressional districts, it was proving less helpful.

By the middle of October, it seemed almost certain that Democrats would win back the House. They had too much verve in their base, too many pickup opportunities, and too much cash not to flip the lower chamber. (Republicans began complaining to donors that fall about a "green wave," citing the record fund-raising sums for congressional challengers across the map.) Trump and his team were highly in tune with this reality. The closer Election Day drew, the clearer the president made it that he did not want to be campaigning with any Republicans who would lose, as it would reflect poorly on him. Strategically, his political team avoided House races almost entirely and stuck to the easier, safer Senate races.

Even in those instances, nothing was guaranteed. Trump could not understand how Democratic senator Jon Tester stood a prayer in Montana, a state the president had carried by 20 points. He was equally miffed by Democratic senator Joe Manchin's staying power in West Virginia, which Trump had won by 42 points. The president insisted on pounding both states all the way to the finish line, certain that his appeal was stronger than that of the incumbent Democrats.

And then there was the curious case of Texas.

For much of the previous year, Ted Cruz was perceived to be sailing to reelection in the Lone Star State. He was a political celebrity with a fat bank account and a proven campaign machine; his Democratic opponent, Robert "Beto" O'Rourke, was a little-known congressman from El Paso, the sixth-biggest media market in the state (behind even the Brownsville/Rio Grande Valley region). Plus, as the conventional wisdom dictated, this was Texas, after all—no place for a Democrat to flourish in the age of Trump.

All this was proving backward. For starters, this was no longer the Texas of George W. Bush. The state's accelerating demographic transformation, paired with the GOP's rightward lurch, was making for an increasingly competitive atmosphere. After four consecutive presidential cycles of landslide double-digit victories for the GOP, Trump carried Texas by 9 points in 2016—a smaller margin than in battleground Iowa. There were warning bells galore, none shriller than the result in Harris County. Anchored by Houston and home to swelling populations of both Hispanics and college-educated whites, Harris County was fought to a virtual tie in 2012, with Obama topping Mitt Romney by fewer than 600 votes. Four years later, Clinton carried the county by 162,000 votes.

For the popular perception of Texas as backcountry, it boasted four of the nation's eleven largest metropolitan areas and was spilling over with the suburbanites who were most hostile to Trump. In a sense, the former Texas governor, Rick Perry, had been *too* successful in luring jobs to the state: By cutting taxes to the bone, he had caused millions of new residents to flood into Texas over the past decade, many of them liberal, college-educated exports from California. This influx, on top of the ever-rising share of Hispanic voters, was dry demographic tinder. The contrast O'Rourke struck with Cruz—and with Trump's GOP—provided the spark.

Young, telegenic, and social media savvy, O'Rourke presented himself as the antidote to the sorry state of American politics. He was fun and authentic, skateboarding on the campaign trail and refusing to hire pollsters or consultants. He also rejected corporate money and super PAC donations, wanting only the aid of small donors. In running this romantic campaign (with a perfect foil in Cruz, viewed as a sort of political Hannibal Lecter by the left), O'Rourke became the darling of the

Resistance. It didn't matter that his platform wasn't fully fleshed out, or that those policies he did embrace (Medicare for All, an assault weapons ban, calling for Trump's impeachment) were tailored more toward national liberals than Texas voters. O'Rourke was a cause more than he was a candidate. And the perks were breathtaking. Drawing mammoth crowds and dotting the state with his signature black-and-white "BETO" signs, O'Rourke raised preposterous sums of money, $38 million in the third quarter alone, a presidential-level haul and the most ever in a U.S. Senate race.[3] (The previous record was $22 million.)

The Cruz campaign was concerned but not flustered. They had expected a comfortable victory in the 10- to 12-point range; as summer turned to fall, and Betomania blew up, they scaled back their projections to the high single digits. Cruz expected his opponent to turn out masses of new Democratic voters; the incumbent would win by mobilizing his own party's base. None of this was terribly worrisome—until the White House started calling.

Trump was delighted upon hearing that summer of Cruz's peril in Texas. Though they both claimed to have moved past their rivalry, with the senator becoming a reliable advocate of the president's agenda, their relationship was no less awkward. Whenever they were together, Trump would recall Cruz's victories in the primary—as well as their attacks on one another. The president had never been defeated by anyone else in politics; because of this, Cruz occupied a space in Trump's psyche that was apparent to their mutual allies. When word came that Cruz was in trouble, then, the president was delighted to play the role of rescuer, joking with aides that he would swoop down to Texas and save Lyin' Ted.

Cruz tried to politely dismiss the president's offers of help, but the phone calls kept coming, at least a half dozen in the month of August alone, with Trump insisting on coming to Texas for what he promised would be the biggest campaign rally of 2018. Cruz was annoyed. He knew what Trump was up to. And the senator didn't want or need his help. Yet he was trapped: If he said yes, then the president's visit could do even more damage with the suburbanites his campaign was bleeding away; if he said no, then Trump might just be liable to do something crazy, such as send a tweet attacking Cruz and hurting his turnout efforts with the GOP base.

The ensuing back-and-forth was a negotiation between competitors masquerading as allies. Cruz, wanting to push the event far away from the major media markets and out into Trump country, recommended they hold the event in Lubbock; the president was adamant that they visit a major city, predicting a capacity crowd. With the discussions at an impasse, Trump took matters into his own hands. "I will be doing a major rally for Senator Ted Cruz in October," he tweeted on August 31. "I'm picking the biggest stadium in Texas we can find."

Cruz was irritated if unsurprised. It took three hours for him to muster a tweet: "Terrific!"

Trump relented on the size of the stadium—Texas has venues holding more than one hundred thousand people, his staff warned, and it would be impossible to hide the empty seats—but he wouldn't budge on the location. This would be the highest-profile event of the election cycle, a demonstration of his mercy and his beneficence. Trump wanted maximum exposure. They settled on the Toyota Center in Houston, filling almost every last seat and drawing vast crowds of protesters outside.

On October 22, two and a half years removed from Trump's accusing Cruz's father of aiding the assassination of JFK and Cruz calling Trump "a pathological liar," the former foes shared the stage in Houston. The president couldn't help but remind everyone of their "nasty" feud in 2016. But that was all behind them now. ("He's not Lyin' Ted anymore," Trump said earlier in the day. "He's Beautiful Ted.") The president credited the Texas senator with leading the charge to pass the GOP agenda, devoting much of the rest of his speech to apocalyptic immigration talk. Democrats, he said, wanted to "give aliens free welfare and the right to vote," and also let in MS-13 gang members, who "like cutting people up, slicing them" instead of using guns. Trump also embraced the term "nationalist," calling himself by that controversial label for the first time.

The Cruz team breathed a sigh of liberation when the event concluded, believing disaster had been avoided. They were right. But the damage was undeniable nonetheless: Cruz's support dropped 5 points overnight in the Houston market, and the local Republican congress-man, John Culberson, saw an even steeper decline.

Then, at the end of October, Trump told Axios in an interview pub-lished one week before Election Day that he planned to end birthright citizenship for the children of illegal immigrants and noncitizens born on American soil.[4] "It's ridiculous. It's ridiculous. And it has to end," Trump said, suggesting he could use an executive order to overturn the promises of the Fourteenth Amendment, enacted at the Civil War's end to protect the rights of newly freed slaves.

Republicans were floored by the president's latest voluntary distrac-tion. "Well, you obviously cannot do that," Ryan responded during an interview with WVLK radio in Kentucky. "You cannot end birthright citizenship with an executive order."

Of course, the Speaker knew better than just about anyone that facts presented no obstacle to Trump. The president made more false claims (1,176) during the two months leading up to Election Day 2018 than he had in all of the previous calendar year (1,011), according to Daniel Dale, a *Toronto Star* reporter who had meticulously chronicled Trump's rela-tionship with the truth. Dale also concluded, "The three most dishonest days of Trump's presidency were the three days prior to the midterms," with a single-day record of 74 false claims being made on Monday, No-vember 5, a little more than three per hour.[5]

Questions of truthfulness and legality and constitutionality not-withstanding, Trump's latest proclamation spelled further political trouble for Republicans with Hispanic constituencies. "It's like he *wants* us to lose!" Cruz bellowed upon hearing of the Axios interview. Launch-ing into his impersonation of Trump, the senator said, "What could I do to *really* antagonize Hispanics? I know! I'll threaten to take away their kids' citizenship!"

If the president was aware of the anger he was incurring within the Republican political class, he didn't show it. Trump was having the time of his life. Earlier in the summer, while he was traveling to South Caro-lina for a rally, storms delayed his arrival by over an hour. The pilots of Air Force One suggested they return to Washington, knowing how far behind schedule they were and seeing no immediate improvement in the weather. Trump wouldn't hear of it. Vowing never to disappoint his thousands of fans waiting on the ground, he grew impatient as Air Force One continued its holding pattern. "Land this fucking plane already!" he bellowed toward the cabin. "Trust me, it's safe! I've been flying longer than you guys have!"

Standing backstage at a boisterous rally in Columbia, Missouri, five days before the election, with Lee Greenwood's "God Bless the USA" pulsing throughout a packed airport hangar, Trump threw his head back and marinated in the moment. Soon enough he would be dazzling a pack of six thousand with his usual riff: Democrats letting the illegals in, Republicans fighting the drugs and criminals, plus the new wrinkle of nixing birthright citizenship. But before any of that, he took a long, introspective pause. Preparing to take the stage, the president seemed to feel it all—the crowd, the music, the energy, the media glare—coursing through his veins.

"I fucking *love* this job!" he howled into the November night.

ONCE USED BY PRESIDENTS AS A BILLIARDS ROOM IN THE FILTHY, FORsaken basement, the Map Room had changed status when Theodore Roosevelt renovated the entire ground floor of the White House. He turned it into multipurpose space, only for later presidents to bring back the pool table and restore the leisurely room's reputation. Franklin D. Roosevelt had another idea: With the onset of World War II, he needed a situation room stocked with records, charts, and maps to track the progress of military engagements across Europe and the Pacific. Hence the Map Room was born, and it has remained ever since, suffering only slight remodels by subsequent administrations careful to keep its integrity and historical value intact.

On the evening of November 6, the Map Room was used to track a different sort of battle—one between Republicans and Democrats, with control of the federal government on the line.

The Office of Political Affairs transformed the space into an impressive Election Night war room. Running down the middle of the floor was one enormous table featuring power strips and docking stations, allowing everyone to project their laptops onto large monitors. A pair of fifty-inch high-definition televisions were situated on either end of the table, resting on roller carts; one showed a four-way split between ABC and the three cable networks; the other was tuned exclusively to Fox News.

True to the room's tradition, maps and charts were plastered across every vacant inch of the walls and table, guiding the president's team through the nearly one hundred races they would be monitoring at the House, Senate, and gubernatorial levels: poll closing times; historical results by state and district; heat maps showing areas of targeted turnout; bellwether counties; contacts for every candidate and campaign; election attorneys by state; and opposition research briefs on Republican candidates whom they suspected might lose, allowing for quick spinning by White House surrogates.

A number of VIPs drifted in and out of the election bunker: Trump's children and their spouses, Pence and his wife, Kellyanne Conway and Sarah Sanders, among others. The vice president took a particular interest in Indiana, asking for maps to be zoomed in so that he could examine county returns and determine whether Democratic senator Joe Donnelly, one of the GOP's top targets, could survive. Other visitors zeroed in on specific states where the president had campaigned, praying for good news to bring him.

In fact, the only prominent official who avoided the Map Room was the president himself. He remained upstairs in the East Room, where a small party was being hosted for friends of the administration, who snacked on pizza and watched the returns come in on Fox News. Trump seemed dour and fatalistic; he made no speech and seemed less conversational than usual, eyeing the bank of televisions and awaiting updates from Stepien, his anxious political director. The president seemed to know that the night would not be one to celebrate.

Downstairs in the Map Room, his team clung to a more optimistic outlook. Word had gotten around that the previous night, Steve Stivers, the chairman of the NRCC, had called Ryan, McCarthy, and Scalise with great news: Republicans were going to keep the House. Ryan was skeptical; all the polling from the Congressional Leadership Fund, the GOP's allied super PAC, showed losses approaching 30. Stivers was adamant that his projections were more accurate, that Republicans would hold on. McCarthy, always eager to deliver good news to Trump, had passed along the message to the president and his team.

In the first three hours following the initial wave of poll closings at 6:00 p.m. Eastern, Republicans were encouraged: Marsha Blackburn had won the Tennessee Senate race, which had been surprisingly competitive all year; and Andy Barr, the GOP incumbent in Kentucky's Sixth District, had fended off a tough Democratic opponent in what was widely viewed as a bellwether race for control of the House of Representatives. Nothing encouraged the president and his team more than the result in Missouri, a state Trump had visited seven times, where Republican challenger Josh Hawley knocked off Democratic senator Claire McCaskill.

On CBS, John Dickerson declared that "Planet House is not spinning the way the Democrats want it to." On ABC, George Stephanopoulos predicted a "disappointing night" for Democrats. On CNN, the liberal commentator Van Jones described the early returns as "heartbreaking." Indeed, by 9:00 p.m. in Washington, things were looking up for the White House.

Thirty-three minutes later, however, a familiar voice pierced through the din. "We are now ready to make one of the biggest calls of the night," announced Bret Baier. "The Fox News Decision Desk can now project the Democrats will take control of the House of Representatives for the first time in eight years."

The Map Room fell silent. Then, after what might have been less than one full second, it ignited with shouted expletives and strewn papers. The president's staff could not process what they had just heard: Early races had broken their way, polls on the West Coast were still open, and no other network or news service had yet made the call for control of the House. Some were sure it was a mistake and shushed their colleagues so they could hear more. "A lot of listeners out there, their heads are exploding," Chris Wallace said, staring straight into the ground floor of the White House. "But this is going to be a very different Washington."

Trump was equally aghast upstairs in the East Room. It felt unreal to everyone watching from 1600 Pennsylvania Avenue. All the "Fake News" organizations were holding back, recognizing the GOP's strong showing early in the night, while Fox News, which typically operated as the president's personal *Pravda,* was sticking a fork in his party.

Trump wanted answers. Sanders, the press secretary, agreed to have White House communications director Bill Shine, a longtime Fox News executive who was still being paid by the network after joining the administration, reach out to his former colleagues for an explanation. What he got was short and sweet: The network was using, for the first time on an Election Night, new and advanced modeling that spliced AP election returns with advanced polling statistics to get a better picture of where certain races were headed. It was clear in their modeling, by the time Fox News made the call at 9:33 p.m., that Democrats would regain control of the House.

It wasn't until 10:22 p.m. that NBC News followed suit. CNN, which had been the first to call the Senate for Republicans (a far less controversial projection), did not join Fox News and NBC News in calling the House for Democrats until after 11:00 p.m.

As the losses for House Republicans piled up, the result was sweet vindication for the election team at Fox News. It was a valuable reminder that, for all the brainwashing practiced by the likes of Hannity, Ingraham, and Jeanine Pirro, the network also employed some outstanding journalists: Baier, whose anchoring is strong and rigidly objective; Chris Wallace, the best interviewer in the business; Chris Stirewalt, the razor-sharp politics editor; Shepard Smith; Bill Hemmer; and several others.

The contrast between the journalism and entertainment wings of the network had been a source of running tension for years, with Baier and Wallace known to apologize to Republican leaders on Capitol Hill for the antics of their colleagues. Wallace, for one, seemed to savor the occasion to shut one of them up on Election Night. When Ingraham opined that the election results showed "the Democrats are going to more of an Ocasio-Cortez party," a reference to the Democratic Socialist in New York, Wallace stopped her.

"I don't think that is a fair thing to say about the Democrats. I think that is a complete mischaracterization," he said, pointing out how the majority was won with moderate Democrats hugging the center. "You know, if you're going to give the Republicans credit for holding on to the

Senate, then I think you have to give Democrats credit for actually flipping the House."

As Trump's staff went into full spin mode, saying the House was lost due to factors beyond the president's control, and crediting Trump with saving the Senate majority, Wallace produced a fresh bucket of cold water.

"I think we are . . . giving too much credit to Donald Trump for holding onto the Senate," he said. "The fact is, this was a historically difficult year for the Democrats. The Democrats had twenty-six seats that they had to defend. The Republicans had nine seats they had to defend."

"So, yes, it's a victory for Donald Trump," Wallace continued, "but this was something he should have been expected to do."

A POLITICAL PARTY CAN ONLY PLAY THE HAND IT IS DEALT, AND REPUBlicans took advantage of the friendly Senate map, knocking off numerous Democratic incumbents: McCaskill of Missouri, Donnelly of Indiana, Heidi Heitkamp of North Dakota, and most important to Trump, Bill Nelson of Florida, in a race he'd been personally invested in because of his friendship with Republican Rick Scott.

There were also a handful of marquee governor's victories to celebrate, including in Florida (where Trump had gone all-in behind Republican Ron DeSantis) and Georgia (where Brian Kemp, the GOP nominee and secretary of state, came under intense scrutiny for purging disproportionate numbers of minority voters from the rolls, only to defeat Democrat Stacey Abrams by 55,000 votes).

More striking were the gubernatorial election results in Maryland and Massachusetts. In Maryland, Republican incumbent Larry Hogan won reelection by a comfortable 12 points; and in Massachusetts, Republican incumbent Charlie Baker won another term by a whopping 34 points. The common denominator: Both men had governed as pragmatic problem solvers in two of the nation's bluest states, setting aside divisive cultural fights and emphasizing kitchen table concerns: schools, roads, housing and health care costs, and the opioid epidemic. Even as the national GOP was addicted to dysfunction for much of the previous decade, some of its state parties were models of competence, and Hogan and Baker were at the forefront.

On the whole, however, Democrats increased their number of governorships. The most celebrated win came in Wisconsin, knocking off Governor Scott Walker, who, friends worried, had gotten greedy in seeking a third term. The left's most symbolic victory came in Kansas, where Democrats toppled the longtime immigration provocateur and Trump ally Kris Kobach, who ran what Republicans described as the worst campaign of the entire election cycle, focusing more on issues of voter fraud and border security than education and health care.

Senate and gubernatorial contests aside, Election Night 2018 was defined by the Democratic takeover of the House of Representatives—the statement it made about the appeal of Trump's party, and the implications for the government moving forward.

When all the votes were counted—which took weeks, considering the molasses mechanics of California—Democrats had won a net gain of 40 seats, well above the preelection forecasts in both parties that ranged between 25 and 30. This result was the culmination of two years of shrewd, self-controlled campaigning that took advantage of the president's unpopularity in the suburbs by straddling the middle of the electorate and not giving in to the tribal nuttiness of the day.

One notable casualty was Dave Brat in the Virginia suburbs. The Freedom Caucus board member lost to Abigail Spanberger, a sharp, centrist former CIA officer who had promised to bridge the divide between the two parties and bring down the decibel level in DC. The symmetry was striking: The man who had slayed Eric Cantor by running an anti-Washington, talk radio–backed campaign pounding the issue of immigration, was ejected from office because of his district's adverse reaction to the president who followed that very blueprint. (Brat was promptly rewarded with a job running the business school at Liberty University.)

While Republicans stood their ground in exurban and rural areas (and in some cases, even grew their support there), the story of the midterms was the Democrats' supremacy in the suburbs. From New York to Philadelphia to Washington to Richmond to Atlanta to Detroit to Chicago to Des Moines to Houston to Oklahoma City to Denver to Salt Lake City to Los Angeles, Republicans bled support in America's suburbs, giving away dozens of districts that had been drawn by GOP lawmakers not long before under the impression that those voters were party lifers.

Driving this transformation of the suburbs was women—and particularly college-educated women. According to national exit polling, Democrats won 59 percent of women overall compared to Republicans' 40 percent. (That 19-point margin was nearly double the 10-point spread by which Republicans lost women in 2016.) And among white women with a college degree, Democrats beat Republicans by 20 points.[6]

The bulk of GOP losses came in districts where Trump's numbers were insurmountably low. In these areas, Republicans witnessed a wipeout of some of their most effective members—including Mike Coffman of Colorado, Barbara Comstock of Virginia, Peter Roskam of Illinois, and Carlos Curbelo of Florida, among others—whose strong individual brands back home were not sufficient to overcome the president's devastating unpopularity. In Orange County alone, Democrats flipped all four of the remaining Republican-held congressional seats. The GOP had been wiped out in the heart of Reagan Country.

Trump could not be held solely responsible for this realignment of the electorate. As the journalist Ron Brownstein has written, the "class inversion" of white-collar suburbanites moving toward the Democrats and blue-collar exurban and rural voters moving toward the Republicans has been under way for a generation. Yet the Trump presidency proved an explosive accelerant: According to the national exit poll of House races, Democrats won whites with a college degree by 8 points; Republicans won whites without a college degree by 24 points.

These demographic splits, statistically speaking, would have been unimaginable one decade earlier.

FOR ALL THE TALK OF TRUMP REMAKING THE ELECTORAL MAP IN 2016 and defying the prescriptions of the Republican National Committee, Reince Priebus and his "autopsy" were haunting the party from the grave. Republicans won white votes overall, 54 percent to 44 percent, according to the exit polls, but Democrats won nonwhites by a margin of 76 percent to 22 percent. Meanwhile, the overall vote share of whites fell 3 percentage points from the 2014 midterms, which itself had been down 3 points from the 2010 midterms.

Trump had won the White House with an inside straight, sweeping the Rust Belt and notching impossibly narrow victories in the three predominantly white states of Pennsylvania, Michigan, and Wisconsin. But repeating that path looked a lot more difficult as the returns came in on Election Night: In the statewide races for governor and U.S. Senate in those states, Democrats went six for six, patching the holes in their "Blue Wall" and regaining control over certain key functions of government, including voting regulations, that could prove decisive in 2020.

Even setting aside the particulars of the Electoral College and Trump's path to reelection, the warnings of the post-2012 RNC autopsy—that the aggregate demographic tradeoffs hurt Republicans long term—rang true once again. Trump's party performed splendidly among the fastest-declining groups of voters and the decreasingly populated parts of the country in 2016, but Democrats dominated among the fastest-growing groups of voters and increasingly populated parts of the country.

This was unsustainable over the long term, and possibly unworkable even in the short term. Exhibit A: Texas.

In a state no Democrat had won statewide since 1994, and where the GOP incumbent, Ted Cruz, was better financed and better organized than almost any Republican in the country, the result was a 2.5-point win for Cruz. If ever there was a moral victory for Democrats, this was it. Cruz won 66 percent of whites, who made up 56 percent of the electorate; but O'Rourke won 69 percent of nonwhites, who made up 44 percent of the electorate. It wouldn't be long before those composition statistics are inverted, with nonwhites comprising the majority of Texas voters; when that happens, the state will cease to be red unless Republicans dramatically improve their performance with minorities.

Harris County, the home of Houston, offered a distillation of the GOP's dilemma. With roughly equal proportions of Hispanics, blacks, and whites—and the whites increasingly college-educated—the county is viewed by both parties as a harbinger of America's demographic future. Cruz lost it by more than 200,000 votes.

It's true that O'Rourke was an exciting candidate who mobilized the Texas Democratic base in unprecedented ways. But it's also true that trend lines in the state bode ominously for the GOP. Back in 2016, not a single House Republican in Texas lost reelection; the only one who came close was Will Hurd, the former CIA agent representing a 71 percent Hispanic district on the southern border. Two years later, a pair of Texas Republicans lost their seats to Democratic challengers, and Hurd, widely viewed as the best campaigner in the House GOP, came dangerously close, winning by fewer than 1,000 votes despite an approval rating in the 70s.

"It was near-presidential turnout in 2018. The Republican base showed up, but it's shrinking as a percentage of the voting population," Hurd said after the election. "In 2016, only one Texas Republican got less than fifty-five percent of the vote; that was me. In 2018, twelve [of us] got less than fifty-five percent of the vote. Two of them lost. The average difference in our margins from 2016 to 2018 was negative fifteen points. This is a trend that has to be stopped, and the only way you stop that trend is by appealing to a broader base of people. If the Republican Party in Texas ceases to look like voters in Texas, there will not be a Republican Party in Texas."

It goes without saying, but if there's not a Republican Party in Texas, there won't be much of a Republican Party in America. The state's 36 Electoral votes have been the foundation of the GOP's path to the White House since 1980. Without Texas, the GOP's presidential math becomes unworkable—period.

It's a similar, albeit less dire situation, in Arizona. Having won sixteen of the past seventeen presidential elections there, Republicans have come to depend on the state's 11 Electoral votes as a red bulwark in the dauntingly blue West Coast. Yet Trump won the state by less than four points in 2016, the smallest margin in two decades, thanks to poor performance with the state's nonwhite population. And unlike in Texas, the Democrats broke through in 2018, winning the Arizona

Senate race by a relatively comfortable margin, with the Democratic nominee, Kyrsten Sinema, carrying 68 percent of the state's nonwhite voters.

Georgia represented a final red flag for the Republican Party. Despite carrying the state in six consecutive presidential elections, Republicans have seen its shifting demographics (increasingly urban, college-educated, and nonwhite) yield shrinking margins over the past decade. In 2018, with Abrams vying to become the nation's first black female governor, Republicans hung on by less than 2 percentage points. The reason: Kemp, the GOP nominee, won 74 percent of whites, who made up 60 percent of the electorate; but he lost 84 percent of nonwhites, who made up 40 percent of the electorate. As in Texas, it's only a matter of time until these composition ratios are reversed.

Waking up to these realities on November 7, Republicans found that no amount of spinning their Senate and gubernatorial wins could mask the three-front war awaiting them in 2020. Democrats were converting the suburbs into political garrisons. They were reasserting themselves in the Rust Belt states, demonstrating the limits of a strategy banking on big margins with working-class whites. And they were creeping closer to parity in three states, Texas, Arizona, and Georgia, that would surely be contested by the party's nominee in all future presidential races.

AS TRUMP PREPARED FOR HIS POSTELECTION PRESS CONFERENCE THAT Wednesday morning, studying the names of his fallen GOP detractors, he envisioned the ways in which he could use the loss to tighten his grip on the party. Many of the defeated Republicans, he told his staff, had been disloyal to him; they deserved to lose. Only by embracing his nationalist, America First policies, he said—and by backing him up in his guerrilla war against the media, the Democrats, and the conventions of Washington—would the Republican Party prosper.

Contemplating the inflection point offered by the midterm results, Trump realized there was good news and bad news.

He had purged the party of perhaps his two most outspoken GOP critics, Arizona senator Jeff Flake and South Carolina congressman Mark Sanford. Despite both those seats being lost to the Democrats, the president took comfort in knowing he was free of those grandstanding, grating, self-righteous traitors who lived to criticize his every utterance.

The bad news: A newly elected Republican was on his way to Washington and intent on filling their shoes. He had little patience for Trump, he told friends, and would not hesitate to police him. He had finally won his first federal campaign after three unsuccessful attempts. He was the incoming junior senator from Utah: Mitt Romney.

CHAPTER TWENTY-FIVE

NOVEMBER 2018

"Some have adapted. And I haven't."

AST NIGHT THE REPUBLICAN PARTY DEFIED HISTORY," PROCLAIMED the president of the United States.

Standing before an overflow audience inside the White House on the afternoon of November 7, his royal-blue tie shimmering and his spray tan exceptionally robust, Donald Trump delivered a postelection press conference the likes of which Americans had never seen.

For an hour and twenty-seven minutes, the president held forth on topics ranging from his popularity among minorities ("I have the best numbers with African American and Hispanic Americans that I've ever had before") to the annexation of Crimea (he blamed Barack Obama instead of Vladimir Putin) to his refusal to release his tax returns ("They're extremely complex, people wouldn't understand them").

Primarily, the appearance was designed as one-part end zone dance and one-part exoneration tour, as he took credit for the contests Republicans had won and distanced himself from those they had lost. Noting how GOP candidates had prevailed in nine of the eleven places he'd visited recently—races picked by his team specifically because of their high likelihood of victory—Trump concluded, "This vigorous campaigning stopped the Blue Wave that they talked about."

Arguing that the odds were stacked against the House GOP due to so many members retiring, without mentioning that plenty of those retirements owed to what lawmakers labeled "Trump fatigue" in Capitol Hill vernacular, the president congratulated those Republicans who had won their elections by remaining loyal to him.

As for the others?

"You had some that decided to 'Let's stay away. Let's stay away.' They did very poorly," the president said. "I'm not sure that I should be happy or sad. But I feel just fine about it."

Trump, the titular head of the Republican Party, then proceeded to namecheck and mock those House Republicans who had campaigned on their own independent brands and who had lost in districts where the president's unpopularity made winning with an *R* next to one's name all but impossible.

"Carlos . . . Cue-bella."

Trump could not pronounce the name of the Miami-area Republican, Carlos Curbelo, one of the GOP's brightest young stars, whose good standing at home was insufficient in the age of Trump.

"Mike Coffman. Too bad, Mike."

Now he was mocking the Colorado congressman, an Army and Marine Corps veteran, who was widely regarded as one of Congress's hardest workers, having taught himself new languages to converse with the constituents in his rapidly diversifying district.

The president continued: "Mia Love."

Trump stopped himself, noting his estranged relationship with the black Utah Republican. "Mia Love gave me no love. And she lost," he said. "Too bad. Sorry about that, Mia."

He went on: "Barbara Comstock was another one. I think she could have won that race, but she didn't want to have any embrace. For that, I don't blame her. But she—she lost. Substantially lost. Peter Roskam didn't want the embrace. Erik Paulsen didn't want the embrace."

After running through his list, Trump paused to reflect on these intraparty traitors.

"Those are some of the people that, you know, decided for their own reason not to embrace—whether it's me or what we stand for," he said. "But what we stand for meant a lot to most people. And we've had tremendous support, and tremendous support in the Republican Party. Among the biggest support in the history of the party. I've actually heard, at 93 percent, it's a record. But I won't say that, because who knows?"

IT WASN'T QUITE A RECORD, BUT TRUMP WAS RIGHT: HE WAS MORE POPular with his party's voters than any president in modern history at the two-year mark, save for George W. Bush's 97 percent approval rating in the aftermath of 9/11. Gallup showed Trump at 88 percent, ahead of Obama at 85 percent, George H. W. Bush at 84 percent, Richard Nixon at 79 percent, Ronald Reagan at 76 percent, Bill Clinton at 75 percent, and Jimmy Carter at 62 percent.[1]

But there was one critical difference: Donald Trump's party lost 40 House seats in 2018, whereas George W. Bush's party *gained* 8 House seats in 2002.

How was this possible? The only mathematical explanation: The party itself was contracting.

In polling, party affiliation is a state of mind, something that is elastic from year to year (and, in this age, from day to day). Fewer voters were identifying as Republicans—or, at least, as Trump Republicans—when pollsters contacted them. It was possible for Trump to maintain a high approval rating with self-identified Republicans while his party was being decimated in red congressional districts because a rising number of voters were vacating their GOP affiliation.

The proof? Consider polling from the Congressional Leadership Fund, the super PAC charged with keeping the House under Republican control. The organization raised and spent roughly $160 million for the election cycle, with an enormous investment in survey work to understand the voters in the dozens of districts they competed in. Every single one of the House Republicans called out by Trump in his postelection press conference (and many more) had higher favorability ratings in their districts as of late October than did the president. And while Trump polled well with GOP voters, the "Generic Republican" option outpaced him on the survey ballots in these districts. For many members—Comstock in Northern Virginia, Coffman in metropolitan Denver, Paulsen in the Twin Cities—Trump's mid-30s favorability was the result of his alienating their suburban, right-of-center voters. It proved insurmountable, despite his support among conservatives.

Just as Trump wrote these members off, so, too, did the professional right wing. "The squishy members who lost their races were the ones who didn't embrace the conservative agenda," David McIntosh, the Club for Growth president, declared at a press conference. David Bozell, the leader of ForAmerica, another activist group, agreed: "Republicans shed a lot of dead weight last night."

Setting aside the fact that most of the aforementioned House Republicans voted with Trump north of 90 percent of the time, and that several self-styled conservatives also lost their races, the dismissiveness was surprising. Two years in the majority had allowed Republicans to accomplish at least some of what they had campaigned on; now, heading into the minority, some on the right seemed strangely enthused about embracing the old Jim DeMint mantra of 30 purists being better than 60 pragmatists.

It simply made no sense.

"Politics are about addition and multiplication, not subtraction and division," as Mississippi governor Haley Barbour used to say. The activist right could be forgiven for ignoring this advice. But of all people, Trump, who stitched together an unlikely coalition en route to winning the White House, would seem to appreciate the imperative of big-tent politics. Instead, all he cared about were displays of loyalty or lack thereof.

One election result pleased the president above all others. "In Jeff Flake's case, it's me, pure and simple. I retired him," Trump boasted on November 7. "I'm very proud of it. I did the country a great service."

Democrats flipped two GOP Senate seats in 2018. One of them was Flake's in Arizona.

His onetime best friend, Mike Pence, sat a few feet away as the president spoke. The two congressmen used to call each other "Butch" and "Sundance," nicknames earned from when they would burst through the swinging doors of the House chamber to register their objection to a Bush-era big-spending proposal. When Flake, on his twentieth wedding anniversary, got stuck in Washington for a busy week of votes, Pence helped fly Flake's wife to DC and arranged a surprise dinner for them on the rooftop of the W Hotel.

Now, as Trump took occasion to spit on his old partner's grave, Pence started straight ahead. If he was angry or upset or bewildered at what the president was saying, he certainly wouldn't, and couldn't, show it.

"Mike is intensely loyal. That's a virtue. And he has never uttered to me one syllable of disagreement with the president, and frankly I admire him for that," Flake says. "We've taken different paths, but I'm not trying to suggest that mine is a more virtuous path than his. He's in a position with considerably more power than I have. And there's something to be said for that. If he can influence the president in a positive direction, then maybe that was a wise choice."

Flake acknowledges that he has changed; that he is not the same hard-line, insurgent-styled conservative he once was. "But the bigger change was the party, which used to be the party of limited government, economic freedom, individual responsibility, free trade," he says. "It has become a more nationalist, nativist, anti-immigration party. That's an unfamiliar standard for most of us. Some have adapted. And I haven't."

THE ADMINISTRATION WAS CONTINUING TO BLEED PERSONNEL. Most notably, back in October, UN ambassador Nikki Haley had abruptly announced her departure. She said all the right things, denying ambitions to challenge Trump in 2020 and saying she simply needed a break, but the perception was reality: The president had lost one of the more respected, competent members of his government.

Jeff Sessions, to the surprise of no one, was the first casualty following the midterms. The attorney general's firing had been a long time coming, having been browbeaten both in public and in private ever since his recusal from the Russia investigation. Sessions wasn't the last to go. By the end of December, a trio of cabinet officials had announced their exits. The first was Ryan Zinke, the secretary of the interior, who faced at least eighteen federal inquiries into his conduct, according to a tally from the left-leaning watchdog group Citizens for Responsibility and Ethics in Washington. He drained himself from the swamp to avoid imminent congressional investigations into his activity once Democrats took control of the House.

The second major departure was that of John Kelly. The White House chief of staff had introduced a modicum of structure and discipline in his early days on the job, impressing the president's friends and giving West Wing staffers fleeting optimism of a change in course. But Kelly had quickly become overwhelmed by Trump's insatiable appetite for disruption. By the fall of 2018, he was disappearing from the White House for lengthy stretches of the day, telling aides only that he was headed to the gym.

As it became clear that Kelly would leave, the search for his replacement turned into something of an open casting call. Mark Meadows publicly lobbied for the job, calling in every favor he could muster, but the president—warned by several friends of the Freedom Caucus leader's star-seeking ways—turned him down. Trump favored Mike Pence's chief of staff, Nick Ayers, a young operative known for his tactical shrewdness and ethical slipperiness. Ayers had, at Pence's instruction, allied himself with Jared Kushner and Ivanka Trump in order to influence the president. This irked West Wing insiders; they called him "Tricky Nicky" and traded whispers about his financial entanglements that the press would feast upon.[2] (He was a multimillionaire in his twenties, having started and then stepped away from lucrative consulting firms whose clients now enjoyed the conspicuous support of Trump.) Sensing the risks, Ayers pulled his name from consideration.

After other options were considered, one was left standing: Mick Mulvaney. Having moonlighted as the interim boss at the Consumer Financial Protection Bureau, taking a wrecking ball to the watchdog agency while simultaneously running the Office of Management and Budget, the former Tea Party congressman had impressed Trump with his tenacity and his talent for multitasking. The president named Mulvaney, who'd joked about safeguarding the Constitution from Trump and had once called him "a terrible human being," as acting chief of staff.

The final resignation—and to official Washington, the most disturbing—was that of Jim Mattis.

More than anyone else in the federal government, the defense secretary had provided peace of mind to those worried that Trump's worst instincts could prove calamitous. The legendary Marine general had also offered a voice of reason in unreasonable times. In an interview with the *New Yorker* after taking over the Pentagon, when asked to cite his biggest concern on the job, he replied, "The lack of political unity in America. The lack of a fundamental friendliness. It seems like an awful lot of people in America and around the world feel spiritually and personally

alienated, whether it be from organized religion or from local community school districts or from their governments."[3]

Weeks after the Charlottesville clashes left the nation shaken and the
president's own party seething at his response, the defense secretary
encountered a group of military officers during a trip to Jordan. "You're
a great example for our country right now. It's got some problems. You
know it and I know it," Mattis told them. "You just hold the line, my fine
young soldiers, sailors, airmen, Marines. You just hold the line until our
country gets back to understanding and respecting each other and being
friendly to one another."

On December 19, 2018, Mattis released a videotaped Christmas
message to his armed forces. The next day, he abruptly announced his
resignation—a stunning rebuke to Trump's decision, made over the
objections of virtually everyone in the administration, to withdraw all
America's troops from Syria. ("If Obama had done this, we would be going nuts right now," said Lindsey Graham. This was a strong statement
given his unmatched metamorphosis from bruising Trump critic to
bald-faced Trump apologist.)

The president, for his part, declared, "We have won against ISIS," a
hyperbolic assertion that no member of either party agreed with.

"My views on treating allies with respect and also being clear-eyed
about both malign actors and strategic competitors are strongly held
and informed by over four decades of immersion in these issues," Mattis
wrote in his resignation letter to Trump. "Because you have the right to
have a Secretary of Defense whose views are better aligned with yours
on these and other subjects, I believe it is right for me to step down from
my position."

Mattis named his departure date of February 28. The president, infuriated by the glowing tributes to the Pentagon chief on television, announced on Twitter that he would be relieved two months earlier.

The turnover in Trump's administration had been nothing short of
staggering: In addition to dozens of lower-level officials, he had lost two
chiefs of staff, two national security advisers, an EPA administrator, a
health and human services secretary, an interior secretary, a secretary
of defense, and a secretary of state. (After Rex Tillerson criticized the

president in December 2018, Trump tweeted that his former secretary of state was "dumb as a rock" and "lazy as hell.")

The previous fall, Bob Corker, the GOP chairman of the Senate Foreign Relations Committee, had said that three men—Mattis, Kelly, and Tillerson—"Help separate our country from chaos." At the dawn of Trump's third year in office, all of them were gone.

"This is like the second half of the second term of a presidency," Paul Ryan said of the administration's staff exodus, "except it's the *second year* of this presidency."

Ryan, who was himself heading for retirement at year's end, felt a heightened level of anxiety over leaving Washington. He had developed close relationships with Mattis and Kelly, talking with them frequently to strategize on ways to insulate the government against Trump's impulses. Now, scanning the administration for moderating influences, he saw far fewer of them. He also saw a president who was harder to influence than he once was.

"Those of us around him really helped to stop him from making bad decisions. *All the time*," Ryan says. "It worked pretty well. He was really deferential and kind of learning the ropes. I think now . . . he sort of feels like he knows the job. He's got it all figured out. He's comfortable in it. And so he's more listening to his own counsel."

Ryan adds, "We helped him make much better decisions, which were contrary to kind of what his knee-jerk reaction was. Now I think he's making some of those knee-jerk reactions."

THE SUNSET OF 2018 PRESENTED A DIFFERENT SET OF CHALLENGES FOR Trump.

The October death of Jamal Khashoggi, a self-exiled Saudi journalist writing for the *Washington Post* and living in America when he disappeared during a visit to Istanbul, had turned into a test of the president's diplomatic and geopolitical priorities. After a lengthy investigation, it was the consensus of U.S. and Turkish intelligence that Khashoggi had been strangled and dismembered inside the Saudi consulate—at the direction of Crown Prince Mohammed bin Salman, the successor to the Saudi throne and a frequent target of Khashoggi's critiques.

This posed a dilemma to Trump's transactional foreign policy doctrine. The crown prince had become an ally of Kushner's and was instrumental in approving a major arms-sale deal in 2017. The butchering of Khashoggi fit the description of a flagrant international offense for which Washington would traditionally have imposed consequences. Instead, Trump decided to cast doubt on the intelligence findings and declined to give the Saudis even a rhetorical slap on the wrist, willfully ceding America's role as a symbolic guardian of human rights.

Just as the furor over Khashoggi was subsiding, like a game of presidential Whack-a-Crisis, another problem popped up: Michael Cohen was back in federal court with a fresh batch of incriminating testimony.

On November 29, the president's former lawyer pleaded guilty to lying to Congress about the plans to build a Trump Tower in Moscow. Contrary to his sworn testimony to congressional investigators, Cohen said the Moscow project was being negotiated well into the summer of 2016—with Trump's awareness and involvement. This meant the GOP front-runner had been pursuing a commercial interest in Russia *while* he was publicly praising Putin and arguing to end Obama's sanctions against the country, all while the Kremlin was waging a sophisticated misinformation campaign aimed at helping him win the presidency.

Trump, aka "Individual 1," accused his onetime fixer of being "weak" for cutting a deal with Mueller in pursuit of a reduced sentence. Indeed, not long after, the special counsel recommended no prison time for Cohen due to his extensive cooperation. But the Southern District of New York felt differently: Cohen was sentenced to three years for the hush money payments to Daniels and, in a separate case, for lying about the Trump Tower Moscow project.

As if Trump didn't have enough to worry about, the one thing that had buoyed him throughout his first two years in office—a booming economy—was showing signs of weakness.

In the fourth quarter of 2018, the Dow and the S&P 500 dropped nearly 12 percent and 14 percent, respectively, while the NASDAQ plunged 17.5 percent in the same period. The final month of the year capped this bearish run: Both the Dow and the S&P 500 plunged by roughly 9 percent, their worst Decembers since 1931. When all was said and done, 2018 was the worst year for the markets since 2008.

This was irksome to a president who had long used the bull markets as a crutch. And while there was blame to go around for the decline—fears of inflation, tech companies facing scrutiny and enhanced regulation—the country's political volatility was undoubtedly a contributor. Trump's game of chicken with China was proving counterproductive: In December, the Commerce Department announced that the U.S. trade deficit had increased to a ten-year high of $55.5 billion, thanks to soybean exports plummeting and imports of consumer goods spiking. The trade deficit had widened for six consecutive months, and much of the damage was being done by China; in November, its trade surplus with the United States reached a record high of $35.6 billion.

Meanwhile, in late November, Trump was incensed to learn that General Motors was slashing 15 percent of its salaried workforce—up to 14,800 jobs—and shuttering five plants. The brunt of the impact would be felt in Ohio and Michigan, two of his Rust Belt strongholds. In the summer of 2017, Trump had promised a raucous crowd in Youngstown, Ohio, that their lost factory jobs were coming back. "Don't move!" the president told them. "Don't sell your house!" Now, little more than a year later, GM was closing its iconic Lordstown Assembly plant in neighboring Warren, Ohio, robbing the area of more than 1,400 jobs.

"I told her, I'm not happy," Trump told reporters of his conversation with GM's CEO, Mary Barra. "The United States saved General Motors, and for her to take that company out of Ohio is not good." In an interview with the *Wall Street Journal*, he talked even tougher. "They better damn well open a new plant [in Ohio] very quickly," Trump said. "I told them, 'You're playing around with the wrong person.'"[4]

Trump was speaking as a president and not a businessman. The reason GM needed a taxpayer bailout to begin with was that it had become bloated with legacy costs and blind to the evolution of the market, producing cars nobody wanted at prices their foreign competitors could beat. If GM was to avoid another bankruptcy, visionary decisions had to be made: eliminating a host of poor-selling smaller cars, doubling down on SUVs, and investing heavily in electric and self-driving vehicles. In business, the short-term pain of such decisions is meant to yield long-term gains, building a stronger company that can hire more workers.

The other dynamic Trump seemed to be ignoring, willfully or otherwise, was the role his own policies had played in decisions such as these. GM had warned, earlier in the year, that his tariffs on imported steel would cost the company $1 billion. Ford Motor Company had said the same. And Harley-Davidson announced back in June that it was moving some of its operations overseas due to the crushing cost of Trump's tariffs, prompting the president to call for a boycott of the motorcycle manufacturer.

It wasn't surprising that the president's interventions in the market didn't work; even in Indiana, where Trump and Pence had made a show of swooping in to rescue Carrier workers with the aid of billions of dollars in state tax incentives, the company continued to jettison jobs throughout 2017 and 2018 as its operations moved to Mexico. "I feel betrayed and deceived—as if President Trump used my pain, and the pain of working-class America, simply to win political points," Quinton Franklin, a laid-off Carrier worker, wrote in a piece for Vox.[5]

What *was* surprising: the silence of conservatives as the American president bullied and threatened private businesses that were making him look bad politically.

It was the best of public service, it was the worst of public service.

In early December, America mourned the passing of former president George H. W. Bush, a son of privilege who enlisted as a fighter pilot, was shot down over the Pacific, and became the patriarch of America's premier political dynasty, all while serving in roles ranging from CIA director to chairman of the Republican National Committee. Like John McCain, the elder Bush had planned his funeral service with precision; but unlike the senator, the former president was insistent on Donald Trump attending. There was no debate to be had: Whatever problems anyone in the family or in the party had with him, the commander in chief would be welcome at his service. This was a final act of class and grace, symbolizing a life spent in service to noblesse oblige.

Less than a week after Bush's funeral brought fond memories of a functional era in politics, Americans were jolted out of their daydream by the spectacle of December 11.

It was supposed to be a three-minute "spray," with reporters and cameramen from the news networks catching a quick glimpse of the meeting inside the Oval Office. In their first formal sit-down since the midterm, the president and vice president were hosting the top Democrats in Congress, Nancy Pelosi and Chuck Schumer. Customarily, the president would offer a few words, something about looking forward to a productive dialogue, and then the media would be shooed from the room so the principals could get down to the business of governing.

Instead, viewers were treated to something that resembled a cold-open *Saturday Night Live* skit. With Trump rocking energetically on the edge of his seat, Pelosi flapping her arm while speaking in short bursts, Schumer failing to suppress his mischievous grin, and Pence observing a solemn vow of silence, the body language was entertaining on its own. And then there was the discourse.

When Pelosi mentioned how House Republicans were packing up their offices after their losses, Trump spoke over her. "And we've gained in the Senate. Nancy. We've gained in the Senate," he said. "Excuse me. Did we win the Senate? We won the Senate."

Schumer turned to reporters. "When the president brags that he won North Dakota and Indiana, he's in real trouble," he said, leaning back, visibly satisfied with himself.

Trump shrugged, cocking his head sideways and looking confused by the insult. "I did. We did win North Dakota and Indiana."

Comparing electoral win-loss records was not the ostensible purpose of the meeting, of course. Washington was careening toward another shutdown: Government funding was due to expire on December 21, and with the Democrats soon to assume control of the House, the president saw this as his final chance to win funding for his promised wall on the southern border. He had already signed certain spending bills into law, but others awaited votes in Congress; one was for the Department of Homeland Security, which would spearhead any wall-building project. Trump was demanding $5 billion. Pelosi and Schumer had already made clear that they wouldn't give a penny more than the $1.3 billion already allocated.

Believing he could rattle his foes under the bright lights, Trump deliberately let the reporters and cameras linger in the room. He pressed his case for border wall funding, claiming that construction was already under way—"A lot of the wall is built"—and accusing Democrats of obstructing Republicans on a project that they had once supported. (It was true that many Democrats had voted for border wall funding in previous Congresses. It was not true that Trump's wall had progressed as promised; in Texas and California, a few dozen miles of fence were being put up, the same type of barrier built under previous administrations.)

The televised scrum could hardly have been more helpful to the Democrats and more harmful to Trump. For starters, the president rattled off a series of easily disproved untruths—that the wall was already under construction, for instance, and that his administration "caught 10 terrorists over the last very short period of time" at the southern border. (There was zero evidence found to substantiate this; Trump's own State Department had previously reported that there was "no credible information that any member of a terrorist group has traveled through Mexico to gain access to the United States."[6])

On top of that, the president strengthened the internal standing of Pelosi. Dozens of House Democrats had campaigned on a promise not to support her return to the Speakership, and in the weeks following Election Day she had worked tirelessly to extinguish any insurrection in the conference. Younger Democrats whispered doubts about the seventy-eight-year-old Pelosi's agility and acuity, probing for vulnerabilities they might exploit to force her from power. Yet those doubts were erased as she went toe to toe with Trump on his turf.

Most consequentially, Trump did what every modern politician, from Ted Cruz to Bill Clinton, Newt Gingrich to Tip O'Neill, had labored to avoid: He accepted the blame for a government shutdown.

Goaded by Schumer, a fellow tough-talking New Yorker who had studied the art of getting under the president's skin, Trump said that he was so committed to building the wall that he would shut down the government if his funding failed to materialize. "You know what I'll say? Yes. If we don't get what we want, one way or the other, whether it's through you, through the military, through anything you want to call, I will shut down the government," Trump said.

"Okay. Fair enough," Schumer said, looking like the cat that had feasted on a whole flock of canaries. "We disagree. We disagree."

Trump's face grew more colorful, his tone sharper, his torso bending so aggressively off his chair that he might have fallen into Schumer's lap.

"I am proud to shut down the government for border security, Chuck. Because the people of this country don't want criminals and people that have lots of problems—and drugs—pouring into our country. So, I will take the mantle. I will be the one to shut it down. I'm not going to blame you for it."

BURSTING THROUGH THE DEEP, OVERBEARING DARKNESS CAME THE faintest hint of light. In the waning days of the 115th Congress, after two years of ruthless polarization and petty score-settling between the two parties, something was happening. It was not a small something; in fact, it was a very big something, something so big, something so important, that it had been unthinkable not long before. Republicans and Democrats were teaming up to pass, on a bipartisan basis, a sweeping, significant piece of legislation. And no one deserved more credit than Trump.

For years, idealists and romantics spanning the partisan spectrum had dared imagine the day in which Congress might finally correct one of its grievous errors: the tough-on-crime legislation of the 1980s and '90s that had led to an explosion in the U.S. prison population. With the onset of mandatory-minimum sentencing and three-strike laws, America had swept up a generation of nonviolent drug offenders, a wildly disproportionate number of them black, into a penal system making lots of profits and doing little rehabilitating.

Subsequent administrations nibbled around the edges of the problem: Bush signed his prisoner reentry program into law in 2008, and Obama passed a law reducing the disparity in sentences between crack and powder cocaine. Yet these policies did little to address the systemic failures of the criminal justice apparatus: the Third World conditions, the absence of training for societal reintegration, the lack of discretion for judges in sentencing drug-related cases, and the financial incentives to build new prisons rather than keep old inmates from returning.

By 2016, according to the Bureau of Justice Statistics, more than two million people were incarcerated in the United States.[7] A bipartisan coalition had emerged during the twilight of the Obama years to push a massive criminal justice reform bill, but it was doomed to failure for a simple reason: Conservative Republicans were reluctant to align themselves with Obama on legislation that could be portrayed as soft on crime. With the election of Trump, a self-described "law and order" candidate, the prospects for reviving the effort looked bleak.

Instead, ever so quietly, the coalition expanded during the new president's first year in office. It was an unlikely crew of conspirators: In the Senate, Democrat Dick Durbin worked alongside Republicans Chuck Grassley and Mike Lee, while on the outside, the Koch brothers teamed with the ACLU and the Center for American Progress. The linchpin: Kushner, the president's son-in-law, who gained a passion for the issue after his father was imprisoned on tax evasion charges.

Kushner didn't exactly have a track record of accomplishment. He had taken on a laughably heavy portfolio, everything from leading the newly created Office of American Innovation, to spearheading a strategy to combat the opioid epidemic, to brokering a peace agreement between the Israelis and the Palestinians. (All in a day's work for a thirty-something with zero government experience.)

And yet, halfway through the second year of Trump's presidency, there were continued signs of progress. The most apparent: Trump's decision in June to commute the sentence of a sixty-three-year-old woman, Alice Johnson, who was two decades into a life sentence for narcotics distribution.

The president's move came after a meeting with Kim Kardashian, the reality TV starlet and wife of Kanye West, that was widely panned in Washington as something of a publicity stunt. Yet the rendezvous had been strategically arranged by Kushner. After inviting Kardashian into the Oval Office to hear more about Johnson's case, Trump found himself agreeing that she should receive clemency. Kushner took this as a clear indication, he told his allies on Capitol Hill afterward, that the president could be persuaded of their broader reform efforts.

Several obstacles remained: John Kelly, a hard-boiled military man who shared much of Trump's sensibilities on law enforcement, had cautioned against loosening sentencing laws. So had Sessions—though, as Kushner calculated, the attorney general's disapproval was likely a net positive given Trump's loathing for the man.

And then there was Tom Cotton. The Arkansas senator, a Harvard Law grad and retired Army captain who had seen tours in Iraq and Afghanistan, was Congress's most vocal opponent of the criminal justice overhaul. He spent the summer and fall of 2018 warning any Republican who would listen that they would get "Willie Hortoned" in their future campaigns by voting to let convicts out early, a reference to the infamous race-baiting ad used against Michael Dukakis in 1988.

Cotton's resistance vexed Kushner and his allies on Capitol Hill. The president, ever a fan of glimmering résumés, had been smitten with Cotton since meeting him in 2015, and the Arkansas senator was one of the few lawmakers who truly had the president's ear. Of all the people susceptible to concerns of being portrayed as soft on crime, the reformers figured, none was more likely to buy Cotton's argument than Trump.

It was a genuine shock, then, when in mid-November the president decided to throw his weight behind the First Step Act. The legislation, scaled back somewhat from the Obama-era effort, would enhance rehabilitation programs for current and former prisoners, provide new funding for antirecidivism initiatives, give judges more discretion in sentencing, reduce certain mandatory-minimum terms, and improve the conditions inside the penal system itself, including an expansion of employment programs.

"Americans from across the political spectrum can unite around prison-reform legislation that will reduce crime while giving our fellow citizens a chance at redemption," Trump announced at the White House. "It's the right thing to do. It's the right thing to do."

Standing behind the president, once again, was Senator Tim Scott. Just as he had a year earlier with the push for "Opportunity Zones" in the tax bill, Scott leveraged his complex alliance with Trump to lobby for the criminal justice legislation.

It was the latest evolution in the schizophrenic relationship between the Republican Party's lone black senator and the man he carefully referred to as its "racially insensitive" president.

That same month, Scott had come under fire from the right for dealing the fatal blow to the nomination of a federal judge, Thomas Farr, who had a long record of disenfranchising African American voters. Predictably, when Scott chose to advance Farr's nomination to a final vote (a procedural move that was in no way reflective of his verdict on Farr), he was met with standard vitriol from the left, with black leaders condemning him on social media and countless people piling on, calling him "Uncle Tom" and worse. When Scott killed Farr's nomination a day later, there were few apologies to be found.

While walking his impossible tightrope, Scott was inching closer to discovering the formula for moving Trump—and, in some sense, decoding the mystery of what made him tick.

"Why would the president of the United States, a guy named Donald J. Trump, take this issue on? It ain't because he's racist. I think he's doing it because he thinks it's the right thing to do," Scott said that month. "When you watch him on TV, it's a presentation. When you sit down and talk with him, sometimes you move the needle. He's not a guy that wants to be attacked. But if you come at him logically, I find that he takes time to listen. If you can get his attention and you've got something meaningful to say, he'll listen to you."

There was one final hurdle to clear: Mitch McConnell.

The Senate majority leader was much closer to Cotton's worldview than Scott's. Forever paranoid about his vulnerabilities on the right, McConnell privately worried that in the two years between passing a bill in 2018 and standing for reelection in 2020, all it would take was a single paroled criminal doing something heinous to end his career.

Trump leaned on McConnell in early December, dismissing the majority leader's claims that there was no time on the legislative calendar for such a vote, and implicitly threatening to make his life miserable if he hung the effort out to dry. Finally, on December 11, McConnell announced his support. After some final tinkering to ensure a maximum number of GOP votes, the bill passed the Senate a week later on a tally of 87 to 12.

It was that rare moment of bonhomie in an age of bitter polarization. Off the Senate floor, Cory Booker, the young, black New Jersey Democrat, embraced Chuck Grassley, the old, white Iowa Republican.

The bill cleared the House on December 20, by a margin of 358 to 36, and was swiftly signed into law. It was rightly viewed as one of the landmark bipartisan achievements of the twenty-first century and it would not have been possible without the active support of the president.

This could have been the safe, feel-good high note that Trump chose to end the year on. Instead, with government funding set to expire on December 21, he decided to gamble.

CHAPTER TWENTY-SIX

DECEMBER 2018

"We had Abraham Lincoln then."

STEVE SCALISE WAS PERCHED INSIDE THE CAPITOL HILL CLUB, EN-joying a glass of red wine on a frosty winter's night, when he was interrupted by a phone call. He hurried to answer: It was his former colleague and the new White House chief of staff, Mick Mulvaney.

"The president is pissed," Mulvaney told Scalise. "He feels like you guys sold him a bill of goods. He's gonna veto this damn thing."

It was Wednesday, December 19, and Washington appeared peaceful—from a distance. Though the government was scheduled to run out of funding on Friday night, there were few outward signs of panic. For the past week, Republicans on both ends of Pennsylvania Avenue seemed to believe that Trump would have no choice but to sign a short-term package to keep the government running into February. Democrats weren't budging on President Trump's request for north of $5 billion to construct a border wall; and even if House Republicans unified behind such a proposal, muscling it through the lower chamber in the twilight of their majority, it stood zero chance of passing the Senate. With these realities in mind, and against the backdrop of Trump preemptively accepting blame for a shutdown, the White House was resigned to the president approving a stopgap measure with just $1.6 billion allocated for border security purposes.

By Wednesday evening, however, the president was worked into a lather. He had spent much of the day talking by phone with Jim Jordan and Mark Meadows. They had convinced Trump of something that was undoubtedly true: He had been played.

Back in the spring, when Paul Ryan persuaded him not to veto the massive omnibus spending bill, the president had been promised that his border wall money would come later that year. It was vitally important to get the increased funding for the military while they could, Ryan said. (The Speaker was joined in this lobbying campaign by a new Montana congresswoman, who told the White House that short-term spending bills were endangering troops' lives. Her name was Liz Cheney, the daughter of the former vice president and an heir to his interventionist foreign policy doctrine.)

Now, with Democrats preparing to take over the House majority, there was one last chance to get that money for the border wall. Instead, Ryan was going home—knowing full well that funding would never, ever materialize—and Trump was getting stiffed.

His wrath swelled as the day went on. Fuming that evening as he watched Freedom Caucus members rail on the House floor against the short-term funding bill, denouncing it as an affront to the president's promises, Trump beckoned his advisers to the residence. He complained that Ryan had lied to him. Mulvaney was both inciting and sympathetic: The chief of staff knew what it was like to be strung along by the House leadership, told to live to fight another day only for that fight never to get picked. He encouraged Trump to stand his ground, to show Congress who was boss, to bend Washington to his will.

The tipping point was Laura Ingraham. Having tuned in to Fox News for her 9:00 p.m. show, the president was mortified listening to the exchange between Ingraham and Jordan as they discussed the dearth of a border wall.

"Jim Jordan, it is true, Congress hasn't done its job on this particular issue," she told him. "I think that not funding the wall is going to go down as one of the worst things to have happened to this administration— forget Mueller."

A moment later, Ingraham aired a clip from Rush Limbaugh's talk radio show from earlier that day, in which Limbaugh compared Trump's failure to build the wall to George H. W. Bush's promise, "Read my lips: no new taxes."

Then, Jordan put the cherry on top. "Four times we promised them that we would build the wall and put it in the spending bill, and now we're saying, 'Oh, no, we're going to kick it to February, when Pelosi's going to be Speaker," he said. "It'll never happen. We've got to do it *now*."

As the president was watching, Meadows called yet again, urging Trump to issue a veto threat that would force Democrats to the negotiating table.

It was at least Meadows's fourth call that day; the Freedom Caucus chairman was using an old trick, phoning Trump late at night while he was in the residence and away from staff, hoping to persuade him of decisions that White House aides disagreed with. On this count, it didn't matter. Mulvaney and Meadows were on the same page. As they spoke, Trump threw his hands in the air, letting fly an impressive string of expletive-deleted remarks about Ryan and Mitch McConnell, wondering how such a gutless party had ever gained power before he came along.

Mulvaney was soon on the horn with Scalise, the House GOP's designated vote-counter, warning him of the president's tailspin. The majority whip quickly contacted both the Speaker and the House majority leader. He conveyed Mulvaney's message and suggested that their plans to pass the stop-gap funding bill on Thursday might be derailed. Ryan and McCarthy did not share his concern. House Republicans, the two agreed, would not tolerate ending the year with a shutdown, especially after having been pummeled in the midterm elections.

As the Senate passed its short-term measure that night, kicking it over to the House for a vote the next afternoon, Ryan went to bed confident and content. It had been a most memorable day: Earlier, inside the Library of Congress, the Speaker had delivered his farewell address, capping his twenty-year congressional run. This was a part of his legacy-burnishing project, along with a widely lampooned six-part, taxpayer-funded video series documenting his career-long pursuit of tax reform.

The point of the speech was threefold. First, Ryan hoped to validate his contributions to the institution, portraying himself as a crusader who sinned only in challenging the status quo. "I acknowledge plainly that my ambitions for entitlement reform have outpaced the political reality, and I consider this our greatest unfinished business," he said, noting nevertheless, "I am darn proud of what we have achieved together to make this a stronger and more prosperous country."

Second, Ryan endeavored to elevate himself above the beleaguered state of American politics. He warned that "genuine disagreement" had given way to "intense distrust" in the age of social media and twenty-four-hour news cycles. "All of this gets amplified by technology, with an incentive structure that preys on people's fears, and algorithms that play on anger," he complained. "Outrage is a brand."

Lastly, most implicitly but most important, the Speaker sought separation from the candidate whose impulses he'd combated and the president whose behaviors he'd ignored. Promising that "our problems are solvable if our politics will allow it," Ryan did not mention Trump by name, yet worked methodically toward persuading the audience that he, Ryan, should not be remembered in the context of the forty-fifth president. "I knew when I took this job, I would become a polarizing figure. It comes with the territory," Ryan said. "But one thing I leave most proud of is that I like to think I am the same person now that I was when I arrived."

In the *personal* sense, this was true: Ryan was still the unfailingly polite, approachable, decent person he'd been all along, the guy House Democrats couldn't bring themselves to dislike even as they accused his policies of killing Grandma.

In the *political* sense, however, Ryan's self-portrait was a mirage. The truth was, he had come to Congress as a Jack Kemp conservative and would depart as a Donald Trump Republican.

It was more complicated than that, certainly. History requires shade and texture. But legacies are reductive by nature. And as the Speaker walked away, closing a messy and mesmerizing chapter in the party's history, the harsh reality was that Ryan would be remembered more for enabling Trump's mischief than for crafting a generational overhaul of the tax code.

It was a political obituary of the Speaker's own writing. His silence in the face of Trump's indignities, and his observance of "exquisite presidential leadership," a line that will live in infamy, would be less remarkable had he not first established himself as one of Congress's good guys, someone whose sense of principle and civility informed his objections to the man in the first place.

Back when he became Speaker, Ryan warned that the Republican

Party's internal fractures threatened to make legislating impossible. "We basically run a coalition government," he complained, "without the efficiency of a parliamentary system."

This was the story of John Boehner's Speakership, certainly. Yet those internecine breakages had largely receded during Ryan's tenure. The party had fallen in line behind Trump; there was no real power struggle within the GOP of 2017 and 2018. This meant that when historians got to asking the obvious questions—How did the party of fiscal sanity become the party of the historic spending increases? How did the party of family values become the party of "grab 'em by the pussy"? How did the party of compassionate conservatism become the party of Muslim bans?—the answers would implicate not just Trump but Ryan as well.

There would be no repairing the tattering of his image. In July of 2016, Ryan's approval among Wisconsin's likely voters was 50 percent favorable and 34 percent unfavorable, according to the Marquette University Law School poll.[1] A month before the 2018 midterm elections, in a Marquette survey done with the same methodology, it had sunk to 41 percent favorable and 49 percent unfavorable.

Politics are cyclical by nature. The war for the future of the Republican Party, he assured himself, would rage on. But in the short term, the battle for the GOP's heart and soul was finished. Trump had won—and Ryan would be remembered as both victim and accomplice.

WILLFULLY IGNORANT TO THE DETERIORATING PERCEPTIONS OF HIS reputation, Ryan was just as blind to the storm brewing inside the GOP conference.

When the House Republicans gathered in the Capitol basement on the morning of Thursday, December 20, Ryan expected a meeting that would be standard for moments such as these: complaints from the conservatives, pushback from the center-right, assurances of lemonade-making from the leadership, and ultimately, no real change in the trajectory of events.

The Speaker was mistaken.

His members were out for blood, and it wasn't just the conservatives.

For more than an hour, one lawmaker after another stood up to make the case for funding the border wall. It wasn't that all of them were on board with the policy; in fact, many of them were not. Yet they all had grown tired of being accused back home of not supporting the president. The last thing they needed was Trump accusing them of treason as they headed home for the holiday recess. If a shutdown amounted to coal in their stockings, a warlike tweet from the president was akin to the Christmas tree catching fire.

There was every reason to fear such an onslaught. Over the past twelve hours, word had spread rapidly throughout the conference of Trump's fury at being stuck with a short-term spending bill that wouldn't fund his wall. Scalise had even warned members that the White House was asking for a copy of his "whip check," an accounting of which way lawmakers were leaning on a vote. Trump was preparing to lash out at any House Republican who didn't stand with him.

If it wasn't clear what needed to be done, Steve Womack's speech erased any doubts. The Arkansas lawmaker was so low-key, so soft-spoken, that his colleagues weren't sure they had ever heard him talk in the House GOP's weekly meeting. This wasn't surprising; it was the hardliners who typically dominated the open-mic portion, and Womack, a longtime ally of the leadership, was nobody's idea of a hard-liner.

On this morning, however, Womack walked to the microphone with a simple message for his colleagues. "We've got to have this fight," he said, "and we've got to have it now."

Much as with Boehner's situation in the fall of 2013, Ryan was trapped: The Speaker would either be blamed for abandoning a core promise to the party's base or vilified for leading the government into a shutdown. And no matter how little leverage Boehner had in those days, Ryan, whose career would expire in two weeks, now had even less. He was the lamest of ducks.

As the House meeting broke up, lawmakers saw their phones sparkle to life with a Twitter alert. "When I begrudgingly signed the Omnibus Bill, I was promised the Wall and Border Security by leadership," Trump wrote. "Would be done by end of year (NOW). It didn't happen!

We foolishly fight for Border Security for other countries—but not for our beloved U.S.A. Not good!"

A few hours later, as the president convened a small summit of lawmakers at the White House, it struck everyone as a most fitting conclusion to the 115th Congress. For the past two years, there had been a tug-of-war for Trump's political soul, pitting Ryan and McCarthy against Jordan and Meadows. Now all four of them, plus a handful of others, were huddled around a coffee table in the Oval Office, pleading their cases to the president.

There wasn't much suspense: They all recognized that Trump's mind had been made up. As it became evident what he wanted them to do—vote on a bill authorizing $5 billion to build a wall on the southern border, then dare the Senate to reject it, somehow believing this would force Democrats to negotiate—the president's staff jumped in on Ryan's behalf. "This is absolutely crazy," said Shahira Knight, the legislative affairs director, who commanded Trump's respect. "It's never going to work."

But the train had left the station. It was nothing if not poetic: Some of the same conservative agitators who had prodded Boehner into a shutdown five years earlier—Meadows, Jordan, Mulvaney—were back at it. At his winter home in Florida, the ex-Speaker was swirling a glass and cackling at the "legislative terrorists" he'd found unfit to serve in Congress who were now running the federal government.

Later that night, after slapping together a bill that met the president's demands, and juicing it with millions of dollars in disaster-relief funds to win over some skeptics, the House passed it by a tally of 217 to 185.

Many senators had already left town for the holidays; now the House was expecting them to come back for a vote on their newly passed bill. They made plans to return, some moving more urgently than others. There was little point to this exercise: The House bill did not have the support of all fifty-one Republicans, much less the additional nine Democratic voters needed to surmount a filibuster. It stood no chance of passing the Senate, and everyone knew it.

Trump made a game effort Friday afternoon to work the phones, trying to sell GOP senators on his master plan to squeeze funding out of red-state Democrats eager to avoid a shutdown over the issue of border security. But he was living in a fantasy land. Nothing resembling the $5.7 billion House package was going to be approved by the Senate. And given how the president had already bragged to America during the surreal Oval Office meeting with Pelosi and Schumer of his willingness to own the shutdown, Democrats weren't fearful of taking the heat.

As the clock struck midnight on Saturday, December 22, two things were apparent. First, this government shutdown was every bit as pointless as its Obamacare-inspired predecessor in 2013; neither one stood a chance of effecting the desired policy change. Second, this one was going to last a lot longer than seventeen days. With Democrats scheduled to seize control of the House on January 3—at which point their leverage would only increase—there was scant chance of a quick resolution. Both sides were digging in, understanding that the fight over Trump's border wall was about a whole lot more.

Lawmakers were told to stand at the ready; a sudden call might come with news of a breakthrough in the negotiations. They scoffed and raced to the airports, hopping flights back home to celebrate the holidays.

The only person stuck in Washington was Trump. Convinced by aides that it would look bad if he left for Mar-a-Lago while tens of thousands of federal employees were being furloughed, the president remained in the White House, spending his days watching Fox News and dialing friends, asking when they thought the Democrats would cave.

IN AMERICA'S TWO-PARTY SYSTEM, ECONOMIC VITALITY HAS TRADItionally acted as the fulcrum for its political swings. When the economy performs well under the president's party, the opposing party is compelled toward the middle; when the economy suffers under the president's party, the opposing party is free to drift toward its base. These rules are not absolute. But particularly in the previous century, the ideological adventurism of a party occurred while out of power and during times of economic turmoil: Democrats in response to Hoover, Republicans in response to Carter, and, in a case historians will study for centuries, Republicans in response to Obama.

How, then, to characterize the Democrats' response to Trump?

To observe the 2018 election season was to witness the party out of power struggling with its very identity, torn between two diverging paths forward. One was to straddle the center, targeting independents and disaffected moderate Republicans. The other was to push unapologetically leftward, courting the energy and activism of the progressive left.

There was no right or wrong answer; in many cases, the calculus depended on the district and its constituencies. But at the dawn of the 116th Congress, with the forty new Democratic members arriving on Capitol Hill, it became apparent that something would have to give.

Consider two incoming members of Michigan's delegation: In the Thirteenth District, a safely blue stronghold anchored in west Detroit, Rashida Tlaib ran on a platform of abolishing Immigration and Customs Enforcement, pursuing a single-payer health care system, and cutting off foreign aid to Israel. (In 2018, Tlaib was one of the first two Muslim women elected to Congress.) An attorney and a member of the Democratic Socialists of America, Tlaib offered a distinct vision for the opposition party.

Twenty minutes away, in Michigan's Eighth District, a longtime Republican lock stretching from Detroit's affluent suburbs westward toward Lansing, Elissa Slotkin offered another. A former CIA analyst who grew up on a farm, Slotkin defeated GOP incumbent Mike Bishop by running as a "midwestern Democrat," emphasizing her support for gun rights, opposition to single-payer health care, and eagerness to secure the southern border, albeit not with a big, beautiful wall.

The weekend before the election, at a rally in Lansing, Slotkin described the ways in which she was ready to work with Trump.

The day she was sworn in, at a party hosted by the group MoveOn, Tlaib declared, "We're gonna go in there and impeach the motherfucker."

There could be no splitting the baby. Even if in agreement on certain issues, this pair of Michigan lawmakers shared as much of a common purpose as Boehner and Jordan, the once-neighboring Ohio Republicans.

The majority-makers were those Democrats like Slotkin, and Jason Crow of Colorado, and Dean Phillips of Minnesota. These freshmen, and others, flipped suburban, culturally moderate GOP districts by presenting themselves as pragmatic centrists. But if the previous ten years had taught Washington anything, it was that the louder, more ideological voices rise to the top.

Not surprisingly, as the new House majority settled in, it was the likes of Tlaib and her fellow Democratic Socialist, Alexandria Ocasio-Cortez, who dominated the conversation about the party's direction. Ocasio-Cortez was especially effective in moving the Overton window, more so than any incoming Democratic lawmaker in at least a generation if not much longer. After Ocasio-Cortez was ridiculed for suggesting a top tax rate of 70 percent on a narrow slice of multimillionaire earners, a number of public polls showed widespread support for the idea, including among Trump's own blue-collar supporters.[2]

"She's got talent. Now, that's the good news," Trump says of Ocasio-Cortez. "The bad news: She doesn't know anything. She's got a good sense—an 'it' factor, which is pretty good, but she knows nothing. She knows nothing. But with time, she has real potential."

Other Republicans were less sanguine at seeing the Democratic Party lurch leftward in response to the GOP's decade-long drift toward the right. "What I hope is that thirty years from now, your children are reading that this was an aberration," Bob Corker, the Tennessee senator, said before leaving office. "But as of now, I am worried both sides of the political aisle are moving toward unacceptable extremes—Republicans toward authoritarianism by not appropriately pushing back on executive overreach and Democrats toward socialism."

It's imperative to assess Trump not as the cause of a revolutionary political climate, but as its consequence; the forty-fifth president's election was the by-product of a cultural, technological, and socioeconomic convulsion that bred disparate yet interconnected strands of populism on both the right (Tea Party) and the left (Occupy Wall Street). Maybe those fatigued Americans pulling for moderate Democrats to take things back to "normal" are fooling themselves. Maybe there's no "normal" to which America can return.

The tactical case for Democrats to hug the middle, in the short term, is the Electoral College: Trump's path to reelection is even narrower than it was in 2016. Once-competitive states such as Virginia and Colorado are off the map; and even states that he carried, such as Arizona and North Carolina, might prove difficult to hold. Trump's strategy is centered on the same trio of Rust Belt states that won him the presidency: Pennsylvania, Michigan, and Wisconsin. Without one, he's in trouble; without two of them, he's almost certainly defeated; without all three, there is no scenario for his reelection.

Recognizing this, Democrats such as Slotkin and Phillips came into office with a message: The party didn't need to veer far left, abandoning the persuadables in Middle America that they won over in 2018. It didn't need to run up its margins in blue states to take back the White House.

"In Michigan, I know a lot of people who voted for Barack Obama and then voted for Donald Trump. And they tell me, 'You know, my life hasn't gotten better from Bill Clinton, George Bush, Barack Obama. I'm like a stage-four cancer patient, and Donald Trump is my experimental chemo,'" Slotkin says. "We need to hear that as Democrats. A lot of people felt like, last cycle, that Donald Trump was the only one talking about the issues that dominate their lives: their job, how much money they make. . . . If we can't address those things, we're not going to win. We don't deserve the Midwest vote if we can't talk about those things."

It sounds good in theory. But the gravitational pull of a party's base can render resistance futile. Phillips, who flipped the suburban Twin Cities district held by Republicans since 1960, said on the Sunday before the election that he wouldn't vote for Pelosi as Speaker—in part to push back against the stereotype of Democrats a coastal party.

And yet, on January 3, in the first vote of their congressional careers, Phillips and numerous other freshmen Democrats supported Pelosi's return to the speakership. In fairness, nobody was running against her, and the promised wave of widespread opposition never materialized. Still, the decision of some Democrats to renege on a core campaign promise in the opening hours of the new Congress reflected the same instinct toward self-preservation that over the previous decade had eroded the institution and left voters feeling cheated by politicians.

"No one here is pure. Every group has their own agenda. When new people come to Congress and ask for my advice, I tell them to do what they told their voters they were going to do," says Raúl Labrador, the retiring Freedom Caucus congressman whose time on the front lines of the Republican civil war holds valuable lessons for the Democrats.

"As long as you keep your promises, you'll at least have your integrity. I think some people lose their soul here. This is a place that just sucks your soul. It takes everything from you."

WHATEVER RELIEF REPUBLICANS FELT AT SEEING THE LEFTWARD TRA-jectory of the opposition party, Democrats were just as elated to witness the GOP's lack of course correction after its mauling in the midterms.

Heads fully submerged in the sand, leading Republicans rejected the notion of a widespread political reorientation in the age of Trump, insisting that the 2018 results foretold no long-term threat to the party. This was voiced most naïvely by Tom Emmer, the new chairman of the National Republican Congressional Committee, who told *National Journal* after the election, "There's a narrative that people are trying to build out there that somehow there's been this shift, this political realignment in the suburbs. That's not true. It isn't there."

To be clear: It *was* true. It *was* there.

Just as the 2010 election saw the purging of Congress's "Blue Dogs," coinciding with Democrats losing their foothold in rural America, 2018 saw the annihilation of Congress's moderate Republicans, coinciding with the GOP's presence fading in the suburbs from coast to coast. This was the definition of a realignment: Democrats flipped two-thirds of the GOP-held House seats with the highest median incomes, according to the *Cook Political Report*'s David Wasserman, while flipping just 6 percent of the GOP-held seats with the lowest median incomes.[3] When the final results had been tabulated, Republicans were left representing just two of the thirty congressional districts with the most college degrees.

For anyone doubting the ramifications of this, or the reality of a "Blue Wave" in 2018, consider that Democrats won nearly 9 million more total votes in House elections than Republicans, breaking the record set in the post-Watergate midterm of 1974.

The GOP's pain was felt far beyond the Beltway. Democrats flipped 7 governor's mansions, 7 legislative chambers, and nearly 400 state legislative seats. The vast majority of these victories were on the strength of a mass resurgence in America's metropolitan expanses.

In Wisconsin, for example, Governor Scott Walker lost his bid for a third term by a single percentage point—after winning by 6 points in 2014. The reason: While his performance improved in the rural middle and northern parts of the state, his numbers dropped by double digits in the suburbs of Milwaukee and Madison.

The implications were straightforward. Republicans already faced an existential threat because of their anemic support from minority voters, and specifically Hispanics, the fastest-growing bloc of the electorate. In a presidential contest, bleeding the support of college-educated white suburbanites to boot would make the party's electoral math unworkable.

The GOP's challenge wouldn't be just with suburbanites writ large, but with women in particular. And its dismal performance among female voters could not be distinguished from its exclusion of female lawmakers: The new Congress convening in 2019 saw a record-setting 102 women serving in the House—but just 13 were Republicans. That number was nearly cut in half from the previous Congress, when 23 women served in the House GOP.

This was a challenge that Elise Stefanik, one of the party's young standouts, was desperate to address. Leading the NRCC's recruitment efforts for the 2018 election, Stefanik, a former Bush administration official representing upstate New York, found it "very, very difficult to recruit women candidates" to run for Congress. "This was a problem pre-Trump, and it's going to be a problem post-Trump," she said, "Although, it's been exacerbated by the president's rhetoric."

Stefanik knew, however, that her party's problems run deeper than its showing with any single demographic group. The congresswoman was witnessing in real time the outgrowth of the "isms" that her former boss, President Bush, once warned of. "There *will* be a post-Trump era," she said. "And I think there's going to be a new generation of voices in the Republican Party that push back on some of the trends we've been seeing—the isolationist, anti-trade, anti-intellectualism trends that are not moving us in the right direction."

The old generation might have its say, too.

On New Year's Day, forty-eight hours before he was sworn in to serve his freshman term, Senator-elect Mitt Romney penned an op-ed in the *Washington Post* that sent shockwaves through the capital city. Explaining that while he agreed with many of Trump's policy decisions, and declaring that he would not "comment on every tweet or fault," Romney warned, "With the nation so divided, resentful and angry, presidential leadership in qualities of character is indispensable. And it is in this province where the incumbent's shortfall has been most glaring."

Trump was disgusted. He recalls a conversation with Romney, during the interview for secretary of state, when he told him, "if only you spent the same energy" against Obama in 2012 as he had opposing Trump in 2016, he would have won the presidency for himself. "But he only wants to play hardball against me," Trump says, rolling his eyes. "Romney had too much respect for Obama."

Trump scolded the incoming senator on Twitter, urging Romney to be a "TEAM player" and help Republicans "WIN!" But for Romney, winning wasn't merely about legislative conquests and electoral triumphs; it was about the government projecting moral leadership, providing an example of comity and dignity for the rest of the country to follow.

For some, these notions were long since irrelevant. To support Trump meant to ignore or justify all that he said and did—period.

This continued to manifest itself most entertainingly on the religious right. As the government shutdown spilled into January, Robert Jeffress, the Dallas pastor who had railed against Romney's Mormonism in 2008 and 2012, told Fox News that the president's tactics were warranted because "The Bible says even heaven itself is going to have a wall around it." Around that same time, Jerry Falwell Jr. told the *Washington Post Magazine* that it was a "distortion" to say America "should be loving and forgiving" because Jesus taught such things. "In the heavenly kingdom the responsibility is to treat others as you'd like to be treated," Falwell Jr. said. "In the earthly kingdom, the responsibility is to choose leaders who will do what's best for your country."

Not all churchgoers and committed Christians were so unblushingly apologetic for Trump. But over his first two years in office, no group had debased itself quite like their foremost clerics.

"These evangelical [leaders] are the biggest phonies of all," says Michael Steele, the former party chairman. "These are the people who spent the last forty years telling everyone how to live, who to love, what to think about morality. And then this motherfucker comes along defiling the White House and disrespecting God's children at every turn, but it's cool, because he gave them two Supreme Court justices. They got their thirty pieces of silver."

There were indicators of progress inside the GOP, however halting and long overdue.

In the middle of January, as the shutdown raged on, McCarthy took a step that should have been taken years earlier: stripping Steve King of his committee assignments.

The Iowa congressman had been making thinly veiled racist comments for at least a decade. And his rhetoric had grown that much bolder since the election of Trump: speaking of "cultural suicide by demographic transformation"; meeting with members of a far-right, Nazi-founded Austrian party; endorsing a self-avowed white nationalist for mayor of Toronto; and warning, "We can't restore our civilization with somebody else's babies."

But it wasn't until he finally, fully removed the veil that Republicans felt compelled to act. "White nationalist, white supremacist, Western civilization—how did that language become offensive?" King said in an interview with the *New York Times*.

When it was published in January, the House of Representatives voted 424 to 1 in favor of rebuking King; the lone dissenter was a Democrat who wanted King formally censured. Meanwhile, the party's leadership removed King from his committees. This would essentially make him useless to his constituents; not taking any chances, Iowa's GOP leaders worked behind the scenes to promote a challenger in the upcoming 2020 primary, a state lawmaker with big donors and deep roots in the Fourth District.

But none of this guaranteed King's defeat: At his first town hall meeting back home after the hullabaloo in Washington, he received a standing ovation.

THE SHUTDOWN WAS A FITTING CONCLUSION TO THE REPUBLICAN PARty's unified ownership of Washington—and a most appropriate beginning to the era of divided government.

"The Wall" had become such an all-eclipsing rhetorical commitment for Trump, both to his base and to the skeptics who questioned his ability to build it, that the president made little effort to understand the policy itself. Nobody who had studied the southern border, Republican or Democrat, thought a physical barrier across the entirety of it, or even much of it, made sense.

"A wall from sea to shining sea is the single most expensive and single least effective way to secure the border," Will Hurd, the Republican congressman and former CIA agent, said after Trump took office.

Hurd would know: His district, stretching from San Antonio to El Paso, includes more of the U.S.-Mexico border, 820 miles, than that of any member of Congress. A national security hawk who studied and lived the border issue every day, Hurd reached the conclusion that a wall simply wasn't going to work. Traffickers would tunnel under or climb over. There were a few urban stretches, perhaps forty or fifty miles in all, where see-through fencing would be effective and necessary. But a physical barrier wasn't remotely the catchall solution Trump claimed it was.

What Hurd offered on behalf of experts on the ground: cutting-edge fiber optic cables and high-definition cameras across the border, monitored by a beefed-up border patrol, that would funnel the flow of drugs and migrants into the legal ports of entry. There, at border inspection stations, the federal government would invest billions of dollars in new technologies capable of screening for the people and products Washington wanted to keep out.

Even as Trump came to understand this argument, the White House preferred to push the dichotomy of The Wall, a symbolic contrast between Republicans who wanted to secure the border and Democrats who didn't. This allowed the two parties to talk past each other, inflating resolvable differences while ignoring easily discovered common ground.

When conservative journalist Byron York pointed out the "supreme weirdness" of the shutdown, indicating the consensus around what many Republicans wanted (fencing in urban sectors, more boots on the ground, technology at ports of entry), George Conway, the Republican lawyer and husband of White House counselor Kellyanne Conway, responded, "Not weird at all. Trump is a master at alienating people he ought to be trying to, and should be able to, persuade. And that's because he can't make a coherent argument. He's incompetent."

Indeed, it should have been no problem for the president to sell Americans on these ideas, reminding them that Democrats once supported most of them. Instead, his hang-up on The Wall, a "manhood thing," as Pelosi suggested to her colleagues, left him thrashing about as the shutdown dragged on. There was no happy ending possible. Trump had gone from promising that Mexico would pay for the Wall, to withholding paychecks from federal employees until the U.S. Congress promised to pay for the Wall.

In a meeting with Democratic leaders in the first week of January, Trump threatened to keep the government closed indefinitely—for months, maybe years—until he got his wall money. Then, speaking from the Rose Garden, he said he might declare a national emergency to procure funding for the project.

It was an absurd idea; aside from shattering the principles of limited government, such a maneuver would be immediately tied up in the courts. And, as some Republicans recognized, it was a slippery slope, constitutionally and otherwise. If a Republican president were to seize funding for the purpose of building a border wall, what was to stop a Democratic president from doing the same, but in pursuit of universal health care, or climate change regulation, or whatever else the left might demand?

Moreover, the idea was deeply insincere at an intellectual and ideological level. A national emergency is for *emergency* scenarios, addressing an urgent problem that can be addressed in no other way. For the previous two years, Republicans had controlled both chambers of Congress. There were a million ways in which appropriators could have shifted numbers around and delivered a steady stream of funding—for the wall, for other security measures, and for the "humanitarian crisis" Trump was now emphasizing. But they didn't. It had never been an urgent priority for GOP lawmakers, and the president was too ineffectual to convince them otherwise. Now that they had incurred his wrath, many of those same lawmakers were signaling their support for an unprecedented power grab.

"Democrats continue to refuse to negotiate in good faith or appropriate any money for border barriers," Meadows tweeted on January 11. "If they won't compromise, POTUS should use asset forfeiture money or other discretionary fees to start construction. If not, he should declare a national emergency. It's time."

This was the Freedom Caucus chairman, a leader of the conservative wing of the pro-liberty, small-government party, urging the president to steal private property to build a wall.

Amazingly, Trump continued to listen to the people who had steered him into this cul-de-sac, Meadows and Mulvaney above the rest. One Republican whom Trump never spoke with throughout the crisis: Hurd. Despite being the only Republican from a border district, and representing more of it than anyone in Congress, the Texas lawmaker waited for a call that never came.

Years earlier, when Hurd left the CIA, he did so because of his outrage at the low caliber of people being sent to Congress. He met many of them while working as an agent, tasked with briefing lawmakers during their trips to the Middle East. Some didn't understand the basic distinction between Sunni and Shia Muslims.

Hurd's departure from the CIA confounded senior national security officials who saw in him a budding superstar. Robert Gates, the former defense secretary who served eight presidents of both parties, issued the first political endorsement of his entire life when Hurd ran for Congress, and suggested that he expected to see Hurd in the White House one day.

For the time being, however, Hurd was stuck dealing with the same low-caliber politicians—and none more so than Trump. This was the state of the modern GOP: As its president shut down the government because of an irrational fight over building a border wall, he did not bother to have a single conversation with the party's lone border-district congressman and its foremost expert on the policies in dispute.

"We had twenty-five hundred furloughed workers at a food bank in San Antonio last week. I just served some more of them at a soup kitchen in DC," Hurd said on January 23. "Here, at the seat of power in the free world, we're forcing federal employees to eat at soup kitchens, while China just launched a mission to the dark side of the moon."

THERE WAS AN ERA IN WHICH THE COUNTRY SEEMED CAPABLE OF DIS-tinguishing its policy battles from its cultural clashes; a time when not every newsworthy development, political or otherwise, was filtered through our preexisting worldviews; a recognition that people were defined far more by their personhood than by their party affiliation.

The *appetite* for this climate remained, as evidenced by the gusher of warm-and-fuzzy responses to the unlikely friendship struck up by George W. Bush and Michelle Obama.

But those days when the default was not to distrust peoples' motives—when the notion of self-selecting into tribes that lived in the same places, shopped at the same stores, and watched the same news shows would have been preposterous—those days were gone.

More enduring than Trump's appointment of judges, or his signing of a tax law, or his deregulating of the energy industry, would be his endorsement of America's worst instincts. The levees were leaky long before he descended his gilded escalator, and certainly other bad actors contributed to the breakage. Yet it was Trump who used his office to flood the national consciousness with fear and contempt, with suspicion and resentment, with ad hominem insults and zero-sum arguments. In so doing, he not only enslaved one half of the country to his callousness, but successfully bade escalation from the other half, plunging all of America and its posterity deeper toward perdition.

Hollywood, naturally, couldn't help but overplay its hand. Its leading men and women lectured on matters of morality while enabling the vilest of predators in their own industry. Comedian Kathy Griffin, the cohost of CNN's New Year's Eve coverage, posted an image of herself holding the fake, bloodied, decapitated head of Trump. Actor Johnny Depp asked aloud of one audience, "When was the last time an actor assassinated a president?" At award shows and galas and film festivals, the pilots of pop culture took turns savaging the president—and, his supporters felt, themselves by extension—in ways that further exacerbated the country's circular firing squad.

Americans were so cantankerously immersing themselves in extraneous debates that the line between reality and parody began to blur. One such dispute broke out over the toxic masculinity addressed, and possibly exaggerated, in an ad from Gillette, the iconic shaving company. (To the future archeologists picking through the ruins of our society, yes, this actually happened.) Inverting the company's traditional slogan from "The best a man can get" to "The best a man can be," the razor empire earned tens of millions of views and sparked a social media cacophony with a spot calling on the male species to evolve. No more bullying. No more whistling at women. No more laughing at sexual humor. And no more . . . boys wrestling with each other in the backyard?

The American fireworks of social indignation were loud and lucent but short-lived, never allowing the aggrieved masses to linger on any given outrage. Sure enough, quicker than you could dial the Dollar Shave Club, a fresh controversy was engulfing the country. It had the three ingredients of a kiss-your-fingertips cultural casserole: race, testosterone, and Make America Great Again hats.

On the evening of Friday, January 18, video emerged online of a strange confrontation from earlier that day. On the site of the Lincoln Memorial in Washington, a group of white teenagers, many of them clad in the president's iconic red baseball caps, encircled an old Native American man. His name was Nathan Phillips. As he sang and beat his drum, the kids whooped and chanted and danced. The video focused on one of them: Sixteen-year-old Nick Sandmann stood across from Phillips, their eyes locked on one another, Sandmann wearing a mysterious smirk.

Not long after, a second video swept through social media, this one of Phillips describing the events thereafter. "As I was singing, I heard them saying, 'Build that wall! Build that wall!'" the Native elder said on camera, his voice choked with emotion. "This is indigenous land. We're not supposed to have walls here."

Judgments of the junior Klansmen were expeditious. By the next morning, America was ablaze. Not since Charlottesville, it seemed, had a story amassed so much attention in so little time. Click-hungry news outlets blasted out reports of the persecution, stressing how Phillips, a Native American *and* a Vietnam veteran, had been accosted by a gang of MAGA-clad teens, while a chorus of celebrities, journalists, politicians, and combinations thereof delivered their damning verdicts.

CNN's chyron called attention to the "Heartbreaking Viral Video," while the *New York Times* published a story headlined, "Boys in 'Make America Great Again' Hats Mob Native Elder at Indigenous Peoples March."

The only problem: Nathan Phillips was lying. Nowhere in the hours of footage reviewed by hundreds of journalists could any of the teens be heard saying, "Build that wall!" And it turned out, he was not a Vietnam veteran, a fact that punctured his other principal claim to sympathy.

As more reporters actually did their job, it became obvious that the Covington Catholic kids had gotten screwed. Video clearly showed that, contrary to being racially charged predators, *they* were the ones preyed upon. The incident had begun with a confrontation between two other groups: Native American activists and members of the Black Hebrew Israelites, an extremist sect and known hate group. As the Covington students looked on, one of the Black Hebrew leaders started calling them "dirty-ass crackers" and threatening to "stick my foot in your little ass." The provocation escalated, all of it one-sided: Black Hebrew members hurled vile insults at the teenagers, calling them "incest babies" and "future school shooters," while mocking the pope ("faggot child-molester!") and Trump ("Your president is a homosexual!")

Granted permission by their chaperones to perform school-spirit chants in the face of the spewing hatred, the lads partook in some synchronized hooting. It was at this point that Phillips, trailed by his fellow indigenous activists, entered the fray, marching toward the students and pounding on his drum. When he came toe to toe with Sandmann, the sixteen-year-old did not move but merely stared straight ahead, wearing the smirk seen 'round the internet. Maybe he meant to intimidate Phillips; perhaps he was just paralyzed by the strangeness of the moment. Either way, the teenager showed zero sign of outward aggression. None of these facts mattered. Most of the do-gooders who impugned Covington Catholic and its students offered no apology. The fire-and-brimstone tweets would remain active, a testament to America's unapologetic rush to judgment circa 2019.

It was all so uniquely Trumpian, a supposed atrocity so perfectly suited to the politics of his reign, that the serendipity went largely overlooked.

The president's ascent had been invited by the right's unresponsiveness to outrage; his ability to get away with political murder owed to the left's gratuitous cries of wolf. Now, nearly one month into a government shutdown, America spent the weekend of January 18, 19, and 20 fixated on faux prejudice by some teenagers while the president of the United States was peddling the real thing.

THAT FRIDAY MORNING, HOURS BEFORE THE COVINGTON CATHOLICS came across the Black Hebrews, Trump fired off a tweet: "Border rancher: 'We've found prayer rugs out here. It's unreal.'" Linking to a story from the *Washington Examiner*, the president annotated his tweet thusly: "People coming across the Southern Border from many countries, some of which would be a big surprise."

This was national security intelligence of epic proportions: Muslims, probably from Syria and Iraq and who knows where else, had traveled to Central America, made the arduous journey north into the United States, and finally crossed over, all the while keeping their prayer rugs in tow, only to clumsily leave them on the American side of the border. Now they were busted: An unknown reporter with the conservative *Washington Examiner*, citing a single unnamed rancher in New Mexico, had blown the story wide open—and the president was reading it.

There were no pictures of the prayer rugs in the *Examiner* story, a slight curiosity in the age of smartphone cameras. But Trump was not dissuaded, and for good reason. There had been photographic proof before; in the summer of 2014, Breitbart.com published a blockbuster: "MUSLIM PRAYER RUG FOUND ON ARIZONA BORDER BY INDE-PENDENT AMERICAN SECURITY CONTRACTORS." That article showed images of a "prayer rug," prompting Texas's lieutenant governor to give notice soon thereafter about Muslim paraphernalia being found on his side of the Rio Grande. He was taken seriously, even though the prayer rug looked a lot like an Adidas soccer jersey. (Upon further examination, it was, in fact, an Adidas soccer jersey.[4])

Trump's tweet was an affront to America herself: The president of the United States was warning citizens that Muslim prayer rugs were being found north of the border, a brazen bit of fearmongering aimed at gaining political advantage amid a legislative fight he was losing in humiliating fashion. The source he consulted before disseminating this information was not the FBI or the CIA or Homeland Security, but a single-sourced *Washington Examiner* piece with no names attached and no photos of the prayer rug in question.

There were no recriminations from his fellow Republicans for this wildly irresponsible statement. And frankly, even the complaints Democrats lodged seemed to slide quietly into the ether. Everyone had grown accustomed to the president casually floating conspiracy theories like a cabin-secluded uncle on a family email chain. This was just another time America would roll its eyes and move on. Besides, there was a race war at the Lincoln Memorial to worry about.

The tweet showed just how desperate Trump had become.

His shutdown was now the longest in U.S. history and there was not a flicker of light to be found in the tunnel. Democrats were holding fast, insisting that they would not negotiate while federal workers were held hostage; and Republicans on Capitol Hill were increasingly agitated, inwardly angry with themselves but outwardly seething at McConnell and McCarthy for having allowed the president to embarass the party like this.

As a last gasp, Trump offered a deal to Democrats: He would grant work permits to certain migrants for three years in exchange for wall funding. But Democrats had no reason to bite: They had all the leverage, and what the president was suggesting fell short of his own previous offers to extend permanent protections for DACA recipients. Accepting something less to end the shutdown, and thus allowing him to claim victory, made no sense politically or policy-wise.

When Pelosi unceremoniously rejected Trump's offer, the dam broke inside the GOP. Senators confronted McConnell and told him in no uncertain terms that the shutdown needed to end; they were spinning their wheels in service of the president's ego while eight hundred thousand federal workers and their families were panicking over the prospect of another missed paycheck.

The majority leader got the message. Assuring the White House that they had no cards left to play, McConnell convinced Trump that they could save face by reopening the government for three weeks. It would give both parties a negotiating window over border security and test whether Democrats were sincere about coming to the table once the government opened up. The president, beleaguered and showing the scantest hint of remorse over the fiasco, agreed.

On January 25, thirty-five days into the shutdown, Trump stood in the Rose Garden and announced a deal to reopen the government for three weeks. "We really have no choice but to build a powerful wall or steel barrier," he said. "If we don't get a fair deal from Congress, the government will either shut down on February 15, or I will use the powers afforded to me under the laws and Constitution of the United States to address this emergency."

The president was determined to project strength. But there was only weakness to be seen. Everyone watching knew the score. Trump had blinked, caved, folded, buckled, lost. The only person who seemed aloof to this reality was the master negotiator himself, the man who ran for president touting his reputation as a winner, a dealmaker, a driver of hard bargains, only to be repeatedly outsmarted by his oppositon once in office.

Later that night, watching television in the White House residence and growing enraged by the universal assessments of his defeat, Trump tweeted, "I wish people would read or listen to my words on the Border

Wall. This was in no way a concession. It was taking care of millions of people who were getting badly hurt by the Shutdown with the understanding that in 21 days, if no deal is done, it's off to the races!"

It was a most forgettable day for the president. That morning, Trump had awoken to news that Roger Stone, his political trickster and hatchet man, was the latest victim of Robert Mueller's investigation. Arrested in the predawn hours at his Florida home, Stone was indicted by a federal grand jury on seven counts. They included obstructing the House Intelligence Committee's investigation of Russian interference in the 2016 election and lying about his communications with WikiLeaks and Trump campaign officials. Stone was also charged with intimidating a witness who was in contact with WikiLeaks's leader, the Kremlin-backed Julian Assange, during the 2016 campaign.

Stone was being accused of the dirtiest word in Republican politics: *collusion.*

Between the indictment of his longest-serving associate and his humiliating defeat at the hands of Pelosi, the president could be excused for feeling low. But his spirits were lifted in no time. The calls came in that night, from Meadows and McCarthy and several other sycophants, cheering Trump and telling him that everything would be fine.

It was just another day in the Republican Party. If anyone was spooked, or distraught, or disgusted, they did their best to hide it.

Some lawmakers were less practiced than others.

Sitting in his new office that Friday afternoon, shortly after the president's Rose Garden speech, Senator Mitt Romney appeared at a loss. He had devoted his life to order and discipline; the only trace of untidiness was the stack of unpacked boxes near his desk. Romney had dealt with Trump enough to know the inborn chaos he wrought, but nothing had prepared the senator for spending his first three weeks in Washington watching the president self-destruct.

It was a job he could have had, Romney thought to himself. A job he *should* have had. A job he would have done with diligence and dignity.

But there was no time to dwell on that. The freshman senator from Utah had work to do. Standing his ground—and standing up to Trump, when the circumstances warranted—would be easier said than done.

Sending tweets and writing op-eds was easy; defying the president, in the face of grinding pressure from party leaders and major donors and voters back home, would be far more difficult. Romney had to tread carefully. Having sold himself as something of a white knight, swooping into Congress to restore balance to the Republican universe, he had two targets on his back: one for the Trump supporters poised to punish his disloyalty, the other for Trump adversaries eager to highlight his hypocrisy the instant he capitulated to Trump.

Over the past decade, Romney had squeezed into different molds to meet different moments. Now he could find freedom in his true political identity—not the full-spectrum conservative or the out-of-touch elitist, but the sincere, pragmatic, well-intentioned statesman who sees that something is wrong and wants to help fix it. Success would be measured at the margins. He understood that. Romney didn't come to the Senate believing he could save the Republican Party from itself. But he *did* take solace in knowing that his six-year term would end in 2024, meaning he would serve at least as long as Trump—and very likely outlast him.

"He will not be president forever," Romney said. "Are we changed forever? In some respects, yes. But we're also going to change again. That's why, in some respects, I think character matters are of such significance. Because policies come and go. But matters of honor, integrity, civility, respect, family orientation, respect for faith, respect for the Constitution—these things are enduring."

Romney had aged. His face was thinner, his presence less commanding, his majestic mane of hair noticeably grayer than it once was. But he still wore that placidly pained expression, the one from when he quit the presidential race at CPAC in 2008, suggesting that something was wrong, and that he had more to say about it.

"Just remember, we've had serious divides in this country before," Romney said, trying to sound reassuring. Then he chuckled. "But, you know, we had Abraham Lincoln then. Now . . ."

The smile slowly vanished, his voice trailing off.

EPILOGUE

HE WEARS A WOOLEN BLUE VEST, A WEEK-OLD BEARD FLECKED WITH early hints of gray, and the look of a man liberated from the cruelest of confinements.

Not long ago, Paul Ryan was the most powerful lawmaker in the United States. Now there is nothing left to denote his significance, just a plainclothes member of his Capitol Police detail leaning against the wall and chomping on a Jimmy Johns sandwich, waiting for his assignment to expire for good in another week. The former Speaker has been content to fade into relative anonymity. Everyone knows him here in Janesville, of course, but they're past the point of treating him any better or worse because of it. Ryan can exhale. No more legislators to babysit; no more presidential Twitter tantrums to abide. He is now a full-time dad who can enjoy his kids' high school years from the comforts of home while collecting six-figure honorariums for forty-five-minute speeches and plum stock options for the hardship of sitting on corporate boards.

Yet Ryan is not at peace. Whatever relief he feels in retirement is tempered by the nagging sense that something is gravely amiss with the government, and the party, he left behind. Revel though he may in the "legal substance that stands a longer test of time"—a restructured tax code, a bigger military, a conservative judiciary—Ryan's grimace gives him away. He knows the foundational tremors that have shaken Washington portend consequences farther reaching than any doubling of the standard deduction. Worse, try as he might to ignore his own agency in the poisoning of our body politic, Ryan knows he could have done more to supply antibodies.

It's this sense of guilt—or fear, perhaps a bit of both—that now animates the former speaker. Sitting in his political office, on the third floor of a brick building on Main Street in Janesville, Ryan attempts to diagnose what went wrong.

"What people will think about, read about, which gets all the attention, is this wave of populism. This disruptive populism, which feeds off identity politics, is what's harmful and hurtful and dark," he says. "But it's more of an indictment on culture and the deinstitutionalization of society. I think technology combined with moral relativism [has] basically blown up norms, including civility in civil society and moral truths. And it's a weaponized system that tears at the institutions that have given us this free society we've enjoyed for over a couple hundred years."

Ryan believes "the real test of our generation" is to "figure out how to re-institutionalize and rebuild these guardrails." This, he insists, cannot be achieved by government. Rather, it falls to the governed, the voters of all partisan affiliations who have shrugged off the coarsening of public life because it reflects their private realities.

"We've gotten so numbed to it all," Ryan says. "Not in government, but where we live our lives, we have a responsibility to try and rebuild. Don't call a woman a 'horse face.' Don't cheat on your wife. Don't cheat on *anything*. Be a good person. Set a good example. And prop up other institutions that do the same. You know?"

Americans may not want to hear these prescriptions from Ryan, believing that he was part of the problem rather than part of the solution. And he may not disagree: Indeed, a principal reason that Ryan quit is that he found it impossible to set that good example from inside Congress. The incentive structures are too warped, the allure of money and fame and self-preservation too powerful, for individuals to change the system from within. Things were trending in that direction long before Donald Trump moved in down Pennsylvania Avenue, and the hastening on his watch has rendered the modern Congress—and the modern GOP—a relic of its former self.

"The old meritocracy is dead," Ryan says. "You can leapfrog good deeds. You can leapfrog earning success. You can leapfrog being a good person, even, and shortcut your way toward the top of the political pile because you're a better entertainer. You're better on Twitter, you have better followers. Hits, clicks, eyeballs, ratings."

He continues, "There were a couple people who did it early, saw the curve, jumped it, tapped the vein. There were a million mini-me's that said, 'Well, shit, if this freshman senator from Texas can do it, I can do it. . . .' And then you had the big gorilla of Donald Trump, the force that he is, just beat them all at the same game."

Regarding the new Democratic majority in the House, and its social-media sensation, freshman New York congresswoman Alexandria Ocasio-Cortez, the former Speaker adds, "They've got this with AOC. They've got the same damn thing. It kind of concerns me. I mean, I sort of wish *they* were being the governing party and holding up standards. But they're going to have the same problem we have."

Well, not necessarily. A renegade rank-and-file member of Congress is hardly the same as a renegade president. Moreover, it's an open question whether the Democratic Party is reborn in the populist, convention-shattering mold of Ocasio-Cortez. On the Republican side, that question has been asked and answered.

For a long stretch of the 2016 campaign, Ryan refused to accept Trump's takeover of the GOP. He traversed the stages of grief: denial (no way can Trump win), anger ("I called him a racist!"), bargaining (the RNC PowerPoint slides), and depression ("This is fatal," he told Reince Priebus) before finally coming to terms with it. This resistance was grounded in a basic belief that the Republican Party was still *his* party.

Looking back, Ryan says, he should have known better. Having considered the converging political, cultural, and socioeconomic events of the twenty-first century and reflected on them in the context of historical intraparty ideological swings, he recognizes now that the American right was primed, even overdue, for revolution.

"Trumpism is a moment, a populist moment we're in, that's going to be here after Trump is gone. And that's something that we're gonna have to learn how to deal with," Ryan says. "I'm a traditional conservative, and traditional conservatives are definitely not ascendant in the party right now. Trump's clearly an indicator of that. But I remember in the early nineties, when I was working for [Jack] Kemp and [Bill] Bennett, we were called neocons back then. And neocons weren't just guys who wanted to invade Iraq; neocons were free-trade, free-market, supply-siders who were also strong on national defense. And then there were the paleocons, which was the Pat Buchanan wing. And the paleocons were kind of what you have now: isolationist, protectionist, and kind of xenophobic, anti-immigrant."

He continues, "We called our wing 'the growth wing,' and we won for a good twenty years. And now their wing is winning. But it's cyclical. We beat the paleocons in the early nineties; they're beating us now.

"The Reagan Republican wing beat the Rockefeller Republican wing," Ryan shrugs. "And now the Trump wing beat the Reagan wing."

ELIZABETH WARREN TRIED TO PLAY THE GAME BY THE PRESIDENT'S rules.

Haunted by her past habit of identifying in academia as a Native American and, more recently, by Trump's jeering cries of "Pocahontas," the Massachusetts senator hoped to neutralize the issue ahead of her campaign for the presidency. Weeks before the 2018 elections, Warren released the results of a Stanford DNA study revealing that she likely had an indigenous ancestor between six and ten generations back, making her anywhere from 1/64 to 1/1,024 Native American. The carefully choreographed rollout—a sleek web video, an exclusive in the *Boston Globe*—reflected a confidence among Warren and her political advisers that they would not only squash the controversy but wield it on the offensive against Trump.

That confidence was sorely misplaced. At a moment of hypersensitivity on the American left regarding matters of identity, Warren's move invited only more skepticism of her past claims—and of her contemporary political judgment. Progressive activists, particularly those of color, hammered her for conflating a genealogist's statistic with a minority's life experience. The Cherokee Nation, which Warren did not consult in advance, slammed her stunt as "inappropriate and wrong." Media outlets that otherwise might have considered the story stale dug deeper; sure enough, days before her February campaign launch, the *Washington Post* unearthed a State Bar of Texas registration card from 1986 listing "American Indian" as Warren's race, in her handwriting.

By this point, Warren had already felt compelled to apologize—first to the Cherokee Nation, then to swarming reporters in the Capitol, and then to voters in Iowa on her pre-launch swing through the state. "I am not a person of color," she said in Sioux City, responding to an irritated caucus-goer who demanded to know why Warren had given Trump "fodder" with the DNA test.

(She wasn't alone in seeking absolution. The opening act of the Democratic race featured a rotating confessional of candidates declaring their past transgressions: Joe Biden for his touchy-feely interactions and his labeling of Mike Pence as a "decent guy"; Kamala Harris for her tough-on-crime policies as a California prosecutor; Bernie Sanders for the sexual harassment complaints against some of his 2016 campaign staffers; Beto O'Rourke for his white privilege and for joking about his wife raising their kids while he campaigned.)

The entire episode—Warren's tone-deaf revelation, the backlash, and her apology—demonstrated once more the folly of fighting Trump on his own turf. A veritable graveyard of Republican presidential candidates stands in testament to his supremacy in the realm of the superficial.

But the line between engaging Trump on substance and being sucked into the vulgarly personal is impossibly thin.

Nearly all the Democratic 2020 hopefuls, for example, envision a more ambitious role for the government in health care than what Barack Obama created with the Affordable Care Act. And many of them support the idea of a single-payer system, eliminating the private insurance market entirely, a position that was considered fringe within the Democratic Party a few years ago. Meanwhile, most of the candidates have also embraced some combination of stances—free college, student loan forgiveness, third-trimester abortion access, reparations for slavery, voting rights for incarcerated felons—that are suddenly and controversially animating the Democratic base.

These are signs of dwindling ideological diversity within the party. And it may not matter: Elections in modern America are won principally by mobilizing the base, not persuading the middle. There is ample reason to believe, then, that Democrats can reclaim the White House by pushing unapologetically leftward.

There is also ample reason to believe that this plays right into Trump's strategy. The president wants nothing more than to "put socialism on trial" in 2020, as Kellyanne Conway says, drawing the brightest possible distinction between the hammer-and-sickle Democrats and the stars-and-stripes Republicans. This isn't about ideas. It's about image. Trump is far less skilled at debating policy than he is at denigrating opponents; the more extreme their policies are perceived to be, the less work he has to do to combat them. It's hard enough to defend a comprehensive government takeover of health care; it's even harder while being slimed as ugly, dumb, or un-American. Keeping one's focus is far easier aspired to than accomplished. Just as he lured Hillary Clinton into talking about "deplorables," just as he baited Warren into releasing a DNA test, just as he provoked Biden into a macho war of words about physical toughness, Trump is planning to make a Democrat beat him at his own game, using the left's political anger and ideological energy against its nominee. This is tactical but also deeply nihilistic: The president knows that even if he loses such a contest, his opponent does, too.

"Within the Democratic Party, I think there is a big debate about how to deal with Trump because he has no boundaries," says David Axelrod, Obama's former chief strategist. "He's willing to do anything and say anything to promote his interests. It's a values-free politics; it's an amoral politics. And so, there is this body of thought that you have to fight fire with fire and so on. But I worry that we'll all be consumed in the conflagration."

TRUMPISM CAN BE UNDERSTOOD AS A CAUTIONARY TALE: THE CYNICISM and the belligerence, the political disruption and the societal wreckage, the heightened distrust of government and the lowered expectations among the governed. These are not the symptoms of a healthy governing entity. Given the toxicity of his time presiding over it, Trump may well be remembered as the president who destroyed the Republican Party.

"Or maybe," says Tony Perkins, the Family Research Council president and onetime Trump skeptic, "he'll be remembered for saving it."

If that sounds crazy, consider the principal complaints about the prior iteration of the GOP. It was pacified. It was insular. It was disconnected from the concerns of everyday voters. It was fragile and apathetic and utterly without conviction.

Enter Trump. He spoke in ways that channeled the angst of forgotten Americans. He campaigned in ways that exposed the impotence and indifference of the ruling class. And he governed in ways that were fearless, prioritizing with single-mindedness his commitments to the few rather than modulating in hopes of gaining approval from the many.

"People were getting tired of the promises being broken. The party was damaged goods, and he has restored its credibility," Perkins says. "Trump is one of the few politicians that I've seen who's actually intent on keeping his promises."

Loath as Democrats will be to acknowledge it, this may be a blueprint. Over the past half century, progressives have repeatedly failed to effect the sweeping changes reflective of their designation. Even in the case of Obama, who remains enormously popular with the left-of-center electorate, many progressives believe his administration fell woefully short on the issues of immigration, climate change, foreign intervention, and even health care, despite his historic shifting of the public policy debate with the passage of the Affordable Care Act.

Meanwhile, in his first two years, Trump has accomplished more for Republicans than any individual in three decades. Setting aside everything else—a tax law whose benefits are not fully demonstrated, a host of executive actions that can be easily unwound by a Democratic administration—Trump's judicial appointments alone have altered the landscape of American life for a generation. As of April 2018, Trump had confirmed one hundred federal judges, far outpacing Obama at the same point in their presidencies. Trump's rapid makeover of the judiciary included two associate justices of the Supreme Court, tipping its balance decidedly to the right, and also 20 percent of *all seats* on the federal appeals courts, as Bloomberg Law reported, a percentage that will climb ever higher throughout 2019 and 2020 thanks to a GOP-controlled Senate with lax confirmation procedures and little else to do legislatively.

It is this "transformation of the courts," as Mitch McConnell describes it, that rationalizes for many Republicans their backing of Trump. Even those who cringe at his autocratic mannerisms, who moan privately at his social media habits, who worry perpetually about the lasting damage done to national institutions and international relationships, see in him someone who has positioned conservatives for long-term victories on myriad issues that will come before the courts: on abortion, gun rights, immigration, religious expression, privacy, voting restrictions, environmental regulations, and virtually everything else that exists along the political fault lines of modern America. Given how bleak things looked for the GOP at the outset of Obama's presidency, with his party in unified control of the government and every expectation that *Democrats* would be the party overseeing this sweeping judicial renaissance, Republicans will take it.

"You have to remember, it was a pretty grim situation at the beginning of this ten-year period," McConnell says. "When I woke up the morning after Election Night 2016, I thought to myself, 'These opportunities don't come along very often. Let's see how we can maximize it.'"

Republicans have, in many ways, maximized their opportunity with Trump. But at what cost?

THE DANGER IN POLITICAL SPASMS TRANSCENDS PARTISAN CONFLICT. What we see in Trump's America is not just two parties repelling one another, but their voters living and thinking and communicating in ways alien to the other side.

Marco Rubio has been preoccupied with this phenomenon since departing the 2016 race. Retroactively analyzing, as all the candidates have, what he missed and what could have been done to counter the appeal of Trump, the Florida senator worries that government-sanctioned polarization has dissolved the nation's basic sense of community.

"Twenty years ago, you and I might disagree strongly on politics, but we're on the board of the same PTA, and our kids go to the same school, they play on the same sports teams, and we go to the same church on Sunday. I knew you as a whole person," Rubio says. "Today, we increasingly know people only by their political views—or we just don't know people unlike [us] at all. And that's particularly pronounced in urban-suburban settings, where you have people who live blocks away from each other but know very little or nothing about each other. And in fact, they have stereotypes about one another that just reinforce it. You add to that the fact that they don't interact socially, they don't interact socioeconomically, they don't interact culturally, they might not even be consuming the same news and information, and the result is you have people living right next to each other who are complete and total strangers."

Rubio says he did not appreciate the depths of our national tribalism until he ran a national campaign. Knowing what he knows now, he wishes he spent more time discussing it, appealing to Americans to step out of their silos and repair the societal bonds necessary for government to begin functioning again. At this stage, he worries, it may be too late.

"History didn't begin in 2001, but for the purposes of this [discussion] it did. Because if 9/11 happened today, I'm not convinced our reaction as a nation would be the same," Rubio says. "If 9/11 happened today, unfortunately, one of the first things you would hear is the assignment of blame through a political lens. People would need some theory as to why this happened. And that's true of any major event: hurricanes, school shootings, pandemics. The immediate reaction is we need a political villain. And so, 9/11 was that last unique period of time."

John Boehner offers a similar analysis. If anything, as dark as it sounds, the former Speaker believes it may take something worse than 9/11 to snap the country out of its self-hatred.

"At some point we're going to have to realize we're Americans first, and Democrats and Republicans and conservatives and liberals second. The country is more important than what each of the parties believe in," he says. "It's going to take an intervening event for Americans to realize that."

An intervening event?

"Something cataclysmic," Boehner responds, gazing upward.

It has been argued that politics is downstream from culture; that elected officials govern in a way that reflects the rhythms of society itself. This is undeniably true. Politicians are reactionaries, not leaders. They achieve and maintain power by responding to public opinion, not by driving it.

Still, it's difficult to see America finding its way out of this predicament of mass polarization without government setting an example. Boehner likes to say that Congress is "nothing more than a slice of America," an institution comprising "some of the smartest people" in the country and "some of the dumbest," "some of the nicest people" and "some that are Nazis." Because of this, lawmakers are every bit as ghettoized as the people and places they represent, projecting onto Congress the anxieties and divisions that stir their constituents back home.

Biden, who spent three and a half decades in the U.S. Senate before becoming vice president, recalls the glory days that predated social media and talk radio and cable news programming. Back then, Biden says, lawmakers in both parties understood that socializing across the aisle was a significant part of doing their jobs. "It was an era that allowed us to get so much done, because we actually got to know one another," he says. "We got to know each other's families; we got to know all about each other. When you know somebody it's awful hard to dislike them, even when you fundamentally disagree with them. . . . When you know, God forbid, that their wife is going through a bout with breast cancer. Or their son has a serious addiction problem. Or their daughter just lost a baby. You know what I mean? It's *hard*."

He remembers the day his rose-tinted version of Washington ceased to exist.

"At the end of the last year of the administration, I decided to go up to the private senators' dining room just to sit and have lunch with some of my Republican friends and Democratic friends," Biden says. "And as I walked in—I realized it doesn't exist anymore. There's no place for Republican and Democratic senators to sit down and eat together. I'm being literal. There used to be two dining rooms: The dining room I can take you into as a guest, and the dining room only a senator can walk into. It had two great big conference tables and a buffet. It's gone."

Biden barks out a question—"What the hell's *happening*, man?"— before answering it himself.

"We've stopped talking to one another."

WITH THE FEDERAL GOVERNMENT REOPENED THREE WEEKS INTO THE new Congress, and cooler heads in both chambers prevailing to *keep* it open, President Trump found himself entering a period of relative calm. No Supreme Court vacancies to fill. No immediate geopolitical crises to confront. No signature policy initiatives to spearhead.

That left him to focus on his reelection efforts. Those efforts, in the opening months of 2019, included declaring a national emergency at the southern border and threatening to close it altogether if Mexico failed to stop illegal crossings; toying with a return to the brutal family-separation policy; proposing that all undocumented immigrants be forcibly sent to so-called sanctuary cities, a stunt that his own aides dismissed as absurd; firing his Homeland Security secretary and purging the department's leadership; endorsing a contentious lower-court decision to invalidate the Affordable Care Act without any readied replacement plan; and responding to a series of controversial (arguably anti-Semitic) remarks from Ilhan Omar, a Muslim freshman congresswoman, by tweeting a video splicing clips of her speaking with footage of 9/11, accompanied by a caption: "WE WILL NEVER FORGET!"

Two common threads emerged. First, for Trump, these are all tried-and-true methods of mobilizing his supporters. The president's renewed emphasis on that which galvanized conservatives in 2016—the dangers of immigration, the evils of Obamacare, the potency of us-versus-them nationalism—suggests that Trump's reelection campaign will look and sound identical to his maiden bid four years earlier.

Second, with a few exceptions, Republicans in Congress did nothing to curb these policy decisions or rebuke the president's behaviors.

Even in declaring his national emergency at the southern border to seize funding that Congress failed to appropriate—a patently unconstitutional power grab—Trump faced little resistance from the purported party of small government. A dozen Senate Republicans joined with Democrats to overturn the declaration, forcing Trump to issue the first veto of his administration. But the other forty-one Senate Republicans went along with Trump, compromising their credibility and inviting a future Democratic president to invoke similar powers to deal with gun violence or climate change or whatever else garners executive enthusiasms. In the case of those senators facing reelection in 2020, such as Thom Tillis, Ben Sasse, and McConnell himself, the reasoning was straightforward: They needed to stay in Trump's good graces.

The elemental prerequisite for GOP lawmakers attempting to keep their job is to stay out of the president's crosshairs, to avoid antagonizing his supporters back in their states and districts. This requires considerable sacrifices, chief among them ideological consistency. But it's a small price to pay for another term with a salary of $174,000; fully funded trips around the world; sprawling staffs catering to their every whim; power-flexing appearances on cable television; black-tie dinners and top-dollar fund-raisers and seats at the table with some of the world's most powerful and well-connected people.

In spite of this culture of allegiance within the Republican Party—enforced through fear, incentivized by proximity to power—Trump still had reason to look over his shoulder.

In late February, America was treated to seven hours of must-see TV when Michael Cohen, the president's former lawyer, testified in front of the House Oversight Committee.

"I am ashamed because I know what Mr. Trump is," Cohen said in his opening statement. "He is a racist. He is a con man. He is a cheat."

The witness did not require much leading. Cohen presented as evidence a personal check from Trump, signed while in office as president, reimbursing him for the hush money paid to Stormy Daniels. He also alleged that Trump told him to lie about the timing of the Moscow building project; that Trump "knew from Roger Stone in advance about the WikiLeaks drop of emails" designed to hurt Hillary Clinton's campaign; that Trump had prior knowledge of his son's meeting with the Kremlin lawyer in the summer of 2016; and that Trump lied about his financials (and potentially committed tax fraud) in the pursuit of bank loans. Cohen also warned that if the president lost his bid for reelection, "there will never be a peaceful transition of power."

Trump's old admonitions were proving prophetic: The "fixer" was causing him a lot of problems.

Republicans on the panel did not challenge these accusations about the president's conduct. In fact, they asked hardly any questions about Trump at all. Instead, they took turns attacking Cohen's credibility, portraying him as a jilted, star-seeking grifter who was headed to jail for lying to Congress already.

They had every reason to do so: The witness was an admitted perjurer, someone whose testimony under normal circumstances wouldn't be taken seriously. Yet these were not normal circumstances. And for all the reasons to remain skeptical of Cohen, here were powerful members of the legislative branch, presented by a witness with damning claims of misconduct by the head of the executive branch, showing not the slightest interest in examining them.

It was a chilling dereliction of duty. And it was rooted in the same motivation that Cohen says kept him shackled to Trump, doing his dirty work, for the previous decade: a fear of disloyalty.

"I did the same thing that you're doing now for ten years. I protected Mr. Trump for ten years," Cohen told the Republicans. "The more people that follow Mr. Trump as I did blindly are going to suffer the same consequences that I'm suffering."

EXPLOSIVE AS IT WAS, COHEN'S TESTIMONY FAILED TO INFLICT TANGIble damage on the president. But it succeeded in further whetting Washington's voracious appetite for something that could: Special Counsel Robert Mueller's report on Russian interference in the 2016 election and Trump's obstruction of the investigation thereof.

On March 22, 2019, twenty-two months following Mueller's appointment, the special counsel delivered his report to the Justice Department. The person responsible for digesting it and producing a summary to the public: William Barr, the new attorney general who had been confirmed to the post just a month earlier. For his first two years in office, Trump complained incessantly that Jeff Sessions did not have his back politically. He would find no such fault with Barr.

After forty-eight hours of frenzied anticipation, the attorney general released a brief synopsis of the special counsel's report. On the question of Russian meddling in 2016—and of potential collusion between Moscow and Trump's team—Mueller's findings were straightforward: "[T]he investigation did not establish that members of the Trump Campaign conspired or coordinated with the Russian government in its election interference activities."

But on the secondary question, regarding Trump's potential

obstruction of justice, Mueller was strikingly less definitive. "[W]hile this report does not conclude that the President committed a crime," the special counsel wrote, "it also does not exonerate him."

Barr felt otherwise. After quoting Mueller's assertion verbatim in his summary memo, the attorney general wrote that he and Rod Rosenstein, the deputy attorney general, "concluded that the evidence developed during the Special Counsel's investigation is not sufficient to establish that the President committed an obstruction-of-justice offense."

The upshot was predictable. "No Collusion, No Obstruction, Complete and Total EXONERATION," the president tweeted that afternoon. "KEEP AMERICA GREAT!"

Widespread confusion over what the special counsel had uncovered, on top of the seemingly warring conclusions reached by Mueller and Barr, prompted an outcry for the Justice Department to release the entire report. It grew deafening when sources close to the special counsel's office told the *New York Times* and the *Washington Post* that Barr's summary did not reflect Mueller's product. (Unbenownst at the time, Mueller himself wrote a letter to Barr complaining that the attorney general's summary "did not fully capture the context, nature and substance" of the investigation, as the *Washington Post* later reported.)

The release of the full, lightly redacted special counsel's report, on the morning of April 18, 2019, could rightly be considered a watershed in presidential history. Drawn from hundreds of under-oath interviews and thousands of documents, digital files, and other investigatory receipts, Mueller's 448-page report portrayed an administration built on corruption and deceit. It illuminated in jaw-dropping detail the web of lies woven by Trump and his team, the chaos and paranoia consuming the White House throughout Mueller's investigation, and the president's multiple efforts to impede its advance.

The probe found that while no Trump campaign officials engaged in a criminal conspiracy with Russia, they were "receptive" to offers of assistance from Moscow—and in fact expected help to arrive.

Mueller found no smoking gun to prove that collusion occurred. There was *attempted* collusion, as with Trump Jr.'s meeting the Kremlin-linked lawyer after promises of dirt on Clinton (and Roger Stone communicating with WikiLeaks), but not *actual* collusion. It was a similar story on the question of obstruction.

According to the report, when then–attorney general Jeff Sessions informed the president of the special counsel's appointment in May 2017, Trump responded, "Oh my God. This is terrible. This is the end of my presidency. I'm fucked."

The president's obsessive fear of Mueller's inquiry prompted him on several occasions to try to thwart it. He pleaded with Sessions to "unre-cuse" himself from the Russia probe and redirect the Justice Department's attention toward investigating Hillary Clinton. He asked his former campaign manager, Corey Lewandowski, to strong-arm Sessions into denouncing the special counsel's investigation. He instructed the White House counsel, Don McGahn, to have Mueller fired.

Had Trump's subordinates not defied these requests, he almost surely would have been charged with obstructing justice. It was the ulti-mate of ironies: Surrounded by people who wielded deception as a polit-ical shield, Trump was likely spared a criminal referral by their refusal to heed his instruction.

"The president's efforts to influence the investigation were mostly unsuccessful," Mueller wrote, "but that is largely because the persons who surrounded the president declined to carry out orders or accede to his requests."

THE SCANDALOUS NATURE OF THE REPORT—"EVEN AS A LONGTIME, quite open critic of Donald Trump," observed *National Review*'s David French, "I was surprised at the sheer scope, scale, and brazenness of the lies, falsehoods, and misdirections detailed by the Special Counsel's Office"—put congressional Democrats in an impossible position.

Launching impeachment proceedings would be a surefire way to en-ergize and unify the Republican Party coming off a thumping in 2018 and heading into a difficult reelection in 2020. Then again, some deci-sions should transcend politics, and constitutional impeachment exists for a reason: to consider the removal of a president who has engaged in conduct, criminal or otherwise, that is detrimental to the republic.

The argument was somewhat academic. President Trump was not

going to be expelled from office—at least, not while Republicans controlled the U.S. Senate.

Even Mitt Romney, who issued the sharpest post-Mueller rebuke of
any Senate Republican—"I am sickened at the extent and pervasiveness
of dishonesty and misdirection by individuals in the highest office of the
land, including the President"—said it was time for Congress to move
on. Predictably, as Romney's GOP colleagues hung him out to dry with
their silence, he came under withering attack from the likes of Rush
Limbaugh, Mike Huckabee, and other high-profile Trump apologists.

In this sense, Mueller's report offered a verdict not just on the integrity of President Trump but on the soul of his Republican Party. No
matter what turns up—in the congressional hearings probing Trump's
financial entanglements, in the Southern District of New York's examination of wrongdoing outside Mueller's purview—the GOP had committed itself to a fully binary view of politics that safeguards Trump's
survival. This was justified not by adherence to principle but by addiction to power: the power to hold office, the power to make laws and influence government, the power to appoint judges, the power to project
ideology onto the culture at large, and the power to deny such powers to
an opposing party.

The question Trump asked two years into his presidency, emerging
from the longest shutdown in government history and awaiting the
findings of Mueller's investigation, was whether Republicans would remain "faithful" and "loyal." It had been answered. Their allegiance to
him, once fleeting and flimsy, had been hardened by fire. The GOP would
belong to Donald Trump for the duration of his presidency.

But what happens when he's gone?

IN THE SPRING OF 2017, NOT THREE MONTHS INTO HIS PRESIDENCY,
Trump was hosting an intimate dinner for some veterans of the transition team when he ambushed them with a most unexpected query.

"Has any president besides Franklin Roosevelt done anything big after their first term?"

It was startling on many levels. For one thing, people around the table were surprised that Trump was demonstrating a textured grasp on

the history of his office. And indeed, while some experts believe the so-called "second term curse" is overstated, there is no question that most multiterm presidents accomplish their major initiatives during their first four years.

But above all, the attendees were taken aback by the implication. Although Trump had been in office a very short time, he was showing signs of misery on the job. The press was unyieldingly critical. His staff was clumsy and ineffectual. Congress was moving like molasses. Democrats were dead-set against him. Even many of the Republicans who smiled in his face, Trump knew, were knifing him in the back.

It all made for a compelling thought experiment at the highest levels of the government, the stuff of whispered fantasy for the likes of Speaker Ryan and Reince Priebus and, later, John Kelly. If Trump achieved a series of major legislative victories in his first term, could he be convinced there was nothing to gain—and everything to lose—by seeking another?

"No, because it's a very big job and there is a lot to do," Trump told the *New York Times* in January 2019, responding to questions about such a scenario. He later added, "Here's the bottom line: I love doing it. I don't know if I should love doing it, but I love doing it."

Whenever Trump does vacate the White House, the Republican Party will face a reckoning. It will have been rebranded as a protectionist, big-spending, anti-immigration entity. Its coalition will be overwhelmingly reliant on exurban and rural working-class whites and less dependent than ever on affluent, diverse suburbanites. Its character on everything from trade to international alliances to entitlement spending will be changed, if not converted entirely, from the turn-of-the-century GOP.

At that point, what will become of Trump's party? Will its identity endure, reshaping the American right for decades or even generations to come? Or will it revert to its Reaganesque roots, embracing once more the concepts of limited government and global integration?

"It's very much an open question," says Karl Rove, the architect of George W. Bush's victories and one of few elder statesmen in today's GOP. "My gut says Trump won't durably change the party. Republicans are free-traders, and this experiment with protectionism is going to end very badly. People in Ohio and Michigan, they're going to see how bad protectionism is. We're anti-communist here; we're not isolationists. I think the American people know that we're five percent of the population and twenty-five percent of the world's economy, and if we want to be prosperous, we can't just wash each other's laundry."

Kellyanne Conway, who is the closest thing to Rove's counterpart in the Trump universe, sees it differently. "It *will* be a Trumpian party," she predicts. "The constant outrage and opprobrium toward Donald Trump miss so much of what he's knitting together in a sustainable way. His version of 'America first' will outlast him. What's the next Republican president going to say? I'm going to raise taxes? I'm going to add regulations? I think getting back in the Paris accords and the Iran nuclear deal and trade deals that screw American workers are great ideas?"

There is, of course, another potential outcome. Republicans abandoned Bush's version of the party when it ceased to align with their needs and attitudes, and they will defect from Trumpism just as quickly. But if and when they do, there is no guarantee of a return to the status quo. While Ryan is right that party politics are cyclical, those cycles don't last forever. The fragmenting of America's two-party system has so accelerated, in such a condensed window, that its implications are impossible to fully appreciate. The 2020 general election could very well pit a Republican nominee who was never a Republican (Trump) against a Democratic nominee who was never a Democrat (Sanders).

With both parties buckling under the weight of extraordinary ideological and cultural pressures, and the electorate as a whole undergoing a sweeping demographic realignment, it's not implausible to envision the post-Trump GOP splitting altogether rather than regressing to an era of paternalistic, top-down party politics. The palace gates were finally broken down by the 2016 election, with Trump's candidacy the chosen battering ram of the populist masses. They may discover that overthrowing the monarchy didn't bring the changes they hoped for—but that doesn't mean they'll reinstall a king.

"The Republican Party is on a pretty thin thread right now," says Raúl Labrador, the former congressman and leader of the Tea Party faction that threatened to break away from the GOP in 2010. "The establishment invited this insurgency by not listening to the American people. It started during the Bush years. It got worse with Boehner. Now [Ryan]. And Trump actually spoke to those people. That's why it's so incumbent on him to listen to them. Because if he doesn't, they will turn on him, too."

What will that look like?

"Right now, they're happy with Trump, but they're going to grow disillusioned if they keep seeing trillions more in debt, if they don't see the immigration problem solved, if they don't see wages go up for everyday Americans," Labrador warns. "I think voters will be looking for a new vehicle to keep those promises to the American people. That's when you're going to see a new political party. And you'll get people from both sides—some of the Bernie Sanders people and some of the Trump people. That's what I see coming."

At the other end of the Republican spectrum, Sara Fagen, White House political director under Bush, offers an identical prediction.

"A lot of people think Trump is a footnote, that he's just here for four or eight years, and then it goes back to normal. But I think that's wrong. I think the party is changed for good," she says. "And it won't be sustainable. We're in a period of incredible change as a country where the extremes of the left and right are going to converge, and you're going to wind up with a third party. Over the next two or four years? No. But in the next twenty? For sure."

In the interim, the jockeying to lead the post-Trump Republican Party has already begun. Marco Rubio and Ted Cruz—both seasoned, shrewd, and afflicted with the presidential bug—are charting their respective policy paths in the Senate. Nikki Haley is taking donor meetings around the country and mapping out her vision for the party. Mike Pence is waiting patiently in the wings, certain that his dutiful subservience will be rewarded. And a crop of ambitious next-generation Republicans is lurking in the shadows. All of these people are sizing up Trumpism and molding themselves to annex some part of its appeal.

They all face the same problem: There is only one Donald J. Trump.

A singular figure in the sweep of American mythology, the forty-fifth president identified a historic convergence of cultural and socioeconomic unrest and used it to remake the political landscape in his image. There can be no imitating Trump's style or replicating his success.

Rarely has a president so thoroughly altered the identity of his party. Never has a president so ruthlessly exploited the insecurity of his people.

I WANT TO KNOW: IS HE TRANSITIONAL OR TRANSFORMATIONAL?

Trump smirks. "I mean, can there be—" he stops abruptly. "*I* don't want to be saying it."

But the president can't help himself. "Can there be a question?" he says, pushing his chair outward and standing up, casting a shadow over the Resolute desk. "Honestly, can there be even a question?"

ACKNOWLEDGMENTS

WHEW.

That was a *bit* long—shocking, no doubt, to everyone who has edited me—so I'll try to be succinct here.

There is no way I could have written this book without the love, encouragement, and unconditional support offered by my wife, Sweta. It's challenging enough to be a working mom with three boys under the age of five. But to carry that load alone for the better part of a year—while her absentee, first-time author husband toiled in his basement study battling anxiety and sleep deprivation—ought to qualify Sweta for some sort of humanitarian award. Naturally, she will shrug this off, her selflessness surpassed only by her humility. The world should know, however, that any praise I may receive for this project rightfully belongs to her. (Your criticisms may still be directed to me.) Sweta, you are the rock of our family, the love of my life, and, as I said on our wedding day, the headline of my heart. I treasure you.

Sitting down one night last November to begin writing, I got as far as typing "Chapter One" before spending an interminable stretch staring at the otherwise barren screen, wondering what I'd gotten myself into and whether it was yet time for a bathroom break. What spurred me to action then, and every night over the next several months, was the thought of my three sons sleeping upstairs. They were far too young—a newborn and two toddlers—to understand what their dad was doing. But they were my inspiration nonetheless. One day my little boys will be men and, like other kids coming of age in that next generation, they will be brimming with questions about this strange national legacy they inherited. My job as a journalist means nothing compared to my duties as a father; fortunately, this book provided me an occasion to wear both hats, putting on paper a decade's worth of reporting that I hope will inform their views of politics, culture, and the human condition. Abraham, Lewis, Brooks: Thank you for clarifying my purpose. Never forget who you are and whose you are. I love the three of you so very much.

Everything I know about parenting, and whatever successes I've had in life, I owe to my mom and dad. They have modeled the virtues of integrity, faithfulness, reliability, perseverance, and trustworthiness that I aspire to in my professional and personal callings. Their devotion—to each other, to their children, and to the Lord—has set an example for which I will be forever grateful. Every child should be so fortunate as to be raised by parents who encourage their every endeavor, commiserate with their every struggle, forgive their every failure, and celebrate their every triumph. Mom and Dad, thank you for everything.

To my three brothers, who collectively showed minimal interest in this project and seemed largely unimpressed with my landing a book deal, thank you for keeping your little brother grounded. The fact that you respect me less for being an author than for being a family man—and for being a better athlete than all of you—speaks volumes about our shared priorities. Chris, J.J., Brian: The next round's on me.

I want to send love to my nieces and nephews (Alexis, Norah, Henry, Madison, Gabriel, Tyler, Isaac, Miles) and my sisters-in-law (Kristi, Steph). And I want to give a special shout-out to Rudy and Kinjal, my *bhai* and *bhen*, for always being here for us and the boys.

There are three people to whom I'm deeply indebted for getting this colossus on wheels. Matt Latimer and Keith Urbahn, my agents at Javelin, went all-in despite having bigger clients and more lucrative projects to pursue. Their guidance was essential from start to finish. And Jonathan Jao, my editor at HarperCollins, could see my vision for telling this story—one not just about Trump, but about the culture that produced him—when many others could not. He also helped whittle down a manuscript that came in 100,000 words long while keeping its soul intact, never losing patience with his author even when he had every right to. Matt, Keith, Jonathan: All three of you guys have been loyal advisers and confidants throughout a process that was daunting and entirely foreign to me. I cannot imagine having smarter, savvier, more responsive people by my side.

Bringing this book to life, particularly on a crash production schedule, would not have been possible without some of the other brilliant minds at HarperCollins. Sarah Haugen deserves special accolades for her capability and good cheer throughout. But even before the manuscript was submitted, the heaviest lifting was done by two people: Derek Robertson, my assistant (and fellow Michigander), who was tireless in his fact checking and researching; and Jim O'Sullivan, my dear friend and dedicated sounding board, who provided real-time feedback on things big and small. Derek and Jim, the peace of mind you provided me was truly invaluable. Please know that I'm exceptionally grateful to the both of you.

Choosing HarperCollins to publish this work came on the advice of a number of industry veterans, all of whom told me some variation of the same thing: There is no better publicity team in the business. Having worked with Tina Andreadis and Theresa Dooley, I cannot overstate just how right they were. While my list of rookie concerns grew longer by the day since signing on that dotted line, I can honestly say that the marketing of this book never concerned me for a moment. Tina and Theresa, you have exuded competence and professionalism every step of the way. I'm so glad to have you in my corner.

Long before I sold myself as someone capable of writing a definitive account of the modern political era, I was a part-time janitor, full-time waiter, and struggling community college student with no real direction in life. It was only after transferring to Michigan State and coming under the tutelage of two men in particular—Eric Freedman and Bill McWhirter—that I realized a love not just for journalism but for political storytelling. Having once believed that no career could be more satisfying than that of a baseball beat writer, Eric and Bill helped me see the thrill of covering a far more important game. I'm not sure where life would have taken me had I not come under their influence—but it certainly wouldn't have been Washington.

It was the work in Eric's program, the Capital News Service, that led to my big break: an internship at the D.C. bureau of the *Wall Street Journal*. For that opportunity I must first thank the late Terry Michael, whose program, the Washington Center for Politics and Journalism, opened the door to a career in the capital. At the *Journal*, it was Mary Lu Carnevale, the eminently gifted (and exceptionally patient) news editor who held this wide-eyed intern's hand when others justifiably wondered what the hell he was doing in her newsroom. Thank you, Mary Lu, for your grace and your kindness.

At that point, I was a penniless intern renting a closet for $400 a month in the not-yet-renovated Petworth neighborhood, driving an Oldsmobile with two plastic-wrapped windows and selling CDs to a pawn shop on 14th Street NW for two dollars apiece to buy ramen noodles. In the decade since, I've outworked a lot of reporters who are smarter and better pedigreed. But I've also benefited tremendously from the risks taken on me by people who could have played it safe: Jim VandeHei, John

Harris, Danielle Jones, Mike Allen, and Ben Smith at *Politico*; Reid Wilson, Josh Kraushaar, Ron Fournier, and Charlie Green at *National Journal*; Rich Lowry and Jack Fowler at *National Review*; and, in my second tour at *Politico*, the amazing Carrie Budoff Brown.

It's imperative at this point that I stop to thank the four colleagues and friends who have meant the most to me in this journey: Kristin Roberts, my erstwhile editor at *National Journal*, who pushed me to maximize my abilities as a source reporter and brought out the best in my stories, though I hated some of her edits; Shane Goldmacher, my old office mate in the *NJ* "locker room," whose polish and precision rubbed off on me every single day, and whose dramatically different background made us an odd yet perfect pairing; Eliana Johnson, who lobbied to bring me aboard *National Review* even as it threatened her status as the publication's top news reporter, and who secured us the autonomy to produce some powerhouse reporting from an opinion journal not traditionally known for it; and Blake Hounshell, the mad tweeter, whose quick-twitch social media instincts belie a storytelling guru whose stewardship of *Politico Magazine* took my feature writing to another level.

All four of you—Kristin, Shane, Eliana, and Blake—have been crucial to the advancement of my career and the refining of my talents, all the while allowing me to remain true to myself. I could ask for nothing more.

As for remaining true to myself, allow me to recognize my original supporting cast—a group of guys who helped make me the man I am today. To Jim Nelson, Bill Duffey, A. J. Lear, Kyle Lamanen, Sean Taylor, Garrett Chapman, Jason Olinik, Joe Powell, Phil Clark, and all my other boys from back home, thanks for having my back. (Honorable mention to the out-of-towners: Andy Geyer, Anthony "Zags" Zagajewski, and the UVA crew that adopted me in Arlington.) And to those mentors who steered me away from trouble as an unruly youth—Sean Carleton, Joe Mackle—I hope you guys take a special satisfaction in seeing that your efforts paid off.

It's also important that I recognize the friendships I've forged along the way—not the D.C. cocktail circuit chums who traffic in mutually beneficial relationships, but the buddies with whom I'll be drinking beers longer after we've escaped the Beltway: Alex Roarty, Scott Bland, Steve Shepard, Raphael Esparza, Sean Sullivan, Patrick Reis, Jim Oliphant, Ben Terris, Charlie Szold, Jonathan Swan, Adam Wollner, Josh Dawsey. You guys have made it much easier to tolerate living and working in the world's most insufferable city. Thanks for the good times, one and all.

On another professional note, I want to express my appreciation to some of the particularly great and generous colleagues I've had the privilege of learning from and working alongside over the past decade: Ron Brownstein, who first suggested this specific angle for a book; Major Garrett, who made time for young reporters and never acted as though we were anything but peers; and Tim Grieve, who made a big bet on my career long before I had done anything to warrant his confidence. I'm also gratified to have called the following people colleagues: Beth Reinhard, Nancy Cook, Alex Burns, Dan Friedman, Richard Just, Nick Tell, Steve Heuser, and the entire *Politico Mag* staff. Also, because I drew from so much of my past work in reconstructing this storyline, I want to make a special point of thanking the people I've shared important bylines with: Billy House, my tutor in covering Congress; Alexis Levinson, my partner on the 2016 debate circuit; Rachael Bade, Elena Schneider, and of course, all my former *Hotline* comrades.

I can't possibly begin to thank the hundreds of sources who fed me information for this book and for so many other stories over the years. (Not that many of you are dying to be named.) That said, I would like to thank two politicians and their staffs in particular: John Boehner and Ted Cruz. While they couldn't be more different—hence one calling the other "Lucifer"—both of these men and their operations, from the top down, have always been accessible and helpful to me, even when my coverage was harsh. They both understand that, even when I've had to kick them in the teeth, I approach my job, and my relationships with their teams, with the utmost objectivity, fairness, and commitment to truth. Speaker Boehner and Senator Cruz, it remains my mission to bring the two of you together over some bottles of red. We could make it a Netflix special, or maybe even pay-per-view. Think it over.

It was some seven years ago that I was first asked to deliver a speech to Georgetown's Government Affairs Institute about the fratricidal nature of Congress and the challenges facing party leadership. What a blessing it has been to go back, again and again, revising and updating that speech and giving it countless times more in the years since. Not only did I discover terrific people along the way—Josh Huder, Worth Hester, Mark Harkins, and Kristin Nicholson, to name a few—but the speeches became an organizing ritual by which critical elements of this book's narrative came together. To my friends at GAI, keep up the great work—and don't expect my fire-and-brimstone routine to lighten up anytime soon.

I'm also grateful to Daniel McCarthy and the Robert Novak Journalism Fellowship Program, which provided grant money to help with my costs pertaining to travel, research, and transcription. It was an honor to be selected for the fellowship, and I highly encourage other young journalists to apply in the years ahead.

An unsung hero in the completion of this book is Clara Martin, our friend and part-time nanny, whose help with the boys was indispensable during some of the dimmest stages of my writing marathon. Thanks a ton, Clara.

Philippians 4:13 reminds us, "I can do all things through Christ which strengtheneth me." This book is proof positive. Coming off an exhausting midterm election stretch and facing some significant personal challenges, there was no reasonable expectation that I could write this manuscript in three months. Looking back, those three months are a blur; I'm not quite sure how it was done. But I do know that Christ strengthened me throughout. And I do know that, despite my straying from the path, He loves me and forgives me.

Lastly, I want to thank the Detroit Lions, my first love and foremost tormenter, for turning in such a wretched performance in the 2018–2019 season. My timetable for producing this book was already unforgiving; had the boys in Honolulu blue put a compelling product on the field, I would have been robbed of two hundred valuable minutes each Sunday last fall. (Also, I found that listening on the radio is considerably less stressful than watching on television.) To the Ford family, Bob Quinn, Matt Patricia, and above all, Matthew Stafford: Now that I'm finished, feel free to go out and get that elusive second playoff win since 1957. *Forward down the field!*

NOTES

CHAPTER ONE: FEBRUARY 2008

1. D. Nowicki and B. Muller, "McCain Profile: Prisoner of War," azcentral .com, 2007, http://archive.azcentral.com/news/election/mccain/articles/2007 /03/01/20070301mccainbio-chapter3.html.
2. Michael Kranish, "Famed McCain Temper Is Tamed," *Boston Globe*, January 27, 2008.
3. Laurie, Goodstein, "Huckabee Is Not Alone in Ignorance on Mormonism," *New York Times*, December 14, 2007.
4. Michael Cooper, "In Christmas-Greeting Style, Huckabee Makes Appeal to Voters," *New York Times*, December 19, 2007.
5. David Folkenflik, *Morning Edition*, transcript, NPR, February 7, 2008.
6. Charles Krauthammer, "The GOP Votes for Safety," *Washington Post*, February 8, 2008.
7. Jonathan Allen, "Bar Is Set High for Immigration Bill in House," *CQ Today* (Washington, DC), June 21, 2007.
8. Richard Gooding, "The Trashing of John McCain," *Vanity Fair*, November 2004.
9. *National Review*, "No on Joe (And Tom Ridge, Too)," *National Review*, August 18, 2008.
10. Elisabeth Bumiller and Michael Cooper, "Conservative Ire Pushed McCain from Lieberman," *New York Times*, August 30, 2008.
11. Henry Blodget, "McCain Staff Not Sure How to Pronounce Sarah Palin's Name," *Business Insider*, September 3, 2008, https://www.businessinsider.com/2008/9 /mccain-staff-not-sure-how-to-pronounce-sarah-palin-s-name.
12. Peggy Noonan, "Palin's Failin'," *Wall Street Journal*, October 17, 2008.
13. Edmund L. Andrews, Michael J. De La Merced, and Mary Williams Walsh, "Fed's $85 Billion Loan Rescues Insurer," *New York Times*, September 16, 2008.
14. Martin Jonathan and Mike Allen, "McCain Unsure How Many Houses He Owns," *Politico*, August 21, 2008, https://www.politico.com/story/2008/08/mccain -unsure-how-many-houses-he-owns-012685.
15. John Bentley, "McCain Says 'Fundamentals' of U.S. Economy Are Strong," CBS News, September 15, 2008, https://www.cbsnews.com/news/mccain-says-funda mentals-of-us-economy-are-strong/.
16. Christopher J. Goodman and Stephen M. Mance, "Monthly Labor Review, 2011," U.S. Bureau of Labor Statistics, Office of Unemployment and Unemployment Statistics.
17. Andrew Clark, "US Bailouts Prevented 1930s-style Great Depression Says New Study," *Guardian*, July 28, 2010, https://www.theguardian.com/business/2010/jul /28/us-bailouts-prevented-1930s-style-great-depression.
18. Scott Kraus, "Local GOP Chief Takes Heat for Use of 'Hussein,'" *Morning Call* (Allentown, PA), October 9, 2008.
19. Leslie Wayne, "Outside Groups Aid Obama, Their Vocal Critic," *New York Times*, January 30, 2008.

20. Ben Smith, "Mount Vernon Statement," *Politico*, February 16, 2010, https://www.politico.com/blogs/ben-smith/2010/02/mount-vernon-statement-025096.

CHAPTER TWO: JANUARY 2009

1. Betsy Klein, "Comparing Donald Trump and Barack Obama's Inaugural Crowd Sizes," CNN, January 21, 2017, https://www.cnn.com/2017/01/20/politics/donald-trump-barack-obama-inauguration-crowd-size/index.html.
2. Gallup, Inc., "Presidential Approval Ratings—Barack Obama," Gallup.com, https://news.gallup.com/poll/116479/barack-obama-presidential-job-approval.aspx.
3. Robert Draper, *Do Not Ask What Good We Do: Inside the U.S. House of Representatives* (New York: Free Press, 2013).
4. Tim Fernholz, "Obama's Republican Hit List," *The American Prospect*, February 9, 2009, https://prospect.org/article/obamas-republican-hit-list.
5. Eric Cantor, "Eric Cantor: What the Obama Presidency Looked Like to the Opposition," *New York Times*, January 14, 2017.
6. Kimberly Amadeo, "Obama's Stimulus Package and How Well It Worked," *The Balance*, January 26, 2019, https://www.thebalance.com/what-was-obama-s-stimulus-package-3305625.
7. Office of the Press Secretary, "Remarks by the President at GOP House Issues Conference," news release, January 29, 2010, The White House: President Barack Obama, https://obamawhitehouse.archives.gov/realitycheck/the-press-office/remarks-president-gop-house-issues-conference.
8. Patrick O'Connor, "At Retreat, Upbeat GOP Looks to 2010," Politico, January 31, 2009, https://www.politico.com/story/2009/01/at-retreat-upbeat-gop-looks-to-2010-018238.
9. Michael Falcone, "New R.N.C. Chairman Wants a 'Hip-Hop' Party," *New York Times*, February 19, 2009, https://thecaucus.blogs.nytimes.com/2009/02/19/new-rnc-chairman-wants-a-hip-hop-party/.
10. Glenn Thrush, "Steele Calls for 'Bling, Bling'-free Stimulus," *Politico*, February 9, 2009, https://www.politico.com/blogs/on-congress/2009/02/steele-calls-for-bling-bling-free-stimulus-015960.
11. Ben Smith, "Steele Offers Jindal 'Slum Love,'" *Politico*, February 26, 2009, https://www.politico.com/blogs/ben-smith/2009/02/steele-offers-jindal-slum-love-016395.
12. Ben Smith, "Steele: Abortion an 'Individual Choice,'" Politico, March 11, 2009, https://www.politico.com/blogs/ben-smith/2009/03/steele-abortion-an-individual-choice-016721.
13. Mike Allen, "Steele to Rush: I'm Sorry," *Politico*, March 2, 2009, https://www.politico.com/story/2009/03/steele-to-rush-im-sorry-019517.
14. Reid Wilson, "RNC Member Calls on Steele to Quit," *The Hill* (Washington, DC), March 5, 2009.
15. "Rahm Emanuel on the Opportunities of Crisis," *Wall Street Journal*, November 19, 2008, https://www.youtube.com/watch?v=_mzcbXi1Tkk.
16. Eric Kleefeld, "FreedomWorks Cuts Estimate for Crowd at Its 9/12 Rally by One Half," *Talking Points Memo*, September 14, 2009, https://talkingpointsmemo.com/dc/freedomworks-cuts-estimate-for-crowd-at-its-9-12-rally-by-one-half.
17. Andy Barr, "Palin Doubles Down on 'Death Panels,'" *Politico*, August 13, 2009, https://www.politico.com/story/2009/08/palin-doubles-down-on-death-panels-026078.

18. "Employment, Hours, and Earnings from the Current Empolyment Statistics Survey," Bureau of Labor Statistics data, U.S. Bureau of Labor, https://data.bls.gov /timeseries/CES0000000001?output_view=net_1mth.

19. "The Employment Situation: January 2009," news release, February 6, 2009, Bureau of Labor Statistics, U.S. Department of Labor, https://www.bls.gov/news .release/archives/empsit_02062009.pdf.

20. "The Employment Situation—August 2009," news release, September 4, 2009, United States Department of Labor, Bureau of Labor Statistics, https://www.bls .gov/news.release/archives/empsit_09042009.pdf.

21. Peter S. Goodman, "U.S. Unemployment Rate Hits 10.2%, Highest in 26 Years," *New York Times*, November 6, 2009.

22. Lydia Saad, "Obama Starts with 68% Job Approval," Gallup.com, January 24, 2009, https://news.gallup.com/poll/113962/obama-starts-job-approval.aspx.

23. Gallup, Inc., "Presidential Approval Ratings—Barack Obama," Gallup.com, https://news.gallup.com/poll/116479/barack-obama-presidential-job-approval.aspx.

CHAPTER THREE: APRIL 2010

1. Sarah Parker, "Poll: Crist Approval High Across the Board," CNN, April 15, 2009, http://politicalticker.blogs.cnn.com/2009/04/15/poll-crist-approval-high across-the-board/.

2. Erick Erickson, "On the NRSC Memo: What About Florida and Texas. And Florina?" *RedState*, September 8, 2009.

3. "Election 2010: Florida Republican Primary for Senate," *Rasmussen Reports*, April 12, 2010, http://www.rasmussenreports.com/public_content/politics/elec tions/election_2010/election_2010_senate_elections/florida/election_2010_flori da_republican_primary_for_senate.

4. Raw data, CNN Opinion Research Poll, Opinion Research Corporation, March 22, 2010.

5. Andy Barr and Ben Smith, "Tea Partiers Give Big Bucks for Palin," *Politico*, January 12, 2010, https://www.politico.com/story/2010/01/tea-partiers-give-big -bucks-for-palin-031409.

6. Robert Gehrke, "Bennett Out; GOP Delegates Reject 18-Year Senate Veteran," *Salt Lake Tribune*, May 8, 2010.

7. Josh Kraushaar, "Reid's Dangerously Low Approval Rating," *Politico*, September 3, 2009, https://www.politico.com/blogs/scorecard/0909/Reids_dangerously _low_approval_rating.html.

8. Bernie Becker, "Club for Growth Endorses in Nevada," *New York Times*, May 19, 2010, https://thecaucus.blogs.nytimes.com/2010/05/19/club-for-growth-endorses -in-nevada/.

9. "Angle to Hispanic Children: 'Some of You Look a Little More Asian to Me,'" *Las Vegas Sun*, October 18, 2010, https://lasvegassun.com/blogs/ralstons-flash/2010 /oct/18/angle-hispanic-children-some-you-look-little-more-/.

10. David Catanese, "Rape Case Haunts Buck in Colorado," *Politico*, October 12, 2010, https://www.politico.com/story/2010/10/rape-case-haunts-buck-in-colorado -043415.

11. "Senate Exit Polls," *New York Times*, https://www.nytimes.com/elections /2010/results/senate/exit-polls.html#colorado.

12. Ben Smith, "O'Donnell Backer Makes Sex Charge in Delaware Race," *Politico*, September 1, 2010, https://www.politico.com/blogs/ben-smith/2010/09/odonnell -backer-makes-sex-charge-in-delaware-race-028961.

13.   Lucy Madison, "Christine O'Donnell Win Net Gain for GOP, Republican Strategists Argue," CBS News, September 15, 2010, https://www.cbsnews.com/news/christine-odonnell-win-net-gain-for-gop-republican-strategists-argue/.

14.   Danny Yadron, "NRSC: No Plans to Fund O'Donnell," *Wall Street Journal*, September 15, 2010, https://blogs.wsj.com/washwire/2010/09/14/nrsc-no-plans-to-fund-odonnell/.

15.   Rachel Weiner, "Michele Bachmann Links Swine Flu to Democrats, Gets History Wrong (VIDEO)," *Huffington Post*, May 25, 2011, https://www.huffingtonpost.com/2009/04/28/michele-bachmann-links-sw_n_192493.html.

CHAPTER FOUR: JANUARY 2011

1.   Andy Barr, "The GOP's No-Compromise Pledge," *Politico*, October 28, 2010, https://www.politico.com/story/2010/10/the-gops-no-compromise-pledge-044311.

2.   Patrick O'Connor and Damian Paletta, "Ryan Is Irked as Obama Picks Apart His Budget Plan," *Wall Street Journal*, April 13, 2011, https://blogs.wsj.com/washwire/2011/04/13/ryan-is-irked-as-obama-picks-apart-his-budget-plan/.

3.   Ralph Z. Hallow, "RNC Faces $20 Million Debt in 2011," *Washington Times*, January 2, 2011.

4.   Jeff Zeleny, "G.O.P. Elects a New Chairman as Steele Drops Out," *New York Times*, January 14, 2011.

5.   Chris Good, "Obama Sticks It to Donald Trump Over Birth Certificate," *The Atlantic* April 27, 2011, https://www.theatlantic.com/politics/archive/2011/04/obama-sticks-it-to-donald-trump-over-birth-certificate/237937/.

6.   Jonathan Capehart, "Donald Trump for President. Seriously?" *Washington Post*, March 17, 2011, https://www.washingtonpost.com/blogs/post-partisan/post/donald-trump-for-president-seriously/2011/03/04/ABvVq4k_blog.html?utm_term=.ab25180c34aa.

7.   Lindsay Powers, "Kim Kardashian, Kris Humphries' Divorce: By the Numbers," *The Hollywood Reporter*, November 1, 2011, https://www.hollywoodreporter.com/news/kim-kardashian-kris-humphries-divorce-by-the-numbers-hoax-255849.

8.   Transcript, *Larry King Live*, CNN, October 8, 1999.

9.   Transcript, *Meet the Press*, NBC, October 24, 1999.

10.   Steve Kornacki, "When Trump Ran Against Trump-ism," NBC News, October 2, 2018, https://www.nbcnews.com/think/opinion/when-trump-ran-against-trump-ism-story-2000-election-ncna915651.

11.   Mitch Daniels, Speech, Ronald Reagan Centennial Dinner, Washington, DC, February 11, 2011.

12.   Joseph E. Stiglitz, "Of the 1%, by the 1%, for the 1%," *Vanity Fair*, May 2011.

13.   David S. Addington, "Obama Couldn't Wait: His New Christmas Tree Tax," Heritage.org, November 8, 2011, http://people.uncw.edu/imperialm/UNCW/PLS_505/Xmass_tree_tax_11_9_11.pdf.

14.   Ann Compton, "Blowback for Obama on Christmas Tree 'Tax,'" ABC News, November 9, 2011, http://abcnews.go.com/blogs/politics/2011/11/blowback-for-obama-on-christmas-tree-tax/.

CHAPTER FIVE: JANUARY 2012

1.   Becca Aaronson, "What Kinds of Jobs Has Texas Created Under Rick Perry?" *Texas Tribune*, August 29, 2011, https://www.theatlantic.com/politics/archive/2011/08/what-kinds-of-jobs-has-texas-created-under-rick-perry/244279/.

2.   Abby Phillip and Dave Levinthal, "Adelson Tally to Gingrich: $20M," *Politico,* April 22, 2012, https://www.politico.com/story/2012/04/gingrich-camp-mired-in -debt-075418.

3.   Gallup, "Media Use and Evaluation," Gallup.com, https://news.gallup.com /poll/1663/Media-Use-Evaluation.aspx.

4.   Gallup, "Americans' Trust in Mass Media Sinks to New Low," Gallup.com, September 14, 2016, https://news.gallup.com/poll/195542/americans-trust-mass-media -sinks-new-low.aspx.

5.   Philip Rucker, "Romney: 'I Like Being Able to Fire People Who Provide Services to Me,'" *Washington Post,* January 9, 2012, https://www.washingtonpost.com /blogs/election-2012/post/romney-sees-need-to-be-able-to-fire-service-providers /2012/01/09/gIQAF18alP_blog.html?noredirect=on&utm_term-.0fe7bdfb1f10.

6.   Richard A. Oppel Jr. and Erik Eckholm, "Prominent Pastor Calls Romney's Church a Cult," *New York Times,* October 7, 2011.

7.   Bryan Hall, "The DeMoss Group—Jay Sekulow vs. Pastor Robert Jeffress," YouTube, October 20, 2010, https://www.youtube.com/watch?v=6Obac8dFNFw.

8.   Alexander Burns, "Perry Backer: Romney Not a Christian," *Politico,* October 7, 2011, https://www.politico.com/story/2011/10/perry-backer-romney-not-a-chris tian-065415.

9.   Alexander Burns, "Woman Mentioned in Priorities Ad Died in '06," *Politico,* August 7, 2012, https://www.politico.com/blogs/burns-haberman/2012/08/woman -mentioned-in-priorities-ad-died-in-06-131332.

10.   David Lightman and Steven Thomma, "Romney Wins in New Hampshire, Strengthens Claim to GOP Nomination," *McClatchyDC,* January 10, 2012, https:// www.mcclatchydc.com/news/politics-government/election/article24721867.html.

11.   Emi Kolawole and Rachel Weiner, "Gingrich Calls Medicare Voucher Proposal 'Right-Wing Social Engineering,'" *Washington Post,* May 16, 2011, https:// www.washingtonpost.com/politics/gingrich-calls-medicare-voucher-proposal -right-wing-social-engineering/2011/05/15/AFHhoR4G_story.html?noredirect=on &utm_term=.cf868c3cc462.

12.   Xuan Thai and Ted Barrett, "Biden's Description of Obama Draws Scrutiny," CNN, February 9, 2007, http://www.cnn.com/2007/POLITICS/01/31/biden .obama/.

13.   Associated Press, "Mourdock Talks Rape, Pregnancy and God's Plan," *Politico,* October 24, 2012, https://www.politico.com/story/2012/10/mourdock-rape-preg nancy-and-gods-plan-082795.

14.   "President: Full Results," CNN, http://www.cnn.com/election/2012/results /race/president/Exit Polls.

15.   Rachel Weiner, "Sean Hannity: I've 'Evolved' on Immigration," *Washington Post,* November 8, 2012, https://www.washingtonpost.com/news/post-politics /wp/2012/11/08/sean-hannity-ive-evolved-on-immigration/?noredirect=on &utm_term=.0a4a81c6f9ac.

CHAPTER SIX: DECEMBER 2012

1.   Jennifer Bendery, "Allen West: I Am the Modern Day Harriet Tubman," *Huffington Post,* October 18, 2011, https://www.huffingtonpost.com/2011/08/18/allen -west-harriet-tubman_n_930052.html.

2.   Marie Diamond, "Allen West: Obama Supporters Are 'a Threat to the Gene Pool,'" ThinkProgress, July 19, 2011, https://thinkprogress.org/allen-west-obama -supporters-are-a-threat-to-the-gene-pool-df4522cc3888/.

3.   Janie Lorber, "Huelskamp Sounds Off on Losing Committee Spots," *Roll Call*, December 13, 2012, https://www.rollcall.com/news/huelskamp_sounds_off_on _losing_committee_spots-219990-1.html.
4.   Justin Sink, "Bobby Jindal: GOP Needs to 'Stop Being the Stupid Party,'" *The Hill*, October 22, 2013, https://thehill.com/video/in-the-news/279243-jindal -republicans-must-stop-being-the-stupid-party.
5.   Dylan Byers, "Rush Limbaugh: 'It's Up to Me and Fox News' to Stop Immigra-tion Reform," *Politico*, January 28, 2013, https://www.politico.com/blogs/media /2013/01/rush-limbaugh-its-up-to-me-and-fox-news-to-stop-immigration-reform -155434.
6.   Kevin Cirilli, "Rubio Wins Big Praise from Limbaugh," *Politico*, January 30, 2013, https://www.politico.com/story/2013/01/rubio-wins-big-praise-from-rush -086873.
7.   Ryan Lizza, "Getting to Maybe," *New Yorker*, June 24, 2013.
8.   Jennifer E. Manning, "Membership of the 113th Congress: A Profile," Washing-ton, DC: U.S. Congressional Research Service, 2014.
9.   Aaron Blake and Sean Sullivan, "The Fix's Week in Politics: Incumbents' Edge, GOP Prioritizing Safety over Civil Liberties," *Washington Post*, January 11, 2011, https://www.washingtonpost.com/national/2014/09/14/4b3fa312-3c4d-11e4 -9587-5dafd96295f0_story.html?noredirect=on&utm_term=.a921e643761f.
10.   Ryan Grim, "Republican Calls Obama 'Very, Very Urban' on House Floor," *Huffington Post*, November 30, 2010, https://www.huffingtonpost.com/2010/11/30 /republican-calls-obama-ve_n_789723.html.
11.   John Bresnahan, "Steve King: Obama 'Favors the Black Person,'" *Politico*, June 14, 2010, https://www.politico.com/blogs/on-congress/2010/06/steve-king -obama-favors-the-black-person-027564.
12.   Aaron Blake, "Rep. Steve King Stands by Controversial Obama Comments," *The Hill*, March 8, 2008, https://thehill.com/homenews/news/14497-rep-steve -king-stands-by-controversial-obama-comments.
13.   James G. Gimpel and Karen Kaufmann, "Republican Efforts to Attract Latino Voters," Center for Immigration Studies, August 1, 2001, https://cis.org/Report/Re publican-Efforts-Attract-Latino-Voters.
14.   Elspeth Reeve, "Steve King Wants to Protect the Border from Cantaloupe-Sized Calves," *The Atlantic*, October 29, 2013, https://www.theatlantic.com/politics /archive/2013/07/steve-king-wants-protect-border-cantaloupe-sized-calves /312984/.

CHAPTER SEVEN: AUGUST 2013

1.   Shane Goldmacher and Daniel Lippman. "When Ted Cruz Wanted to Be Part of the Establishment," *Politico*, January 26, 2016, https://www.politico.com/mag azine/story/2016/01/ted-cruz-2016-establishment-george-bush-213561#ixzz3yM CyqjrZ.
2.   Shane Goldmacher, "The Push to Defund Obamacare Is Running Out of Steam," *National Journal*, August 19, 2013, https://www.theatlantic.com/poli tics/archive/2013/08/the-push-to-defund-obamacare-is-running-out-of-steam /278815/.
3.   Eduardo Porter, "For G.O.P., Health Care as a Scare," *New York Times*, October 2, 2013.
4.   Mark Barrett, "Mark Meadows Has Taken Chances in Rapid Rise to Power," [Asheville, NC] *Citizen-Times*, April 1, 2017, https://www.citizen-times.com/story

/news/local/2017/04/01/mark-meadows-has-taken-chances-rapid-rise-power
/99865648/.

5. Matt Fuller, "Norquist Has Leadership's Back Against Heritage, Club for
Growth," *Roll Call*, September 11, 2013, https://www.rollcall.com/news/norquist
-emerges-as-key-gop-leadership-ally.

6. Susan Davis, "House GOP Votes to Delay Obamacare for One Year," *USA Today*,
September 29, 2013, https://www.usatoday.com/story/news/politics/2013/09/28
/house-saturday-shutdown-battle-senate-obama/2887537/.

7. "Boehner: 'This Isn't Some Damn Game,'" CNN, October 4, 2013, https://www
.cnn.com/videos/politics/2013/10/04/bts-dc-boehner-shutdown-debt-ceiling-fri
day.cnn.

8. Susan Davis and Richard Wolf, "Mitch McConnell: Democrats Will Rue Filibuster Vote," *USA Today*, November 21, 2013, https://www.courier-journal.com
/story/news/local/2013/11/21/mitch-mcconnell-democrats-will-rue-filibuster
-vote/3665463/.

9. Club for Growth, "Club for Growth Opposes Ryan-Murray Budget Proposal,"
news release, December 11, 2013, The Club for Growth, https://www.clubforgrowth
.org/club-for-growth-opposes-ryan-murray-budget-proposal/.

10. Joel B. Pollak, "Mark Levin to Paul Ryan: Budget Deal Is 'Mickey Mouse,'"
Breitbart, December 11, 2013, https://www.breitbart.com/politics/2013/12/10
/mark-levin-to-ryan-budget-deal-is-mickey-mouse/.

11. Michael A. Needham, "Budget Deal a Step Backward: Opposing View," *USA
Today*, December 11, 2013, https://www.usatoday.com/story/opinion/2013/12/10
/budget-deal-heritage-action-for-america-editorials-debates/3975653/.

12. Emma Dumain, "RSC Members Reflect on Paul Teller's Departure," *Roll Call*,
December 13, 2013, https://www.rollcall.com/news/rsc-members-reflect-on-paul
-tellers-departure.

CHAPTER EIGHT: APRIL 2014

1. Elizabeth Titus, "McConnell Aide: 'Holding My Nose,'" *Politico*, August 8, 2013,
https://www.politico.com/story/2013/08/mitch-mcconnell-campaign-chief-hold
ing-his-nose-095346.

2. Nia-Malika Henderson, "Mark Udall Has Been Dubbed 'Mark Uterus' on the
Campaign Trail. That's a Problem," *Washington Post*, October 13, 2014, https://
www.washingtonpost.com/news/the-fix/wp/2014/10/13/mark-udall-has-been
-dubbed-mark-uterus-on-the-campaign-trail-thats-a-problem/?utm_term=.5b0
4e2497797.

3. Anna Wolfe, "Media Buying Firm Purchases Pro-Cochran Ads, Funded by
NRSC," *Jackson Free Press*, July 10, 2014, http://www.jacksonfreepress.com/news
/2014/jul/10/cochran-friendly-pac-refuses-release-donor-list-cr/.

4. Susan Ferrechio and Eric Gay, "How the Dream Act May Have Led to Eric Cantor's Loss," *Washington Examiner*, June 11, 2014, https://www.washingtonexaminer
.com/how-the-dream-act-may-have-led-to-eric-cantors-loss.

5. Michael Patrick Leahy, "Laura Ingraham: We Should Have Traded Cantor for Bergdahl," Breitbart, June 4, 2014, https://www.breitbart.com/politics
/2014/06/04/Laura-Ingraham-and-David-Brat-Blast-Eric-Cantor-on-Amnesty
-Flip-Flops-at-Campaign-Rally/.

6. Dara Lind, "The 2014 Central American Migrant Crisis," Vox, October 10, 2014,
https://www.vox.com/2014/10/10/18088638/child-migrant-crisis-unaccompanied
-alien-children-rio-grande-valley-obama-immigration.

7.   Trymaine Lee, "Eyewitness to Michael Brown Shooting Recounts His Friend's Death," MSNBC, August 19, 2014, http://www.msnbc.com/msnbc/eyewitness-mi chael-brown-fatal-shooting-missouri.

8.   Michelle Ye Hee Lee, "'Hands Up, Don't Shoot' Did Not Happen in Ferguson," *Washington Post*, March 19, 2015, https://www.washingtonpost.com/news /fact-checker/wp/2015/03/19/hands-up-dont-shoot-did-not-happen-in-ferguson /?noredirect=on&utm_term=.ecc18dbf1a92.

9.   "Sharp Racial Divisions in Reactions to Brown, Garner Decisions," Pew Research Center for the People and the Press, December 8, 2014, http://www.people -press.org/2014/12/08/sharp-racial-divisions-in-reactions-to-brown-garner-deci sions/.

10.   Katie Glueck, "Poll: Race Relations Same or Worse," *Politico*, September 15, 2014, https://www.politico.com/story/2014/09/politico-poll-race-relations-under -obama-110924.

CHAPTER NINE: JANUARY 2015

1.   Ed O'Keefe, "Jeb Bush: Many Illegal Immigrants Come out of an 'Act of Love,'" *Washington Post*, April 6, 2014.

2.   "Democracy in Action Transcript of Donald Trump Speech at Iowa Freedom Summit," P2016.org, January 24, 2015, http://www.p2016.org/photos15/summit /trump012415spt.html.

3.   Mark Sappenfield, "Why It Matters That Donald Trump Is Attacking Mitt Romney," *Christian Science Monitor*, January 25, 2015, https://www.csmonitor.com /USA/Politics/DC-Decoder/2015/0125/Why-it-matters-that-Donald-Trump-is -attacking-Mitt-Romney.

4.   David Lightman, "Does GOP Take Palin and Trump Seriously?" McClatchy DC, January 28, 2015, https://www.mcclatchydc.com/news/politics-government /election/article24779314.html.

5.   Lucy McCalmont, "Hannity, Puzzled, Quizzes Palin on Iowa Speech," *Politico*, January 28, 2015, https://www.politico.com/story/2015/01/sean-hannity-sarah -palin-speech-iowa-114675.

6.   "Massie Hinted That He Wouldn't Vote for Boehner with Tweet from Drive-Thru," *River City News*, January 5, 2015, http://www.rcnky.com/articles/2015 /01/05/massie-hinted-he-wouldnt-vote-boehner-tweet-drive-thru.

7.   Tarini Parti, "Jeb Super PAC Raises $103 Million," *Politico*, July 31, 2015, https://www.politico.com/story/2015/07/jeb-bush-superpac-103-million-2016 -120853.

8.   Michael Kruse, "How Marco Slew His 'Mentor,'" *Politico*, February 19, 2016, https://www.politico.com/magazine/story/2016/02/marco-rubio-slew-jeb-bush -mentor-2016-213652.

9.   Philip Rucker, "Rubio Pollster Calls the White House Contender 'the Michael Jordan of American Politics,'" *Washington Post*, March 31, 2015, https://www.wash ingtonpost.com/news/post-politics/wp/2015/03/31/rubio-pollster-calls-the -white-house-contender-the-michael-jordan-of-american-politics/?utm_term=.f2 548d8ffce6.

10.   "2016 Republican Presidential Nomination," RealClearPolitics, https://www .realclearpolitics.com/epolls/2016/president/us/2016_republican_presidential _nomination-3823.html.

11.   James R, Carroll, "Poll: Rand Paul 2016 Frontrunner in IA, N.H.," *Louisville Courier-Journal*, July 18, 2014, https://www.courier-journal.com/story/politics

-blog/2014/07/17/rand-paul-leads-2016-republican-presidential-field-poll-finds/12777025/.

12. Nicholas Confessore, "Network of 'Super PACs' Says That It Has Raised $31 Million for Ted Cruz Bid," *New York Times*, April 8, 2015.

13. "Donald Trump's Presidential Announcement Speech," *Time*, June 16, 2015, http://time.com/3923128/donald-trump-announcement-speech/.

14. John Siciliano, "Trump Declares the 'silent Majority' Is Back in Business," *Washington Examiner*, October 29, 2016, https://www.washingtonexaminer.com/trump-declares-the-silent-majority-is-back-in-business.

15. Jennifer Agiesta, "Poll: Majority Sees Confederate Flag as Southern Pride-CNNPolitics," CNN, July 2, 2015, https://www.cnn.com/2015/07/02/politics/confederate-flag-poll-racism-southern-pride/index.html.

16. "Changing Attitudes on Gay Marriage," Pew Research Center's Religion and Public Life Project, June 26, 2017, http://www.pewforum.org/fact-sheet/changing-attitudes-on-gay-marriage/.

17. John Parkinson, "Boehner Warns Obama Not to 'Burn Himself' on Immigration Reform," ABC News, November 6, 2014, https://abcnews.go.com/blogs/politics/2014/11/boehner-warns-obama-not-to-burn-himself-on-immigration-reform/.

18. Jake Sherman and John Bresnahan, "GOP Group Targets House Conservatives on DHS Fight," *Politico*, March 2, 2015, https://www.politico.com/story/2015/03/american-action-network-targets-house-conservatives-dhs-funding-115668.

19. Lauren French and Jake Sherman, "House Conservative Seeks Boehner's Ouster," *Politico*, July 29, 2015, https://www.politico.com/story/2015/07/house-conservative-john-boehner-ouster-120742.

20. Andrew Kaczynski, "Trump Isn't into Anal, Melania Never Poops, and Other Things He Told Howard Stern," BuzzFeed News, February 16, 2016, https://www.buzzfeednews.com/article/andrewkaczynski/trump-isnt-into-anal-melania-never-poops-and-other-things-he#.geR1xNzqpX.

CHAPTER TEN: SEPTEMBER 2015

1. Jonathan Martin and Maggie Haberman, "G.O.P. Seeks Pledges of Loyalty to Nominee," *New York Times*, September 3, 2015.

2. Brett LoGiurato, "Donald Trump Just Handed the Republican Party a Major Gift," *Business Insider*, September 3, 2015, https://www.businessinsider.com/donald-trump-press-conference-independent-pledge-2015-9.

Ali Dukakis, "Rick Perry Calls Donald Trump a 'Cancer on Conservatism,'" ABC News, July 22, 2015.

3. Jonathan Dornbush, "'South Park' Season 19 Premiere Recap: Stunning and Brave," *Entertainment Weekly*, September 16, 2015, https://ew.com/article/2015/09/16/south-park-season-19-premiere-react/.

4. Theresa Vargas and Annys Shin, "President Obama Says, 'I'd Think about Changing' Name of Washington Redskins," *Washington Post*, October 5, 2013.

5. Cindy Boren, "Senators Urge NFL to Act on Redskins' Name, Citing NBA Action with Donald Sterling (updated)," *Washington Post*, May 22, 2014, https://www.washingtonpost.com/news/early-lead/wp/2014/05/22/senators-urge-nfl-to-act-on-redskins-name-citing-nba-action-with-donald-sterling/?utm_term=.7a8f367a425c.

6. Anne Steele, "Washington Redskins Racist? What Do the Navajo Really Think?" *Christian Science Monitor*, October 13, 2014, https://www.csmonitor.com

/USA/Sports/2014/1013/Washington-Redskins-racist-What-do-the-Navajo-really
-think.

7.　Elena Schneider, "McCarthy Rumors Follow Ellmers to North Carolina," *Politico*, October 19, 2015, https://www.politico.com/story/2015/10/ellmers
-house-north-carolina-republicans-214903.

8.　Mike DeBonis and Robert Costa, "'Supermajority' of House Freedom Caucus to Back Paul Ryan's Speaker Bid," *Washington Post*, October 21, 2015.

9.　Peter Sullivan and Cristina Marcos, "House Passes Boehner-Pelosi Medicare Deal in Resounding Vote," *The Hill*, March 26, 2016, https://thehill.com/policy
/healthcare/237068-house-passes-medicare-deal-in-overwhelming-392-37-vote.

CHAPTER ELEVEN: OCTOBER 2015

1.　"In Their Own Words: Obama on Reagan," *New York Times*, January 21, 2008, https://archive.nytimes.com/www.nytimes.com/ref/us/politics/21seelye-text.ht
ml?module=inline.

2.　Hans Johnson, "California's Population," Public Policy Institute of California, March 2017, https://www.ppic.org/publication/californias-population/.

3.　"Donald Trump: We'll Be Saying 'Merry Christmas' At Every Store If I'm President," CBS DC, October 23, 2015, https://washington.cbslocal.com/2015/10/23
/donald-trump-well-be-saying-merry-christmas-at-every-store-if-im-president/.

4.　Eugene Scott, "Trump Believes in God, but Hasn't Sought Forgiveness-CNNPolitics," CNN, July 19, 2015, https://www.cnn.com/2015/07/18/politics
/trump-has-never-sought-forgiveness/index.html.

5.　"Trump Appears to Mock Reporter's Disability," *Los Angeles Times*, https://
www.latimes.com/85160455-157.html.

6.　Sabrina Siddiqui and Ben Jacobs, "Trump Faces Backlash from Both Parties after Call to Bar Muslims Entering US," *Guardian*, December 8, 2015, https://www
.theguardian.com/us-news/2015/dec/07/donald-trump-muslim-ban-backlash
-jeb-bush-chris-christie.

7.　Avi Selk, "Pence Once Called Trump's Muslim Ban 'Unconstitutional.' He Now Applauds the Ban on Refugees," *Washington Post*, January 28, 2017, https://
www.washingtonpost.com/news/the-fix/wp/2017/01/28/mike-pence-once-called
-trumps-muslim-ban-unconstitutional-he-just-applauded-the-order/?utm_term
=.4c6f1142d134.

8.　Tim Hains, "Trump on Defeating ISIS: We Have to Take Out Their Families," RealClearPolitics, December 2, 2015, https://www.realclearpolitics.com/video
/2015/12/02/trump_isis_is_our_1_threat_—_we_cant_be_fighting_everybody_at
_the_same_time.html.

9.　Eric Bradner, "Donald Trump Praises Alex Jones' 'Amazing' Reputation," CNNPolitics, December 2, 2015, https://www.cnn.com/2015/12/02/politics/donald
-trump-praises-9-11-truther-alex-jones/index.html.

10.　"Speaking Time for Each Candidate," *New York Times*, October 28, 2015, https://www.nytimes.com/live/republican-debate-cnbc-boulder/speaking-time
-for-each-candidate.

11.　Matt Flegenheimer, "Ted Cruz Adds Two More Endorsements from Social Conservatives," *New York Times*, December 9, 2015, https://www.nytimes.com
/politics/first-draft/2015/12/09/ted-cruz-adds-two-more-endorsements-from
-social-conservatives/.

12.　"Donald Trump on the Day He Took the Pledge," Transcript, *The Hugh Hewitt Show*, September 3, 2015.

13. Eric Bradner, "Donald Trump: I'd 'Get Along Very Well with' Putin," CNNPolitics, October 11, 2015, https://www.cnn.com/2015/10/11/politics/donald -trump-vladimir-putin-2016/index.html.
14. David Jackson, "Putin: Trump Is 'Absolute Leader' of U.S. Presidential Race," *Atlanta Journal Constitution*, September 4, 2016, https://www.ajc.com/news/national -govt—politics/putin-trump-absolute-leader-presidential-race/rHxZ27Joxp4S1X nrPeyGDJ/.
15. Callum Borchers, "Timeline of Trump's Praise for Putin While Trump Tower Moscow Was in the Works," *Washington Post*, August 28, 2017, https://www .washingtonpost.com/news/the-fix/wp/2017/08/28/timeline-of-trumps-praise -for-putin-while-trump-tower-moscow-was-in-the-works/.
16. Jeremy Diamond and Greg Botelho, "Putin Praises 'Bright and Talented' Donald Trump," CNNPolitics, December 18, 2015, https://www.cnn.com/2015/12/17 /politics/russia-putin-trump/index.html.
17. Marin Cogan, "The Secret American Subculture of Putin-Worshippers," *National Journal*, September 19, 2013, https://www.nationaljournal.com/s/70126 /secret-american-subculture-putin-worshippers.
18. Victor Davis Hanson, "The Value of Putin," *National Review*, February 11, 2014, https://www.nationalreview.com/2014/02/value-putin-victor-davis-hanson/.
19. Tim Hains, "Trump vs Stephanopoulos: Has Anybody Proven Putin Has Killed Anybody?" RealClearPolitics, December 20, 2015, https://www.realclearpolitics .com/video/2015/12/20/trump_vs_stephanopoulos_has_anybody_proven_putin _has_killed_anybody.html.

CHAPTER TWELVE: JANUARY 2016

1. "Ted Cruz Trounces Marco Rubio in Third Quarter Fundraising," CBS News, October 9, 2015, https://www.cbsnews.com/news/ted-cruz-trounces-marco-rubio -in-third-quarter-fundraising/.
2. Editorial Board, "Marco Rubio Should Resign, Not Rip Us Off," *Sun Sentinel* (Fort Lauderdale, Florida), October 29, 2015.
3. Maggie Haberman and Matt Flegenheimer, "Cruz, at Fund-Raiser, Is Said to Question If Trump Has 'Judgment' to Be President," *New York Times*, December 11, 2015.
4. Eli Stokols, "Trump Takes on Cruz, but Lightly," *Politico*, December 12, 2015, https://www.politico.com/story/2015/12/trump-cruz-returns-fire-216708.
5. "Trump in 'FNS' Exclusive: Cruz a 'Maniac,' Hillary's 'Stupidity Killed Hundreds of Thousands,'" Fox News, December 14, 2015, https://insider.foxnews.com /2015/12/13/trump-fns-calls-cruz-maniac-claims-hillary-killed-hundreds-thou sands-people.
6. Jeremy Diamond, "Donald Trump: Hillary Clinton Got Schlonged in 2008," CNNPolitics, December 22, 2015, https://www.cnn.com/2015/12/21/politics/don ald-trump-hillary-clinton-disgusting/index.html.
7. David Wright, "Ted Cruz: Donald Trump 'Embodies New York Values,'" CNNPolitics, January 13, 2016, https://www.cnn.com/2016/01/13/politics/ted -cruz-donald-trump-new-york-values/index.html.
8. Ryan Teague Beckwith, "Republican Debate: Read the Full Transcript," *Time*, January 15, 2016, http://time.com/4182096/republican-debate-charleston-tran script-full-text/.
9. Paul Bond, "Leslie Moonves on Donald Trump: "It May Not Be Good for America, but It's Damn Good for CBS," *Hollywood Reporter*, March 1, 2016,

https://www.hollywoodreporter.com/news/leslie-moonves-donald-trump-may
-871464.

10. Mike McIntire, "Cruz Neglected to Report Loan from '12 Race," *New York Times*, January 13, 2016, https://www.nytimes.com/2016/01/14/us/politics/ted
-cruz-wall-street-loan-senate-bid-2012.html.

11. Mike Allen, "What Ted Cruz Said Behind Closed Doors," *Politico*, December 23, 2015, https://www.politico.com/story/2015/12/ted-cruz-gay-marriage-secret
-audio-217090.

12. Michael Barone, "Probing for Clues in the Iowa Caucus Numbers," RealClear-Politics, February 5, 2016, https://www.realclearpolitics.com/articles/2016/02/05
/probing_for_clues_in_the_iowa_caucus_numbers_129564.html.

13. Chad Merda, "So Many People Are Booing at the GOP Debate, It's Trending on Google," *Sun-Times*, February 13, 2016, https://web.archive.org/web/20
160215060124/http://national.suntimes.com/national-world-news/7/72/259
6882/many-people-booing-gop-debate-trending-google/.

14. Team Fix, "The CBS News Republican Debate Transcript, Annotated," *Washington Post*, February 13, 2016, https://www.washingtonpost.com/news/the-fix
/wp/2016/02/13/the-cbs-republican-debate-transcript-annotated/?utm_term
=.1c89a3133504.

15. Michelle Hackman, "Donald Trump Called Ted Cruz a 'Pussy' and the Media Won't Repeat It," Vox, February 9, 2016, https://www.vox.com/2016/2/9/10950182
/donald-trump-ted-cruz.

16. Amy Davidson Sorkin, "Donald Trump's Scalia-Conspiracy Pillow Fight," *New Yorker*, June 19, 2017, https://www.newyorker.com/news/amy-davidson/donald
-trumps-scalia-conspiracy-pillow-fight.

17. Team Fix, "The CNN-Telemundo Republican Debate Transcript, Annotated," *Washington Post*, February 25, 2016, https://www.washingtonpost.com/news/the
-fix/wp/2016/02/25/the-cnntelemundo-republican-debate-transcript-annotated
/?utm_term=.908e1062553e.

18. Jeremy Diamond, "Donald Trump on Protester: 'I'd Like to Punch Him in the Face,'" CNNPolitics, February 23, 2016, https://www.cnn.com/2016/02/23/poli
tics/donald-trump-nevada-rally-punch/index.html.

19. Daniel White, "Donald Trump Tells Crowd to 'Knock the Crap Out of' Hecklers," *Time*, February 1, 2016, http://time.com/4203094/donald-trump-hecklers/.

20. Alexandra Jaffe, "Donald Trump Has 'Small Hands,' Marco Rubio Says," NBC News.com, February 29, 2016, https://www.nbcnews.com/politics/2016-election
/donald-trump-has-small-hands-marco-rubio-says-n527791.

21. Ryan Teague Beckwith, "Read Mitt Romney's Speech About Donald Trump," *Time*, March 3, 2016, http://time.com/4246596/donald-trump-mitt-romney-utah
-speech/.

22. Team Fix, "The Fox News GOP Debate Transcript, Annotated," *Washington Post*, March 3, 2016, https://www.washingtonpost.com/news/the-fix/wp
/2016/03/03/the-fox-news-gop-debate-transcript-annotated/?utm_term=.b20
c8db42b69.

23. "Michigan Exit Polls," CNN, March 8, 2016, https://www.cnn.com/election
/2016/primaries/polls/mi/Rep.

24. "Michigan Exit Polls," CNN, March 8, 2016, https://www.cnn.com/election
/2016/primaries/polls/mi/dem.

CHAPTER THIRTEEN: MARCH 2016

1. Evan Thomas, "The Slickest Shop in Town," *Time*, March 3, 1986.
2. Kenneth P. Vogel, "Paul Manafort's Wild and Lucrative Philippine Adventure," *Politico*, June 10, 2016, https://www.politico.com/magazine/story/2016/06/2016 -donald-trump-paul-manafort-ferinand-marcos-philippines-1980s-213952.
3. Joe Palazzolo, Nicole Hong, Michael Rothfeld, Rebecca Davis O'Brien, and Rebecca Ballhaus, "Donald Trump Played Central Role in Hush Payoffs to Stormy Daniels and Karen McDougal," *Wall Street Journal*, November 9, 2018, https://www .wsj.com/articles/donald-trump-played-central-role-in-hush-payoffs-to-stormy -daniels-and-karen-mcdougal-1541786601?mod=searchresults&page=1&pos=1.
4. "Homewrecker Carly Fiorina Lied About Druggie Daughter," *National Enquirer*, September 23, 2015, https://www.nationalenquirer.com/real-life/homewrecker -carly-fiorina-lied-about-druggie-daughter/.
5. "Bungling Surgeon Ben Carson Left Sponge in Patient's Brain!" *National Enquirer*, October 7, 2015, https://www.nationalenquirer.com/celebrity/bungling -surgeon-ben-carson-left-sponge-patients-brain/.
6. "Senator Marco Rubio's Cocaine Connection!" *National Enquirer*, December 31, 2015, https://www.nationalenquirer.com/true-crime/marc-rubio-cocaine-con nection-drug-scandal/
7. "Ted Cruz Sex Scandal: 5 Secret Mistresses," *National Enquirer*, March 30, 2016, https://www.nationalenquirer.com/photos/ted-cruz-sex-scandal-cheating -affairs-mistresses/.
8. L. V. Anderson, "In Which Donald Trump Mocks Ted Cruz for Not Having a Trophy Wife," *Slate*, March 24, 2016, https://slate.com/human-interest/2016/03 /trump-mocks-ted-cruz-for-not-having-a-trophy-wife.html.
9. "Cruz to Trump: 'You're a Sniveling Coward and Leave Heidi the Hell Alone,'" NBCNews.com, March 24, 2016, https://www.nbcnews.com/video/cruz-to-trump -you-re-a-sniveling-coward-and-leave-heidi-the-hell-alone-651661379665.
10. Andrew Kaczynski, "Mike Pence Argued in an Op-Ed That Disney's "Mulan" Was Liberal Propaganda," BuzzFeed News, July 17, 2016, https://www.buzzfeed news.com/article/andrewkaczynski/mister-ill-make-a-man-out-of-you.
11. "Pence (Finally) Endorses Cruz," *The Rush Limbaugh Show*, April 29, 2016, https://www.rushlimbaugh.com/daily/2016/04/29/pence_finally_endorses_cruz/.
12. Sean Gregory, "Bob Knight Talks Donald Trump, Hillary Clinton, Border Wall," *Time*, April 29, 2016, http://time.com/4312558/bob-knight-donald-trump -indiana/.
13. Tony Cook, Brian Eason, Stephanie Wang, and Chelsea Schneider, "Primary Blog: Cruz Will Be in Indy for Election Returns," *Indianapolis Star*, May 3, 2016, https://www.indystar.com/story/news/politics/2016/05/01/trump-cruz-sanders -make-final-push-indy-monday/83797800/.
14. Ian Schwartz, "Trump Ties Ted Cruz's Father to Lee Harvey Oswald, JFK As-sassination," RealClearPolitics, May 3, 2016, https://www.realclearpolitics.com /video/2016/05/03/trump_ties_ted_cruzs_father_to_lee_harvey_oswald_jfk_as sassination.html.
15. David Wright, Tal Kopan, and Julia Manchester, "Cruz: Trump 'Is a Pathological Liar,'" CNN, May 3, 2016, https://www.cnn.com/2016/05/03/politics/donald -trump-rafael-cruz-indiana/index.html.

CHAPTER FOURTEEN: MAY 2016

1.   Calvin Woodward, "Trump and Ryan Pledge Unity but Remain Divided on Policy," AP News, May 12, 2016, https://apnews.com/f576403e7dfc49579f8e28 65c25bbb31.

2.   Benjy Sarlin, "Donald Trump Warns Supporters Could Riot If He Doesn't Get GOP Nomination," NBCNews.com, March 17, 2016, https://www.nbcnews.com /politics/2016-election/trump-warns-supporters-may-riot-fears-violence-esca late-n540516.

3.   Libby Nelson, "Donald Trump Says Women Don't like Hillary Clinton. They Dislike Him Even More," Vox, April 27, 2016, https://www.vox.com/2016 /4/26/11514948/trump-clinton-women.

4.   Rove, Karl, "What Donald Trump Needs Now," *Wall Street Journal*, May 4, 2016, https://www.wsj.com/articles/what-donald-trump-needs-now-1462399988.

5.   Jenna Johnson, "In General Election, Donald Trump Plans to Focus on 15 States," *Washington Post*, May 26, 2016, https://www.washingtonpost.com/news /post-politics/wp/2016/05/26/in-general-election-donald-trump-plans-to-focus -on-15-states/?utm_term=.a846c18fea19.

6.   M. J. Lee, Dan Merica, and Jeff Zeleny, "Bernie Sanders Endorses Hillary Clin- ton," CNN, July 12, 2016, https://www.cnn.com/2016/07/11/politics/hillary-clinton -bernie-sanders/index.html.

7.   "Trump Wants a 'Deportation Force,'" The Daily Beast, November 11, 2015, https://www.thedailybeast.com/trump-wants-a-deportation-force.

8.   Alexandra Jaffe, "Pence Meets with Trump, Says 'Nothing Was Offered' on VP Front," NBCNews.com, July 4, 2016, https://www.nbcnews.com/politics/2016-elec tion/mike-pence-meets-trump-nothing-was-offered-vp-front-n603271.

9.   Theodore Schleifer, "Trump: Judge with Mexican Heritage Has an 'Inherent Conflict of Interest,'" CNN, June 3, 2016, https://www.cnn.com/2016/06/02/poli tics/donald-trump-judge-mexican-heritage-conflict-of-interest/index.html.

10.   Amanda, Terkel, "Sorry, Brown People, Donald Trump Doesn't Believe You're Real Americans," *Huffington Post*, June 3, 2016, https://www.huffingtonpost.com /entry/donald-trump-mexicans_us_5751e972e4b0c3752dcda87f.

11.   Mohammad Zargham and Richard Cowan, "Trump Should Stop Attacking Minority Groups: Senate's McConnell," Reuters, June 7, 2016, https://www.reuters .com/article/us-usa-election-mcconnell-idUSKCN0YT2CK.

12.   Manu Raju, Eugene Scott, and Deirdre Walsh, "Graham: Trump's Judge Com- ments 'Un-American,'" CNN, June 7, 2016, https://www.cnn.com/2016/06/07/poli tics/lindsey-graham-donald-trump/index.html.

13.   Nick Gass, "Christie: Trump 'Not Racist,' Has 'Right' to Express Opinions on Judge," *Politico*, June 7, 2016, https://www.politico.com/story/2016/06/chris -christie-trump-not-racist-223994.

14.   Aaron Blake, "Chris Christie's Poll Numbers Are Absolutely Brutal," *Wash- ington Post*, May 24, 2016, https://www.washingtonpost.com/news/the-fix/wp /2016/05/19/this-poll-is-brutal-for-potential-trump-vice-presidential-candidate -chris-christie/.

15.   Ryan Lovelace, "Pence: Electing Hillary Clinton Would Be 'Extremely Care- less.'" *Washington Examiner*, July 13, 2016, https://www.washingtonexaminer.com /pence-electing-hillary-clinton-would-be-extremely-careless.

16.   Ashley Parker, Alexander Burns, and Maggie Haberman, "A Grounded Plane and Anti-Clinton Passion: How Mike Pence Swayed the Trumps," *New York Times*,

July 16, 2016, https://www.nytimes.com/2016/07/17/us/politics/donald-trump -mike-pence.html.

CHAPTER FIFTEEN: JULY 2016

1. U.S. Census Bureau, Census, Raw data, U.S. Census Bureau, Washington, DC.
2. Emily Jashinsky, "Rep. Massie's Theory: Voters Who Voted for Libertarians and Then Trump Were Always Just Seeking the 'Craziest Son of a Bitch in the Race,'" *Washington Examiner*, May 7, 2018, https://www.washingtonexaminer.com /rep-massies-theory-voters-who-voted-for-libertarians-and-then-trump-were-al ways just-seeking-the-craziest-son-of-a-bitch-in-the-race.
3. Ryan Teague Beckwith, "Republican Convention: Read Ted Cruz Speech Transcript," *Time*, July 21, 2016, http://time.com/4416396/republican-convention-ted -cruz-donald-trump-endorsement-speech-transcript-video/.
4. Dara Lind, "Ted Cruz Savors Victory in Iowa Caucus Speech," Vox, February 2, 2016, https://www.vox.com/2016/2/2/10892740/ted-cruz-iowa-caucus-speech.
5. Isaac Arnsdorf, "Cruz Super PACs Spent $10 Million in January," *Politico*, February 21, 2016, https://www.politico.com/story/2016/02/ted-cruz-super-pac -backers-january-spending-219575.
6. Haberman, Maggie, "Influential Donors Criticize Ted Cruz for His G.O.P. Convention Speech," *New York Times*, July 23, 2016, https://www.nytimes .com/2016/07/24/us/politics/influential-donors-criticize-ted-cruz-for-his-gop -convention-speech.html.
7. Louis Nelson, "Trump Campaign: Kasich 'Embarrassing His State' by Skipping Convention," *Politico*, July 18, 2016, https://www.politico.com/story/2016/07/paul -manafort-john-kasich-convention-225692.
8. "Full Text: Donald Trump 2016 RNC Draft Speech Transcript," *Politico*, July 21, 2016, https://www.politico.com/story/2016/07/full-transcript-donald-trump -nomination-acceptance-speech-at-rnc-225974.
9. "FULL TEXT: Khizr Khan's Speech to the 2016 Democratic National Convention," ABC News, August 1, 2016, https://abcnews.go.com/Politics/full-text-khizr -khans-speech-2016-democratic-national/story?id=41043609.
10. Donna Brazile, "Inside Hillary Clinton's Secret Takeover of the DNC," *Politico*, November 2, 2017, https://www.politico.com/magazine/story/2017/11/02/clinton -brazile-hacks-2016-215774.
11. Nick Corasaniti and Maggie Haberman, "Trump Suggests Gun Owners Act Against Clinton," *New York Times*, August 10, 2016.
12. Andrew Kirell, "Trump Fan: 'I Would've Taken Him to the Shed' for Joking About Hillary's Murder," The Daily Beast, August 10, 2016, https://www.the dailybeast.com/trump-fan-i-wouldve-taken-him-to-the-shed-for-joking-about -hillarys-murder.
13. Kopan Tal, "Donald Trump: I Meant that Obama Founded ISIS, Literally," CNN, August 12, 2016, https://www.cnn.com/2016/08/11/politics/donald-trump -hugh-hewitt-obama-founder-isis/index.html.
14. Seth Masket and Jennifer Victor, "Clinton Has Vastly More Campaign Offices than Trump: How Much of an Advantage Is This?" Vox, October 5, 2016, https:// www.vox.com/mischiefs-of-faction/2016/10/5/13174624/trump-field-offices.
15. Clare Foran, "The Curse of Hillary Clinton's Ambition," *The Atlantic*, September 19, 2016, https://www.theatlantic.com/politics/archive/2016/09/clinton -trust-sexism/500489/.

16. "A Clear Lead for Clinton after the Conventions," *Washington Post*, August 9, 2016, https://www.washingtonpost.com/page/2010-2019/Washington Post/2016/08/07/National-Politics/Polling/release_435.xml?tid=a_inl.

17. Abigail Tracy, "Clinton Accuses Trump of Supporting Racism in Scathing Campaign Speech," *Vanity Fair*, August 25, 2016, https://www.vanityfair.com /news/2016/08/hillary-clinton-donald-trump-breitbart-alt-right.

18. Steve Wyche, "Colin Kaepernick Explains Why He Sat During National Anthem," NFL.com, August 27, 2016, http://www.nfl.com/news/story/0ap3 000000691077/article/colin-kaepernick-explains-why-he-sat-during-national -anthem.

19. Nick Wagoner, "Transcript of Colin Kaepernick's Comments About Sitting During National Anthem," ESPN, August 28, 2016, http://www.espn.com/blog /san-francisco-49ers/post/_/id/18957/transcript-of-colin-kaepernicks-comments -about-sitting-during-national-anthem.

20. Sean Sullivan, "Trump Slams Colin Kaepernick: 'Maybe He Should Find a Country That Works Better for Him,'" *Washington Post*, August 29, 2016, https:// www.washingtonpost.com/news/post-politics/wp/2016/08/29/trump-slams-col in-kaepernick-maybe-he-should-find-a-country-that-works-better-for-him/.

21. Ashley Killough, "Trump: 'You Can Get the Baby out of Here,'" CNN, August 3, 2016, https://www.cnn.com/2016/08/02/politics/donald-trump-ashburn-virginia -crying-baby/index.html.

22. *National Review* Staff, "Conservatives Against Trump," *National Review*, January 28, 2016, https://www.nationalreview.com/magazine/2016/02/15/conserva tives-against-trump/.

23. Editorial Board, "USA TODAY's Editorial Board: Trump Is 'Unfit for the Presidency,'" *USA Today*, September 30, 2016, https://www.usatoday.com/story /opinion/2016/09/29/dont-vote-for-donald-trump-editorial-board-editorials-de bates/91295020/.

CHAPTER SIXTEEN: OCTOBER 2016

1. David A. Fahrenthold, "Trump Recorded Having Extremely Lewd Conversation About Women in 2005," *Washington Post*, October 8, 2016, https:// www.washingtonpost.com/politics/trump-recorded-having-extremely-lewd -conversation-about-women-in-2005/2016/10/07/3b9ce776-8cb4-11e6-bf8a -3d26847eeed4_story.html?noredirect=on.

2. Jake Sherman, "Ryan 'Sickened' by Trump, Joint Appearance Scrapped," *Politico*, October 8, 2016, https://www.politico.com/story/2016/10/paul-ryan-donald -trump-comments-women-wisconsin-229307.

3. Cristiano Lima, "'I'm Out': Rep. Chaffetz Withdraws His Endorsement of Trump," *Politico*, October 8, 2016, https://www.politico.com/story/2016/10/rep -chaffetz-withdraws-his-endorsement-of-trump-229335.

4. Aaron Blake, "Three Dozen Republicans Have Now Called for Donald Trump to Drop Out," *Washington Post*, October 9, 2016, https://www.washingtonpost .com/news/the-fix/wp/2016/10/07/the-gops-brutal-responses-to-the-new-trump -video-broken-down/.

5. Matthew DeFour, "Paul Ryan Heckled on Home Turf after Donald Trump, Mike Pence Scratched from GOP Event," Madison.com, October 9, 2016, https:// madison.com/wsj/news/local/govt-and-politics/paul-ryan-heckled-on-home -turf-after-donald-trump-mike/article_8e0acf67-1f2f-56a4-b42d-e8ebcbd0bc67 .html.

6. Matthew Boyle, "Exclusive-Audio Emerges of When Paul Ryan Abandoned Donald Trump: 'I Am Not Going to Defend Donald Trump—Not Now, Not in the Future,'" Breitbart, March 14, 2017, https://www.breitbart.com/politics/2017/03/13/exclusive-audio-emerges-of-when-paul-ryan-abandoned-donald-trump-i-am-not-going-to-defend-donald-trump-not-now-not-in-the-future/.
7. Jim Rutenberg, Marilyn W. Thompson, David D. Kirkpatrick, and Stephen Labaton, "For McCain, Self-Confidence on Ethics Poses Its Own Risk," *New York Times*, February 21, 2008.
8. Brandon Byron, "Mulvaney, Person Debate in York Just Days Before 5th Congressional District Election," *The State*, November 2, 2016, https://www.thestate.com/news/politics-government/article112195367.html.
9. Hugh Hewitt, "It's the Supreme Court, Stupid," *Washington Examiner*, August 1, 2016, https://www.washingtonexaminer.com/its-the-supreme-court-stupid.
10. "Full Transcript: Third 2016 Presidential Debate," *Politico*, October 20, 2016, https://www.politico.com/story/2016/10/full-transcript-third-2016-presidential-debate-230063.

CHAPTER SEVENTEEN: OCTOBER 2016
1. "2012 U.S. General Election," CNN, http://www.cnn.com/election/2012/results/state/MI/president/.
2. "Official 2016 Presidential General Election Results," Federal Election Commission, shttps://transition.fec.gov/pubrec/fe2016/2016presgeresults.pdf
3. "Exit Polls 2016," CNN, https://www.cnn.com/election/2016/results/exit-polls.
4. "Transcript: Donald Trump's Victory Speech," *New York Times*, November 9, 2016, https://www.nytimes.com/2016/11/10/us/politics/trump-speech-transcript.html.

CHAPTER EIGHTEEN: NOVEMBER 2016
1. Lindsey McPherson, "Ryan Re-Elected Speaker with Only 1 GOP Defection," *Roll Call*, January 3, 2017, https://www.rollcall.com/politics/ryan-re-elected-Speaker-115th.
2. Gabriella Paiella, "Lena Dunham: 'I Still Haven't Had an Abortion, But I Wish I Had,'" *The Cut*, December 20, 2016, https://www.thecut.com/2016/12/lena-dunham-i-havent-had-an-abortion-but-i-wish-i-had.html.
3. "Democrats' Platform," *New York Times*, August 18, 1996, https://www.nytimes.com/1996/08/18/opinion/l-democrats-platform-070246.html.
4. David Wasserman, "Senate Control Could Come Down to Whole Foods vs. Cracker Barrel," FiveThirtyEight, October 8, 2014, https://fivethirtyeight.com/features/senate-control-could-come-down-to-whole-foods-vs-cracker-barrel/.
5. Jamie Gangel and Elise Labott, "First on CNN: Trump Nixes Elliott Abrams at State," CNN, February 10, 2017, https://www.cnn.com/2017/02/10/politics/elliott-abrams-trump-state-department-tillerson/index.html.
6. Max Fisher, "16 Most Hair-Raising General Mattis Quotes," *The Atlantic*, July 9, 2010, https://www.theatlantic.com/politics/archive/2010/07/16-most-hair-raising-general-mattis-quotes/340553/.
7. Jonathan Swan, "Exclusive: Trump Campaign CEO Wanted to Destroy Ryan," *The Hill*, June 12, 2017, https://thehill.com/blogs/ballot-box/presidential-races/300445-exclusive-trump-campaign-ceo-wanted-to-destroy-ryan.
8. Lisa Hagen, "Trump Compares Ryan to 'a Fine Wine,'" *The Hill*, December 14,

2016, https://thehill.com/homenews/news/310307-trump-compares-ryan-to-a-fine-wine.

9.   Rachael Bade and Elena Schneider, "Freedom Caucus Agitator Who Pushed Out Boehner Pleads for Ryan's Help," *Politico*, July 12, 2016.

10.   Michael Wolff, "Ringside with Steve Bannon at Trump Tower as the President-Elect's Strategist Plots 'an Entirely New Political Movement,'" *Hollywood Reporter*, December 19, 2016, https://www.hollywoodreporter.com/news/steve.

11.   Rosalind S. Helderman, Tom Hamburger, Kevin Uhrmacher, and John Muyskens, "Timeline: The Making of the Christopher Steele Trump-Russia Dossier," *Washington Post*, February 6, 2018, https://www.washingtonpost.com/graphics/2018/politics/steele-timeline/?utm_term=.e13f4da287d7.

12.   Evan Perez, Jim Sciutto, Jake Tapper, and Carl Bernstein, "Intel Chiefs Presented Trump with Claims of Russian Efforts to Compromise Him," CNN, January 10, 2017, https://www.cnn.com/2017/01/10/politics/donald-trump-intelligence-report-russia/index.html.

13.   Ben Guarino, "Shaking Hands Is 'Barbaric': Donald Trump, the Germaphobe in Chief," *Washington Post*, January 12, 2017, https://www.washingtonpost.com/news/morning-mix/wp/2017/01/12/shaking-hands-is-barbaric-donald-trump-the-germaphobe-in-chief/?utm_term=.8562afcdf43e.

CHAPTER NINETEEN: JANUARY 2017

1.   "The Inaugural Address," The White House, January 20, 2017, https://www.whitehouse.gov/briefings-statements/the-inaugural-address/.

2.   Glenn Kessler and Michelle Ye Hee Lee, "Fact-Checking President Trump's Inaugural Address," *Washington Post*, January 20, 2017, https://www.washingtonpost.com/news/fact-checker/wp/2017/01/20/fact-checking-president-trumps-inaugural-address/?utm_term=.2fc20523c9c3.

3.   "AAPC Statement on U.S. Auto Sector's Doubling of Exports," AAPC, February 27, 2015, http://www.americanautocouncil.org/content/aapc-statement-us-auto-sectors-doubling-exports.

4.   Federica Cocco, "Most US Manufacturing Jobs Lost to Technology, Not Trade," *Financial Times*, December 2, 2016, https://www.ft.com/content/dec677c0-b7e6-11e6-ba85-95d1533d9a62.

5.   Serena Marshall, "Obama Has Deported More People than Any Other President," ABC News, August 29, 2016, https://abcnews.go.com/Politics/obamas-deportation-policy-numbers/story?id=41715661.

6.   Yashar Ali, "What George W. Bush Really Thought of Donald Trump's Inauguration," *New York Magazine*, March 29, 2017, http://nymag.com/intelligencer/2017/03/what-george-w-bush-really-thought-of-trumps-inauguration.html.

7.   Elias Groll, "Trump Goes to CIA to Attack Media, Lie About Crowd Size, and Suggest Stealing Iraq's Oil," *Foreign Policy*, January 21, 2017, https://foreignpolicy.com/2017/01/21/trump-goes-to-cia-to-attack-media-lie-about-crowd-size-and-suggest-stealing-iraqs-oil/.

8.   Glenn Kessler, "Spicer Earns Four Pinocchios for False Claims on Inauguration Crowd Size," *Washington Post*, January 22, 2017, https://www.washingtonpost.com/news/fact-checker/wp/2017/01/22/spicer-earns-four-pinocchios-for-a-series-of-false-claims-on-inauguration-crowd-size/?utm_term=.61cd8cb84f74.

9.   Erica Chenoweth and Jeremy Pressman, "This Is What We Learned by Counting the Women's Marches," *Washington Post*, February 7, 2017, https://www.wash

ingtonpost.com/news/monkey-cage/wp/2017/02/07/this-is-what-we-learned-by
-counting-the-womens-marches/?noredirect=on&utm_term=.03da22c1a2c9.

10.   Lydia Saad, "Trump Sets New Low Point for Inaugural Approval Rating," Gal
lup.com, January 23, 2017, https://news.gallup.com/poll/202811/trump-sets-new
-low-point-inaugural-approval-rating.aspx.

11.   Meridith McGraw, "A Timeline of Trump's Immigration Executive Order
and Legal Challenges," ABC News, June 29, 2017, https://abcnews.go.com/Politics
/timeline-president-trumps-immigration-executive-order-legal-challenges/story
?id=45332741,

12.   Sean Sullivan, "Leading Republican Senators Criticize Trump's Refugee
and Travel Ban-48 Hours Later," *Washington Post*, January 29, 2017, https://
www.washingtonpost.com/news/powerpost/wp/2017/01/29/leading-republican
-senators-criticize-trumps-refugee-and-travel-ban-48-hours-later/?utm_term
=.37444a7077df.

13.   Leif Reigstad, "Here's Where Texas Politicians Stand on Donald Trump's Refugee Ban," *Texas Monthly*, February 2, 2017, https://www.texasmonthly.com/the
-daily-post/heres-texas-politicians-stand-donald-trumps-refugee-ban/.

14.   Amber Phillips, "Rand Paul, a Copy Machine and a 'secret' Obamacare Bill,"
Washington Post, March 2, 2017, https://www.washingtonpost.com/news/the-fix
/wp/2017/03/02/rand-paul-a-copy-machine-and-a-secret-obamacare-bill/?utm
_term=.1e54b311fe27.

15.   Glenn Thrush and Maggie Haberman, "Trump the Dealmaker Projects Bravado, but Behind the Scenes, Faces Rare Self-Doubt," *New York Times*, March 23,
2017, https://www.nytimes.com/2017/03/23/us/politics/trump-health-care-bill
-regrets.html.

16.   Madeline Conway, "Ryan: 'Obamacare Is the Law of the Land' for Foresee-
able Future," *Politico*, March 24, 2017, https://www.politico.com/story/2017/03
/obamacare-repeal-failed-paul-ryan-reaction-236478.

17.   Matt Apuzzo, Maggie Haberman, and Matthew Rosenberg, "Trump Told Russians That Firing 'Nut Job' Comey Eased Pressure from Investigation," *New York
Times*, May 29, 2017, https://www.nytimes.com/2017/05/19/us/politics/trump-rus
sia-comey.html

18.   Miller Greg and Greg Jaffe, "Trump Revealed Highly Classified Information to
Russian Foreign Minister and Ambassador," *Washington Post*, May 15, 2017, https://
www.washingtonpost.com/world/national-security/trump-revealed-highly-clas
sified-information-to-russian-foreign-minister-and-ambassador/2017/05/15
/530c172a-3960-11e7-9e48-c4f199710b69_story.html?utm_term=.af67f15f565f.

19.   Michael S. Schmidt, "Comey Memo Says Trump Asked Him to End Flynn Investigation," *New York Times*, May 16, 2017, https://www.nytimes.com/2017/05/16
/us/politics/james-comey-trump-flynn-russia-investigation.html.

CHAPTER TWENTY: JUNE 2017

1.   Stephanie Armour, "Medicaid Expansion Gains Popularity in Red States,"
Wall Street Journal, June 14, 2018, https://www.wsj.com/articles/medicaid-expan
sion-gains-popularity-in-red-states-1528974001.

2.   Letitia Stein, "Angry over U.S. Healthcare Fail, Trump Voters Spare Him
Blame," Reuters, March 26, 2017, https://www.reuters.com/article/us-usa
-obamacare-trump-supporters-idUSKBN16X005.

3.   Max Siegelbaum and David Steen Martin, "In Trump Country, Voters Don't

Blame President for the Health Care Bill Debacle," PBS, March 28, 2017, https://
www.pbs.org/newshour/health/trump-country-voters-dont-blame-president
-health-bill-debacle.
4.   Gary Langer, ABC News, July 16, 2017, https://abcnews.go.com/Politics
/months-record-low-trump-troubles-russia-health-care/story?id=48639490.
5.   *Legal View with Ashleigh Banfield*, Transcript. CNN, February 18, 2016.
6.   Jo Becker, Matt Apuzzo, and Adam Goldman, "Trump Team Met with Lawyer
Linked to Kremlin During Campaign," *New York Times*, July 8, 2017, https://www
.nytimes.com/2017/07/08/us/politics/trump-russia-kushner-manafort.html.
7.   Jo Becker, Matt Apuzzo, and Adam Goldman, "Trump's Son Met with Russian
Lawyer After Being Promised Damaging Information on Clinton," *New York Times*,
July 9, 2017.
8.   Ryan Lizza, "Anthony Scaramucci Called Me to Unload About White House
Leakers, Reince Priebus, and Steve Bannon," *New Yorker*, July 27, 2017, https://
www.newyorker.com/news/ryan-lizza/anthony-scaramucci-called-me-to-un
load-about-white-house-leakers-reince-priebus-and-steve-bannon.

CHAPTER TWENTY-ONE: AUGUST 2017

1.   Shawna Thomas, "Sen. Scott Says Trump's Moral Authority Was Compromised
by His Tues. Comments on Charlottesville," *Vice News*, August 17, 2017, https://
news.vice.com/en_us/article/j5dab3/tim-scott-trump-charlottesville-race.
2.   Jamelle Bouie, "The Secret of Tim Scott's Success? White Voters," *Slate*, November 11, 2014, https://slate.com/news-and-politics/2014/11/tim-scott-won
-south-carolinas-senate-seat-the-first-black-republican-senator-elected-from
-the-south-since-reconstruction.html.
3.   Jodi Kantor and Megan Twohey, "Harvey Weinstein Paid Off Sexual Harassment Accusers for Decades," *New York Times*, October 5, 2017, https://www.ny
times.com/2017/10/05/us/harvey-weinstein-harassment-allegations.html.
4.   Dan Balz and Scott Clement, "Poll: Trump's Performance Lags Behind Even
Tepid Public Expectations," *Washington Post*, November 5, 2017, https://www
.washingtonpost.com/politics/poll-trumps-performance-lags-behind-even-tepid
-public-expectations/2017/11/04/25d2a912-bf4d-11e7-959c-fe2b598d8c00_story
.html?noredirect=on&utm_term=.b061c8826598.
5.   Jugal K. Patel and Alicia Parlapiano, "The Senate's Official Scorekeeper Says
the Republican Tax Plan Would Add $1 Trillion to the Deficit," *New York Times*, No-
vember 28, 2017, https://www.nytimes.com/interactive/2017/11/28/us/politics/tax
-bill-deficits.html.

CHAPTER TWENTY-TWO: JANUARY 2018

1.   Michael Rothfeld and Joe Palazzolo, "Trump Lawyer Arranged $130,000 Pay-
ment for Adult-Film Star's Silence," *Wall Street Journal*, January 12, 2018, https://
www.wsj.com/articles/trump-lawyer-arranged-130-000-payment-for-adult-film
-stars-silence-1515787678.
2.   Maggie Haberman, "Michael D. Cohen, Trump's Longtime Lawyer, Says He
Paid Stormy Daniels Out of His Own Pocket," *New York Times*, February 14, 2018,
https://www.nytimes.com/2018/02/13/us/politics/stormy-daniels-michael-cohen
-trump.html.
3.   Michael S., Schmidt and Adam Goldman, "'Shaken' Rosenstein Felt Used by
White House in Comey Firing," *New York Times*, June 29, 2018, https://www.ny
times.com/2018/06/29/us/politics/rod-rosenstein-comey-firing.html.

4. "Vice President Pence Delivers Naval Academy Commencement Address," C-SPAN, May 26, 2017, https://www.c-span.org/video/?429079-1/vice-president-pence-delivers-naval-academy-commencement-address.
5. E. J. Montini, "Will All-Talk Jeff Flake Finally Take Action on Supreme Court Nominee?" Azcentral, August 21, 2018, https://www.azcentral.com/story/opinion/op-ed/ej-montini/2018/08/20/jeff-flake-brett-kavanaugh-supreme-court-do-something/1038948002/.

CHAPTER TWENTY-THREE: JULY 2018
1. Eli Watkins and Clare Foran, "EPA Chief Scott Pruitt's Long List of Controversies," CNN, July 5, 2018, https://www.cnn.com/2018/04/06/politics/scott-pruitt-controversies-list/index.html.
2. Philip Bump, "A Summary of the Fruit of the Mueller Investigation, to Date," *Washington Post*, July 13, 2018, https://www.washingtonpost.com/news/politics/wp/2018/07/13/a-summary-of-the-fruit-of-the-mueller-investigation-to-date/?utm_term=.09721b623365.
3. George Stephanopoulos, ABC News, July 2, 2018, https://abcnews.go.com/Politics/michael-cohen-family-country-president-trump-loyalty/story?id=56304585.

CHAPTER TWENTY-FOUR: SEPTEMBER 2018
1. Elena Schneider, "'Something Has Actually Changed': Women, Minorities, First-time Candidates Drive Democratic House Hopes," *Politico*, September 11, 2018, https://www.politico.com/story/2018/09/11/white-men-democratic-house-candidates-813717.
2. Rachel Ventresca, "Clinton: 'You Cannot Be Civil with a Political Party that Wants to Destroy What You Stand For,'" CNN, October 9, 2018, https://www.cnn.com/2018/10/09/politics/hillary-clinton-civility-congress-cnntv/index.html.
3. Zach Montellaro, "O'Rourke Raises Record-Smashing $38 Million in Third Quarter," *Politico*, October 12, 2018, https://www.politico.com/story/2018/10/12/orourke-raises-38-million-896833.
4. Jonathan Swan and Stef W. Kight, "Exclusive: Trump Targeting Birthright Citizenship with Executive Order," Axios, October 30, 2018, https://www.axios.com/trump-birthright-citizenship-executive-order-0cf4285a-16c6-48f2-a933-bd71fd72ea82.html.
5. Daniel Dale, "815 False Claims: The Staggering Scale of Donald Trump's Pre-Midterm Dishonesty," *Toronto Star*, November 15, 2018, https://www.thestar.com/news/world/analysis/2018/11/15/815-false-claims-the-staggering-scale-of-donald-trumps-pre-midterm-dishonesty.html.
6. "2018 Midterms: Exit Polling," CNN, November 6, 2018, https://www.cnn.com/election/2018/exit-polls.

CHAPTER TWENTY-FIVE: NOVEMBER 2018
1. "Presidential Job Approval Center," Gallup.com, https://news.gallup.com/interactives/185273/presidential-job-approval-center.aspx.
2. Kenneth P. Vogel and Katie Rogers, "Nick Ayers Is Rising Fast in Trump's Washington: How Far Will He Go?" *New York Times*, November 21, 2018, https://www.nytimes.com/2018/11/21/us/politics/nick-ayers-white-house.html.
3. Dexter Filkins, "James Mattis, a Warrior in Washington," *New Yorker*, June 20, 2017, https://www.newyorker.com/magazine/2017/05/29/james-mattis-a-warrior-in-washington.

4.   "Transcript of President Trump's Interview with the *Wall Street Journal*," *Wall Street Journal*, November 26, 2018, https://www.wsj.com/articles/transcript-of -president-trumps-interview-with-the-wall-street-journal-1543272091.

5.   Quinton Franklin, "I Lost My Job at Carrier After President Trump Promised to Save It," Vox, February 6, 2018, https://www.vox.com/first-person /2018/2/6/16976176/carrier-president-trump-layoffs.

6.   Aaron Blake, "The GOP's Pernicious Link Between Terrorism and the Border Wall," *Washington Post*, January 4, 2019, https://www.washingtonpost.com/poli tics/2019/01/04/gops-pernicious-link-between-terrorism-border-wall/?noredi rect=on&utm_term=.6d8fcd3488a7.

7.   Danielle Kaeble and Mary Cowhig, "Correctional Populations in the United States, 2016," Bureau of Justice Statistics, U.S. Department of Justice, Washington, DC.

CHAPTER TWENTY-SIX: DECEMBER 2018

1.   Charles Franklin, "New Marquette Law School Poll Finds Clinton Widening Lead over Trump in Wisconsin," Marquette Law School Poll, August 10, 2016, https://law.marquette.edu/poll/2016/08/10/new-marquette-law-school-poll -finds/.

2.   Matthew Sheffield, "Poll: A Majority of Americans Support Raising the Top Tax Rate to 70 Percent," *The Hill*, January 15, 2019, https://thehill.com/hilltv /what-americas-thinking/425422-a-majority-of-americans-support-raising-the -top-tax-rate-to-70.

3.   David Wasserman, "Five Takeaways from Democrats' House Triumph," *The Cook Political Report*, November 8, 2016, https://cookpolitical.com/analysis /five-takeaways-democrats-house-triumph.

4.   Aaron Rupar, "Trump's Unfounded Tweet Stoking Fears About Muslim 'Prayer Rugs,' Explained," Vox, January 18, 2019, https://www.vox.com /2019/1/18/18188476/trump-muslim-prayer-rugs-tweet-border.

INDEX